Social Justice and the City

This special collection aims to offer insight into the state of geography on questions of social justice and urban life. While using social justice and the city as our starting point may signal inspiration from Harvey's (1973) book of the same name, the task of examining the emergence of this concept has revealed the deep influence of grassroots urban uprisings of the late 1960s, earlier and contemporary meditations on our urban worlds (Jacobs, 1961, 1969; Lefebvre, 1974; Massey and Catalano, 1978) as well as its enduring significance built upon by many others for years to come. Laws (1994) noted how geographers came to locate social justice struggles in the city through research that examined the ways in which material conditions contributed to poverty and racial and gender inequity, as well as how emergent social movements organized to reshape urban spaces across diverse engagements including the U.S. Civil Rights Movement, anti-war protests, feminist and LGBTQ activism, the American Indian Movement, and disability access.

This book was originally published as a special issue of *Annals of the American Association of Geographers*.

Nik Heynen is a Professor in the Department of Geography at the University of Georgia, Athens, USA. His research interests include urban political ecology and the politics of race, class, and gender.

Social Justice and the City

Edited by
Nik Heynen

Routledge
Taylor & Francis Group

LONDON AND NEW YORK

First published 2019
by Routledge
2 Park Square, Milton Park, Abingdon, Oxon, OX14 4RN, UK

and by Routledge
52 Vanderbilt Avenue, New York, NY 10017, USA

First issued in paperback 2020

Routledge is an imprint of the Taylor & Francis Group, an informa business

British Library Cataloguing in Publication Data
A catalogue record for this book is available from the British Library

ISBN 13: 978-0-367-66355-1 (pbk)
ISBN 13: 978-1-138-32274-5 (hbk)

Typeset in Goudy
by Out of House Publishing

Publisher's Note
The publisher accepts responsibility for any inconsistencies that may have arisen during the conversion of this book from journal articles to book chapters, namely the possible inclusion of journal terminology.

Disclaimer
Every effort has been made to contact copyright holders for their permission to reprint material in this book. The publishers would be grateful to hear from any copyright holder who is not here acknowledged and will undertake to rectify any errors or omissions in future editions of this book.

Contents

CONTENTS

CONTENTS

Citation Information

The chapters in this book were originally published in the journal *Annals of the American Association of Geographers*, volume 108, issue 2 (March 2018). When citing this material, please use the original page numbering for each article, as follows:

CITATION INFORMATION

CITATION INFORMATION

CITATION INFORMATION

Chapter 24
Datafying Disaster: Institutional Framings of Data Production Following Superstorm Sandy
Ryan Burns
Annals of the American Association of Geographers, volume 108, issue 2 (March 2018) pp. 569–578

Chapter 25
Cultivating (a) Sustainability Capital: Urban Agriculture, Ecogentrification, and the Uneven Valorization of Social Reproduction
Nathan McClintock
Annals of the American Association of Geographers, volume 108, issue 2 (March 2018) pp. 579–590

Chapter 26
From "Rust Belt" to "Fresh Coast": Remaking the City through Food Justice and Urban Agriculture
Margaret Pettygrove and Rina Ghose
Annals of the American Association of Geographers, volume 108, issue 2 (March 2018) pp. 591–603

For any permission-related enquiries please visit:
www.tandfonline.com/page/help/permissions

Introduction: The Enduring Struggle for Social Justice and the City

Nik Heynen, Dani Aiello, Caroline Keegan, and Nikki Luke

Myriad questions related to social justice have shaped urban geographic scholarship, among which two things remain clear: geographers maintain fidelity to the idea that the discipline should keep working to understand unjust processes within urban life and simultaneously seek solutions to make cities more just. Beyond this, few geographers today would come to the same set of defining characteristics of what a just city would look like, or agree on the right questions to ask toward its realization. What the concept of social justice lacks in terms of facilitating intellectual and political consensus, it makes up for in centering heterodox efforts at generating relevant theory and practice that can change the social circumstances of people living in cities, regardless of how these terms are defined.

It is out of these enduring commitments, demands, and possibilities that the theme of this special issue emerged: Social Justice and the City. This special collection offers insights into the state of the discipline on questions of social justice and urban life. Although using social justice and the city as our starting point might signal inspiration from Harvey's (1973) book of the same name, the task of examining the emergence of this concept has revealed the deep influence of grassroots urban uprisings of the late 1960s and earlier and contemporary meditations on our urban worlds (Jacobs 1961, 1969; Lefebvre [1974] 1991; Massey and Catalano 1978) as well as its enduring significance built on by many others for years to come. Laws (1994) described how geographers came to locate social justice struggles in the city through research that examined the ways in which material conditions contributed to poverty and racial and gender inequity, as well as how emergent social movements organized to reshape urban spaces across diverse engagements including the U.S. civil rights movement, antiwar protests, feminist and lesbian, gay, bisexual, transgender, and queer (LGBTQ) activism, the American Indian Movement, and (dis)ability access. The twenty-six essays that make up this special issue collectively offer a theoretically robust and empirically astute picture of how far this idea has evolved since the radical turn in geography in 1968 and 1969 and more specifically within a flagship journal of the discipline. There was broad support across the journal's editorial board for this theme as well as broad interest, evidenced by the fact that the initial call for abstracts yielded 131 considered submissions. When inviting full paper submissions, special emphasis was given to capturing the breadth of how scholars and teams of scholars push the ways we can envision and talk about social justice and the city. Thirty-six full papers were invited and of those we include twenty-six articles here. Only seven of the essays in the collection either explicitly engage or reference Harvey's (1973) *Social Justice and the City*, another indication that the idea has roots and a trajectory that moves much beyond this more often cited origin.

What follows in this article is an effort to trace the genealogy of urban social justice within the *Annals* to understand its origins since the journal's first publication in 1911 and gesture at where it might be going. To frame the articles that follow, we work through the archives of the *Annals* starting with the first published issue, mapping changes in the definition of social justice in three cuts. In the first section, we consider the political discussions of justice and injustice up to the radical turn in the discipline that prefigured what would become social justice as a dominant theme of investigation in geography. We then show, in selected ways, the rapid theoretical development of social justice in its variegated forms after the turn up to this special issue. Over time, we note how the empirical emphasis of articles widens to consider a broad range of geographies, identities, and political aims with a greater preponderance of specifically urban studies. Third, we discuss the ways in which articles published in the *Annals* have treated "the city" and urban geographical processes more broadly. Following this deeper context, we offer some summary of the twenty-six special issue articles. The shift across the journal's disciplinary history is quite extraordinary, with much of the early research drawing from racist, sexist, colonial, and environmentally determinist thought and transitioning into much more socially engaged and progressive, sometimes radical, scholarship.

Although the collective insights across this special issue suggest that there has been substantially more engagement with politically relevant and timely geographic questions of social justice over the span of the journal's history, we feel it necessary to point out the slow and painful pace of change with regard to significant inequalities in the discipline that shape our geographic knowledge production, our mutual intellectual thriving, and publishing within geography. A very recent and profound testament to this reality is Kelsky's (2017) compilation of gender-based discrimination and violence in the academy more broadly. On this subject, a tweet from Dr. Carrie Mott (who has an article in this special issue) on 10 December 2017 said, "I would love to see men in Geography take seriously a survey and responses that are being gathered by @ProfessorIsIn. For every response on there (includes some in Geog) there are MANY more. @theAAG." Kelsky's crowdsourced survey had, the last time we looked, 2,213 cases of sexism, sexual harassment, and sexual assault listed across all academic disciplines and ranks. Unlike other important surveys (see Hanson and Richards 2017; Webley Adler 2017), Kelsky's survey includes geographers sharing a range of oppressive and violent incidents they have experienced.

In the face of slow but steady progress in the content of geographic scholarship, why is it that the practice of doing that scholarship is still fraught and more difficult for many women-identified, non-white, LGBTQ, and gender-nonconforming people who continue to endure unequal workplace conditions? How is it that our collective record of sustained attention to socially unjust geographies can continue to be accompanied by such deep betrayals of justice in our professional and personal relations? Of course, part of the answer to this is that those geographies of injustice we examine also exist within the patriarchal conditions of our intimate social worlds. Beyond these ongoing and dehumanizing incidents of gender-based violence, a fidelity to social justice requires us to contend with the additional (intersectional) reality that racist, classist, heteronormative, transphobic, anti-immigrant, anti-Muslim, and anti-Semitic dynamics (among others) continue to exist within the embodied practices of our discipline despite our ostensible aims to address them in our work. As we venture into the archives of geography within the *Annals*, marking our progression, it seems important to maintain urgency for social justice oriented work, but this cannot be meaningfully pursued without the work of dismantling the uneven social relations that shape our professional relationships, our research, and our publishing. How can we bring a renewed commitment beyond the published pages to guarantee that our departmental hallways, conference halls, and other spaces where geographers interact are more socially just also? When will we come to realize that without ethical, everyday social relations embedded within our practice of publishing about social justice, our work will always lack in emancipatory potential?

Prefiguring Social Justice

Normative approaches to social justice were absent in the early issues of the *Annals*. There were theoretical precursors, however, that prefigured what ultimately became some of the most important political ideas in those early issues. As a way of providing a foundation for this special issue, we marked the radical turn as a threshold to cultivate a sense of how discussions of equity, truth, ethics, and principles were framed beforehand, prior to when social justice emerged as a central preoccupation. Preceding these prefigurative discussions of justice, in the earliest decades of the *Annals*, we see many examples of environmental determinist analyses that exemplify the ways in which geography contributed to the development of racist and colonial typologies (see Woods [1998]; Byrd [2011]; Tuck and Yang [2012], for counternarratives and context).

Whitbeck (1912) offered the earliest insight into the terrain of the political for the *Annals* in its second volume, stating that "when the white man came to Wisconsin, thirty million acres of timber lay untouched before him. In those forests stood more than one hundred billion feet of white pine and as much more of other merchantable timber. At the close of the Civil War, enough white pine stood in Wisconsin to liquidate the national debt which that war had piled up" (59). Where Indigenous people are excluded from Whitbeck's discussion of the opportunities for economic gain envisioned in the extraction of timber, Gregory (1915) evoked a colonizing sentiment that infantilizes Indigenous peoples when he said, "To the Hopis the coming of the white man was welcome. Under the protection of a stronger race, the farms of their ancestors, practically abandoned for 250 years, were reoccupied" (117).

That same year, Brigham (1915) questioned how and whether geographers should consider race, asking, "Our references to the race problem might seem superfluous, for if this field belongs essentially to the anthropologists, what right has the geographer there" (14). Brigham (1915, 18) revealed, however, the environmental determinist roots of his orientation by commending the development of Ratzel's "Darwinism" in the field of geography. Other key examples of determinism include Semple (1919) on the influence of tree species and specific forest resources to people and their ship-building enterprises and on irrigation and reclamation in the Mediterranean (Semple 1929). Just as influential as Semple's work in the early twentieth century is Huntington (1924), whose determinist scholarship helped shape the discipline and was supported by scholars such as Bowman (1932) and Whittlesey (1945), among others. As the determinist trend within geography waned, scholars nevertheless continued to rescue Ratzel's ideas, for example, when Sauer (1971), in his essay "The Formative Years of Ratzel in the United States," reflected on "the grave and unresolved crisis of destructive exploitation and urban malaise" (254), pointing to Ratzel's prophetic insights into the development of the U.S. landscape.

As early as the late 1940s, an intervention by Platt (1948) suggested the beginning of a pushback against determinism within the discipline: "If we avoid a deterministic approach and give our best efforts to the pursuit and use of knowledge, we can rightly hope to bend our common course in the direction of our desire, and to cause a trend of events (cause in a true philosophic and not in a pseudo-scientific sense) toward greater human welfare" (132). In descriptive essays that predate this more obvious shift, however, geographers observing Indigenous social worlds also subtly worked against the dominant determinism of their time. For instance, Haas (1926, 172, 175) refused to measure the "advancement" of Indigenous people against Western norms, yet speaks highly of the "progress," "skill," community-organized public projects, and governance structures among the "Cliff-Dwelling" peoples of the Southwestern United States. In the first mention we found of (in)justice in the *Annals*, Haas (1926) observed, "Had our colonial history been written by the red man, or even by unbiased minds, the story would read quite differently" (171). Across the years that followed, Trewartha (1938) discussed the history of Indigenous conflict in North American French settled regions and Martin (1930) and Meyer (1956) reflected on the negotiation in the courts and with the federal government over Indigenous land claims.

Early scholarship on race and racism in geography is now easily critiqued, thanks to the sophistication with which Black Geographies have matured (see McKittrick and Woods 2007); nevertheless, there are some notable essays from those early years. Several articles comment on racialized conflict and integration between Indigenous and black peoples with the introduction of the slave trade: Hans (1925) described how in the West Indies and Brazil "the Indian could not and would not adapt himself to slave conditions" (91), Parsons (1955, 52) noted how Miskito peoples were conscripted to put down slave rebellions, and Price (1953, 155) discussed mixed-race settlements. Although still steeped in a determinist logic, Parkins (1931) aimed to critique the institutional roots of slavery in the United States and mentioned the importance of the urban for fostering social movements, when he suggested that "the very poor, landed or landless, were inaudible then as now. There were no large urban groups to contest the control of the planter classes" (8). Waibel (1943, 119) discussed the transformative impacts of abolishing the slave trade and how planters sought to spread the "American principle" of the plantation economy into West Africa. Meanwhile, much later on, Nostrand (1970) offered the first discussion of "Hispanic" as an ethno-racial category in the *Annals*.

Even during the apex of the determinist trend, more progressive currents of thought began to consider questions of rights, governance, and organizational structures able to improve human lives. Colby (1924), for instance, discussed cooperative marketing in the formation of a raisin trust that suggests collective organization for social benefit, and Visher (1925) evaluated what standards guaranteed a respectable livelihood in terms of homestead allotments—perhaps an early exploration of equality in housing and habitability. Although environmental justice and urban political ecology are relatively newer domains within the *Annals*, earlier studies also nod to these ideas. McMurry (1930) recorded the enclosure of natural areas, "bought up by wealthy sportsmen and developed into private game preserves for the exclusive use of the owners. Individuals and clubs have leased considerable acreage for hunting purposes, and numerous farmers derive appreciable income from this source" (12). Whitaker (1941) discussed inequity of natural resource depletion and the need for governing authorities to intervene. Twenty years later, McNee (1961) published a precursor to work done on petro-economics that considered the multidirectional forces at work in regions dominated by international petroleum companies. This was followed quickly by the first extensive discussion of Marxist geography by Matley (1966) that interestingly worked to emphasize the ongoing importance of environmentalism in Soviet geography and their interpretations of Marx.

Especially relevant to this special issue are discussions of housing rights. Hance (1951) published one of the first papers to explicitly discuss this, describing crofting settlements in the Outer Hebrides that "make a perplexing problem for the public health officials. With little means to pay taxes except when employed, owning no land to be taxed, they create congestion and often possess the poorest of the houses in the community" (87). Sometime following this, both Ward (1968) and Holzner (1970) touched on different housing problems related to urban "blight," its psychosocial repercussions, and the right to adequate housing. Perhaps presaging a broader right to the city discussion, these publications were reflective of debates in other regions around land reform and rights to land (see Chardon 1963).

It was not until 1959 that Naylon (1959, 361) made the first explicit mention of "social injustice" in the *Annals* in his discussion of land fragmentation in Spain. Highlighting the need for geographers to pay more careful attention to land as a domain of inequality, Naylon explained that the Spanish example "is of interest in showing that under-employment, low productivity, and social discontent are not related only to large estates and their associated monocultures, although the glaring social injustices of the latifundios [large estates] have in the past received the most attention from reformers and academic observers." Bushman and Stanley (1971), in another early and explicit reference, similarly reflected on the political possibilities present in a social justice coalition as they discussed political trends in the U.S. Southeast. They noted, "The Democratic Party in the region undoubtedly will continue its shift toward a more liberal position on racial matters and issues of social justice" (666).

Toward the latter half of the 1960s, the *Annals* begin to evidence increasingly sophisticated analyses of the local geographies of political struggles to address poverty, colonial dispossession, and racial inequity. Whereas Meinig (1972) located the movements of "American Indians, New Mexican Hispanos, Mexican Americans, and Black Americans, [who were] never accorded full social integration" (182) in rural settings, other scholars considered how these struggles specifically articulate within the city. Lowry (1971) unequivocally stated, "Negroes and whites unquestionably did not enjoy equally the benefits of economic advances

in and around urban places" (586). Finally, perhaps one of the most provocative related essays we found in the archives was Murray's (1967) "The Geography of Death in the United States and the United Kingdom," in which he detailed the correlation between mortality rates and socioeconomic conditions: "Areas of large minority group populations, such as Indian reservations, Spanish-American districts, and most particularly, the high-proportion Negro areas of the South, portray rates higher than the average" (310). He continued that the health effects of poverty and racial inequity extend into urban areas: "Some of the highly urbanized counties in the Northeast also reveal above average rates, partly because of their attraction for minority and deprived groups whose longevity is jeopardized" (310).

Social Justice after the Radical Turn

In her retrospective of the *Annals*, Kobayashi (2010) suggested, "As the 1970s rolled around, contributions to the *Annals* began to reflect a larger concern for geography's role in society, responding to larger societal concerns about the ongoing Vietnam War, the advent of activism over environmental degradation, the second-wave feminist movement, and the burgeoning human rights movement" (1099). Although development of social justice research (broadly defined) quickly proliferated after the radical turn in geography, there are some key essays we believe offer an especially interesting context for this special issue. In his presidential address delivered at the sixty-eighth annual meeting of the Association of American Geographers (AAG), published as "From Colonialism to National Development: Geographical Perspective on Patterns and Policies," Ginsburg (1973) highlighted the growing attention to issues of social justice: "The increasing concern with relevance in academia is shared by most geographers, who recognize themselves to be citizens and human beings as well as scientists and scholars" (1). He went on to write, "The equity principle involves much more than regional equity. It also involves the distribution of benefits among all the people, not merely some of them. ... Everyone recognizes the enormous disparities in income and welfare even within given metropolitan areas and the social injustices associated with that maldistribution" (16).

The 1970s and early 1980s in the *Annals* saw more heterodoxy than is commonly attributed to this period of geographic scholarship. In his piece, "The Geography of Human Survival," Bunge (1973) suggested, "The penance of the new wave of exploration is to undo the exploitation of the early geographers. The world is not some vast treasure trove of unlimited wealth, human and mineral, to be carted off to the homeland as booty" (290). We also see a continued thread of Marxist geography, with Peet (1975) writing just a few years later, "There is little point, therefore, in devoting political energies to the advocacy of policies which deal only with the symptoms of inequality without altering its basic generating forces. Hence the call for social and economic revolution, the overthrow of capitalism, and the substitution of a method of production and an associated way of life designed around the principles of equality and social justice" (564).

Although there might have seemed to exist a broad consensus that geography's relevance increasingly hinged on its attention to such broader social questions, there were many fault lines already evident within and across these politics. Some, like King (1976), Ley (1980), and Helburn (1982), added differently nuanced and theoretically positioned ways of considering questions of social justice. Toward the end of the decade as these debates unfolded, Bunge (1979) wrote of these differences, "I am short tempered with academic geographers, even Marxist ones. The campus geographers tend to separate theory from practice. They read too much and look and, often, struggle not at all. They cite, not sight. In the heady atmosphere of all theory and no practice all sorts of objections are raised to our work, but the one that is most fearful is an ideological Marxist reductionism" (171).

Long-standing questions about land and Indigenous struggles over land took on much more emancipatory trajectories in the 2000s as evidenced by Wolford's (2004) "This Land Is Ours Now," Harris's (2004) "How Did Colonialism Dispossess?," and Radcliffe (2007), who argued, "Articulated as alternatives to neoliberalism, indigenous geographies of hope are grounded in critiques of racism, colonial legacies, and particular forms of economic political power" (393). More recently we can see Blomley (2014) furthering these discussions when he argued, "Property is an instrument of sociospatial justice, whether in relation to colonialism or other social and political settings. As a set of relations, powerfully constitutive of space, property can serve both as an instrument of dispossession and as a ground for resistance" (1303).

If determinist, racially insensitive, and outrightly unjust research was published in the earliest decades of the Annals, after a relative silence during the mid- and late 1980s, the 1990s and onward have seen a flourishing of research working to extend an explicitly antiracist, social justice orientation in publishing. Dominant within this thread have been questions about housing discrimination (Holloway 1998) and segregation (Ellis et al. 2012). To this end, geographers have continued to better connect processes of racialization to other spatial practices as evidenced by Hoelscher's (2003) suggestions that "the culture of segregation that mobilized such memories, and the forgetting that inevitably accompanied them, relied on performance—ritualized choreographies of race and place, and gender and class, in which participants knew their roles and acted them out for each other and for visitors" (677). Other key publications extended the political horizons on which geographers continued to take questions of race and racialization seriously, including their relation to immigration (Leitner 2012), public space (Tyner 2006), antihunger politics and social reproduction (Heynen 2009), racial violence (Inwood 2012), white supremacy (Inwood and Bonds 2016), prison politics (Bonds 2013), neoliberal regulation (Derickson 2014), and mobility (Alderman and Inwood 2016; Parks 2016). Much of this work has helped expand on the driving insights from Kobayashi and Peake (2000), who argued that "strategies of resistance are also diverse. They are expressed through distinctive racialized identities, and take many forms that may range from everyday cultural practices to political movements, and may cover the ideological spectrum" (398; see Kobayashi 2014 for a more in-depth overview).

Although feminist scholarship should now be considered a key pillar of Annals publications around questions of social justice, it was slow to develop. One of the most notable contributions to opening up feminist scholarship in the 1990s was Jones and Kodras's (1990) "The Feminization of Poverty in the U.S." in which they argued, "First, raising the national minimum wage above the poverty level will end the injustice of working full-time yet remaining poor. Second, a re-evaluation of the worth of women's work, through pay equity legislation, would be a major step towards eliminating gender-based wage differentials" (180). Meanwhile, Katz's (1991) "Sow What You Know" was a landmark paper extending discussions of social reproduction in the field. England's (1993) "Suburban Pink Collar Ghettos" and Wright's (2004) "From Protests to Politics: Sex Work, Women's Worth, and Ciudad Juárez Modernity" worked crucially to widen the spectrum of feminist scholarship the Annals published.

Other key feminist geographic interventions include Hovorka's (2005) "The (Re) Production of Gendered Positionality in Botswana's Commercial Urban Agriculture Sector" and Brickell's (2014) "'The Whole World Is Watching': Intimate Geopolitics of Forced Eviction and Women's Activism in Cambodia." The growth of feminist scholarship was extended into more specific research on sexuality and LGBTQ studies.

We have seen a slow, although steady, increase in publications situating social justice within LGBTQ geographies including Waitt's (2006) "Boundaries of Desire: Becoming Sexual through the Spaces of Sydney's 2002 Gay Games" and Schroeder's (2014) "(Un)holy Toledo: Intersectionality, Interdependence, and Neighborhood (Trans)formation in Toledo, Ohio." Collins, Grineski, and Morales (2017) most recently helped to show the increased cross-cutting developments within LGBTQ geographies in their "Sexual Orientation, Gender, and Environmental Injustice" by emphasizing that "it is important to recognize that same-sex partnering in households is a highly visible expression of minority sexual orientation (in contrast to being LGBT single or in the closet) and is thus an important residential indicator of the status of the LGBT community in social justice terms" (89).

Though there has been important evolution in geographic work published on race, gender, and sexuality, there has also been continued evolution of innovative intersectional and internationalist approaches to staging more inclusive and comprehensive efforts at social justice oriented questions. Gilbert's (1998) "'Race,' Space, and Power: The Survival Strategies of Working Poor Women" highlights "the significance of place and context in shaping the relationship between space and multiple relations of power, in this case, racism and gender. Therefore it becomes important to ask *how* mobility and immobility are related to historically and geographically situated constellations of power relations" (616). Mullings's (1999) "Sides of the Same Coin?" was another important publication in this context. This effort at thinking across subject positions was also explored in internationalist contexts by Hodder (2016) and Featherstone (2013), who argued, "The various forms of internationalism associated with labor, Pan-Africanism, anti-colonialism and feminism . . . were not homogenizing universalisms but built on the mutual constitution of gender, class, race and national subordination to create agendas for struggle and visions of social equality and justice" (1408). Indeed, much of this work has gone into making

legible the invisible connections that so powerfully make places, as Elwood et al. (2015) explained: "When relational place-making involves engaged struggle with difference and inequality, actors might begin to recognize, articulate, and question race and class norms or poverty politics that were previously invisible and taken for granted" (136).

During this period, environmental questions about justice and, by extension, political ecology, began to take social justice questions in innovative and new directions for geographers. Lawrence's (1993) "The Greening of the Squares of London" is one of the earliest discussions in the *Annals* that is explicitly interested in "nature and (in) the city." Likewise, the *Annals* was host to one of the earliest environmental justice papers, in which Bowen and Salling (1995) argued, "Environmental justice is the policy rubric within which issues such as environmental equity, environmental discrimination, and environmental racism are embedded" (641). Only a few years following came one of the most important papers published to extend questions of social and environmental justice, especially related to race, in Pulido's (2000) "Rethinking Environmental Racism." She explained, "The issue of racism itself raises both scholarly and political concerns. I believe that as geographers, we need to diversify and deepen our approach to the study of racial inequality. Our traditional emphasis on mapping and counting needs to be complemented by research that seeks to understand what race means to people and how racism shapes lives and places" (33). Other work by Boone et al. (2009) and Holifield (2012) only further helped to solidify the journal's content on research focusing on environmental justice.

The last special issue in the *Annals* (edited by Braun in 2015) contained a number of essays that extended social justice oriented questions through socionatural empirical contexts. In that issue, Mansfield et al. (2015) noted, "In a world of massive and ubiquitous socioecological change, it is time to rally not around the tired environmentalisms of 'protecting nature' but around protecting and fostering the social natures that lead to the most just outcomes for humans and nonhumans alike" (292). Likewise, Derickson and MacKinnon (2015) and Wainwright and Mann (2015) extended important political arguments related to climate change, as did Rice, Burke, and Heynen (2015), who argued for new, more egalitarian forms of knowledge production around these important questions: "In the case of climate, organic intellectuals could articulate the knowledge of ordinary people and subalterns in place-based, culturally attuned ways that spark more inclusive and just climate actions, thus replacing traditional intelligentsia with a more egalitarian politics of knowledge" (255). In that same special collection, Collard, Dempsey, and Sundberg (2015) articulated how central decolonial approaches are to this overall project of social justice: "Orienting toward abundant futures requires walking with multiple forms of resistance to colonial and capitalist logics and practices of extraction and assimilation. Decolonization is our guide in this process" (329).

"The City"

Given the breadth and sophistication of urban studies today, it is interesting to assess the ways in which "the city" and urban social processes evolved slowly within the early pages of the *Annals*. At first, much like the wider breadth of urban literature at the time, the city is staged merely as the site in which economic processes agglomerate. Whitbeck (1912), who offered the first political discussion in the *Annals*, also published the first explicit reference to a city, describing the industrial urban geography of the city of Sheboygan (Wisconsin), where over half of its wage earners were employed in furniture factories, "chiefly chair factories" (62). Just over a decade later in an essay examining similar themes of industrialization in Kentucky, Davis (1925) published the first explicit reference to "urban development." Yet, it was not until 1951 that Branch (1951) offered the first discussion of "planning" or "city planning" when he wrote, "Every actual situation in community planning combines physical and socio-economic considerations in inseparable combination" (281).

Gregory (1915) wrote an article about the settlement of Tuba, which is today an incorporated city with a population of only about 9,000 but was in 1915 a heavily used route across northern Arizona. This article matter-of-factly details the history of dispossession of local residents related to the establishment of the town. Two years later, Jefferson (1917) wrote, "A great country population cannot exist to-day among civilized men without bringing cities into existence. Neither Norway nor Ecuador, on the other hand, can have a great city because they have no great country population" (6). In this sense, Jefferson's work reveals early thinking that

structures a city–country binary within early colonial and determinist logics.

In the early years of the Great Depression, an urban geographic nuance emerged to understand the development of U.S. cities, much of which throughout the 1930s was predictably influenced by Chicago school emphasis on zones, functions, and an organism understanding of urban form. For instance, Colby (1933) wrote, "The modern city is a dynamic organism constantly in process of evolution. This evolution involves both a modification of long established functions and the addition of new functions. Such functional developments call for new functional forms, for modification of forms previously established, and for extensions of, and realignments of, the urban pattern" (1). Just a few years later, Kellogg (1937) described a relational logic between cities and urban economic system: "In the city invention replaces philosophy, the cathedral replaces the church, the delicatessen the garden, the night club the home; in short the city is sterile, biologically and spiritually. Later these cities, built by the wealth and sons of the soil become dominant and make economic and social arrangements to their own liking" (147).

In the longest urban geographic essay in the *Annals* to this point, Taylor (1942) employed an explicitly environmental determinist analysis in his sixty-seven-page "Environment, Village and City" to survey how the local environment shapes some twenty case studies of urban developments. In a section subtitled "Possibilism Applied to Race, Nation and City," Taylor wrote, "As most of my readers know, I have always been a rather definite environmentalist. In concluding this address, I wish to consider whether determinism or possibilism is of more importance in connection with the three types of human groups which I have studied with some thoroughness" (65). Here we capture the sort of questions that geographers were considering related to urban social process and spatial form that link directly over time as a problematic contribution to environmentally determinist understandings of the built form and urban marginality, in conversation with ideologies of the ghetto and culture of poverty to emerge later from the urban literature, broadly defined.

Murphey (1954) offered an early instance of the language of revolutionary change and the urban (in Western Europe and China), which is, interestingly, also the *Annals'* first reference to Marx, when he suggested, "The industrial revolution has emphasized the economic advantages of concentration and centrality. But is it true to say that change, revolutionary change, has found an advantage in urbanization; in concentration and numbers?" (349). Murphey foreshadowed questions taken up later in urban political ecology when he suggested, "The whole economical history of society is summed up in the movement of this ... separation between town and country" (350). Out of this language of revolutionary change, and similarly foreshadowing fundamental changes to urban growth and structure in the United States, came Nelson's (1962) "Megalopolis and New York Metropolitan Region." The following year, Burton (1963) published the first article explicitly discussing "urban sprawl." Although scale was mobilized in different ways in reference to "the city," Ulack (1978) published the first mention of "neighbourhood change" as well as "urban squatters." Ultimately, attention to the dynamics of scalar, urban change led to framing problems and their solutions in far more nuanced ways that considered linkages across processes and practices at multiple scalar levels as evidenced when Dingemans (1979) published the *Annals'* first analysis of "redlining" and "urban design."

Throughout the 1980s and 1990s, the *Annals* published vibrant urban theoretical papers that significantly advanced ongoing discussions, such as Soja (1980), who, responding to Harvey (1973), argued for the importance of space and extended Lefebvre's work on everyday urban life and the concept of uneven development in generative ways. Harvey (1990) also continued to push urban theory toward more comprehensive formulations. Dear and Flusty's (1998) "Postmodern Urbanism" offered the first reference to this theme in the *Annals* by way of comparative analysis of the Chicago and L.A. School approaches of urban geography.

At the same time, we saw urban geography expand from "the city" to other spatial configurations as evidenced by Walker and Heiman's (1981) claim that a "major response of Great Society liberals to the black and poor people's movements of the 1960s was to attack suburban exclusion as the cause of lack of access to jobs and housing" (74). They went on to note, "The social and land use control reform movements thus coincided in a program to 'open up the suburbs' which helps open up emergent work of suburban geographies focused of social justice questions" (74). As the concept of suburbia was developing for urban geographers so too was "global cities" research. Although there were a number of descriptive urban papers about cities in the Global South throughout these decades, it was not until Mitchelson and Wheeler's (1994) first reference to the idea of

"global cities" that we see it as a concept and object of analysis explicitly connected to work on globalization.

One of the most important themes within this special issue to follow relates to processes of displacement and gentrification, making a deeper context of this theme in the *Annals* important. Price and Young (1959) offered some of the earliest discussion of housing markets presaging cycles of urban change, gentrification, and industrial decline: "Whether the land should have been built up to avoid such a problem is a question this generation does not ask, and the next one, not faced with our choices, really will not be able to answer. ... With continued growth the old core will demand attention, if for no other reason than the great space it occupies" (112).

Another landmark publication is that of Schaffer and Smith (1986), discussing "The Gentrification of Harlem," in which they captured the state of discussion twenty-seven years after Price and Young by stating, "Debate over gentrification has emerged around three main questions: the significance of the process (or its extent), the effects of gentrification, and its causes. ... It will quickly become obvious that these three issues are closely interrelated" (348). That same year Pratt (1986) published "Housing Tenure and Social Cleavages in Urban Canada," showing a strong spatial correlation between home ownership and access to capital gains across all classes of homeowners. The next year the *Annals* published Smith's (1987) landmark "Gentrification and the Rent Gap," in which he expanded on his earlier idea, stating, "If the early literature tended narrowly to emphasize either consumption-side or production-side explanations (such as the rent gap), it should now be evident that the relationship between consumption and production is crucial to explaining gentrification. The restructuring of the city, of which gentrification is only a part, involves a social and economic, spatial and political transformation" (464). To this, Ley (1987) replied, "The discussion of the rent gap thesis occupied only a small portion of my paper," going on to suggest that Smith "is anxious to defend it, particularly as his theoretical framework has not fared well in recent reviews" (465).

Dubois's (1903) prophetic declaration that "the problem of the twentieth century is the problem of the color-line" (41) was clear in the pages of the *Annals* related to "the city." In Bogue (1954) we see race used descriptively to examine urban structure and evolution: "Literally thousands of nonwhites (mostly

Negroes) left the soil to settle in urban centers. The result was almost a doubling of the nonwhite population of cities within the decade" (130). Hart's (1960) "The Changing Distribution of the American Negro" details the urbanization of black populations, and Lewis (1965) discussed the impact of African American migration on electoral geographies of Flint, connecting political geographies to the expansion of African American neighborhoods.

Importantly, given the urban uprisings that in part ushered in geography's broader radical turn, 1970 is the first year we see an article using the language of "racial segregation" in the city. In a hugely important essay, Rose (1970) outlined early understandings of these dynamics: "The Negro ghetto represents an expanding residential spatial configuration in all of the major metropolitan areas in the United States. The process of ghetto development is essentially related to the refusal of Whites to share residential space with Blacks on a permanent basis, and to the search behavior employed by Blacks in seeking housing accommodations" (1). Related to this theme, the early 1970s saw a significant increase in attention to racial urban geography, including Brunn and Hoffman (1970), Hartshorn (1971), and Bennett (1973).

At this same time, Bunge published on his (and his collaborators') Fitzgerald project, extending the discussion of race in the city but adding an explicit and important engagement with youth geographies: "Urban exploration, the use of geography in the protection of children, survival geography, because of the overwhelming need, cannot wait for a gradual acceptance. Those of us who are convinced must plunge ahead even if it upsets our fellow tradesmen" (Bunge 1974, 485). Bunge cross-cut this discussion by talking explicitly about race: "There is only 'racism,' and only a racist would not know it. To use a racist term like 'white racism' implies that there is a 'black racism,' which there is not. If you bridle at this logic you are a racist, so follow the following logic, over and over, if necessary, until you have purged yourself of your racism. It is true that blacks often hate whites, but this is not racism, this is a reaction to racism" (485). Thus, Bunge's expounding on the importance of language and categories here suggests that for him, urban geographic research should always be linked to both anti-racist praxis and writing.

The mid-1970s and 1980s saw a smattering of important work connecting racialized processes to the structure of the city. Another landmark essay in this regard was Anderson's (1987) "The Idea of Chinatown," which opened up discussion about the social construction of race and ethnoracial urban districts for geographers when she argued, "Racial categories are cultural ascriptions whose construction and transmission cannot be taken for granted" (580). The next year Marston (1988) published "Neighborhood and Politics: Irish Ethnicity in Nineteenth Century Lowell, Massachusetts," further expanding discussions of race and ethnicity, followed soon after by Aiken's (1990) "A New Type of Black Ghetto in the Plantation South." Specifically, Aiken thought through the localized dynamics of race across scalar processes, and noted that "[a]t the regional scale, unequal changes in white and black municipal populations produced an increase in segregation among the municipalities of the Yazoo Delta. At the local scale, a pattern of residential desegregation has emerged in particular municipalities" (223).

The connection between (urban) space, property, and vulnerable bodies was a thread that began to develop in earnest in the 1990s. Rowe and Wolch's (1990) "Social Networks in Time and Space: Homeless Women in Skid Row, Los Angeles" is one of the earliest ethnographic projects in the city with street-based vulnerable populations. Related to this, Mitchell (1995) helped raise the visibility of the politics of public space by suggesting, "So long as we live in a society which so efficiently produces homelessness, spaces like these will be—indeed *must* be—always at the center of social struggle. For it is by struggling over and within space that the natures over 'the public' and of democracy are defined" (128). Others such as Blomley (2003) opened up discussions on urban property relations and the spatial power therein, for example, by arguing that "[p]hysical violence, whether realized or implied, is important to the legitimation, foundation, and operation of a Western property regime" (121). This marks a key turn within urban geography specifically, toward understanding the underlying colonial dynamics of urban land, no doubt indebted to ongoing Indigenous struggles Blomley witnessed within the city.

Although Mitchelson and Wheeler (1994) opened the language of "global cities" in the *Annals*, there is a string of detailed papers published early in the journal about the development of cities and regions outside of North America, including Hall (1934) on Japan and Hoselitz (1959) on cities in India. The tone and tenor of these early articles offer important context for appreciating the evolution of work in cities across the world, especially postcolonial cities. McFarlane and Graham's (2014) "Informal Urban Sanitation:

Everyday Life, Poverty, and Comparison" and Doshi and Ranganathan's (2017) "Contesting the Unethical City: Land Dispossession and Corruption Narratives in Urban India" are exemplars for thinking through this growth.

Unearthing the evolution of the city in the *Annals* yielded substantial anti-colonial and anti-racist interventions, as did going through the back issues with an eye toward excavating what came before social justice. The different cuts we took through the archives offer up a range of different kinds of colonial, gendered, racialized, and environmental determinist sentiments that help contextualize how far the discipline has evolved. This context is especially important for highlighting the innovative and politically astute research within this special issue.

This Special Issue

The more contemporary archives of the *Annals* show the continued interest of geographers in big questions about justice (Mitchell 2004), ethics (Kearns 1998), democracy (Barnett and Bridge 2013), and political relevance of geographic scholarship (Staeheli and Mitchell 2005). Many of the themes that have long been central to papers published in the *Annals* are represented in this special issue, albeit often in ways critical of the approach of many of the earliest publications. There are several essays included that take on the social theoretical connections between democracy and theories of justice. For instance, Barnett engages notions of social justice within human geography and urban studies by tracing the recurrent disavowal of "liberalism" in debates on social justice and the city, the just city, and spatial justice. Through a feminist approach, Wright explores the way in which the forced disappearances of Mexican students in 2014 open up new approaches to link democratic theories of justice with the creation of counterpublics that become necessary amidst the absence of those activists who cannot stand for their rights. Examples embodying these counterpublics, such as mass graves and prisons, paint a picture of a necessary restoration to versions of democracy attempting to operate amidst disappearances. Lake uses a relational approach to social justice to investigate the larger democratic ramifications for social movement collective action against housing displacement.

The centrality of gentrification and displacement starting in the 1980s within the *Annals* continues in

this special issue. Lees, Annunziata, and Rivas-Alonso insist that notions of survivability are key to understanding the ongoing ravages of planetary gentrification and that organizing against it should be composed of both overt opposition and everyday (often invisible) resistance. Shin's paper seeks to show how grassroots people come to realize notions of rights in their struggles against urban redevelopment and displacement amidst efforts around urban speculative accumulation in Seoul. Muñoz's research into housing justice in Buenos Aires shows that when precarious housing is understood at the scale of the home, as opposed to broader urban spatial scales, it can offer new ways of seeing how the right to the city is endeavored, challenged, and denied. Maharawal and McElroy discuss the Anti-Eviction Mapping Project active within the San Francisco Bay Area to demonstrate how countermapping, inspired by feminist and decolonial science studies, can be mobilized through robust data visualization and oral historical analysis for posing challenges to the proliferation of gentrification and eviction. Through sophisticated storytelling methodologies, they embody social justice work in innovative and compelling ways.

In considering contradictions of social justice and the city, several articles in the special issue investigate questions of law, policy, and policing. Swanson investigates how zero tolerance policing as exported from New York to Ecuador led to transnational displacements of street vendors from Ecuador back to New York, demonstrating the perverse role of policy mobilities exacerbating gentrification and socioscalar disruptions at the global scale. Hamilton and Foote use police torture of hundreds of black men in Chicago to help us understand how far theorizing race and violence have evolved since the beginnings of the radical turn in geography and particularly since the publication of *Social Justice and the City*. They show the urgent necessity of vigilance in their theorizing amidst ongoing violence done to black men in cities across the United States as well as the rise of Black Lives Matter. Brownlow investigates the underpolicing of rape in U.S. cities to show how race continues to bias where rape is likely to be undercounted and hidden.

In the rich, if recent, tradition of feminist scholarship in the *Annals*, several articles in this special issue extend different forms of feminist theorization to questions of social justice. Mott's article centered on organizing around Arizona's racial profiling legislation SB 1070 offers thought-provoking social movement insights into how anarcha-feminist solidarity work rooted in horizontal praxis can negotiate across race, class, gender, language, and documentation status. Dowler and Ranjbar use the lens of a feminist ethics of care to look at how efforts at "just praxis" and "positive security" can be mobilized to overcome vulnerability to political and environmental violence in both Belfast, Northern Ireland, and Orumiyeh, Iran. Arpagian and Aitkin open up possibilities of social justice in the city in their discussion of Roma dispossession in Bucharest that foregrounds the emotive politics of care as foundational to politics against dispossession and exclusion.

Two of the articles help build the growing focus on LGBTQ struggles in the city. Roberts and Catungal investigate a public–private partnership between an LGBTQ-focused center, a private philanthropist, and the City of Toronto to show the ways social justice ideals continue to be limited and thwarted through the proliferation of urban processes of neoliberalization. Goh investigates how "unjust geographies" that cross-cut race, class, and gender are central to LGBTQ activism in New York and how queer community organizers and activists are fighting for social and spatial change.

Questions of citizenship are also taken up in the special issue. Ye and Yeoh investigate how narratives of multiracialism help explain how the diversification of peoples in the global city is also paralleled by the diversification of precarity. Richardson's investigation of Occupy Hong Kong offers insight into postcolonial ideas of social justice through a focus on citizenship and whiteness for better understanding the contested politics of universal suffrage in a centrally important global city.

The emphasis on property and vulnerability present in the *Annals* is also extended by several articles in this special issue. Safransky builds from scholars of critical race studies, critical Indigenous studies, and decolonial theory to investigate underresearched questions related to the cultural and racial politics of land and property dispossession in Detroit. She asserts land justice to be a historical diagnostic for thinking about similar questions in other North American cities. Barraclough investigates the urban valences of settler colonialism by showing how ideologies of the cowboy and the frontier continue to limit the abilities of Indigenous and other marginalized people to realize social justice in the U.S. West.

Although also related to questions of property and land, several essays in the collection attend to environmental justice, urban political ecology, and disaster response. Grove et al. investigate the deep historical connections between environmental inequality and segregation in Baltimore. Using the Baltimore Ecosystem Study (BES) as a foundation, the article shows the indelible patterns left by decades of urban social processes on socioracial patterns of the city that help understand the long-term notions of environmental justice. De Lara mobilizes urban political ecology and theories of racial capitalism to show how sustainability discourse is conceptualized as a way of challenging green capitalism in Southern California. Valdivia expands discussions of environmental justice and urban political ecology, putting these approaches into tighter conversation around the politics of social reproduction amidst the ravages of frontier-style petro capitalism. Simpson and Bagelman work to decolonize urban political ecology through unpacking and deconstructing narratives of settler colonialism. They show how agricultural politics around Indigenous food production in British Columbia become central to the reproduction of urban political ecologies in settler colonial cities. Burns brings in a digital humanitarian discussion of data collection amidst urban disasters, using the case of Superstorm Sandy to show the struggles and motivations of emergency providers and other interested researchers who struggle to capture the needs of individuals and communities during emergency disaster events.

Finally, and related to and springing from questions about environmental justice, two articles examine food justice. McClintock explores strategic resistance to gentrification in different ways than others in the special issue by focusing on the connections between food justice and social reproduction, asserting that "ecogentrification is not only a contradiction emerging from an urban sustainability fix, but is central to how racial capitalism functions through green urbanization." Pettygrove and Ghose examine the neoliberalization of food activism in Milwaukee to show the ongoing complexities of urban agricultural production across public–private partnerships.

Conclusion

This special issue reflects on the trajectory of urban social justice oriented work within geography, the compelling and important research undertaken to represent and contribute to these struggles, and the work that remains to be done to enact just practices across our various engagements within and beyond the academy. It is troubling to us to review the colonial, racist, and masculinist history of geography that endured in at least the metaphorical language used to summarize its early progress. After fifty years of publication for the *Annals*, Whitaker (1954) suggested, "The first fifty years of this Association have seen the pioneering tasks completed, the land taken up, the clearing done, the seed planted. Our task is to produce more abundant harvests as the years go by" (244). We agree with Blaut's (1979) view on the long history of geographic scholarship in his "The Dissenting Tradition" when he said, "Geography is not socially and politically neutral. It never has been such, and it never can be such" (158). In a similar way, Lave (2015) argued that within this moment of waning scientific influence, "lending our authority, however reduced, to the production of knowledge for progressive, justice-focused ends," geographers might collectively "achieve more of our political and intellectual goals by embracing the progressive aspects of our reduced authority than by fighting its erosion" (245). The range of problems and approaches those political questions provoke continue to become evermore nuanced and robust. This is evidenced both in the wide interest in the framing of this special issue as well as the thought-provoking and important scholarship published herein. To this end, we share Barkan and Pulido's (2017) sentiment that "all of this suggests a great role for geographic knowledge in the pursuit of justice" (38), and we think that this special issue conveys that this sentiment well. At the same time, a more emancipatory focus on social justice as it relates to how we produce geographic knowledge necessitates renewed and vigilant attention to the embodied practices, cross-positional interactions, and often outright oppressive conditions some of us continue to enact at the expense and harm of others.

Acknowledgments

Producing this special issue required many committed reviewers who we would like to thank. Special thanks to Richard Wright, James McCarthy, Kate Derickson, Mat Coleman, and Jamie Peck. Extra special thanks go to Jennifer Cassidento and Lea Cutler, without whom this special issue, or the high-quality publishing in the *Annals* more generally, would not be possible.

References

Aiken, C. S. 1990. A new type of black ghetto in the plantation South. *Annals of the Association of American Geographers* 80 (2):223–46.

Alderman, D. H., and J. Inwood. 2016. Mobility as antiracism work: The "hard driving" of NASCAR's Wendell Scott. *Annals of the American Association of Geographers* 106 (3):597–611.

Anderson, K. J. 1987. The idea of Chinatown: The power of place and institutional practice in the making of a racial category. *Annals of the Association of American Geographers* 77 (4):580–98.

Barkan, J., and L. Pulido. 2017. Justice: An epistolary essay. *Annals of the American Association of Geographers* 107 (1):33–40.

Barnett, C., and G. Bridge. 2013. Geographies of radical democracy: Agonistic pragmatism and the formation of affected interests. *Annals of the Association of American Geographers* 103 (4):1022–40.

Bennett, D. C. 1973. Segregation and racial interaction. *Annals of the Association of American Geographers* 63 (1):48–57.

Blaut, J. M. 1979. The dissenting tradition. *Annals of the Association of American Geographers* 69 (1):157–64.

Blomley, N. 2003. Law, property, and the geography of violence: The frontier, the survey, and the grid. *Annals of the Association of American Geographers* 93 (1):121–41.

———. 2014. Making space for property. *Annals of the Association of American Geographers* 104 (6):1291–1306.

Bonds, A. 2013. Economic development, racialization, and privilege: "Yes in my backyard" prison politics and the reinvention of Madras, Oregon. *Annals of the Association of American Geographers* 103 (6):1389–1405.

Boone, C. G., G. L. Buckley, J. M. Grove, and C. Sister. 2009. Parks and people: An environmental justice inquiry in Baltimore, Maryland. *Annals of the Association of American Geographers* 99 (4):767–87.

Bouge, D. J. 1954. The geography of recent population trends in the United States. *Annals of the Association of American Geographers* 44 (2):124–34.

Bowen, W. M., and M. J. Salling. 1995. Toward environmental justice: Equity in Ohio and Cleveland. *Annals of the Association of American Geographers* 85 (4):641–63.

Bowman, I. 1932. Planning in pioneer settlement. *Annals of the Association of American Geographers* 22 (2):93–107.

Branch, M. C. J. 1951. Physical aspects of city planning. *Annals of the Association of American Geographers* 41 (4):269–84.

Brickell, K. 2014. "The whole world is watching": Intimate geopolitics of forced eviction and women's activism in Cambodia. *Annals of the Association of American Geographers* 104 (6):1256–72.

Brigham, A. P. 1915. Problems of geographic influence. *Annals of the Association of American Geographers* 5 (1):3–25.

Brunn, S. D., and W. L. Hoffman. 1970. The spatial response of negroes and whites toward open housing: The Flint referendum. *Annals of the Association of American Geographers* 60 (1):18–36.

Bunge, W. 1973. The geography of human survival. *Annals of the Association of American Geographers* 63 (3):275–95.

———. 1974 Fitzgerald from a distance. *Annals of the association of American Geographers* 64 (3):485–88.

————. 1979. Perspectives on theoretical geography. *Annals of the Association of American Geographers* 69 (1):169–74.

Burton, I. 1963. A restatement of the dispersed city hypothesis. *Annals of the Association of American Geographers* 53 (3):285–89.

Bushman, D. O., and W. R. Stanley. 1971. State senate reapportionment in the Southeast. *Annals of the Association of American Geographers* 61 (4):654–70.

Byrd, J. A. 2011. *Transit of empire: Indigenous critiques of colonialism.* Minneapolis: University of Minnesota Press.

Chardon, R. E. 1963. Hacienda and Ejido in Yucatán: The example of Santa Ana Cucá. *Annals of the Association of American Geographers* 53 (2):174–93.

Colby, C. C. 1924. The California raisin industry: A study in geographic interpretation. *Annals of the Association of American Geographers* 14 (2):49–108.

————. 1933. Centrifugal and centripetal forces in urban geography. *Annals of the Association of American Geographers* 23 (1):1–20.

Collard, R., J. Dempsey, and J. Sundberg. 2015. A manifesto for abundant futures. *Annals of the Association of American Geographers* 105 (2):322–30.

Collins, T. W., S. E. Grineski, and D. X. Morales. 2017. Sexual orientation, gender, and environmental injustice: Unequal carcinogenic air pollution risks in greater Houston. *Annals of the American Association of Geographers* 107 (1):72–92.

Davis, D. H. 1925. Urban development in the Kentucky mountains. *Annals of the Association of American Geographers* 15 (2):92–99.

Dear, M., and S. Flusty. 1998. Postmodern urbanism. *Annals of the Association of American Geographers* 88 (1):50–72.

Derickson, K. D. 2014. The racial politics of neoliberal regulation in post-Katrina Mississippi. *Annals of the Association of American Geographers* 104 (4):889–902.

Derickson, K. D., and D. MacKinnon. 2015. Toward an interim politics of resourcefulness for the Anthropocene. *Annals of the Association of American Geographers* 105 (2):304–12.

Dingemans, D. 1979. Redlining and mortgage lending in Sacramento. *Annals of the Association of American Geographers* 69 (2):225–39.

Doshi, S., and M. Ranganathan. 2017. Contesting the unethical city: Land dispossession and corruption narratives in urban India. *Annals of the American Association of Geographers* 107 (1):183–99.

Du Bois, W. E. B. 1903. *The souls of black folk.* Chicago: A. C. McClurg.

Ellis, M, S. R. Holloway, R. Wright, and C. S. Fowler. 2012. Agents of change: Mixed-race households and the dynamics of neighborhood segregation in the United States. *Annals of the Association of American Geographers* 102 (3):549–70.

Elwood, S., V. Lawson, and S. Nowak. 2015. Middle-class poverty politics: Making place, making people. *Annals of the Association of American Geographers* 105 (1):123–43.

England, K. 1993. Suburban pink collar ghettos: The spatial entrapment of women? *Annals of the Association of American Geographers* 83 (2):225–42.

Featherstone, D. 2013. Black internationalism, subaltern cosmopolitanism, and the spatial politics of antifascism. *Annals of the Association of American Geographers* 103 (6):1406–20.

Gilbert, M. R. 1998. "Race," space, and power: The survival strategies of working poor women. *Annals of the Association of American Geographers* 88 (4):595–621.

Ginsburg, N. 1973. From colonialism to national development: Geographical perspective on patterns and policies. *Annals of the Association of American Geographers* 63 (1):1–21.

Gregory, H. E. 1915. The oasis of Tuba, Arizona. *Annals of the Association of American Geographers* 5 (1):107–19.

Haas, W. H. 1926. The cliff-dweller and his habitat. *Annals of the Association of American Geographers* 16 (4):167–215.

Hall, R. B. 1934. The cities of Japan: Notes on distribution and inherited forms. *Annals of the Association of American Geographers* 24 (4):175–200.

Hance, W. A. 1951. Crofting settlements and housing in the Outer Hebrides. *Annals of the Association of American Geographers* 41 (1):75–87.

Hans, W. H. 1925. The American Indian and geographic studies. *Annals of the Association of American Geographers* 15 (2):86–91.

Hanson, R., and P. Richards. 2017. Sexual harassment and the construction of ethnographic knowledge. *Sociological Forum* 32 (3):587–609.

Harris, C. 2004. How did colonialism dispossess? Comments from an edge of empire. *Annals of the Association of American Geographers* 94:165–82.

Hart, J. F. 1960. The changing distribution of the American. *Annals of the Association of American Geographers* 50 (3):242–66.

Hartshorn, T. A. 1971. Inner city residential structure and decline. *Annals of the Association of American Geographers* 61 (1):72–95.

Harvey, D. 1973. *Social justice and the city.* Baltimore: Johns Hopkins University Press.

————. 1990. Between space and time: Reflections on the geographical imagination. *Annals of the Association of American Geographers* 80 (3):418–34.

Helburn, N. 1982. Geography and the quality of life. *Annals of the Association of American Geographers* 72 (4):445–56.

Heynen, N. 2009. Bending the bars of empire from every ghetto for survival: The Black Panther Party's radical antihunger politics of social reproduction and scale. *Annals of the Association of American Geographers* 99 (2):406–22.

Hodder, J. 2016. Toward a geography of black internationalism: Bayard Rustin, nonviolence, and the promise of Africa. *Annals of the American Association of Geographers* 106 (6):1360–77.

Hoelscher, S. 2003. Making place, making race: Performances of whiteness in the Jim Crow South. *Annals of the Association of American Geographers* 93 (3):657–86.

Holifield, R. 2012. Environmental justice as recognition and participation in risk assessment: Negotiating and translating health risk at a superfund site in Indian Country. *Annals of the Association of American Geographers* 102 (3):591–613.

Holloway, S. R. 1998. Exploring the neighborhood contingency of race discrimination in mortgage lending in

Columbus, Ohio. *Annals of the Association of American Geographers* 88 (2):252–76.

Holzner, L. 1970. The role of history and tradition in the urban geography of West Germany. *Annals of the Association of American Geographers* 60 (2):315–39.

Hoselitz, B. F. 1959. The cities of India and their problems. *Annals of the Association of American Geographers* 49 (2):223–31.

Hovorka, A. J. 2005. The (re) production of gendered positionality in Botswana's commercial urban agriculture sector. *Annals of the Association of American Geographers* 95 (2):294–313.

Huntington, E. 1924. Geography and natural selection: A preliminary study of the origin and development of racial character. *Annals of the Association of American Geographers* 14 (1):1–16.

Inwood, J. 2012. Righting unrightable wrongs: Legacies of racial violence and the Greensboro truth and reconciliation commission. *Annals of the Association of American Geographers* 102 (6):1450–67.

Inwood, J., and A. Bonds. 2016. Confronting white supremacy and a militaristic pedagogy in the U.S. settler colonial state. *Annals of the American Association of Geographers* 106 (3):521–29.

Jacobs, J. 1961. *The death and life of great American cities.* New York: Vintage.

———. 1969. *The economy of cities.* New York: Vintage.

Jefferson, M. 1917. Some considerations on the geographical provinces of the United States. *Annals of the Association of American Geographers* 7 (1):3–15.

Jones, J. P., III, and J. E. Kodras. 1990. Restructured regions and families: The feminization of poverty in the U.S. *Annals of the Association of American Geographers* 80 (2):163–83.

Katz, C. 1991. Sow what you know: The struggle for social reproduction in rural Sudan. *Annals of the Association of American Geographers* 81 (3):488–514.

Kearns, G. 1998. The virtuous circle of facts and values in the new western history. *Annals of the Association of American Geographers* 88 (3):377–409.

Kellogg, C. E. 1937. Soil and the people. *Annals of the Association of American Geographers* 27 (3):142–48.

Kelskey, K. 2017. Sexual harassment in the academy: A crowdsource survey. The professor is in. Accessed August 3, 2017. https://docs.google.com/spreadsheets/d/1S9KShDLvU7C-KkgEevYTHXr3F6InTenrBsS9yk-8C5M/htmlview#gid=1530077352

King, L. J. 1976. Alternatives to a positive economic geography. *Annals of the Association of American Geographers* 66 (2):293–308.

Kobayashi, A. 2010. People, place, and region: 100 years of human geography in the *Annals. Annals of the Association of American Geographers* 100 (5):1095–1106.

———. 2014. The dialectic of race and the discipline of geography. *Annals of the Association of American Geographers* 104 (6):1101–15.

Kobayashi, A., and L. Peake. 2000. Racism out of place: Thoughts on whiteness and an antiracist geography in the new millennium. *Annals of the Association of American Geographers* 90 (2):392–403.

Lave, R. 2015. The future of environmental expertise. *Annals of the Association of American Geographers* 105 (2):244–52.

Lawrence, H. W. 1993. The greening of the squares of London: Transformation of urban landscapes and ideals. *Annals of the Association of American Geographers* 83 (1):90–118.

Laws, G. 1994. Social justice and urban politics: An introduction. *Urban Geography* 15 (7):603–11.

Lefebvre, H. [1974] 1991. *The production of space,* trans. D. Nicholson-Smith. Oxford, UK: Basil Blackwell.

Leitner, H. 2012. Spaces of encounters: Immigration, race, class, and the politics of belonging in small-town America. *Annals of the Association of American Geographers* 102 (4):828–46.

Lewis, P. F. 1965. Impact of negro migration on the electoral geography of Flint, Michigan, 1932–1962: A cartographic analysis. *Annals of the Association of American Geographers* 55 (1):1–25.

Ley, D. 1980. Lineral ideology and the postindustrial city. *Annals of the Association of American Geographers* 70 (2):238–58.

———. 1987. Reply: The rent gap revisited. *Annals of the Association of American Geographers* 77 (3):465–68.

Lowry, M., II. 1971. Population and race in Mississippi, 1940–1960. *Annals of the Association of American Geographers* 61 (3):576–88.

Mansfield, B., K. McSweeney, L. Horner, D. K. Munroe, C. Biermann, J. Law, and C. Gallemore. 2015. Environmental politics after nature: Conflicting socioecological futures. *Annals of the Association of American Geographers* 105 (2):284–93.

Marston, S. A. 1988. Neighborhood and politics: Irish ethnicity in nineteenth century Lowell, Massachusetts. *Annals of the Association of American Geographers* 78 (3):414–32.

Martin, L. 1930. The Michigan–Wisconsin boundary case in the Supreme Court of the United States, 1923–26. *Annals of the Association of American Geographers* 20 (3):105–63.

Massey, D. B., and A. Catalano. 1978. *Capital and land: Landownership by capital in Great Britain.* London: Edward Arnold.

Matley, I. M. 1966. The Marxist approach to the geographical environment. *Annals of the Association of American Geographers* 56 (1):97–111.

McFarlane, C. R. Desai, and S. Graham. 2014. Informal urban sanitation: Everyday life, poverty, and comparison. *Annals of the Association of American Geographers* 104 (5):989–1011.

McKittrick, K., and C. A. Woods. 2007. *Black geographies and the politics of place.* Toronto: Between the Lines.

McMurry, K. C. 1930. The use of land for recreation. *Annals of the Association of American Geographers* 20 (1):7–20.

McNee, R. B. 1961. Centrifugal–centripetal forces in international petroleum company regions. *Annals of the Association of American Geographers* 51 (1):124–38.

Meinig, D. W. 1972. American Wests: Preface to a geographical interpretation. *Annals of the Association of American Geographers* 62 (2):159–84.

Meyer, A. H. 1956. Circulation and settlement patterns of the Calumet Region of northwest Indiana and northeast Illinois (the second stage of occupance–pioneer settler and subsistence economy, 1830–1850). *Annals of the Association of American Geographers* 46 (3):312–56.

Mitchell, D. 1995 The end of public space? People's Park, definitions of the public, and democracy. *Annals of the Association of American Geographers* 85 (1):108–33.

———. 2004. Geography in an age of extremes: A blueprint for a geography of justice. *Annals of the Association of American Geographers* 94 (4):764–70.

Mitchelson, R. L., and J. O. Wheeler. 1994. The flow of information in a global economy: The role of the American urban system in 1990. *Annals of the Association of American Geographers* 84 (1):87–107.

Mullings, B. 1999 Sides of the same coin? Coping and resistance among Jamaican data-entry operators. *Annals of the Association of American Geographers* 89 (2):290–311.

Murphey, R. 1954. The city as a center of change: Western Europe and China. *Annals of the Association of American Geographers* 44 (4):349–62.

Murray, M. A. 1967. The geography of death in the United States and the United Kingdom. *Annals of the Association of American Geographers* 57 (2):301–14.

Naylon, J. 1959. Land consolidation in Spain. *Annals of the Association of American Geographers* 49 (4):361–73.

Nelson, H. J. 1962. Megalopolis and New York metropolitan region: New studies of the urbanized eastern seaboard. *Annals of the Association of American Geographers* 52 (3):307–17.

Nostrand, R. L. 1970. The Hispanic-American borderland: Delimitation of an American culture region. *Annals of the Association of American Geographers* 60 (4):638–61.

Parkins, A. E. 1931. The antebellum South: A geographer's interpretation. *Annals of the Association of American Geographers* 21 (1):1–33.

Parks, V. 2016. Rosa Parks redux: Racial mobility projects on the journey to work. *Annals of the American Association of Geographers* 106 (2):292–9.

Parsons, J. J. 1955. The Miskito pine savanna of Nicaragua and Honduras. *Annals of the Association of American Geographers* 45 (1):36–63.

Peet, R. 1975. Inequality and poverty: A Marxist-geographic theory. *Annals of the Association of American Geographers* 65 (4):564–71.

Platt, R. S. 1948. Determinism in geography. *Annals of the Association of American Geographers* 38 (2):126–32.

Pratt, G. 1986. Housing tenure and social cleavages in urban Canada. *Annals of the Association of American Geographers* 76 (3):366–80.

Price, E. T. 1953. A geographic analysis of white-negro-Indian racial mixtures in eastern United States. *Annals of the Association of American Geographers* 43 (2):138–55.

Price, E. T., and R. N. Young. 1959. The future of California's southland. *Annals of the Association of American Geographers* 49 (3):101–17.

Pulido, L. 2000. Rethinking environmental racism: White privilege and urban development in Southern California. *Annals of the Association of American Geographers* 90 (1):12–40.

Radcliffe, S. A. 2007. Latin American indigenous geographies of fear: Living in the shadow of racism, lack of development, and antiterror measures. *Annals of the Association of American Geographers* 97 (2):385–97.

Rice, J. L., B. J. Burke, and N. Heynen. 2015. Knowing climate change, embodying climate praxis: Experiential knowledge in Southern Appalachia. *Annals of the Association of American Geographers* 105 (2):253–62.

Rose, H. M. 1970. The development of an urban subsystem: The case of the negro ghetto. *Annals of the Association of American Geographers* 60 (1):1–17.

Rowe, S., and J. Wolch. 1990. Social networks in time and space: Homeless women in skid row, Los Angeles. *Annals of the Association of American Geographers* 80 (2):184–204.

Sauer, C. O. 1971. The formative years of Ratzel in the United States. *Annals of the Association of American Geographers* 61 (2):245–54.

Schaffer, R., and N. Smith. 1986. The gentrification of Harlem? *Annals of the Association of American Geographers* 76 (3):347–65.

Schroeder, C. G. 2014. (Un)holy Toledo: Intersectionality, interdependence, and neighborhood (trans)formation in Toledo, Ohio. *Annals of the Association of American Geographers* 104 (1):166–81.

Semple, E. C. 1919. Climatic and geographic influences on ancient Mediterranean forests and the lumber. *Annals of the Association of American Geographers* 9 (1):13–40.

———. 1929. Irrigation and reclamation in the ancient Mediterranean region. *Annals of the Association of American Geographers* 19 (3):111–48.

Smith, N. 1987. Gentrification and the rent gap. *Annals of the Association of American Geographers* 77 (3):462–65.

Soja, E. W. 1980. The socio-spatial dialectic. *Annals of the Association of American Geographers* 70 (2):207–25.

Staeheli, L. A., and D. Mitchell. 2005. The complex politics of relevance in geography. *Annals of the Association of American Geographers* 95 (2):357–72.

Taylor, G. 1942. Environment, village and city: A genetic approach to urban geography; with some reference to possibilism, *Annals of the Association of American Geographers* 32 (1):1–67.

Trewartha, G. T. 1938. French settlement in the driftless hill land. *Annals of the Association of American Geographers* 28 (3):179–200.

Tuck, E., and K. W. Yang. 2012. Decolonization is not a metaphor. *Decolonization: Indigeneity, Education & Society* 1 (1):1–40. Accessed August 3, 2017. http://decolonization.org/index.php/des/article/view/18630.

Tyner, J. A. 2006. "Defend the ghetto": Space and the urban politics of the Black Panther Party. *Annals of the Association of American Geographers* 96 (1):105–18.

Ulack, R. 1978. The role of urban squatter settlements. *Annals of the Association of American Geographers* 68 (4):535–50.

Visher, S. S. 1925. Regional geography of southeastern Wyoming from the viewpoint of land classification. *Annals of the Association of American Geographers* 15 (2):65–85.

Waibel, L. 1943. The political significance of tropical vegetable fats for the industrial countries of Europe. *Annals of the Association of American Geographers* 33 (2):118–28.

Wainwright, J., and G. Mann. 2015. Climate change and the adaptation of the political. *Annals of the Association of American Geographers* 105 (2):313–21.

Waitt, G. R. 2006. Boundaries of desire: Becoming sexual through the spaces of Sydney's 2002 gay games.

15

Annals of the Association of American Geographers 96 (4):773–87.

Walker, R. A., and M. K. Heiman. 1981. The quiet revolution. *Annals of the Association of American Geographers* 71 (1):67–83.

Ward, D. 1968. The emergence of central immigrant ghettoes in American cities: 1840–1920. *Annals of the Association of American Geographers* 58 (2): 343–59.

Webley Adler, K. 2017. Female scientists report a horrifying culture of sexual assault. *Marie Claire*. Accessed December 15, 2017. http://www.marieclaire.com/career-advice/a14104684/sexual-harassment-assault-in-science-field/

Whitaker, J. R. 1941. Sequence and equilibrium in destruction and conservation of natural resources. *Annals of the Association of American Geographers* 31 (2):129–44.

Whitaker, J. R. 1954. The way lies open. *Annals of the Association of American Geographers* 44 (3):231–44.

Whitbeck, R. H. 1912. Industries of Wisconsin and their geographic basis. *Annals of the Association of American Geographers* 2 (1):55–64.

Whittlesey, D. 1945. The horizon of geography. *Annals of the Association of American Geographers* 35 (1): 1–36.

Wolford, W. 2004. This land is ours now: Spatial imaginaries and the struggle for land in Brazil. *Annals of the Association of American Geographers* 94 (2): 409–24.

Woods, C. 1998. *Development arrested: The blues and plantation power in the Mississippi Delta.* London: Verso.

Wright, M. W. 2004. From protests to politics: Sex work, women's worth, and Ciudad Juárez modernity. *Annals of the Association of American Geographers* 94 (2):369–86.

NIK HEYNEN is a Professor in the Department of Geography at the University of Georgia, Athens, GA 30601. E-mail: nheynen@uga.edu. His research interests include urban political ecology and the politics of race, class, and gender.

DANI AIELLO is a PhD Candidate in the Department of Geography at the University of Georgia, Athens, GA 30601. E-mail: daiello@uga.edu. Her research interests include housing inequality, critical race, settler colonial studies, and qualitative geographic information systems. She is currently working on a comparative project on evictions in Atlanta, Georgia, and Vancouver, British Columbia, as ongoing forms of racialized dispossession and violence.

CAROLINE KEEGAN is a PhD Student in the Department of Geography at the University of Georgia, Athens, GA 30601. E-mail: caroline.keegan@uga.edu. Her research interests include labor, social reproduction, and critical race and ethnic studies. She is currently working on a project examining farmworker labor and drought.

NIKKI LUKE is a PhD Student in the Department of Geography at the University of Georgia, Athens, GA 30601. E-mail: knluke@uga.edu. Her research examines labor, environmental, and environmental justice politics around energy transition in the United States and draws from urban political ecology and feminist and antiracist scholarship.

1 Geography and the Priority of Injustice

Clive Barnett

This article considers the challenges that follow from giving conceptual priority to injustice in the analysis of political life. Human geography, urban studies, and related fields of spatial theory meet this challenge halfway, insofar as expressions of injustice through social movement mobilizations are given primacy over philosophical elaborations of justice. The privileging of practice over theory, however, reproduces a structure of thought in which justice continues to be understood as an egalitarian ideal against which injustice shows up as an absence or deviation. The practical primacy accorded to expressed claims of injustice inadvertently displaces a model of authoritative, monological reasoning about the meaning of justice from ideal theory onto explanatory accounts and ontologies of space. Basic assumptions about how spatial theory matters to questions of justice are disclosed by tracing the recurrent disavowal of "liberalism" in debates on social justice and the city, the just city, and spatial justice. Thinking about claims of injustice in a double sense—as involving demands on others that require vindication—calls into question the value of inherited ideals of the political significance of the "the city," by drawing attention to the enactment of distributed public spaces of claims-making, reasoning, and accountable action. *Key Words: the city, democracy, injustice, justice, public reason.*

本文考量在分析政治生活中, 赋予不正义概念上的优先性所带来的挑战。人文地理学、城市研究, 以及空间理论的相关领域在途中遭遇了此般挑战, 因为相较于正义的哲学阐述而言, 透过社会运动动员所表达的不正义被赋予了优先性。但偏好实践而非理论, 却再生产了正义不断被理解为对抗不正义的自发性理想之思想结构, 其中不正义展现作为一种缺乏或非常态。赋予表达有关不正义的宣言实践上的优先性, 无意间将有关正义的意义之威权且单一逻辑的说理模型, 从理想化的理论置换成解释性的说明与空间本体。本文藉由追溯在社会正义与城市、正义城市和空间正义的辩论中对 "自由主义" 不断重复的否定, 揭露空间理论如何关乎正义问题的基本预设。透过关照製定提出宣称、说理和应负责任的行动中的分派公共空间, 以双重意义思考有关不正义的宣称——作为涉及要他人进行辩白的需求——质疑 "城市" 的政治显着性的内在理想之价值。 关键词: 城市, 民主, 不正义, 正义, 公共说理。

Este artículo se refiere a los retos que se desprenden de conceder prioridad conceptual a la injusticia en el análisis de la vida política. La geografía humana, los estudios urbanos y los campos relacionados de la teoría espacial enfrentan este reto a medias, en la medida en que a las expresiones de injusticia a través de las movilizaciones del movimiento social se les da primacía sobre las elaboraciones filosóficas de la justicia. Privilegiar la práctica sobre la teoría, sin embargo, reproduce una estructura de pensamiento en la que la justicia sigue entendiéndose como un ideal igualitario contra el cual la injusticia se manifiesta como ausencia o desviación. La primacía práctica acordada para las reclamaciones expresas de injusticia desplaza inadvertidamente un modelo de razonamiento autoritario y monológico acerca del significado de la justicia desde la teoría ideal en los relatos explicativos y ontologías del espacio. Los supuestos básicos sobre cómo importa la teoría espacial en cuestiones de justicia son revelados trazando la recurrente negación del "liberalismo" en debates sobre justicia social y ciudad, la ciudad justa y la justicia espacial. Pensar en las reclamaciones de injusticia en un doble sentido—implicando las demandas de otros que requieren vindicación—pone en duda el valor de los ideales heredados del significado político de "la ciudad", llamando la atención hacia la reinstauración de los espacios públicos distribuidos, de acción responsable, generadora de reclamaciones y razonable. *Palabras clave: la ciudad, democracia, injusticia, razón pública.*

Social justice has been a central theme of critical scholarship in human geography, urban and regional studies, and planning theory since the 1970s. Spatial theorists are often reluctant to specify the content of justice or other normative standards, however. Although some scholars do argue that it is necessary to justify the substance of normative standards, it is more common to assert the primacy of practice as the arena in which the value of justice arises, so that analytical attention is given to the investigation of explicit struggles against injustice. Both of these pathways remain faithful to a "normal model" of theorizing justice (see Shklar 1990). In both cases, it is assumed that justice is a positive ideal from which injustice is a deviation, an ideal that is only ever empirically registered as an absence. The paradigm case of the normal model is the work of Rawls (1971, 1993), who held fast to the view that injustice is recognized as a departure from a prior ideal of justice. In spatial disciplines, commitment to the normal model of thinking about justice is most evident in the stylistic convention of running justice and injustice together as one word—as "(in)justice" (Soja 2010, 5).

The idea that injustice appears against the background of an egalitarian ideal, from which injustice is a departure, is a shared feature of what are otherwise quite different traditions of analysis. In strands of radical geography, the insights of alternative ontologies or explanatory theories are meant to supersede what are regarded as the inherently individualizing and universalizing tendencies of normative reasoning about the justifiability of particular social arrangements. In strands of radical empiricism, it is presumed that amassing empirical data on patterns of inequality is equivalent to mapping injustice, the persistence of which is attributed to a set of "beliefs" (e.g., Dorling 2011). More generally, critical social scientists often assume that injustice becomes visible by comparing actual patterns of disadvantage against ideal theories. For example, Smith's (1994) account of social justice as a process of equalization presupposes that injustice is identified by reference to an ideal of equality: "Social justice is manifest in reductions in inequality: a process of returning to equality" (118). From a somewhat different perspective, Wright (2009) invoked Derrida's account of the aporia that separates the idea of justice from its impure expression in any given form of law to argue that justice is an ideal that although never fully realizable in practice nevertheless orients action in the present.

I argue here that by continuing to think of injustice and justice as lying on either side of a division between practice and theory, spatial theorists inadvertently reproduce the form of authoritative third-person reasoning that is the most problematic feature of normative political philosophies of justice. My suggestion is that giving conceptual priority to injustice in understanding the dynamics of political life—where this refers to theorizing injustice independently from a prior formulation of an ideal principle of justice—requires a more fundamental reconsideration of the conventions of critical theorizing about the spatialities of political action (see Barnett 2017). Giving priority to injustice is associated with an emphasis on claims-making as the dynamic through which relations of domination are challenged. Thinking of justice in terms of claims-making might appear, at first sight, to support a view of the spaces of politics that privileges accessibility to classically defined public spaces, not least those of "the city." It seems to invite us to think of claims as being expressed in particular ways, through more or less spectacular performance of presence in physical space. In contrast, I argue that giving conceptual priority to injustice as a norm for critical analysis requires a rather more provisional account of the spatial configurations through which claims of injustice can be articulated and vindicated than has become established in contemporary geographical thought. It also challenges the prevalent conventions of critique in these fields. Critical analysis often presents the exposure of the facts of geographical entanglement and the contingencies of spatial patterns as if this carried automatic normative force (e.g., Whatmore 2002; Massey 2004). By contrast, from the perspective associated with giving conceptual priority to injustice, the demonstration of relations of spatial interconnectivity and interdependence should serve as the occasion for further reflection on the parameters of practical reasoning about political action—parameters that remain somewhat taken for granted in critical spatial thought (see also Young 2011).

Justice Emergent

The idea that questions of social justice are best approached by closely cleaving to "the plurality of injustice" is a long-standing convention in human geography and urban studies (Merrifield and Swyngedouw 1995, 3). Spatial disciplines have become adept at pluralizing the norms of justice to include issues of distributive equity, recognition, interactional justice, procedural justice, and care (e.g., Fincher and Ivesen 2012; Low and Ivesen 2016). Research on environmental justice is perhaps the most advanced field of geographical inquiry in which primacy is accorded to the practical emergence of expressions of injustice. The guiding ethos in this field is a vision of movement-oriented scholarship that is "more focused on understanding and addressing the problem than on constructing an ideal" (Schlossberg 2013, 47). On this understanding, the focus of attention is on how various standards of justice are strategically developed and deployed by social movements. Likewise, research guided by the idea of "the right to the city" is framed by an assumption that values arise from engaged political action. So, for example, Soja's (2010, 8) discussion of "spatial justice" tracks how this notion "is being used politically and strategically in social movements of all kinds." The emphasis on the emergent qualities of normative criteria is associated with a view of claims-making as the medium through which expressions of injustice are articulated, as in Lefebvre's (1996, 158) presentation of the idea of the right to the city as a "cry and demand."

The recommendation that critical analysis be responsive to the immanent values that express shared senses of injustice should certainly be taken seriously. There are two temptations, however, that the practical privileging of injustice can lead us into, and both should be resisted. The first is the impression that one could reconstruct the meaning of justice by tracing the contents of visible expressions of the sense of injustice. The second and related temptation is to assume that critical analysis necessarily involves an elective identification with favored activist voices or with the interests of victims. Both of these temptations threaten to distract from the conundrum that arises from any assertion of the primacy of expressions of injustice: Attending only to expressed claims of harm or injury or exploitation can lead us to pass over the ways in which the dampening of victims' capacities to express their own experiences of harm and injury and exploitation is often a central feature of unjustifiable power relations. This problem of epistemic injustice—that is, the systematically skewed distributions of believability and self-interpretability (Fricker 2007)—complicates any straightforward assertion that analytical attention and normative primacy should be given to expressed claims of injustice.

To be clear, I am not suggesting that the reluctance to explicitly theorize about the content of justice necessarily undermines the normative stance of critical thinking in spatial disciplines (compare Olson and Sayer 2009). Despite appearances, in fact, this reluctance leaves in place the forms of normative authority that characterize more explicit constructions of ideal criteria of evaluation. My argument is that giving only practical priority to injustice leaves in place a set of assumptions about modes of justification that are most in need of revision. In discussions of the urbanization of injustice,

the right to the city, or grammars of injustice, visible expressions of injustice are routinely interpreted as responses to what remain largely taken-for-granted sources of wrong: capitalist exploitation of labor and the environment; neoliberal governance; intersectional formations of gendered, sexualized, and racialized oppression; and accumulation by dispossession. In this recurring form of analysis, the authoritative apprehension of the meaning of injustice is displaced from normative philosophies of justice onto the revelatory force ascribed to explanatory theories of space and alternative ontologies of spatiality.

The practical primary accorded to claims of injustice in critical spatial theory is, then, associated with a form of justification that actually retains many of the most problematic features associated with normative philosophies of ideal justice. This is evident in a particular view of what spatial theory can do for critical analysis. Alternative ontologies and explanatory narratives are supposed to be normatively compelling because they demonstrate that things could be different, that things are not as they first appear, and that current arrangements are crossed by histories and contingencies that mean that they could be reconfigured, performed differently, or imagined afresh. Revealing the processes shaping the production of space, the assemblage of constituted orders, and the formation of provisional settlements goes alongside an affirmation of the dynamics of becoming, contradiction, performativity, and paradox that both explain the reproduction of fixed patterns and relationships and also offer the possibility of transforming them. Spatial theory is therefore meant to assist in laying bare the devices through which overarching structures of injustice are reproduced, revealing fundamental sources of injustice by unmasking the exclusionary, naturalizing, or essentializing effects of flat, absolute, fixed concepts of space or identity. The authority for this revelatory maneuver sometimes relies on access to a superior epistemology capable of determining the difference between significant and insignificant differences (Harvey 1996); or sometimes it relies on ontological accounts of the necessarily contingent formation of settled orders that give implicit normative priority to the relative openness and contestability of those orders (Massey 2005); or sometimes it relies on the idea that the processes of spatialization through which the world is constituted as knowable and actionable in the first place are themselves sources of injustice (e.g., Dikeç 2010). Across different forms of theoretical discourse, it is presumed that the task of spatial theory is to demonstrate "the ontological case for social justice" (Gleeson 1996, 233).

Questions of social justice are not open to either epistemological or ontological resolution, and to think that they are is a kind of category error. Matters of justice are essentially contested, for sure, but this is only to say that their significance is unavoidably a matter of appraisal and of judgment. This implies that political hope should not be hinged on demonstrating the possibility of change per se, supported perhaps by proclaiming the need for a renewal of utopian vision. My suggestion here is that the primacy accorded to expressions of injustice does not go far enough and that it needs to be freed from its continuing subservience to the monological styles of reasoning associated with both normative political philosophy and critical theories of space and spatiality. To elaborate on this claim, in the next section I consider the significance of the recurrent gesture of disavowal toward "liberal formulations" that defines the engagement with issues of justice in critical spatial thought.

Dogmas of Egalitarianism

In spatial disciplines, Rawls's (1971) account of justice as fairness functions as a kind of obligatory passage point for discussions of social justice. It is an account that many writers feel they must mention without ever quite wanting to use in any fundamental fashion. In his critique of utilitarian approaches to justice, Rawls proposed two principles that, he claimed, deserved universal assent: a principle that everyone should be accorded as wide a range of basic political liberties as plausible and a principle to regulate socioeconomic distributions. It is here, in the second principle, where Rawls elaborated on what he called the "difference principle," according to which just distributions of wealth and income need not necessarily be equal "but must be to everyone's advantage" (Rawls 1971, 61). The two principles are arranged in "a serial order," so that departures from the first principle cannot be justified by greater social and economic equality (Rawls 1971, 60–61). Rawls summarized the significance of the two principles of justice by saying that "injustice, then, is simply inequalities that are not to the benefit of all" (62).

The real originality of Rawls's theory lies not in the identification of the two principles themselves but in his account of the universal justifiability of those two principles. Here, too, lies the source of the most sustained criticism of his account. Rawls presented the

notion of "the original position"—the set of contractual obligations people would agree to if placed behind "a veil of ignorance"—as the device to establish that these two principles of justice are indeed those that people would and should rationally adopt. Across his own refinements of his theory of justice, Rawls remained committed to a vision of depersonalized public reason, guided by a conviction that value pluralism is a threat to peaceable politics and therefore that "comprehensive doctrines" should be sequestered away from the public sphere. In this conviction and its theoretical sublimation via an ideal model of disembodied third-person reasoning, Rawls seems to place questions about unequal, unfair, and discriminatory social relations of class, gender, race, or sexuality beyond the scope of the conversation of justice. It is this model of rational justification as a guide to institutional design and as a mode of philosophical reasoning that defines "methodological Rawlsianism" (Pateman 2002, 39–40).

Critical analysis in human geography and urban studies has tended to focus attention on that range of issues covered by Rawls's difference principle—questions about the distributive equity of social and economic outcomes. For example, Smith (1977, 7) operationalized the Rawlsian account of justifiable "constraints on equality" in relation to a series of geographical patterns and processes. Importantly, Smith's elaboration of a welfare approach to geographical inquiry is shaped by a presumption that inequality is not the same as injustice (Smith 1977). His proposition that "an unequal distribution is not necessarily unjust" (Smith 1979, 40) follows from a geographical imagination that is sensitive to the value of difference, particularity, and pluralism, one that foregrounds the question of under what circumstances differential treatment can be justified (see also Walker 2012; Davoudi and Brooks 2014).

In Smith's engagement with the significance of the difference principle, however, Rawls's ordering of two principles is not registered. More generally, in human geography, urban studies, and other spatialized social sciences, the significance of Rawls's first principle is not given much attention at all (see Katznelson 1995). Writing in the 1970s, Smith presumed a functioning practice of "planning" as the medium through which principles of justice and welfare would be delivered. A similar assumption, about the promise of a decentralized planning system, underwrites Harvey's (1973) pivotal treatment of the principle of territorial justice in *Social Justice and the City*. Thinking of justice as a matter pertaining to the "the division of benefits and the allocation of burdens" (97) that arise from the division of labor and social cooperation, Harvey (1973) transposed Rawls's difference principle onto relations between cities and regions, addressing the problem of how to maximize "the prospects of the least fortunate region" (110).

Harvey's deployment of the difference principle to establish the content of a just distribution was a prelude to his conclusion that markets are wholly incompatible with the achievement of a just distribution, however that is formulated (Harvey 1973). This proposition is just one example of a broader habit of mind that has shaped radical spatial thought ever since, in which "liberal" philosophies of justice are dismissed as remaining at the surface level of "distribution," when what is needed is an attention to the causal dynamics of "production" (Harvey 1973, 15). From such a perspective, one always already knows in advance that distributive concepts of justice and liberal theories of rights obscure and sustain the underlying structural sources of injustice. These are properly located in processes of class power, property relations, accumulation by dispossession, and exploitation, mediated by dynamics of gender, race, or sexuality and state formation. Elaborating on the "production" of such formations is given both diagnostic preeminence and normative primacy. It is a move evident, for example, in Harvey's own treatment of his original proposition that theorizing justice involves specifying the forms of "a just distribution justly arrived at" (Harvey 1973, 98), which he later refined so that the idea of justice refers to the "just production of just geographical differences" (Harvey 1996, 5). The latter formula expresses simultaneously the idea that an analysis of the production of phenomena always trumps concerns about fair distribution and that grasping the dynamics of a particular order of production provides the key to discriminating between forms of difference that are to be valued, on the one hand, and patterns of disadvantage that are expressions of unjust inequalities on the other. The same form of reasoning informs Soja's (2010) treatment of the idea of spatial justice. He presented Rawls as overemphasizing distributive "outcomes" at the cost of the analysis of "process." In so doing, the focus of Rawlsian theory on the rationality of principles of justice—on the *process* of justifying the two principles—is completely elided, so that the question of the adequacy of that account does not even arise. In the same move, Soja also conflated a concern with process in terms of procedural justice—with the fair application of rules—with a social

scientific notion of grasping the causal processes behind the production of inequality.

The preference for identifying fundamental causes has the unfortunate effect of reducing the meaning of distributive justice to substantive equality in the allocation of divisible socioeconomic goods and services. The specifically political meaning of the difference principle in Rawls's original formulation, as one part of an account that is governed by a norm of democratic equality, is barely mentioned in the engagement with questions of social justice and equality by spatial theorists. The foreclosure of the problem of the democratic justification of normative standards in spatial theories of social justice is illustrated by Fainstein's (2011) account of the just city. She found theories of communicative planning and deliberative democracy to be inadequately attentive to the underlying social relations of inequality that will always ensure that outcomes of putatively fair procedures will fall short of creating just outcomes. Fainstein set up a stark contrast between democracy (defined as a set of procedures of deliberation and inclusive participation) and justice (defined as a substantive principle of equity). In drawing up this contrast, Fainstein (2011, 15) was explicit in asserting the "precedence of justice," presented as a standard of equitable outcomes that has precedence over norms of democratic process. She thereby effectively reverses Rawls's serial ordering of principles. Fainstein's elevation of substantive justice as a principle of evaluation illustrates a structure of thought that is perfectly able to pluralize the criteria to be mobilized—extending these to include ideals of recognition as well as equality, need as well as merit—without addressing the democratic limits of this way of reasoning about justice as an ideal.

I have dwelt on the marginal reference to Rawls in debates about justice in human geography, urban studies, and planning theory not because I think we should cleave closely to his style of philosophical reasoning. I have done so because the gesture of disavowal through which Rawls shows up in this tradition of thought reveals some of what is involved in being "in the true" of critical spatial theory (see Bennett 1993). First, Rawls is often mentioned as offering an overly consensual view of justice, so that spatial theorists are able to claim special sensitivity to issues of antagonism, conflict, and hostility. Liberalism, however, is concerned with nothing so much as issues of antagonism, conflict, and hostility (it just thinks about these in particular ways). Refusing to recognize this concern leads to the rather one-sided view that what properly counts as "the political" is only to be found in acts of disruption, resistance, and contestation (see Barnett 2012). Second, Rawls is condemned for proposing a merely distributive understanding of justice, so that spatial theorists can assert their superior grasp of the generate dynamics that reproduce inequalities—a grasp that is availed them by complex theories of the production of space, ontologies of relational spatiality and the constitutive spatialization of the political, or spatial dialectics. Third, it is common to claim that Rawls was concerned with the maximization of individual liberty, as if his liberalism was essentially a variation of a Lockean tradition of limited government. Whatever its various faults, though, Rawls's theory of justice is a rather serious attempt to think through problems of cooperation and coexistence—of sharing, of being-with others—without reducing these problems to the ideal of self-owning selves; or to higher order abstractions such as community, hegemony, the public, or the state; or to hopes invested in ethics or personal virtue. The kind of individualism central to Rawls's theory of justice—one that seeks to honor the principle of treating persons "not as means but as ends in themselves" (Rawls 1971, 179–80)—is not invalidated by simply invoking facts about interdependence or ontologies of relational entanglement. It starts off from both an acknowledgment of and a worry about relationships of dependence, vulnerability, and unintended consequences. Keeping this in mind helps us to see that to engage critically with the problem of justice does not depend on access to superior explanatory or ontological insight. It requires a consideration of the full implications for spatial theory of thinking of practices of justification as intersubjective affairs, undertaken in second-person registers.

From Geographies of Justice to Democratic Inquiry

My contention is that the ways in which Rawls is referred to—either as a representative of a liberalism to be dismissed as hopelessly parochial in its unreflective universalism or as a source of an ideal theory to be used as an evaluative yardstick—indicate a recurring misrecognition of the central concerns that animate a great deal of the political theory that does now show up in geographical research. Human geography, planning theory, and urban studies have drawn on ideas from a range of political theorists in the last two or three decades, including writers such as Chantal Mouffe, Nancy Fraser,

Seyla Benhabib, Jacques Rancière, James Tully, John Dryzek, William Connolly, and Iris Marion Young. What is usually found in the work of these thinkers is a willingness to insist on the importance in political life of power, or difference, or antagonism, or affect. This is a discovery made without much reckoning with the various pictures of critical reasoning presented by these thinkers, however. If liberalism is a problem for the political theorists most popular in spatial disciplines, then it is above all because of the difficulties identified in the account of public reason of the sort presented by Rawls.

So, for example, Iris Marion Young's work has certainly served as an important reference point in geographical thought for recentering the analysis of justice on the shared apprehension of injuries and harms such as exploitation, marginalization, powerlessness, violence, and cultural denigration (Young 1990). Young's work seems to have much in common with spatial theorists' own attunement to the worldly sources of injustice (see Fainstein 2007). In this respect, as with the appeal made to other political thinkers, spatial theorists most often find in Young's work a confirmation of the things that, as social scientists, they already know anyway. The real significance of Young's work, though, lies not just in what she theorized about but how she theorized about normative issues (see Jaggar 2009). Young did not start from a prior model of an ideal society, beginning instead from particular expressions of injustice. She did not seek to control for the messy pluralism of commitments and passions that differently placed people bring to the conversation of justice; she treated such differences as epistemic resources rather than obstacles or impediments to shared reasoning. She did not appeal to ideal images of society to assess real situations, instead looking for the modes of evaluation that are articulated by struggles against injustice. Each of these three departures from methodological Rawlsianism would, no doubt, find affirmation in the characteristic privileging of struggles for justice over ideal theory in critical spatial theory. Crucially, though, Young did not rest content with the simple affirmation of emergent feelings of justice. Questions of justice, for Young (2006) as much as for Rawls (1999), always remained a political matter, rather than a matter of epistemology or metaphysics. Saying "political" here certainly implies a reference to matters of basic institutional design (see Waldron 2016). Perhaps more fundamentally, it also refers to questions of how to imagine the forms of reasoning and justification through which collective living might be coordinated.

Two decades ago in the midst of debates in geography about difference, relativism, and universal values, Young (1998, 40) asserted, "To invoke the language of justice and injustice is to make a *claim.*" It is here that we can begin to specify just what giving conceptual priority to injustice implies for critical inquiry (see Barnett 2017). There are two initial points that need to be underscored. First, giving conceptual priority to injustice requires shifting away from an idea that the currency of justice is some form of divisible set of primary goods, or resources, or rights, or even capabilities. It requires instead thinking of matters of injustice and justice as primarily to do with questions of "the relative social and political standing of persons with respect to each other—the issue, in short, not of *what you have* but of *how you are treated*" (Forst 2007, 260). Giving conceptual priority to injustice involves thinking about how social relations, institutional arrangements, and norms systematically disadvantage some persons as participants in shared practices of public life (see Fraser 2003). This leads to the second point about giving conceptual priority to injustice: It involves focusing on the ways in which questions of justice always involve claims made by one party on other agents, claims about the justifiability of a state of affairs that call for vindication of certain sorts (see Anderson 2010).

The significance of claims-making to the view of the priority of injustice I am recommending here lies in the presumption that the meaning of injustice is necessarily arrived at through intersubjective engagement. This emphasis on reasoning about justice dialogically rather than monologically is a feature of Storper's (2013) recent revival of the theme of justice and territorial development. He focused on the difficult relationship between thinking of justice in procedural terms—in terms of "fair interactions"—and in substantive terms—in terms of "good consequences." Storper's recommendation is that we need "more social choice conversations" (228); that is, a concerted and systematic search for better ways of finding out what people in one place consider to be just ways of relating to other places. In critical spatial theory, this is one of the few examples of theorizing not only about how to achieve a just conception in a just manner but about how to arrive democratically at conceptions of what is just in the first place, where this involves respecting the pluralism of peoples' concerns.

Grasping the difference between two ways of thinking about the justification of ideas of justice—a third-person, monological style and a second-person, dialogical style—is therefore central to fully appreciating the challenges of giving conceptual priority to injustice. Young once argued, against Harvey, that saying a situation is unjust does not necessarily imply a reference, implicit or explicit, to a standard of ideal justice (Young 1998). To insist that it must do so is to presume that justice is a label of some sort, which can be more or less appropriately applied to various phenomena. Thinking that we must in advance have a standard against which to judge affairs gets the grammar of justice the wrong way around, and it always leads back to the idea that justice is some kind of ideal. Justice talk is, however, in a strong sense, rhetorical: "The meaning of 'justice' has to do with what people intend to convey in saying it, not with the features of the phenomena they say about it" (Pitkin 1972, 173). In this view, injustice is understood not as an absence of justice, just as illness is not an absence of health. Injustice is a positive condition, arising from an experience of injury of some kind or, more broadly, from an "abhorrence of wrong" (Wolgast 1987, 194). The strong implication of this simple-sounding proposition is that justice is something developed not to satisfy an ideal standard, either an a priori principle or an emergent one, but is a universalizing response to situated expressions of injustice: "We craft responses to wrong, our purpose being not to satisfy some preconceived picture of justice but to address the snares of injustice" (Wolgast 1987, 145). To put it another way, justice is not an ideal at all. It is a condition that is approached through processes of repair, recognition, redress, reparation, and redistribution.

It is worth noting that by this understanding of the grammar of justice, giving conceptual priority to claims of injustice involves affirming the passionate dynamics through which political action is generated as a response to varied forms of harm, injury, or mistreatment (see Barnett 2016). This, too, might resonate with current interests in spatial theory in political affect, emotions, and feelings. Two difficulties arise, however, from any simplistic affirmation of the passions, both of which seem to undermine the normative value of attending to experiences of injustice that the affirmation is meant to support. First, the emphasis that is placed on forms of passionate expression in arguments for the priority of injustice seems to make judgment a purely capricious matter. This appearance might be only heightened when one notices that the sense of injustice often starts out from negative emotions, such as anger, indignation, resentment, or vengeance, rather than empathetic or sympathetic ones (see Shklar 1990). This is related to the second difficulty that arises from privileging passionate claims of injustice. Recalling the phenomena of epistemic injustice should temper any straightforward valorization of explicit cries of injustice precisely because such valorization risks obscuring the structures of harm that stifle the expressions of some actors.

These two difficulties suggest that giving priority to claims of injustice runs the risk of merely lending a normative sheen to the grievances of those with the loudest voices. Taken together, both issues therefore demand a further clarification of the double significance of the idea of claims that is so central to the project of giving conceptual priority to injustice. First, the idea of claims at work here implies that matters of justice arise in contexts in which existing patterns of power are contested through the voicing of objections of one form or another. In this sense, claims are asserted against felt injustices. Second, though, the idea of claims of injustice also refers to the notion that these claims are, indeed, assertions; that is, they are claims made on the attention of others and, as such, are subject to a democratic test by being passed through the medium of argument and debate. It should be said that the proposition that claims of injustice can be assessed as to whether they are warranted is not just a matter of determining epistemological certainty or even normative validity. It follows instead from a view in which practices of justification are made central to the experience and articulation of injustice as injustice. This is not to be mistaken for a linear process in which individual, subjective, emotional feelings are subsequently given cognitive rationalization. Giving conceptual priority to injustice involves, instead, thinking of the sense of injustice as arising from and being processed through intersubjectively mediated, shared inquiry. Injustice is, in short, a thoroughly public phenomenon.

Conclusion

What makes a state of affairs unjust to those immediately on the receiving end of domination, exploitation, or violence, as well as to those called on to act in response to such states of affairs, is not the reference to a prior construction of what counts as a properly just arrangement. Saying this does not cast us adrift from the safety accorded by clinging to universal principles into the depths of relativism. Giving priority to the double sense of claims of injustice outlined earlier involves treating other persons democratically; that is, as free and equal citizens. The conceptual prioritization of injustice in critical theory rests on a commitment to the idea that harms and violations and wrongs are experienced, felt, expressed, assessed, and vindicated in situations of intersubjective interaction.

Giving conceptual priority to injustice in critical analysis, in the sense I have outlined here, should compel us to adjust the normative assumptions through which geographical thought continues to apprehend the spatialities of political action. The double significance of claims as assertions alerts us to the fact there is more to political claims-making than practices of assembly, dissent, encounter, and protest (see Parkinson 2012). The scope of claims-making extends beyond the mere expression of a demand to include the processes by which claims are evaluated, adjusted, accorded recognition and validity, and acted on. The emphasis on practices of justification as the mediums in which injustice is experienced and articulated therefore disrupts the political significance usually ascribed to the figure of the city. Accounts of the dynamics of capitalist urbanization or more generalized accounts of processes of exclusionary spatialization often posit a homology between urban space as both the arena in which injustice is produced and the stage on which it is best resisted. The spaces in which injustice is apprehended and justice is enacted, however, are stretched out over space and time; they are distributed spaces of recognition, justification, and vindication. They are not well modeled on the images of copresence evoked by ideas of the city or the forum or the street. Critical inquiry into the geographies of injustice requires an appreciation of the variable relations among three analytically distinct dimensions of political action: the spatial dynamics involved in the generation of inequalities and injuries, the spaces through which those patterns are translated into expressions of injustice, and the spatialities of the practices that seek to vindicate claims of injustice by crafting just courses of legitimate action (see Barnett 2014). There is no reason to suppose that the articulated geographies of these three distinct aspects of claims-making will always coalesce into a shape that conforms to the inherited associations between politics and the city found across various strands of spatial thought.

In short, the focus on claims-making involved in giving not only practical but conceptual priority to injustice requires an acknowledgment of all of the ways in which the spatialities of political action exceed the romantic preference for performative models of assembly and demonstration and protest. The geographies of justice emerge through the combination of spaces of mobilization and agitation, deliberation and compromise, bargaining and deal making, decision and delivery, accountability and revision; that is, across all of the spaces through which the "full ritualization of conflicts" is enacted (Hampshire 1993, 6).

Acknowledgments

Thanks to Nick Gill and Karen Bickerstaff for comments on an earlier draft of this article, as well as to three anonymous referees and Nik Heynen for incisive criticism and advice on how to improve it.

ORCID

Clive Barnett ⓘ http://orcid.org/0000-0002-1291-1421

References

Anderson, E. 2010. The fundamental disagreement between luck egalitarians and relational egalitarians. *Canadian Journal of Philosophy* 40 (Suppl. 1):1–23.

Barnett, C. 2012. Situating the geographies of injustice in democratic theory. *Geoforum* 43:677–86.

———. 2014. What do cities have to do with democracy? *International Journal of Urban and Regional Research* 38 (1):1625–43.

———. 2016. Toward a geography of injustice. *Alue ja Ympäristö* 2016 (1):111–18.

———. 2017. *The priority of injustice: Locating democracy in critical theory.* Athens: University of Georgia Press.

Bennett, T. 1993. Being "in the true" of cultural studies. *Southern Review* 26 (2):217–38.

Davoudi, S., and E. Brooks. 2014. When does unequal become unfair? Judging claims of environmental injustice. *Environment and Planning A* 46:2686–2702.

Dikeç, M. 2010. Justice and the spatial imagination. *Environment and Planning A* 33 (10):1785–1805.

Dorling, D. 2011. *Injustice: Why social inequality still persists.* Bristol: Policy Press.

Fainstein, S. 2007. Iris Marion Young (1949–2006): A tribute. *Antipode* 39 (2):382–87.

———. 2011. *The just city.* Ithaca, NY: Cornell University Press.

Fincher, R., and K. Ivesen. 2012. Justice and injustice in the city. *Geographical Research* 50 (2):231–41.

Forst, R. 2007. Radical justice: On Iris Marion Young's critique of the "distributive paradigm." *Constellations* 14 (2):260–65.

Fraser, N. 2003. Social justice in the age of identity politics: Redistribution, recognition, and participation. In *Redistribution or recognition: A politico-philosophical exchange,* ed. N. Fraser and A. Honneth, 7–109. London: Verso.

Fricker, M. 2007. *Epistemic injustice: Power and the ethics of knowing* Oxford: Oxford University Press.

Gleeson, B. 1996. Justifying justice. *Area* 28 (2):229–34.

Hampshire, S. 1993. Liberalism: The new twist. *New York Review of Books* 40 (14):44–46.

Harvey, D. 1973. *Social justice and the city.* London: Edward Arnold.

———. 1996. *Justice, nature and the geography of difference.* Oxford, UK: Blackwell.

Jaggar, A. 2009. *L'Imagination au pouvoir*: Comparing John Rawls's method of ideal theory with Iris Marion Young's method of critical theory. In *Dancing with Iris: The philosophy of Iris Marion Young,* ed. A. Ferguson and M. Nagel, 95–101. Oxford, UK: Oxford University Press.

Katznelson, I. 1995. Social justice, liberalism, and the city. In *The urbanization of injustice,* ed. A. Merrifield and E. Swyngedouw, 45–64. London: Lawrence and Wishart.

Lefebvre, H. 1996. *Writings on cities.* Oxford, UK: Blackwell.

Low, S., and K. Ivesen. 2016. Propositions for more just urban public spaces. *City* 20 (1):10–31.

Massey, D. 2004. Geographies of responsibility. *Geografiska Annaler: Series B, Human Geography* 86 (1):5–18.

———. 2005. *For space.* London: Sage.

Merrifield, A., and E. Swyngedouw. 1995. Social justice and the urban experience. In *The urbanization of injustice,* ed. A. Merrifield and E. Swyngedouw, 1–17. London: Lawrence and Wishart.

Olson, E., and A. Sayer 2009. Radical geography and its critical standpoints: Embracing the normative. *Antipode* 41 (1):180–98.

Parkinson, J. 2012. *Democracy and public space: The physical sites of democratic performance.* Oxford, UK: Oxford University Press.

Pateman, C. 2002. Self-ownership and property in the person: Democratization and the tale of two concepts. *Journal of Political Philosophy* 10 (1):20–53.

Pitkin, H. F. 1972. *Wittgenstein and justice: On the significance of Ludwig Wittgenstein for social and political thought.* Berkeley, CA: University of California Press.

Rawls, J. 1971. *A theory of justice.* Oxford, UK: Oxford University Press.

———. 1993. *Political liberalism.* New York: Columbia University Press.

———. 1999. Justice as fairness: Political not metaphysical. In *John Rawls: Collected papers,* ed. S. Freeman, 388–420. Cambridge, MA: Harvard University Press.

Schlossberg, D. 2013. Theorising environmental justice: The expanding sphere of a discourse. *Environmental Politics* 22 (1):37–55.

Shklar, J. 1990. *The faces of injustice.* Cambridge, MA: Harvard University Press.

Smith, D. M. 1977. *Human geography: A welfare approach.* London: Edward Arnold.

———. 1979. *Where the grass is greener: Living in an unequal world.* London: Croom Helm.

———. 1994. *Geography and social justice: Social justice in a changing world.* Oxford, UK: Blackwell.

Soja, E. 2010. *Seeking spatial justice*. Minneapolis: University of Minnesota Press.

Storper, M. 2013. *Keys to the city: How economics, institutions, social interaction, and politics shape development*. Princeton, NJ: Princeton University Press.

Waldron, J. 2016. *Political political theory: Essays on institutions*. Cambridge, MA: Harvard University Press.

Walker, G. 2012. *Environmental justice: Concepts, evidence, politics*. London: Routledge.

Whatmore, S. 2002. *Hybrid geographies: Natures, cultures and spaces*. London: Sage.

Wolgast, E. 1987. *The grammar of justice*. Ithaca, NY: Cornell University Press.

Wright, M. 2009. Justice and the geographies of moral protest: Reflections from Mexico. *Environment and Planning D: Society and Space* 27:216–33.

Young, I. M. 1990. *Justice and the politics of difference*. Princeton, NJ: Princeton University Press.

———. 1998. Harvey's complaint with race and gender struggles: A critical response. *Antipode* 30:36–42.

———. 2006. Taking the basic structure seriously. *Perspectives on Politics* 4 (2):91–97.

———. 2011. *Responsibility for justice*. Oxford, UK: Oxford University Press.

CLIVE BARNETT is Professor of Geography and Social Theory at the University of Exeter, Exeter, UK. E-mail: c.barnett@exeter.ac.uk. His research interests include work on the geographies of public life and spaces of democracy, social theory, and the urbanization of responsibility in global policy discourses. He is the author most recently of *The Priority of Injustice: Locating Democracy in Critical Theory* (University of Georgia Press, 2017).

2 Against the Evils of Democracy

Fighting Forced Disappearance and Neoliberal Terror in Mexico

Melissa W. Wright

On 26 September 2014, Mexican police forces in Iguala, Guerrero, attacked and abducted four dozen students known as *normalistas* (student teachers); some were killed on the spot and the rest were never seen again. Within and beyond Mexico, rights activists immediately raised the alarm that the *normalistas* had joined the country's growing population of "the disappeared," now numbering more than 28,000 over the last decade. In this article, I draw from a growing scholarship within and beyond critical geography that explores forced disappearance as a set of governing practices that shed insight into contemporary democracies and into struggles for constructing more just worlds. Specifically, I explore how an activist representation of Mexico's *normalistas* as "missing students" opens up new political possibilities and spatial strategies for fighting state terror and expanding the Mexican public within a repressive neoliberal and global order. I argue that this activism brings to life a counterpublic as protestors declare that if disappearance is "compatible" with democracy, as it appears to be within Mexico, then disappeared subjects demand new spaces of political action. They demand a countertopography where the disappeared citizens of Mexico make their voices heard. Activists demonstrate such connections as they compose countertopographies for counterpublics across the Americas landscape of mass graves, prisons, and draconian political economies, mostly constructed in the name of democracy and on behalf of securing citizens. Understanding how Mexico's activists confront the intransigent problems of state terror, spanning from dictatorships to democracies, offers vital insights for struggles against policies for detaining and disappearing peoples there and elsewhere in these neoliberal times. *Key Words: feminism, forced disappearance, Mexico, neoliberalism, social movements, state terror.*

2014年九月二十六日, 墨西哥在格雷罗州伊瓜拉的警力, 攻坚并劫持了四十位名为 *normalistas* (学生教师) 的学生; 有些人当场被击毙, 其馀则消失无踪。在墨西哥之内与之外的维权行动者, 立即警告 *normalistas* 已成为该国逐渐增加的 "失踪" 人口的一部分, 而在过去十年间, 失综人口数已超过两万八千人。我们于本文中, 运用批判地理学之中与之外探讨被迫失踪的逐渐增加之学术研究, 该研究将被迫失踪视为对当代民主与打造更为公平世界的奋斗提供洞见的一组治理实践。我特别探讨墨西哥的 *normalistas* 作为 "失踪的学生" 之社会行动再现, 如何在压迫的新自由主义与全球秩序中, 开启对抗国家暴政并扩大墨西哥公众的崭新政治可能与空间策略。我主张, 随着抗议者宣称, 若失踪能够见容于民主的话——如同在墨西哥一般——失踪的主体则需要崭新的政治行动空间, 此一社会行动主义换醒了反对大众。他们要求失踪的墨西哥市民能够被听见声音的反製图学。社会行动者透过为反对大众构成横跨美洲的大型墓园、监狱与残暴的政治经济地景——多半是以民主和保护市民安全之名建造——的反製图学, 显示上述连结。理解墨西哥社会行动者如何对抗从独裁到民主政体皆存在的国家恐怖主义之顽固问题, 对于对抗新自由主义时期在各处监禁或迫使人民失踪的政策, 提供了重要的洞见。 *关键词: 女权主义, 被迫失踪, 墨西哥, 新自由主义, 社会运动, 国家暴政。*

El 26 de septiembre de 2014 las fuerzas policiales mejicanas de Iguala, Guerrero, atacaron y secuestraron a cuatro docenas de estudiantes *normalistas* (estudiantes de docencia); a algunos los mataron allí mismo y del resto nunca se volvió a saber. De inmediato, dentro y fuera de México, activistas de derechos humanos proclamaron que los normalistas habían pasado a ser parte de la creciente población de "desaparecidos," que ahora suma más de 28.000 personas desde la última década. Me baso en este artículo en un creciente cuerpo de erudición dentro y fuera de la geografía crítica que explora la desaparición forzada como un conjunto de prácticas de gobierno repudiadas en las democracias contemporáneas y dentro de las luchas que buscan construir mundos más justos. Específicamente, exploro la manera como una representación activista de los normalistas de México como "estudiantes desaparecidos" abre nuevas posibilidades políticas y estrategias espaciales para combatir el terror del estado y expandir el público mejicano dentro de un orden global represivo y neoliberal. Sostengo que este activismo le da vida a un contra-público cuando quienes protestan declaran que si la desaparición es "compatible" con la democracia, como parece serlo en México, entonces los sujetos desaparecidos demandan nuevos espacios de acción política. Ellos reclaman una contra-topografía donde los ciudadanos desaparecidos en México puedan hacer oír sus voces. Los activistas demuestran tales conexiones a medida que componen contra-topografías para los contra-públicos a través del paisaje de las Américas representado en fosas comunes, prisiones y economías políticas draconianas, generalmente construidas a nombre de la democracia y con el pretexto de dar seguridad a los ciudadanos. Entender cómo confrontan los activistas de México los problemas intransigentes del terror del estado, que van desde las dictaduras hasta las democracias, ofrece posibilidades vitales para las luchas contra las políticas de detener y desaparecer pueblos, allí y en otras partes, en estos tiempos neoliberales. *Palabras clave: feminismo, desaparición forzada, México, neoliberalismo, movimientos sociales, terror de estado.*

On 26 September 2014, city and federal police forces in the Mexican city of Iguala, Guerrero, fired on several buses that carried students, known as *normalistas* (student teachers), from a rural teachers' college in Ayotzinapa. The students, all young men from the mainly indigenous and mountainous reaches of southern and central Mexico, had commandeered the buses, as they had in previous years, to make the almost 200-km journey to Mexico City and join the annual 2 October commemoration of the 1968 government massacre of students and other prodemocracy demonstrators. They never arrived. Instead, they were ambushed by police forces who killed several, along with some bystanders in the street, before loading the remaining forty-three into police vehicles. These *normalistas* were never seen again.

In the days immediately following the September 2014 massacre and disappearance of the *normalistas*, their families and friends sought information from an obstructionist police department. Student videos surreptitiously taken during the police ambush began circulating on social media, as did texts sent on 26 September 2014 from student phones exposing their fear and vulnerability as military and police officials rounded them up, failed to tend to their injuries, and bullied them in the police station. The silencing of the phones and termination of texts on that date spurred more concern. A few days later, when the *normalistas* did not return to the Aytozinapa school and to their adjacent cornfields bursting with mature crop, their families and friends knew that something was terribly wrong.

Some days later, officials declared that the *normalistas* had been "infiltrated" by drug gangs who instigated the attack. After immediate public rebuke of that first official version, the government provided, some weeks later, a second one: The *normalistas* had been ambushed by corrupt police officers working in cahoots with a drug gang, allied with the then-mayor and his wife, who then murdered, incinerated, and tossed the *normalistas'* ashes into a nearby river. With this final official version, the federal government declared the *normalista* case "closed." Since then, the Mexican National Commission of Human Rights (CNDH) and major international rights groups have debunked the government's story and have accused the Mexican state of perpetrating the crimes and covering them up. By 2015, these same groups raised an alarm that not only had the forty-three *normalistas* joined the country's growing population of "the disappeared" (people forcibly detained and killed by the state) but that Mexico was becoming "a country of the disappeared" (Campa 2015), with numbers reaching above 20,000 since the onset of a drug war declared in 2006 by the federal government and with support from the United States. As the realization spread over the ensuing months that the *normalistas* were "disappeared," millions of people took to Mexico's streets with the consistently coupled accusation and demand: "It was the state!" (*¡Fue el Estado!*) and "Return them alive! (*¡Regrésenlos!*)."[1]

In this article, I explore how these demands, as part of a broader protest against forced disappearance, open up new political possibilities and spatial strategies for fighting state terror within a repressive neoliberal and global order. In my discussion, I draw from a number of scholarly traditions: interdisciplinary work within and beyond critical geography, human rights and activist scholarship across the Americas, and feminist and Marxist perspectives that study state terror and this particular manifestation of forced disappearance (see Borland 2006; Ross 2008; Taylor 2012; DeJesus, Bosco, and Humaydan 2014). Taken together, they expose a politically and economically complex set of practices that provide insight into contemporary governance and transnational capitalism. Important for my exploration is a contextualizing of the *normalista* disappearance and massacre within the current "democratic" moment of Latin America (see Smith 2009; Aristegui and Trabulsi 2010; Pastrana 2016). The state terror of forced disappearance, after having subsided during the region's "third wave of democracy," is seeing a comeback that is provoking much debate across the Americas over the meaning and integrity of democratic governance (see also Fazio 2016; Lomnitz 2017). As a prominent international human rights lawyer and Argentinian judge poignantly observed, "No one was prepared for ... forced disappearance during democracy" (Bertoia 2014). Now, as Mexico leads among countries experiencing a population explosion of "the forcibly disappeared" (Human Rights Watch 2013), a pressing question emerges: If "democratization" was the answer to the evils of dictatorship in the modernizing Americas, what is the answer to the evils of democracy?[2] In the following, I argue that the protestors in Mexico today confront that question as they demand a radical opening to democracy that includes an expansion of the Mexican public sphere to include the detained and disappeared as part of the political landscape.

This expansion emerges when seen through the combination of theoretical frames that connect grievability to humanity and, in turn, humanity to the idea of the political "public" and its spaces of action. A burgeoning literature on grief as social action in recent years derives from a convergence of theoretical concerns with the politics of death (i.e., necropolitics) and the "grievability" of the most vulnerable and precarious populations in the world, who also provide the most exploitable labor within a globalized capitalist system that treats them as disposable (Mbembe 2003; Butler 2009). Butler's work on grievability and precarity has inspired wide-reaching scholarship into the politics of grief and its meaning for ontological understandings of "the human." As she writes, "One way of posing the question of who 'we' are in these times of war is by asking whose lives are considered valuable, whose lives are mourned, and whose lives are considered ungrievable" (Butler 2009, 38). Particularly influential for my approach here is the modification of these insights by feminist and postcolonial scholars working "beyond" the Global North where grievability emerges through decades-long resistance to state terror that operates within the webbing of imperialist and transnational capitalist interests (e.g., see R. A.

Hernández 2008; Stephen 2013; Lloyd 2017). Among other critical insights, this diverse scholarship clearly illustrates how such public performances of grief constitute "dissent" by asserting the "grievability" of an "ungrievable subject" (as defined by the status quo). This research creates dialogue with critical race studies in the northern reaches of the continent that, for instance, has demonstrated the centrality of grief within strategies for resisting systemic racism, including genocide and the racist underpinnings of capitalist exploitation (see McKittrick 2014; Weheliye 2014; Davis 2017). As such work indicates, understanding how social movements mobilize across the webbing of social hatreds (i.e., racism, misogyny, and disgust of the poor) is essential for providing insights useful to those whose fights for the right to grieve are fights for social justice more broadly. This position finds backing with current work on social justice in Mexico. As anthropologist Maiana Mora (2017) has recently written in her critical engagements with the *normalista* mobilizations, studies that ask how racism contributes to forced disappearance can help "generate new juridical-political strategies and open other demands" (29) as protestors assert the grievability of those who are missing.

Drawing from this rich and varied scholarship, I maintain that the protests for the *normalistas* generate a radical politics of grievability by refuting state attempts to lump the victims into the "ungrievable" abyss of the "narco" (Wright 2017). As their protests illustrate, such state representations of the ungrievable draw directly from long-standing racist and antirural sentiments that render the nonmestizo and rural working poor as disposable and peripheral to the country (see Ochoa Muñoz 2015). Taking on this long and dehumanizing legacy, protestors refuse to allow the government to represent the *normalistas* as drug punks and thugs unworthy of public grief and therefore unworthy of public justice. Instead, the protestors launch a radical politics over the representation of the *normalistas* as they declare them, against state efforts to the contrary, to be the country's best and the brightest: "Los Estudiantes" (The Students). As the much-beloved Mexican author and public intellectual Elena Poniatowska (2014) declared in a massive protest in the *Mexico City Zócalo*, "A government that kills its students, kills its future!" In response to her declaration, the hundreds of thousands who overflowed in the hemisphere's largest public plaza roared their grief for the disappeared *normalistas* as an expression of their outrage over Mexico's lost "future."

Such proclamations create what Nancy Fraser (1992) has termed a "subaltern counter-public," an "arena where members of subordinated social groups invent and circulate counter-discourses, which in turn permit them to formulate oppositional interpretations of their identities, interests, and needs" (81). In short, as protestors declare that grieving the *normalistas* represents nothing less than the grieving of Mexico's lost future, they force a reconceptualization of those who count in Mexico's public sphere and of those who represent the nation itself. This subaltern public is not the prospering, urban, white and *mestizo* population but the rural and indigenous poor who suffer in poverty, endure the worst violence of its drug wars, and immigrate as economic exiles to the United States. To assert the centrality of these subalterns to Mexican public life, the protestors use creative spatial strategies for demonstrating that the disappeared students deserve their rights and privileges as part of the Mexican "public"—in the streets, institutions, and imagination of the Mexican *pueblo*. In other words, they demonstrate how this counterpublic invigorates a countertopography (to borrow from Katz 2001) where the disappeared *normalistas*, and all of those marginalized populations they represent, stand for and in the center of the Mexican nation.

Students, Dictators, and Democratic Terror

To appreciate the revered meaning of *students* in Mexico and the radical politics represented by the current protests that align *normalistas* with them requires examining the tenacity of forced disappearance as a tool of state terror that bridges the country's former dictatorship with its current democratic era. Although Mexico is an outlier in Latin America for its twentieth-century experience with a civilian, rather than a military-led, dictatorship, the country has, like its southern neighbors, coupled modernity with overt political oppression under the seventy-year reign of the governing Institutional Revolutionary Party (PRI). From 1929 to 2000, during what would come to be known as "The Perfect Dictatorship,"[3] the federal state governed through a heavily centralized system of political and economic repression exercised in concert with regional elites and global capitalist financiers.

By the mid-1960s, resistance to oppression had grown across Mexico with the formation of guerrilla organizations in both rural and urban areas. In Mexico City, university students at the country's flagship public university, the Universidad Nacional Autónoma de México (UNAM), along with students and civilians from other parts of the country, openly organized pro-democratization and leftist networks. As political oppression targeted such efforts in what would come to be known as Mexico's own "Dirty War," students consolidated their efforts across a span of ideological ideas calling for democratic reforms to violent revolution. On 2 October 1968, ten days before the opening ceremonies of the Olympics in Mexico City, the Mexican government attacked student and civilian groups in a central residential area known as "The Plaza of the Three Cultures." Several hundred died and another thousand were detained, many never to be seen again. Government suppression of information about the event was ferocious, but despite the risks, a young journalist at the time, Elena Poniatowksa (1991), compiled a benchmark account in her still widely read book, *Massacre in Mexico*. Through Poniatowska's reporting, and related efforts, Mexico's students came to represent the heart of a country that was struggling for democracy and its revolutionary ideals and against a rapacious capitalism craving its natural resources and human labor.

In the ensuing years, urban and rural college campuses provided places of ongoing resistance to state terror as students mobilized for the release of those detained and for political and economic changes at the national level (Aguayo Quezada 2001). As many of these events occurred contemporaneously with the Chilean and Argentinian military coups and their vicious targeting of universities in those countries, many Mexican students formed or joined underground resistance movements, some arming themselves along with other organizations. Although the student activists came from all walks of Mexican life, many of them from the rural and indigenous communities, and many of them women, the dominant portrait of the student martyr came to be seen as a young male who represented the mestizo protagonist of the country's ongoing "revolutionary nationalism" and modern struggles (see Alonso 1988). Consequently, women and racial minorities, especially those in rural areas, continued to stand as those "outside" of political Mexico. The legacy of that representation directly contributes to the context in which the *normalistas* challenge the lasting hegemony of the virile, urban mestizo student martyr as symbolic of Mexico's future and progress.

Also important for understanding the current context of the 2014 Iguala massacre is the historical evolution of the rural *normal* schools such as the Raúl Isidro Burgos Rural School in Ayotzinapa, Guerrero. Established by indigenous communities working with the fledging Revolutionary governments in the 1910s and 1920s, the *normales* are teaching colleges whose mission has been to train teachers from within the very communities they seek to educate (see "Who Are the *normalistas?*" 2014). Yet, as the country's revolutionary commitments quickly yielded to the pressures of expanding global capitalism and the demands of the country's long-standing racial and economic elites, the *normales*' mission included, by necessity, a political education that included Marxist and proindigenous commitments and that, relative to the dominant social context, came to be seen as "radicalization." The Ayotzinapa *normal* boasts a particularly notable part of this radicalized history as the educational site for a prominent *guerrillero*, Lucio Cabañas, who had studied there in the 1950s and who, in the 1960s, sought to bring about revolutionary change. In addition to political, cultural, and economic critique, the Ayotzinapa school, like the country's other rural *normales*, also stresses self-reliance (through farming combined with study) and commitments to a modest life of teaching in the country's rural and indigenous communities (Solano 2014).

In recent years, especially during the implementation of neoliberal political economic reforms beginning in the 1980s, the rural *normales* expanded their educational mission to include public resistance to the privatization of public entities and the reorganization of the country's rural economy, which culminated in massive changes to communal landholding in 1992 and, then, the implementation of the North American Free Trade Agreement in 1994. The divisions between a rural and indigenous resistance movement and the urban student protests came out forcefully with the Zapatista rebellion in the southern state of Chiapas. Although both rural and urban students were involved in supporting and participating in the Zapatista movement, the Zapatistas presented an indigenous profile to "revolutionary" Mexico that stood in contrast to the urban mestizo student martyr (see Harvey 1998). The government's violent response to the rebellion targeted the rural areas and the *normales* like that in Ayotzinapa and the communities they served. In response, the Raúl Isidro Burgos Rural School and others intensified their open rebellion to neoliberalism and the state terror that accompanied it during the 1990s.

When the PRI finally ceded to pressures and allowed a rival party to assume the presidency in 2000,

the transition from dictatorship to something heralded as "democracy" did not mean an end to state terror, especially in the rural and indigenous areas. Notably, the federal government's extension in the 1990s of a "war ideology" that criminalized much of the agricultural economy, within a drug war, and that also criminalized those working in the informal economy took direct aim at rural Mexico. Consequently, within a few years of the official end of the global Cold War in 1989 and its justification of domestic warfare against the internal enemies of communism, Mexico's Dirty War machinery changed its focus—again in synergy with its powerful funder and ally to the north—to fix its sights on "drug-terrorists" (Paley 2015). By 2006, Mexico, like Colombia, had orchestrated its own nation-wide "drug war" that, with massive U.S. funding, wrought economic and social devastation on the country's rural and indigenous communities and created a humanitarian crisis in many parts of the country (Shirk 2011; Human Rights Watch 2013). Many rural *normales*, such as the Raúl Isidro Burgos Rural School in Ayotzinapa, raised the public alarm that the "drug war" was really a war being waged against their communities to make them ever more vulnerable to a vicious state that worked in the interest of global capitalist and racist elites. Meanwhile, their students became increasingly known throughout the region as "radical trouble-makers."

These dynamics were clearly at play in the events surrounding the September 2014 massacre and disappearance of the *normalistas* and the months following as journalists, activists, and academics struggled against the governing logic of declaring the victims to be ungrievable "narcos." Immediately declaring the students to be "infiltrated" by the gang *Los Rojos* (the Reds), the government attempted to remove the crimes and the *normalistas* from public debate by representing the murdered and disappeared students as not worthy of larger public concern. As the government's well-documented logic had it, the *normalistas* were just more dead *narcos* in a drug war that needed to run its course (see Wright 2017). Therefore, when in October 2014 some million people marched on Mexico City and demanded the *normalistas* live return—as the return of the country's best and brightest (its "students")—they directly challenged the government's position. Indeed, in their rebuke of the logic that sought to minimize the significance of the *normalista* deaths by dismissing the significance of their lives (as just more dead *narcos*), the protests challenged the status quo within Mexico even further as they subverted the racial and urban order of who counted as Mexico's *pueblo* (its people)—for the students of the Aytozinapa *normal* do not fit the profile of the urban and mestizo population initially lionized as the nation's students out of the 1968, and subsequent, massacres and mobilizations. Certainly, and as many scholars and analysts of these protests have since noted, the politics on behalf of the *normalistas* radically challenge the politics of racial and regional representation within even the country's politically progressive sectors (Figueroa 2015; Lomnitz 2017). They demand recognition of a Mexican public that centers its formerly subaltern subjects within the very soul of the nation to proclaim that when members of this public are injured, the nation is injured. In other words, their demands for the live return of the *normalistas* are demands for a vibrant "counterpublic," as Fraser (1992) conceives it, "where members of subordinated social groups invent and circulate counter discourses to formulate oppositional interpretations" (123) of who counts as the people and their government. And this counterpublic must, by necessity, come to life in the public places of the public sphere.

The Counterplaces of Counterpublics

The massive mobilizations across the country that demand the live return of the students frequently take place behind the symbolic leadership of the *normalistas'* families, most notably of their "fathers," who stand in front of the marches, blockades, and press conferences as the official figureheads of the movement. In these events, the public grief of fathers for their sons materializes as a public expression of outrage in the country's public sphere.[4]

Through a careful examination of the politics of representation behind their symbolic leadership,[5] we can see how these father-activists, like many "mother-activists" across the Americas, create the spaces of action for the counterpublics who not only defy the Mexican state but that also raise new possibilities for political subjects and their spaces of action. I refer to these spaces of action as countertopographies in the sense intended by Katz (2001); that is, as the spatial processes through which emerges "a political logic that both recognizes the materiality of cultural and social difference and [that] can help mobilize transnational and internationalist solidarities to counter the imperatives of globalization" (709). Recognizing these countertopographies behind the apparently

conservative positioning of fathers as those leading the search for their student-sons requires peeling back the conservative facade of patrilineal symbolism to expose a multiscalar politics of gender, racism, and neoliberalism within a globalized struggle over human rights and state terror in democratizing Americas.

As the father-activists extend a long legacy of family-organized activism within Mexico and across the Americas since the dictatorship era, they stand as exceptions from the more common mother-led movements that have been so effective in bringing attention to forced disappearance from Argentina's military junta to Mexico's *feminicidio* (the killing of women with impunity) in recent years (see Monárrez Fragoso 2002; Lagarde y de los Ríos 2008; Wright 2011). As one Mexican political analyst, Ileno Semo, has written, "In the [Argentinian] movement of the Plaza de Mayo, as in others in Colombia, Chile, and Guatemala, the mothers, as one expects, defined the political challenges. In the case . . . of Ayotzinapa, fathers have been added (men), which adds an unprecedented dimension to the special place that the father figure occupies in our culture" (Semo 2015).

In both cases, mother- and father-activists use a set of socially conservative concepts regarding gender roles within family structures to represent their standing as respectable families within a heteronormative and patriarchal social order that upholds the cultural and political traditions of Roman Catholicism. Whereas mother-activists over the decades of activism against state terror in the Americas typically present their political demands within a representation of themselves as "mothers looking for their children," the father-activists of Ayotzinapa stand for "the family" and its demands, and rights, in relation to the state. The gendered contrast of fathers representing the family and of mothers looking for children reflects the politics of sexual difference within patriarchal organizations of the family—as mothers tend to children and fathers represent the whole domestic unit within the public sphere. The contrast, however, bears teasing apart more fully to appreciate the radical potential that extends from this conservative representation of political subjects and their radical political demands.

The spatial strategies of the mother-activists who have taken on Latin America's dictatorships have drawn a tremendous amount of attention over the last decades. Most notably, the occupation of women as mother-activists in the public square has refashioned the meaning of "public women," from a historically pejorative connotation of prostitution, to include their standing as political women, who through a public expression of motherhood and grief challenge state terror and those who benefit from it (Wright 2006; Taylor 2012). Behind their protest lies the understanding that the state is the reason why good family women are out on the street in the first place, as they are out looking for their children whom the state has abducted (Bosco 2006). In other words, the power of these protests depends on an understanding of the spatial logics behind conservative practices. As women take to the streets as "mothers," they create countertopographies of action across the scales linking conservative ideologies of patriarchal families to social justice movements calling for radical political and economic changes. In this way, and across such apparently differentiated political and spatial domains, they knit together local and transnational alliances around the idea that these "mothers" promise to return to their rightful place in the home once their children return, alive and well. Of course, the revolutionary potential of their message is not lost on the governing elites (or anyone else) who knows that these mother-activists will never return home (as their disappeared children are not alive and well): Their protest is eternal (see Escobar and Álvarez 1992).

Likewise, the "fathers" of the *normalistas* have drawn on this legacy of mother-activists searching for their disappeared "children." In this case, the fathers use the socially and politically conservative tenants of patriarchal authority to assert their moral legitimacy in presenting their families' demands, and those of their communities by extension, to the state. They articulate their demands for their sons' return through a patriarchal logic that represents these sons, as their patrilineal descendants, who, in turn, represent the basis of Mexican society—young men who are studying to establish their professional trajectories and become national leaders. Such symbolism reaffirms the declarations of these missing sons as the missing students of Mexico's future: They are its future fathers and patriarchs.

Central to the radicality of this otherwise conservative message is the subversive politics of race and class within the assertion by these particular father-activists, and the multitudes marching within them, that their indigenous and brown-skinned rural and very economically humble families indeed represent "the families" of Mexico and that, by extension, "their sons" are the country's missing "future." In other words, to assert that these fathers and these sons as

representative of Mexico's future is to challenge the racial order of Mexico in which indigenous families are regularly relegated to its cultural, political, and economic margins (see Harvey 1998; R. A. Hernández 2008; Saldaña-Portillo 2016). Consequently, behind the conservative symbolism of fathers demanding justice for their sons, this activism defies the racist political economy of a neoliberal order that has further impoverished indigenous and rural Mexico on the basis that this part of country, and its populations, are not central to the nation's priorities. Therefore, as fathers hold up posters of their sons' official school photos, organize desks to represent their sons' absences from classrooms, and refer to their missing sons as the country's "missing future," they generate a counterpublic that spatially reconfigures the politics of race and region that have been a key underpinning to globalized politics of Mexico's neoliberal restructuring in the late twentieth century.

The countertopographies of these counterpublics extend further through the alliances formed among populations and places typically differentiated or even dualistically opposed. For instance, as urban and rural students collaborate collectively as "students," they bring together the indigenous and mestizo populations, who attend extremely different kinds of educational institutions (e.g., urban universities vs. rural teachers' colleges) that long characterize the country's rural–urban and class-riven divides. Urban professors also take a leading role as many, during the first year of protests, held mock classes outside of their buildings, with forty-three empty desks adorned with photos of the missing *normalistas*. In October 2014, I was witness to one such "classroom" at the Universidad Autónoma Metropolitana-Xochimilco (UAM-X) campus in Mexico City when a group of professors took turns sitting silently at a desk in front of an "empty classroom," held outside of the administrative building. With unoccupied blue plastic desk chairs arranged, seminar-style, in a circle, the professors stared ahead, their mouths covered with tape. Behind them, a banner declared their purpose: "There is no class today. Forty-three students are missing. I do not want you to be the next one" (author observation, October 2014). As one student who was watching the professor demonstration explained to me, "Before Iguala, we never really thought of the *normalistas* as students like us. But they are." The many photos of indigenous faces on the desks stared out at the largely mestizo crowd that stood in support of the protest. Some students from this urban campus helped the Ayotzinapa families harvest the autumn corn that had been planted by their student-sons alongside their schools that was at risk of rotting with their disappearance.

In efforts to fortify the transnational and multiscalar potential of these politics, as a way not only to pressure the Mexican government but also to provide solidarity to movements beyond the country, Mexican activists (many of them rural and urban students) in the early weeks of the protests tweeted: #AyotzinapaFerguson. Their efforts were to draw out the topographies binding the massacre and disappearance of the *normalistas* with the police violence that targets racial minorities, and especially African Americans, in the United States. Although this tweet did not generate as much attention in the United States as it did in Mexico, it reflects an ongoing effort to generate countertopographies for the counterpublics emerging out of the social justice movements that link racial, sexual, and economic justice. These countertopographies take shape, everyday, through the brave mobilizations of people who defy the daily threats of terrorizing states (who collude, for instance, in the making of drug wars). Through the courageous politics of grief and radical representations of the *normalistas* as Mexico's "families, students, and its future," a new public seeks to create a new Mexican geography that includes rather than excludes the rural, indigenous, and working poor. As a result of the power of this politics, and the new social and spatial relations it generates, many Mexican activists (and the journalists and scholars who support them) suffer harassment, death, and disappearance as the Mexican state, in the name of democracy and drug war, continues to terrorize its protesting citizens.

Further Reflections on Democratic Evils

A recent Amnesty International Report (2017) on Mexico describes the ongoing interconnections of state terror, a U.S.-supported drug war that evolved out of the Cold War, and neoliberal policies that impoverish the most economically and socially vulnerable:

> Ten years since the start of the so-called "war on drugs and organized crime", the use of military personnel in public security operations continued and violence throughout the country remained widespread. There continued to be reports of torture and other ill-treatment, enforced disappearances, extrajudicial executions and arbitrary detentions. Impunity persisted for human rights violations and crimes under international law. . . .

Human rights defenders and independent observers were subjected to intense smear campaigns; journalists continued to be killed and threatened for their work.

As activists expose themselves to enormous risks in this context, the social justice struggles continue to expand through collaborations and creative approaches to the typical divisions that emerge during social justice work. Needed within such efforts, of course, are collaborations from scholars who, working alongside activists, provide practical insights into the complexities and risks that these movements encounter. Critical geography, with its emphasis on the spatial processes inherent to these kinds of mobilizations, has much to offer. The call for counterpublics requires an understanding of the geographic processes through which these publics come to life. Certainly, within movements such as those on behalf of Mexico's missing students, there is an urgent need for geographers who commit to producing knowledge about the making of solidarity, including the generation of counterplaces for counterpublics, and who strive to make that knowledge accessible and widely understandable.

Acknowledgments

I am especially grateful to Dr. Mónica Inés Cejas for her helpful insights on the topics I discuss in this article. I am also indebted to Dr. Hector Padilla, Dr. Juanita Sundberg, and Leobardo Álvarado, with whom I have worked on a related project on militarization along the Mexico–U.S. border. Our collaborations have helped me in innumerable ways as I work through these ideas. I am solely responsible for any errors.

Funding

This project has received funding from The National Science Foundation under award number 1023266. Any opinions, findings, and conclusions or recommendations expressed in this material are those of the author and do not necessarily reflect the views of the National Science Foundation.

Notes

1. For a carefully and bravely documented account of these events, please see A. Hernández (2017).

2. My framing of this question comes from the intersection of academic research, such as Mendoza's (2006) "The Undemocratic Foundations of Democracy" and the implicit logic of the social movements discussed here.

3. So dubbed by the prized Chilean author Mario Vargas Llosa and broadcast to the world more recently by filmmaker Luis Estrada with his film, *The Perfect Dictator (La Dictadura Perfecta)*.

4. Even as the word for *parents* in Spanish is *los padres* (as the feminine *mothers* is subsumed by the masculine referent), in the *normalista* campaigns, *los padres* refers specifically to the fathers.

5. I refer to their leadership as "symbolic" as there are many who lead and organize these campaigns, including women and those not related to the victims as family members. Important to note also is the diversity of these campaigns even as they consistently demand the live return of the disappeared.

References

Aguayo Quezada, S. 2001. *La Charola: Una historia de los servicios de inteligencia en México*. México City, Mexico: Grijalbo.

Amnesty International. 2017. Mexico 2016/2017. Accessed August 29, 2017. https://www.amnesty.org/en/countries/americas/mexico/report-mexico/.

Aristegui, C. 2014. Caso Iguala: Hace un mes y no aparecen los 43 estudiantes. Accessed July 5, 2015. http://aristeguinoticias.com/2410/mexico/caso-iguala-1-mes-y-no-aparecen-los-43-estudiantes/.

Aristegui, C., and R. Trabulsi. 2010. *La transicion: Conversaciones y retratos de lo que se hizo y se dejo de hacer por la democracia en Mexico*. Grijalbo, Mexico City: Penguin Random House Grupo Editorial.

Atach Zaga, L. 2015. Carteles por Ayozinapa: Del mensaje a la acción. In *Carteles por Ayotzinapa*, 7–8. Mexico City, Mexico: The Museum of Memory and Tolerance.

Bertoia, L. 2014. No one was prepared for a forced disappearance during democracy. *Buenos Aires Herald* March 16. Accessed June 6, 2017. http://www.buenosairesherald.com/article/154540/%27no-one-was-prepared-for–a-forced-disappearance–during-democracy%27.

Borland, E. 2006. The mature resistance of Argentina's Madres de Plaza de Mayo. In *Latin American social movements: Globalization, democratization and transnational networks*, ed. H. Johnston and P. Almeida, 115–30. Lanham, MD: Rowman and Littlefield.

Bosco, F. 2001. Place, space, networks and the sustainability of collective action: The Madres de Plaza de Mayo. *Global Networks: A Journal of Transnational Affairs* 1 (4):307–29.

———. 2006. The Madres de Plaza de Mayo and three decades of human rights activism: Embeddedness, emotions and social movements. *Annals of the Association of American Geographers* 96 (2):342–65.

Butler, J. 2009. *Frames of war: When is life grievable?* Ann Arbor: University of Michigan Press.

Campa, H. 2015. En este sexenio, 13 desaparecidos. Reporte Especial. *Proceso*. Accessed July 3, 2015. http://hemeroteca.proceso.com.mx/?page_id=278958&a51dc26366d99bb5fa29cea4747565fec=395324&rl=wh.

Davis, A., ed. 2017. *Policing the black man*. New York: Pantheon.

DeJesus, K., F. Bosco, and I. Humaydan. 2014. Enforced disappearance: Spaces, selves, societies, suffering. *ACME: An International E-Journal for Critical Geographies* 13 (1):73–78.

Escobar, A., and S. Álvarez. 1992. *The making of social movements in Latin America*. Boulder, CO: Westview.

Fazio, C. 2016. *Estado de emergencia*. Mexico City, Mexico: Grijalbo.

Figueroa, B. 2015. Officially unofficial: The evolution of Mexico's student protests. Accessed April 3, 2017. http://www.coha.org/officially-unofficial-the-evolution-of-mexicos-student-protests/.

Fraser, N. 1992. Rethinking the public sphere: A contribution to the critique of actually existing democracy. In *Habermas and the public sphere*, ed. C. J. Calhoun, 109–42 . Cambridge, MA: MIT Press.

Fregoso, L.-R., and C. Bejarano, eds. 2010. *Terrorizing women: Feminicide in the Americas*. Durham, NC: Duke University Press.

Gilmore, R. W. 2007. *Golden gulag: Prisons, surplus, crisis and opposition in globalizing California*. Berkeley: University of California Press.

Gould, K. 2014. Framing disappearance: H.I.J.@.S., Public Art and the making of historical memory of the Guatemalan civil war. *ACME: An International E-Journal for Critical Geographies* 13 (1):100–34.

Grandin, G. 2007. *Empire's workshop: Latin America, the United States, and the rise of the new imperialism*. New York: Holt.

Harvey, N. 1998. *The Chiapas rebellion*. Durham, NC: Duke University Press.

Heinle, K., C. Molzahn, and D. Shirk. 2015. Drug violence in Mexico: Data and analysis through 2014. Justice in Mexico Report. Accessed September 25, 2017. https://justiceinmexico.org/2015-drug-violence-in-mexico-report-now-available/

Hennessey, K., and T. Wilkinson. 2013. Obama, visiting Mexico, shifts focus from drug war. *The Los Angeles Times* May 2. Accessed September 25, 2017. http://articles.latimes.com/2013/may/02/world/la-fg-mexico-obama-20130503

Hernández, A. 2017. *La verdadera noche de Iguala: La historia que el gobierno quiso ocultar*. New York: Vintage Español.

Hernández, R. A., ed. 2008. *Etnografías e historias de resistencia*. Mexico City, Mexico: CIESAS.

Human Rights Watch. 2013. Mexico's disappeared: The enduring cost of a crisis ignored. February 20. Accessed July 3, 2015. https://www.hrw.org/report/2013/02/20/mexicos-disappeared/enduring-cost-crisis-ignored.

Johnson, T. 2014. A college of missing Mexican students. *The Christian Science Monitor* October 13. Accessed September 29, 2016. http://www.csmonitor.com/World/Americas/2014/1013/At-college-of-missing-Mexican-students-history-of-revolutionary-zeal.

Jones, O. 2014. Body pictographs and the disappeared: Ghosting (through) city spaces. A short essay with photographs. *ACME: An International E-Journal for Critical Geographies* 13 (1):135–52.

Katz, C. 2001. On the grounds of globalization: A topography for feminist political engagement. *Signs* 26 (4):1213–34.

Keller, R. 2015. *Mexico's Cold War: Cuba, the United States, and the legacy of the Mexican Revolution.* New York: Cambridge University Press.

Koonings, K., and D. Kruijt, eds. 2009. *Societies of fear: The legacy of civil war, terror and violence in Latin America.* London: Zed.

Lagarde y de los Ríos, M. 2008. Antropología, feminismo y política: Violencia feminicida y derechos humanos de las mujeres. In *Retos teóricos y nuevas prácticas,* ed. M. Bullen and M. C. Díez Mintegui, 209–39. Donostía, Spain: Ankulegi.

La Jornada Jalisco. 2015. No están vivos, no están muertos: Están desaparecidos. March 30. Accessed July 3, 2015. http://lajornadajalisco.com.mx/2015/03/no-estan-vivos-no-estan-muertos-estan-desaparecidos/.

Lloyd, M. 2017. Naming the dead and the politics of the human. *Review of International Studies* 43 (2):260–79.

Lomnitz, C. 2017. *La nación desdibujada.* Mexico City, Mexico: Malpaso.

Mbembe, A. 2003. Necropolitics. *Public Culture* 15 (1):11–40.

Mbembe, J. A., and L. Meintjes. 2003. Necropolitics. *Public Culture* 15:11–40.

McKittrick, K. 2014. *Sylvia Wynter: On being human as praxis.* Durham, NC: Duke University Press.

Mendoza, B. 2006. The undemocratic foundations of democracy: An enunciation from postoccidental Latin America. *Signs: Journal of Women in Culture and Society* 31 (4):932–39.

Monárrez Fragoso, J. E. 2002. La cultura del feminicidio en Ciudad Juárez. 1993–1999. *Frontera Norte* 12 (23):87–111.

Mora, M. 2017. Desaparición forzada, racismo institucional y pueblos indígenas en el caso Ayotzinapa, México. *LASAFORUM* XLVIII (2):29–31. Accessed June 4, 2017. http://lasa.international.pitt.edu/forum/files/vol48-issue2/Debates-AntiRacistStruggles-3.pdf

Novo, C. M. 2003. The culture of exclusion. *Bulletin of Latin American Research* 22 (3):249–68.

Ochoa Muñoz, K. 2015. (Re)pensar el derecho y el sujeto del indio desde una mirada descolonial. *Revista Internacional de Comunicación y Desarrollo* 4:47–60.

Paley, D. 2014. *Drug war capitalism.* Oakland, CA: AK Press.

Pastrana, D. 2016. Mexico, a democracy where people disappear at the hands of the state. Interpress News Agency August 26. Accessed June 2, 2017. http://www.ipsnews.net/2016/08/mexico-a-democracy-where-people-disappear-at-the-hands-of-the-state/.

The Perfect Dictator (La Dictadura Perfecta). 2014. Dir. L. Estrada. Mexico City: Bandidos Films.

Poniatowska, E. 1991. *Massacre in Mexico.* Columbia: University of Missouri Press.

———. 2014. Regrésenlos" (discurso de Elena Poniatowska en el Zócalo). *La Jornada* October 26. Accessed July 5, 2015. http://www.jornada.unam.mx/ultimas/2014/10/26/201cmexico-se-desangra201d-dice-elena-poniatowska-en-el-zocalo-4330.html.

Ross, A. 2008. The body counts: Civilian casualties and the crisis of human rights. In *Human rights in crisis,* ed. A. Bullard, 35–48. New York: Ashgate.

Saldaña-Portillo, R. J. 2016. *Indian given: Racial geographies across Mexico and the U.S.* Durham, NC: Duke University Press.

Semo, I. 2015. Los Padres de Aytozinapa. Accessed September 25, 2017. http://www.estudioscriticosdelacultura.com/articulos/index/14

Shirk, D. 2011. *The drug war in Mexico: Confronting a shared threat.* New York: Council on Foreign Relations Press.

Smith, W., ed. 2009. *Latin American democratic transformations.* Miami, FL: University of Miami Press.

Solano, P. 2014. Los Maestros Rurales. *La Jornada.* Accessed September 26, 2016. http://www.jornada.unam.mx/2014/10/12/politica/010n1pol.

Stephen, L. 2013. *We are the face of Oaxaca: Testimony and social movements.* Durham, NC: Duke University Press.

Stevenson, M. 2017. Mexico has so many mass graves. Associated Press. Accessed April 3, 2017. http://bigstory.ap.org/article/ba51de3dcba842baa059dfdecd7b2aad/mexico-has-so-many-graves-lacks-space-bodies-exhumed.

Taylor, D. 2012. *Disappearing acts: Spectacles of gender and nationalism in Argentina's Dirty War.* Durham, NC: Duke University Press.

Weheliye, A. G. 2014. *Habeas viscus: Racializing assemblages, biopolitics, and black feminist theories of the human.* Durham, NC: Duke University Press.

Weld, K. 2014. *Paper cadavers: The archives of dictatorships in Latin America.* Durham, NC: Duke University Press.

Who are the *normalistas*? Radicals? Revolutionaries? Communists? 2014. *Borderland Beat.* October 18. Accessed July 3, 2015. http://www.borderlandbeat.com/2014/10/who-are-normalistas-radicals.html.

Wright, M. W. 2006. *Disposable women and other myths of global capitalism.* London and New York: Routledge.

———. 2011. Necropolitics, narcopolitics, and femicide: Gendered violence along the Mexico–U.S. border. *Signs* 36 (3):707–31.

———. 2017. Epistemological ignorances and fighting for the disappeared: Lessons from Mexico. *Antipode* 49 (1):249–69.

MELISSA W. WRIGHT is Professor of Geography and Women's, Gender, and Sexuality Studies at The Pennsylvania State University, University Park, PA 16802. E-mail: mww11@psu.edu. Her current research studies social justice movements, feminist and antiracist politics, and state terror in modern Mexico and across the Americas.

3 Locating the Social in Social Justice

Robert W. Lake

A concept of social justice in which the social names a subset of justice suggests that the social constitutes a distinct sphere within which a distinctively social justice is produced and experienced and within which a specifically social injustice can be addressed. Theorists from Dewey to Latour to Foucault, however, have questioned the conceptualization of the social as a separate substantive domain within which a distinctively social justice can be found. I seek to move from a substantive to a relational conceptualization of the social in social justice, drawing from Dewey's concept of the social as an associative rather than an aggregative relation. A relational approach situates the social not in a delimited substantive domain within which justice can be assessed but as a mode of collective association through which justice is performed and produced. Relocating the social from a substantive sphere to a relational practice transforms the problem of social justice. Rather than assessing the justice of outcomes within a specifically social sphere, the problem of social justice addresses the interactive practices of social actors engaged in the collective project within which justice is dialectically and simultaneously a process and an outcome, a means and an end. I illustrate the challenges of practicing a relational conception of social justice in an antidisplacement protest against a neighborhood redevelopment proposal in Camden, New Jersey. The case study suggests that furthering the goal of social justice focuses on everyday practices of associative interaction in which relations of democratic equality are undermined or encouraged. *Key Words: democratic justice, relational justice, social justice.*

社会作为正义的一个子集之"社会正义",意味着社会构成了一个特殊的领域。在这个领域中,社会正义进行生产与被经历,且一种特定的社会不正义能够获得处置。但从杜威到拉图以至福柯等理论家,质疑将社会视为分离的实质领域、且特定的社会正义可于其中寻得的概念化。我运用杜威将社会视为联合而非集合关系的概念,寻求将社会正义中的社会,从实质转为关系性的概念化。关系性取径,并非将社会定位为可从中获得正义的限定实质领域,而是将社会置于正义进行展演并生产的集体关联模式。将正义从实质领域再安置于关系性实践之中,改变了社会正义的问题。与其评估一个特定社会领域中的正义结果,社会正义的问题,应对参与至集体计画的社会行动者的互动实践,而正义于该计画中是辩证且自发的过程与结果、方法与目的。我描绘在新泽西坎顿一个针对邻里再开发计画的反迫迁抗议行动中,实践关系性的社会正义概念之挑战。该案例研究主张,推进社会正义的目标,需聚焦关联性互动的每日生活实践,其中民主平等的关系,可能会受到侵蚀抑或鼓励。*关键词: 民主正义, 关系性正义, 社会正义。*

Un concepto de justicia social en el que lo social califica un subconjunto de la justicia sugiere que lo social constituye una esfera distintiva dentro de la cual una justicia distintivamente social es producida y experimentada, y dentro de la cual puede abocarse una injusticia específicamente social. Sin embargo, teóricos desde Dewey a Foucault, pasando por Latour, han cuestionado la conceptualización de lo social como un dominio sustantivo separado dentro del cual pueda hallarse una justicia distintivamente social. Por mi parte, busco moverme de la conceptualización sustantiva de lo social hacia una relacional, apoyándome en el concepto de Dewey de lo social, más como una relación más asociativa que de agregación. Un enfoque relacional sitúa lo social no en un dominio sustantivo delimitado dentro del cual pueda evaluarse la justicia, sino como un modo de asociación colectiva a través de la cual la justicia es ejercida y producida. Relocalizar a lo social desde una esfera sustantiva a una práctica relacional transforma el problema de la justicia social. Más que evaluar la justicia de los resultados dentro de una específica esfera social, el problema de la justicia social aboca las prácticas interactivas de actores sociales comprometidos en un proyecto colectivo dentro del cual la justicia es dialéctica y simultáneamente un proceso y un resultado, un medio y un fin. Ilustro los retos de practicar una concepción relacional de la justicia social en una protesta anti-desplazamiento hecha contra una propuesta de redesarrollo vecinal de Camden, Nueva Jersey. El estudio de caso sugiere que fortaleciendo eso el fin de la justicia social se enfoca en las prácticas cotidianas de interacción asociativa en la cual las relaciones de igualdad democrática se socaban o promueven. *Palabras clave: justicia democrática, justicia relacional, justicia social.*

The forty residents of the Lanning Square neighborhood of Camden, New Jersey, who crowded into the basement of the New Mickle Baptist Church on an evening in June 2007 had come to attend a "redevelopment update meeting" called by the Camden Department of Development and Planning to explain the city's plan to redevelop their neighborhood. A cloud of anger and uncertainty hung over the meeting as residents fearing displacement sought information regarding the plan for the neighborhood's future. Tensions were high because the redevelopment plan listed 350 properties, including eighty occupied parcels, comprising more than a third of the entire neighborhood, that the city would acquire in implementing its proposed plan (Camden, NJ Department of Development and Planning 2008b). Yet despite the scale of property acquisition and displacement, residents lacked information on the content of the plan, the timing of property acquisitions, the method of compensation for acquired properties, provisions for relocation assistance and replacement housing, and many other details of the plan that would have direct and enduring consequences on their lives, their neighbors, and their neighborhood. Further impeding the redevelopment of Lanning Square was a legacy of distrust and disbelief built up over fifty years of failed urban renewal projects and fruitless attempts to rejuvenate the city of Camden (Lake et al. 2007). Once a major center of manufacturing employment,[1] the city experienced massive deindustrialization, disinvestment, and white flight in the decades after World War II, with the loss of 157,000 manufacturing jobs between 1950 and 1970 and a 40 percent drop in total population by 2000 (Cowie 1999; Gillette 2003, 2005; Sidorick 2009; Hedges and Sacco 2012). By 2010, Camden's population was 95 percent African American and Hispanic; the city's median household income was 36 percent of the statewide average; and Camden ranked among the poorest cities in the country, with 42 percent of households in poverty (U.S. Census Bureau 2010, 2012). The New Jersey state legislature took control of Camden's city government in 2002 by appointing a chief operating officer empowered with override authority over the democratically elected mayor and city council and charged with overseeing the city's redevelopment.[2] After decades of housing abandonment, private and public disinvestment, land clearance, and urban renewal, only 2,896 residents remained in the Lanning Square neighborhood by 2010, down from a peak of 17,347 in 1940 (Camden,

NJ Department of Development and Planning 2008a; Camden Redevelopment Agency 2014).

On its face, the displacement of Lanning Square's residents in the name of urban redevelopment constitutes a condition of injustice by virtually any extant formulation of the concept (Barnett 2016). The proposed redevelopment plan continued a persistent practice that concentrated the costs of neighborhood renewal and urban revitalization on vulnerable, marginalized, powerless, and disenfranchised residents who were excluded from decisions that affected them and from which they would reap few if any benefits. The inequitable distribution of costs and benefits violates Rawls's (1971) difference principle, which holds that unequal distributions should most benefit the least advantaged. The redevelopment plan fails on the three dimensions of justice—equity, democracy, and diversity—that Fainstein (2010), following Rawls, identified as criteria for the "just city." If compensating residents at the preredevelopment value of their property constitutes exploitation, displacing residents from the neighborhood to be redeveloped reproduces marginalization, excluding residents from the planning process reflects their powerlessness, clearing away a working-class community of color to make way for a middle-class housing market constitutes cultural imperialism, and disrupting long-standing ties of neighboring and community is a form of violence, then the situation in Lanning Square and across most of the city of Camden arguably characterizes all five "faces of oppression" named by Young (1990, 40) as the hallmarks of injustice.[3]

In what sense, though, is Lanning Square a case of a specifically "social" conceptualization of justice? What is gained or lost by characterizing the situation in Lanning Square as a matter of social justice? A concept of social justice in which the social names a subset or domain of justice suggests that the social constitutes a distinct sphere within which a specifically social kind of injustice can be identified and addressed and a distinctively social sort of justice can be produced and experienced.[4] The characterization of Lanning Square's residents using racial, ethnic, income, and other demographic and social categories, for example, seems to locate the observed injustices within an implicit sphere of the social described by those categories, corresponding to a discrete disciplinary ontology that distinguishes between social and nonsocial (i.e., economic, environmental, juridical, psychological, etc.) domains of justice (Walzer 1983). Alternatively, the situation in Lanning Square might implicate a specifically social kind of justice if the observed injustice is determined by the structure of society or is extruded by an inequitable and exploitative social structure understood once again as located behind, and existing prior to, the observed practices (Harvey 1973; MacLeod and McFarlane 2014). In both the demographic and the structural formulations, social justice tautologically resides in outcomes (e.g., income inequality, racial disparity) implicitly situated within a social sphere of concern.

Theorists from a variety of perspectives, however, have questioned the ontological presupposition of a social sphere, domain, or structure that might serve as the setting or context of an explicitly social justice. Taking his inspiration from Dewey,[5] Latour (2005) observed that the "social is not a place, a thing, a domain, or a kind of stuff but a provisional movement of new associations" (238). By resituating the social in this way, Latour signals a shift from a substantive to a relational conception of the social understood as an encounter in the constant process of enactment. My purpose in this article is to move from a substantive to a relational concept of social justice that situates the social not in a delimited substantive domain within which (in)justice is located but as a mode of collective engagement through which justice or injustice is both performed and (re)produced. Within a relational conception of social justice, the social refers to a process of collective interaction grounded in an ontology of association and encounter rather than referring to a substantive (yet nonetheless abstract) sphere of the social with indeterminate boundaries and a dubious ontological presence.

Relocating the social from a substantive arena to a relational process transforms the problem implicated in the idea of social justice. Rather than assessing the justice of outcomes within a specifically social sphere, the problem of social justice within a relational frame addresses the practices of social (i.e., collective) actors engaged in the collective (i.e., social) project of social interaction within which justice is dialectically and simultaneously a process and an outcome (Lake 2016). In the remainder of this article, I examine a relational concept of social justice that locates the social in the collective process forming individual subjectivities that thereby constitute the collective conditions on which the possibility of justice depends. I return to Lanning Square in the final section of the article to assess the justice of that collective social process as it unfolded in a particular place and time.

From Social Sphere to Social Relation

In his well-known book *Reassembling the Social*, Latour (2005) was explicit in his rejection of "the social" as a separately discernable context or domain of social life. "[T]here is nothing specific to social order," he argued, "no social dimension of any sort, no 'social context,' no distinct domain of reality to which the label 'social' or 'society' could be attributed" (4). To ground one's conceptualization of the social in a preexisting sphere is to "believe the social to be always already there, ... whereas the social is not a type of thing either visible or to be postulated" (Latour 2005, 8). Society, Latour asserted, "is not the whole 'in which' everything is embedded" (241). Although postulating the social as a separate sphere such that "the social could explain the social" (3) has become "the default position of our mental software" (4), Latour (2005) pointed out that "the 'social' as in social problems, the social question, is a nineteenth-century innovation" (6) adopted as a disciplinary convenience to "stabilize" and simplify the complexity of our experience of multiply interacting actors comprising the empirical world (Foucault 1970).[6]

In place of the default conceptualization of the social as a preexisting substantive domain, Latour offered a relational concept of the social as a process of association or connection among actors arising by virtue of their mutual encounter and interaction in the world. The "social does not designate a thing among other things ... but *a type of connection* between things" (Latour 2005, 5). A consequence of the relational and processual understanding of the social is that society always remains "a task to be fulfilled" (184) rather than the context of action already established behind the practice: "The body politic has always been, as John Dewey put it ... a *problem*, a ghost always in risk of complete dissolution. Never was it supposed to become a substance, a being, a sui generis realm that would have existed beneath, behind, and beyond political action" (162). An a priori assumption of the social, furthermore, undermines the possibility of politics because "if there is a society, *then no politics is possible*" (Latour 2005, 250). To the contrary, Latour insisted, politics constructs the possibility and the characteristics of the social as "politics is defined ... as the progressive composition of collective life" (41). To be social, on this relational account, is to be political because relations must always be performed and how they are performed is a political act.[7]

Latour's relational approach to the social echoes two core principles of Deweyan pragmatism: an anti-foundational rejection of a priori abstract categories and a commitment to a processual over a substantive ontology (Lake 2016, 2017b, 2017c). Dewey insisted on the precedence of empirical observation over abstract conceptualization, considering the latter as ontologically unverifiable and a hindrance to problem solving (Dewey [1920] 2004).[8] Abstract foundational concepts such as the state, the social, and the atomistic liberal individual constitute, for Dewey, merely theoretical constructs that simplify and thus obscure empirical complexity while relying on a self-referential ontology for their conceptual validation. Considering the origins of the state, for example, Dewey warned against the power of abstract constructs to deflect attention away from the empirical world of praxis and onto the conceptual plane of theoretical discourse:

> The concept of the state, like most concepts which are introduced by "The," is both too rigid and too tied up with controversies to be of ready use. ... The moment we utter the words "The State" a score of intellectual ghosts rise to obscure our vision. Without our intention and without our notice, the notion of "The State" draws us imperceptibly into a consideration of the logical relationship of various ideas to one another, and away from facts of human activity .... Conceptions of "The State" as something *per se*, something intrinsically manifesting a general will and reason, lend themselves to illusions. (Dewey [1927] 2012, 44, 77)

On the same account, Dewey dismissed Lockean formulations of liberal individualism reliant on a doctrine of natural rights "inherent in the very structure of the individual" (Dewey [1927] 2012, 87), because the assertion of divinely endowed natural rights remains an abstract construct unsusceptible to empirical verification. If the atomistic liberal individual is merely a construct of eighteenth-century political philosophy that has been naturalized and internalized to appear as a universal principle, however ("as something *given*, something already there"; Dewey [1920] 2004, 111; see also Lake 2017a]), then the understanding of society as the aggregation of autonomous individuals is equally a conceptual convenience rather than an empirical fact.[9]

Dewey disparaged the aggregate conception of the social as merely a "numerical" agglomeration of atomistic individuals in which "relations to other individuals were as mechanical and external as those of Newtonian atoms to one another" (Dewey [1939]

1991, 92). Against the contractarian view of society as the aggregation of autonomous individuals who choose to congregate for reasons of mutual self-interest, Dewey proposed an associational idea of the social in which individual identity derives from the collective of which the individual is a part. Pragmatism's adherence to observable practice derives the associational view from the inevitability of encounter, interaction, and communication among individuals-in-association, as against the aggregative assumption of a mythical (because it is unobservable) Rousseauian social contract established by presocial individuals (Rasmussen 2011). "Association in the sense of connection and combination is a 'law' of everything known to exist," Dewey ([1927] 2012) observed, and "nothing has been discovered which acts in entire isolation. ... The behavior of each," he concluded, "is modified by its connection with others" (51). In the aggregative view of liberal individualism, individuals bring their already-formed selves into confrontation with others and a common interest emerges from the aggregation of self-interests. In the associative view, in contrast, individuals are formed through their association with others, in connection with whom they derive their interests, values, beliefs, and actions. Society in Dewey's associative view constitutes what Sandel (1998) described as a constitutive conception that "describes not just what (members) *have* as fellow citizens but also what they *are*, not a relationship they choose ... but an attachment they discover, not merely an attribute but a constituent of their identity" (150). Whereas individuals form a community in the aggregative model, community in the associative model is "constitutive of the shared self-understandings of the participants" (Sandel 1998, 173).[10]

If society constitutes the individuals of which it is comprised, then attention turns to the relational process through which subjects are formed and individuality is constructed. Here Dewey deserves to be quoted at length:

> When self-hood is perceived to be an active process it is also seen that social modifications are the only means of the creation of changed personalities. Institutions are viewed in their educative effect:—with reference to the types of individuals they foster. ... We are led to ask what the specific stimulating, fostering and nurturing power of each specific social arrangement may be. ... Just what response does *this* social arrangement, political or economic, evoke? ... Does it release capacity? If so, how widely? Among a few, with a corresponding depression in others, or in an extensive and equitable way? ...

Are men's senses rendered more delicately sensitive and appreciative, or are they blunted and dulled by this and that form of social organization? Are their minds trained so that the hands are more deft and cunning? Is curiosity awakened or blunted? ... Such questions as these ... become the starting-points of inquiries about every institution of the community when it is recognized that individuality is not originally given but is created under the influences of associated life. ... But instead of leading us to ask what it does in the way of causing pains and pleasures to individuals already in existence, it inquires what is done to release specific capacities and coordinate them into working powers. What sort of individuals are created? (Dewey [1920] 2004, 113–14)

Considering the constitutive effects of "associated life" foregrounds the relational practices through which association is performed and experienced, including relations that are regularized in formal institutions and those enacted through the micropractices of everyday life. A relational and associational concept of the social reflects what Fraser (1996, 2004) called a politics of recognition, in which misrecognition "means social subordination in the sense of being prevented from participating *as a peer* in social life" (Fraser 2004, 129). Subject positions are established and identities are formed, on this account, through the diffuse, dispersed, multilocational, and multivocal relational practices of the everyday and it is through those relations, rather than in the hidden hand of determinative structures, that the politics of social justice is practiced and experienced.[11] As Healey (2012) observed:

Those struggling for political change tend to focus on capturing control of key arenas and on the installation of new institutional designs and/or policy programs through which to redirect and re-regulate the way resources flow. Yet it is in the micro-dynamics of practices that such ambitions are brought into being, brought to life and experienced by both those involved in their "implementation" and by those who are supposed to experience their outcomes. (24; see also Davis 1989)

Social structures must be performed, as noted earlier, and it is through the micropractices of everyday life that structural relations are reproduced and structural change can be enacted.

Return to Lanning Square

The micropractices of relational (in)justice were starkly in evidence in the series of redevelopment update meetings convened in Lanning Square in the summer of 2007. Meetings routinely began with a recitation of "ground rules" for behavior (e.g., one person speaks at a time; each person who wants to speak will have a chance; be respectful; stay on the agenda).[12] Unremarkable on their face, the rules valorized norms of rational, dispassionate speech, implicitly characterized the issues at stake as susceptible to rational-technical discourse, and delegitimized residents' subjective experiences of anger and passion. Although redevelopment officials represented the meetings as a form of community engagement, the rules for discussion constituted a process of internal exclusion (Young 2000) that established the relational microdynamics of that engagement in ways that limited residents' abilities to express their views.

Planning officials and consultants repeatedly self-identified as experts when presenting information to meeting attenders. A planning consultant introduced a presentation on community demographics by enthusing that

We've done endless surveys, endless meetings, we've explored and explored. We've looked at endless data sources [and] conducted deep data analysis of who you are. We have reams and reams and boxes and boxes of information. The city and state authorized us to do that work for you so we can get the right priorities for redevelopment.

Planners established a hierarchy of legitimacy by claiming a monopoly on knowledge that simultaneously elevated their own expertise and devalued residents' experiential knowledge of themselves and their local knowledge of the neighborhood. That expertise, in turn, validated the consultant's descriptions of residents over residents' own self-identification, by virtue of the consultant's superior ability to inform residents of "who you are." Indeed, the denigration of resident identities began long before the convening of redevelopment update meetings, in the compilation of properties to be acquired by the city in implementing its redevelopment plan. An appendix to the Lanning Square Redevelopment Plan lists properties "To Be Acquired" by tax parcel block and lot numbers (Camden, NJ Department of Development and Planning 2008b), a compilation that omits any reference to the identity of the individuals whose homes are to be acquired in a process of dehumanization and erasure that begins long before residents are physically displaced.

City officials were adept at controlling the meeting agenda as a strategy of disempowerment. As a redevelopment official made clear:

The plan is done. We're not here to redo the plan. The only thing to focus on is some difficult implementation issues. How to get started. What would make sense as the first phase of redevelopment.

While foreclosing discussion of the substance of the plan, the official invited participation in deciding on the plan's implementation:

We want to present the plan to City Council . . . but only if the community has reached agreement on how the plan will be implemented.

That invitation appeared short-lived, however, when the official declared that, at the next meeting:

I will present to you a recommended process for implementing the plan.

The meeting agenda also became a tool for channeling discussion away from residents' concerns, as illustrated in the following exchange:

Resident: Mr. [. . .] said no homes would be taken.

Official: No, that's not what he said. He said he had an idea about an approach that would minimize the number of acquired properties.

Resident: Are you going to take my property?

Official: What's on that list is what's going to be acquired.

Resident: Why not talk about that now?

Official: Because there's a lot of other things to talk about.

Resident: We've been having meetings forever. We need some action.

Official: Let's talk about the section on the agenda about [neighborhood] assets and challenges.

Resident: Have you made these people aware that they're not going to be living in the community?

Official: Please, not tonight!

Resident: Yes, tonight! That's why all these people are here.

Official: I'm here to finish the agenda we've been working on.

Although this exchange reflects substantially unequal power relations between residents and city officials, residents were not without a voice in contesting the city's practices. As one resident responded:

You talk about this game that you're playing but what good will it do if we're not here to take advantage of it? You

cannot mess around with people's lives. [You're] here to take the focus away from what's real. We bought our homes here because we thought this was where we would be. We bought these homes because this is where we want to be. How can you come in and tell these older people: "We want your home?" The city's idea of civic engagement is to ask: "When would you like your house torn down? Would you like it torn down this month or next month?"[13]

Locating Social Justice

The displacement of residents from Lanning Square violates standards of justice on which many would agree. Injustice is not indexed merely by the number of people or households displaced, though, and the problem of social justice emerged long before the distributive outcome of displacement became manifest in the material landscape of Camden. A relational approach to the social locates social justice in what Fraser (1996, 2004) called participatory justice and Anderson (1999) called democratic equality, a relational rather than a solely distributional practice that situates justice in relational practices of mutual recognition and respect rather than in a fair distribution of goods, opportunities, capabilities, or something else. A relation of democratic equality is the location of a specifically social justice because, as Anderson (1999) concluded, it "conceives of equality as a relationship among people rather than merely as a pattern in the distribution of divisible goods" (336; see also Lake 2014).

Social justice was little in evidence in Lanning Square because the relational encounter between officials and residents failed the test of democratic equality. Injustice began in the dehumanization and erasure of residents' identities in the city's list of properties to be acquired; in the silencing of residents' voices in meetings ostensibly called to promote resident participation in neighborhood redeveloping; in the denigration and disparagement of residents' expertise and local knowledge; and in the construction of residents' subjectivities as bodies to be controlled and manipulated to further the city's objectives.

The practices of encounter and association that constructed individual identities in Lanning Square are simultaneously procedural and consequential, transcending the traditional dualism of process and outcome within the relational practice. In a relational understanding of social justice, the dialectic of collective encounter comprises both the process of its performance and the situation of its outcome, presenting in turn the condition for the continuation of the relational encounter (Lake 2006; Harman 2014). The relational process of subject formation answers the question that Dewey posed—"What sort of individuals are created?"—and it is the individuals thus constituted whose resulting practices construct the world as it is produced and experienced on the ground.

Acknowledgments

I presented an earlier version of this article at the New York Pragmatist Forum at Fordham University and I have benefited greatly from that group's constructive engagement with pragmatist ideas. I am extremely grateful for helpful comments on previous drafts by Bob Beauregard, Nik Heynen, Kathe Newman, Stephanie Pincetl, and two anonymous reviewers for the journal. I am solely responsible for any shortcomings remaining in the article.

Notes

1. Camden's industrial history dates to the early nineteenth century, facilitated by the city's location at the confluence of the Cooper and Delaware rivers and its proximity to Philadelphia (Camden, NJ Department of Development and Planning n.d.; Gillette 2005). As described by Cowie (1999): "By the 1920s, the South Jersey city contained a variety of textile mills and leather processors, a huge ironworks, the Campbell's soup cannery, cigar factories, a pen manufacturer, paint and chemical processing plants, ... the bustling docks, shipyards, and four thousand workers of the New York Shipbuilding Company [and] the Victor Talking Machine Company's [later Radio Corporation of America, and then RCA] sprawling complex of factories" (12).
2. See the Municipal Rehabilitation and Economic Recovery Act (MRERA) of 2002 (S428/A2054) at http://www.njleg.state.nj.us/bills/BillView.asp (accessed 28 January 2017). For a description of the legislation and subsequent amendments, see http://www.camconnect.org/resources/MRERALegislation.html.
3. Young (1990, 39–65) listed exploitation, marginalization, powerlessness, cultural imperialism, and violence as comprising the "five faces of oppression," which, together with domination constitute her definition of injustice.
4. The conception of social justice as a distinct social sphere derives from Rawls (1971), for whom justice refers to "the way in which the major social institutions distribute fundamental rights and duties and determine the division of advantages from social co-operation" (7). A review of the vast literature that adopts Rawls's

position is beyond the scope of this discussion but see, for example, Hayward and Swanston (2011).

5. Expressing his rejection of abstract foundational assumptions, Latour (2005) observed that "it's only the freshness of the results [i.e., the avoidance of foundational presuppositions] of social science that can guarantee its political relevance. And no one has made the point as forcefully as John Dewey did" (261). In Dewey's ([1929] 1984) words, "We are given to thinking of society in large and vague ways. We should forget 'society.' ... There is no society at large, no business in general" (120).
6. Scott (1998, 91) considered "the discovery of society as a reified object that ... could be scientifically described" as a legacy of the Enlightenment and a precondition for the ascendancy of authoritarian high modernism (see also Beauregard 2015).
7. A relational and processual approach to the social recognizes that power is unequally distributed across structural positions but it also insists that structural relations must be performed and enacted and that it is in the indeterminacy of such praxis that a politics of possibility is situated. See also Arendt's (1958, 2005) distinction between behavior and action, contrasting ritualized, habitual social behavior in accordance with norms with the agentic unpredictability of political action.
8. Dewey was unrelenting in his denigration of "the teleological character" of abstract theorizing: "Men looked at the work of their own minds and thought they were seeing realities in nature. They were worshipping under the name of science, the idols of their own making. ... The social philosopher, dwelling in the region of his concepts, 'solves' problems by showing the relationship of ideas instead of helping men solve problems in the concrete. ... Social theory ... exists as an idle luxury rather than as a guiding method of inquiry and planning" (Dewey [1920] 2004, 20, 110–11).
9. "Natural rights and natural liberties," Dewey asserted, "exist only in the kingdom of mythological social zoology" (Dewey [1935] 2000, 27). "The only thing which imports obscurity and mystery ... is the effort to discover alleged, special, original, society-making causal forces, whether instincts, fiats of will, personal, or an immanent, universal, practical reason, or an indwelling, metaphysical, social essence and nature. These things do not explain, for they are more mysterious than are the facts they are evoked to account for" (Dewey [1927] 2012, 53). As Festenstein (1997) concluded, "It is an abstractionist error of old individualism to envisage individuals as prior to their social world" (69).
10. According to Young (1990), whereas "the aggregate model conceives the individual as prior to the collective," in the association model "groups, on the other hand, constitute individuals" (44–45; see also Young 2011).
11. For a discussion of the similarities and differences between Dewey and Foucault on the relationality of subject formation, see Marshall (1995) and Rabinow (2011).
12. All meeting excerpts are from the author's contemporaneous field notes.

13. The Camden Planning Board approved the Lanning Square Redevelopment Plan by unanimous vote on 12 June 2008 and City Council approval followed shortly thereafter. The redevelopment plan called for construction of 400 single-family homes and declared the entire neighborhood eligible for public taking through eminent domain throughout the project's officially designated twenty-five-year life span.

References

Anderson, E. 1999. What is the point of equality? *Ethics* 109 (2):287–337.

Arendt, H. 1958. *The human condition.* Chicago: University of Chicago Press.

———. 2005. *The promise of politics.* New York: Schocken.

Barnett, C. 2016. Towards a geography of injustice. *Alue Ja Ympäristö* 45 (1):111–18.

Beauregard, R. 2015. *Planning matter: Acting with things.* Chicago: University of Chicago Press.

Camden, NJ Department of Development and Planning. n.d. History. City of Camden. Accessed January 28, 2017. http://www.ci.camden.nj.us/history/.

———. 2008a. Lanning Square: A study to determine the need for redevelopment. Accessed January 28, 2017. http://www.ci.camden.nj.us/wp-content/uploads/2013/04/Lanning-Square-Redevelopment-Study.pdf.

———. 2008b. Lanning Square redevelopment plan. Accessed January 28, 2017. http://www.ci.camden.nj.us/wp-content/uploads/2013/04/Lanning-Square-Redevelopment-Plan.pdf.

Camden Redevelopment Agency. 2014. Crossing over, CDC presentation, December 2. Accessed January 28, 2017. http://camdenredevelopment.org/getattachment/8cc7283c-7489-4449-8f07-e8940beefcf0/Test-Plan.aspx.

Cowie, J. 1999. *Capital moves: RCA's seventy-year quest for cheap labor.* New York: New Press.

Davis, P. 1989. Law as micro-aggression. *Yale Law Journal* 98:1559–77.

Dewey, J. [1920] 2004. *Reconstruction in philosophy.* Mineola, NY: Dover.

———. [1927] 2012. *The public and its problems: An essay in political inquiry.* University Park, PA: Penn State University Press.

———. [1929] 1984. Individualism, old and new. In *John Dewey, The later works, 1925–1953: Vol. 5. 1929–1930,* ed. J. Boydston, 41–123. Carbondale: Southern Illinois University Press.

———. [1935] 2000. *Liberalism and social action.* Amherst, NY: Prometheus.

———. [1939] 1991. I believe. In *John Dewey, The later works, 1925–1953: Vol. 14. 1939–1941,* ed. J. Boydston, 91–97. Carbondale: Southern Illinois University Press.

Fainstein, S. 2010. *The just city.* Ithaca, NY: Cornell University Press.

Festenstein, M. 1997. *Pragmatism and political theory: From Dewey to Rorty.* Chicago: University of Chicago Press.

Foucault, M. 1970. *The order of things: An archaeology of the human sciences.* London: Tavistock.

Fraser, N. 1996. Social justice in the age of identity politics: Redistribution, recognition, and participation. Tanner Lectures on Human Values, Tanner Humanities Center, University of Utah. Accessed October 10, 2017. https://tannerlectures.utah.edu/lecture-library.php

———. 2004. Institutionalizing democratic justice: Redistribution, recognition, and participation. In *Pragmatism, critique, judgment,* ed. S. Benhabib and N. Fraser, 123–47. Cambridge, MA: MIT Press.

Gillette, H. 2003. The wages of disinvestment: How money and politics aided the decline of Camden, New Jersey. In *Beyond the ruins: The meanings of deindustrialization,* ed. J. Cowie and J. Heathcott, 139–59. Ithaca, NY: Cornell University Press.

———. 2005. *Camden after the fall: Decline and renewal in a post-industrial city.* Philadelphia: University of Pennsylvania Press.

Harman, G. 2014. *Bruno Latour: Reassembling the political.* London: Pluto.

Harvey, D. 1973. *Social justice and the city.* Oxford, UK: Blackwell.

Hayward, C., and T. Swanstrom, eds. 2011. *Justice and the American metropolis.* Minneapolis: University of Minnesota Press.

Healey, P. 2012. Re-enchanting democracy as a mode of governance. *Critical Policy Studies* 6 (1):19–39.

Hedges, C., and J. Sacco. 2012. *Days of destruction, days of revolt.* New York: Nation Books.

Lake, R. 2006. Recentering the city. *International Journal of Urban and Regional Research* 30 (1):194–97.

———. 2014. Methods and moral inquiry. *Urban Geography* 35 (5):657–68.

———. 2016. Justice as subject and object of planning. *International Journal of Urban and Regional Research* 40 (6):1206–21.

———. 2017a. Big data, urban governance, and the ontological politics of hyper-individualism. *Big Data and Society* 4 (1):1–10.

———. 2017b. A humanist perspective on knowledge for planning: Implications for theory, research, and practice. *Planning Theory and Practice* 18 (2):291–319.

———. 2017c. On poetry, pragmatism, and the urban possibility of creative democracy. *Urban Geography* 38 (4):479–94.

Lake, R., K. Newman, P. Ashton, R. Nisa, and B. Wilson. 2007. *Civic engagement in Camden, New Jersey: A baseline portrait.* New York: MDRC. Accessed August 29, 2017. http://www.mdrc.org/publication/civic-engagement-camden-new-jersey-baseline-portrait.

Latour, B. 2005. *Reassembling the social: An introduction to actor-network-theory.* Oxford, UK: Oxford University Press.

MacLeod, G., and C. McFarlane. 2014. Introduction: Grammars of urban injustice. *Antipode* 46 (4):857–73.

Marshall, J. 1995. On what we may hope: Rorty on Dewey and Foucault. *Studies in Philosophy and Education* 13:307–23.

Rabinow, P. 2011. Dewey and Foucault: What's the problem? *Foucault Studies* 11:11–19.

Rasmussen, C. 2011. *The autonomous animal: Self-governance and the modern subject.* Minneapolis: University of Minnesota Press.

Rawls, J. 1971. *A theory of justice*. Cambridge, MA: Harvard University Press.

Sandel, M. 1998. *Liberalism and the limits of justice*. 2nd ed. Cambridge, UK: Cambridge University Press.

Scott, J. 1998. *Seeing like a state: How certain schemes to improve the human condition have failed*. New Haven, CT: Yale University Press.

Sidorick, D. 2009. *Condensed capitalism: Campbell Soup and the pursuit of cheap production in the twentieth century*. Ithaca, NY: Cornell University Press.

U.S. Census Bureau. 2010. *Quick facts, Camden City, New Jersey*. Accessed January 28, 2017. http://www.census.gov/quickfacts/table/PST045215/341000.

———. 2012. Household income for states: 2010 and 2011. Accessed January 28, 2017. https://www.census.gov/prod/2012pubs/acsbr11-02.pdf.

Walzer, M. 1983. *Spheres of justice*. New York: Basic Books.

Young, I. 1990. *Justice and the politics of difference*. Princeton, NJ: Princeton University Press.

———. 2000. *Inclusion and democracy*. Oxford, UK: Oxford University Press.

———. 2011. *Responsibility for justice*. Oxford, UK: Oxford University Press.

ROBERT W. LAKE is Professor in the Edward J. Bloustein School of Planning and Public Policy and a member of the Graduate Faculties in Geography and Urban Planning at Rutgers University, New Brunswick, NJ 08901. E-mail: rlake@rutgers.edu. His research interests include the politics of urban land markets, collaborative and community-based planning, the financialization of public policy, and pragmatist approaches to the politics of knowledge production.

4 Resisting Planetary Gentrification

The Value of Survivability in the Fight to Stay Put

Loretta Lees, Sandra Annunziata, and Clara Rivas-Alonso

In-depth studies of and attempts to theorize or conceptualize resistance to gentrification have been somewhat sidelined by attention to the causes and effects of gentrification in the now rather extensive gentrification studies literature. Yet resistance to gentrification is growing internationally and remains a (if not the) key struggle with respect to social justice in cities worldwide. In this article, we address this gap head on by (re)asserting the value of survivability for looking at resistance to gentrifications around the globe. U.S. urban scholars have been at the forefront of writing about resistance to gentrification, especially in cities like San Francisco and New York City, but in a situation of planetary gentrification it is imperative that we learn from other examples. Critically, we argue that practices of survivability can be scaled up, down, and in between, enabling the building of further possibilities in the fight against gentrification, the fight to stay put. There needs to be a stronger and more determined international conversation on the potential of antigentrification practices worldwide and here we argue that survivability has a lot to offer these conversations. *Key Words: planetary gentrification, resistance, survivability.*

针对反抗贵族化的深度研究与理论化或概念化的企图，在今日相当广泛的贵族化研究之文献中，多少因其对贵族化的导因与效应之关注而被排除在外。但反抗贵族化的行动，在全世界中皆逐渐增加，并维持作为全世界城市中有关社会正义的主要 (若非唯一重要的) 斗争。我们于本文中，透过 (重新) 评估存活性在检视全球反抗贵族化中的价值，直接应对上述阙如。美国的城市研究者，身处书写反抗贵族化的前沿，特别是在旧金山与纽约市中，但在全球性的贵族化境况下，我们必须从其他的案例中学习。我们批判性地主张，存活性的实践，能够上、下调整尺度，或位居其中，从而推进打造对抗贵族化的未来可能——一个为了留下来的斗争。对于反贵族化的全球实践，必须有更强大且更为坚定的国际对话，而我们于此主张，存活性对此般对话贡献良多。关键词: 全球贵族化, 反抗, 存活性。

Los estudios a profundidad de la resistencia al aburguesamiento, y los intentos para teorizarla o conceptualizarla, han sido marginados en cierta medida por la atención que se concede a las causas y efectos del aburguesamiento o gentrificación en la literatura de estudios de este fenómeno, ciertamente muy extensa en este momento. Pero la resistencia a la gentrificación está creciendo internacionalmente y se mantiene como una lucha clave, por no decir *la* lucha clave, con respecto a la justicia social en las ciudades de todo el mundo. En este artículo abordamos de frente esta brecha (re)afirmando el valor de la supervivencia para observar la resistencia a la gentrificación alrededor del globo. Los eruditos urbanos de los EE. UU. han estado en la línea de avanzada de la producción de artículos acerca de la resistencia a la gentrificación, especialmente en ciudades como San Francisco y Nueva York, pero en una situación de gentrificación planetaria es imperativo que aprendamos de otros ejemplos. Críticamente, sostenemos que las prácticas de supervivencia pueden escalarse hacia arriba, hacia abajo o en el medio, facilitando la construcción de mayores posibilidades en la lucha contra la gentrificación, la pelea por permanecer en el sitio. Se necesita una conversación internacional más fuerte y determinada sobre el potencial de las prácticas anti-gentrificación a escala mundial, y aquí sostenemos que la supervivencia tiene mucho por ofrecer a estas conversaciones. *Palabras clave: gentrificación planetaria, resistencia, supervivencia.*

Over the past fifty years, gentrification scholars have produced one of the largest literatures in urban studies, yet until more recently there have been relatively few academic studies of resistance to gentrification. Detailed studies of antigentrification protests, struggles, and activism have been sidelined by attention to the causes and effects of gentrification. Academic writings on resistance to gentrification are now growing, perhaps not surprising given the fact that resistance to gentrification is growing internationally and remains a (if not the) key struggle with respect to social justice in cities worldwide. In this growing literature, though, there has been little consideration of what constitutes (successful) resistance and how gentrification scholars conceptualize resistance. In this article, we mull over these issues, focusing specifically on the value of survivability as a practice of resistance that we think deserves much more attention from gentrification scholars.

Survivability is a critical concept, we argue, that holds real promise for a properly global gentrification studies. Vinthagen and Johansson (2013) discussed how survivability is constantly negotiated in and through informality, invisibility, temporalities, and the limits to solidarity. Given that informality is a new area in global gentrification studies, it also makes good sense to draw on work from development studies[1] in a Global South context where survival is a matter of daily life. In addition, the concept of survivability introduces a welcome perspective of individual action into the field of gentrification studies, which has perhaps tended to make assumptions about the collective nature of resistance. Indeed, in this article we make the crucial point that any understanding of resistance to gentrification needs to be tempered by the fact that individuals need to focus foremost on their individual survival and welfare, in addition to that of their families. In reality, planetary resistance to gentrification is composed of both overt opposition and everyday (often invisible) resistances, which are entangled and in a constant process of becoming.

Like Harvey (1973), we see social justice as contingent on the nature of urbanization and urbanism and something that is inherently geographical. Harvey wrote *Social Justice and the City* spurred on by events in U.S. cities in the 1960s and by the work of Marxists interested in community-based urban social movements; forty years later, in *Rebel Cities*, Harvey (2013), like ourselves, was spurred on by similar yet different events. Our focus on social justice and the city in this article is specific to escalating processes of planetary gentrification and resistance to them. Unlike Harvey (1973), we take our lens further than Anglo-American cities and, in so doing, pay proper attention to more cosmopolitan readings of gentrification and resistance to it. Unlike Harvey (2013), we seek to provide a deeper framework for researching social struggles and their internal dynamics. We are also three female scholar-activists writing in a sea of male urban geographical scholarship, so the result is perhaps a different reading of resistance to gentrification and the fight for social justice in the city (cf. Gibson-Graham 1996, 2006). Ours is a neo-Marxist reading of gentrification and resistance to it that harnesses the power of Marxist analysis at the same time as enabling the epistemic authority that comes out of marginalized people's everyday lives. Following Koopman (2015), we look at resistance through critical engagement with the politics of everyday life.

It is interesting to note that Harvey (1973) said a lot about gentrification without actually mentioning the word. He talked about how the spatial structure of the city will change if the preferences of richer groups change. Indeed, he stated, "They can with ease alter their bid rent function and move back into the centre of the city" (135). His discussion of the elimination of ghettos, polarization, Hausmannization, and what he called "urban renew" are all questions that are at the center of twenty-first-century gentrification studies if in a different way, even if in 1973 he did not consider the spread of gentrification beyond the central city and beyond the Global North. What Harvey (1973, 2013) did not do was investigate urban social movements fighting for social justice in any detail or, for that matter, individuals fighting for the survival of themselves and their families. Revisiting resistance to gentrification and (re)asserting the value of survivability is especially important in the context of the everyday, visceral realities of eviction and displacement or threat of eviction and displacement due to gentrification globally.

Studies of resistance to gentrification usually talk about it in relation to eviction, yet eviction is a process that has been described as the most understudied mechanism of reinforcing inequality (Desmond 2016) and it remains a hidden housing problem (see Hartman and Robinson 2003). Urban scholars have sought to conceptualize the right to the city, the right to stay put, but they have spent less energy on conceptualizing the actual fight to stay put in the face of gentrification. In focusing on the fight to stay put, in this article, we hope not only to put research on the everyday resistances of ordinary people at the center, not the margins, of gentrification studies but also to inform that literature by attention to practices of survivability.

Resistance to Planetary Gentrification

At the turn of the twenty-first century, Hackworth and Smith (2001) proclaimed that resistance to gentrification was all but dead, but since the global financial crisis and Arab Spring, this is no longer the case. In recent years, antigentrification resistance has made international headline news, as Gezi Park in Istanbul, Occupy London, and the Tsunami Tour in Rome, among others, testify to. Not since the Tompkins Square Park riots in New York City (see Smith 1996) had antigentrification resistance made headline news. Resistance to gentrification has also transcended the neighborhood and indeed city scale to become national; for example, the Abahlali baseMjondolo movement in South Africa. There is also now recognition that antigentrification resistance outside of the Global North is not new, for resistance to gentrification was happening in South Korea in the 1980s before gentrification authors in the West even began to discuss a global gentrification. It began to organize systematically and was supported by other social movements such as the democracy movement and the labor movement (Lees, Shin, and Lopez-Morales 2016).

Outside of the detailed discussions of Chester Hartman's scholar-activism in San Francisco (see Hartman 1974, 1984; Hartman, Keating, and LeGates 1982), until more recently, discussions of resistance in the gentrification literature have tended to be sketchy, with little to no in-depth research involved. There is evidence now that this is changing, however, as the recent special issue of *Cities* (Goetz 2016) on resistance to the gentrification of public housing around the globe attests to. The literature on resistance to gentrification has also, until recently, been dominated by European-American case studies, when in a situation of planetary gentrification it is imperative that we learn from examples outside of Europe and North America. The gentrification literature has said even less about successful resistance (one exception is NION in Hamburg; see Novy and Colomb 2013), although it would seem that "success" is on the increase as the new gentrification tax in Vancouver and the Milieuschultz Law in Berlin attest to. In London, campaigners fighting against the gentrification of Europe's largest public housing estate also had a rare win (Braidwood and Dunton 2016). There has, however, been discussion of successful resistances in the Global South, and it is here that the Global North would do well to learn.

In an antidisplacement campaign run in the Coyoacan neighborhood in Mexico City, artisans and street vendors successfully practiced antigentrification strategies by organizing outdoor exhibitions aimed at tourists and the media alike (see Crossa 2013). In Chacao, Venezuela, women mobilized against the

gentrification of their barrio (see Velásquez Atehortúa 2014). Their resistance was also peaceful, but it was helped by the support of the Socialist majority in the National Assembly, who then passed a series of reforms that supported the People's Power (*El Poder Popular*). The government allowed the barrio women to build a pioneers' camp (a *Campamento de Pioneros*) on the land to begin the process of building a Socialist community for 600 families. This became a new model of social policy development that involved people contesting neoliberalism by the marginalized being involved in executing government programs. In partnership with the Socialist state, they could stand up to the power of real estate elites, bankers and developers, and so on. In so doing, they successfully fought off an urban renewal project that was to gentrify the Old Market in Caracas. Betancur (2014) argued that unlike in the Global North, gentrification in Latin America has run into stubborn resistance from the (informal) self-help and self-employment spaces in which the lower classes live. As a result, gentrification has been much more limited than expected. Indeed, in Lima, residents organized around the Comite Promotor para la Renovacion Urbana with *Renovacion urbana sin Desalojos* (Urban Renewal without Evictions; International Alliance of Inhabitants 2008) and in Colombia opponents of gentrification named the Office of Urban Development the Office of Urban Displacement! What does success really mean in terms of resistance to gentrification, though? Is success purely about winning the fight to stay put? What if the fight to stay put is lost, but the fight has mobilized national or international attention? A struggle might lose on one level but obtain incredible visibility able to inform other levels of action. Defining successful resistance is both important and strategic.

Samara, He, and Chen (2013) claimed that consciousness of the "right to the city" and cross-class alliances are increasingly being formed in newly industrialized countries. We would add that this is also the case in less industrialized countries also being affected by the speculation in the secondary circuit of capital that is the defining feature of twenty-first-century planetary gentrification. For example, the Mahigeer Tahreek (indigenous coastal fisherfolk communities of Pakistan) movement successfully fought off attempts to gentrify Karachi's coastline (and privatize its public beaches) from global capital and Dubai- and Malaysia-based real estate companies (see Hasan 2015; see also Hasan 2012). In 2007, they wrote a letter, "Development to Destroy Nature and Displace

People," the outcome of discussions between various stakeholders but especially local communities. As well as outlining the destruction of nature—from green turtles, to mangroves, to fish and birds—they also were clear that it would displace people, the fishing communities who had been living on the coast for centuries. The project, it was claimed, would affect their livelihoods, which were based on subsistence fishing and beach leisure activities. Despite more than 100 villages being in the project area, their future was not mentioned at all in the project proposal. The letter also claimed that given that lower- and lower-middle-class Karachiites would not be able to go to the beach, this would increase the divide between the rich and poor in society. The letter was followed up by public demonstrations and a press campaign. Meetings were held with the Chief Secretary along with prominent civil society individuals, and because of opposition from all segments of society, the developer, Limitless, backed out of the project in 2009.

Resistance to gentrification is not a singular entity; there are many different forms and practices, and these need to be researched in context. Furthermore, the concept of resistance itself can be highly relative and context dependent, and there is an urgent need to unpack it further. There have been a number of recent reviews of the literature on resistance to gentrification (e.g., Gonzales 2016; Lees and Ferreri 2016; Annunziata and Rivas-Alonso forthcoming). In their detailed review of the academic literature,[2] Annunziata and Rivas-Alonso (forthcoming) usefully identified the main practices as *institutional prevention*, the implementation of public housing policies, enforcing tenants' protections, and community planning tools (e.g., Newman and Wyly 2006); *mitigation measures*, delaying eviction, compensation (e.g., Kolodney 1991; Gallaher 2016); plus legal strategies and counternarratives (e.g., Blomley 2004); and *the production of alternatives* (see Holm and Kuhn 2011; Janoschka 2015). Resistance to gentrification, of course, can encompass a number of these different practices enacted simultaneously or consequently by the same or different groups. Much less attention, however, has been paid to practices of resistance that draw on the strategic mobilization of identity and cultural practices deeply rooted in the everyday (see Soymetel 2014). Practices that are not overtly antagonistic and not very visible can produce resistance and indeed can demonstrate more innovative approaches to survival in the face of gentrification.

There is a tendency for global scholars to articulate resistance at an abstract level; for example, Leitner, Sziarto, Sheppard, and Maringanti (2007) and Mayer (2009) seek solidarity across different classes to challenge the uneven spatiality created by neoliberal governance and globalization. This can appear grounded at times; for example, Routledge (2012) discussed their performances on the ground, but grounded is not contextualized and being against capitalism and neoliberalism per se is not necessarily the same as being against gentrification. Other work has tried to connect the local with the global context; for example, Maeckelbergh (2012) connected neoliberalism, the outbreak of the subprime mortgage crisis in the United States, and its three by-products—the housing crisis, gentrification, and foreclosure—through East Harlem–based social movements' autonomous struggle.

There is also a significant difference between the storming the barricades type of antigentrification battle and the everyday practices of resistance (Lees 1999). Fighting gentrification does not always have to be confrontational; indeed, direct confrontation is too dangerous (or even less likely to succeed for cultural and political reasons) in some parts of the world, as seen in the case of Chinese resisters adopting "rightful resistance" (Erie 2012). When faced with rent hikes or eviction from their homes, displacees often simply prioritize the moment. Considering all of the different practices people employ to stay put is important if we want to escape analysis that merely describes landscapes of despair and offers little more than blanket statements about neoliberal hegemony. The reality is that everyday millions of people faced with gentrification and threatened by displacement and eviction face situations that are not as black and white as some of the gentrification scholarship would have us believe; delicate decisions have to be made in relation to the present and presumptions about the future. These decisions are more often within a world of shrinking possibilities as the paths for capital accumulation are stabilized further. Following Koopman (2015), we turn now to look at resistance to gentrification through critical engagement with the politics of everyday life. We argue that "staying put" is not just a seductive slogan; critically, it is a matter of survivability, and that survivability is part of the fight to stay put.

(Re)asserting the Value of Survivability in Resisting Gentrification

The value of survivability in gentrification studies was noted in a discussion of the differentiated ideas of resistance, reworking, and resilience with respect to state-led gentrification in London (Lees 2014). Drawing on Katz (2004), it was suggested that we consider "an oppositional consciousness that achieves emancipatory objectives (resistance), an impact on the organisation of power relations if not their polarised distribution (reworking), and an enabling of survival in circumstances that do not allow changes to the causes that dictate survival (resilience)" (Cloke, May, and Johnsen 2010, 12). DeVerteuil's (2016) recent work on resilience to gentrification in Los Angeles, Sydney, and London has also recognized the utility of the notion of survivability. Deeper conceptual work needs to be done, however. In thinking more about how gentrification scholars might conceptualize survivability, we can learn from resistance studies; for example, Vinthagen and Johansson's (2013) epistemological framework for the study of resistance that includes (1) repertoires of everyday resistance practices, (2) the relationships of agents, (3) spatialization, and (4) temporalization of everyday resistance. They also suggested intersectionality as the way forward, for it allows us "to capture the construction of multiple and shifting identities of agents of resistance and the interplay between these, as well as the contradictory positions of being both dominant and subordinate, depending on which system/context/relationship subjects are positioned and position themselves in" (Vinthagen and Johansson 2013, 424).

In our Introduction, we mentioned that Vinthagen and Johansson (2013) discussed how survivability is constantly negotiated in and through informality, invisibility, temporalities, and the limits to solidarity. Informality is an essential part of everyday survival, and studies of informality in the gentrification literature have looked beyond ambiguous homeownership situations to the eviction of street vendors and other informal activities from central cities (see Lees, Shin, and Lopez-Morales 2015, 2016). The way in which informality is enacted allows for different escape routes, and the most effective networks of support are embodied in informality; in connections, whether acquaintances or family relations, social capital is mobilized when there is a need for help and where the promise of a future leverage widens the possibilities available. Regulating visibility is a key tool to stay put. In more authoritarian settings, the more visible someone is, the bigger the risk of being made to disappear. In this context, those affected by gentrification fine-tune their visible involvement depending on the circumstances, what there is to gain or lose. Those in more precarious positions might decide to step back, get involved in movements in subtler ways, and eventually make use of more visible tools of protest if momentum is gained. Making the invisible visible is a political act (Lees and Ferreri 2016).

Vinthagen and Johansson (2016) were interested in "how everyday resistance in the form of activities, social relations and identities, is spatially organized and how everyday resistance is practised in and through space as a central social dimension" (425). The issue of the spatiality of resistance opens up a whole set of issues around positionality, marginality, and scale, for resistance is "localised, regionalised and globalised at the same time that economic globalisation slices across geopolitical borders" (Chin and Mittelman 1997, 35). Inasmuch as resistance is spatialized, however, it is also temporarily organized. Hartman, Keating, and LeGates (1982) pointed out how acting timely is crucial in stopping demolition, eviction, and displacement. Solidarity networks can solidify or dissipate depending on how well a position is negotiated, and when survival is compromised, solidarity among those under threat can be limited.

Harvey (1973, 2013), among others, would argue that resistance is oriented toward the change of a larger system that perpetuates injustice, but this is not always possible, and it pays little attention to the smaller, more intimate scale of resistance to gentrification. This is where the value of survivability as related to everyday practices comes in. The concept of survivability, we would argue, can be used to scale up from the micro to macro scales of resistance (from the individual to the neighborhood, city, nation, and internationally; see Smith 1992); in addition, scaling can be done in between. The value of a scaled survivability is that it enables us to focus both on the survival of the collective and also critically of the individual. The potential of individual practices of resistance, which have often been much more successful than big organized resistance (Vinthagen and Johansson 2013, 2016), have been overshadowed in the gentrification studies literature. Indeed, it is important to study resistance at the micro scale because we are confronted by a coherent and hegemonic urban neoliberal order that pushes people into vulnerability and survivability. Survivability allows us to talk radically about geography, the focus being on the fundamental, material need to survive. Heynen (2006, 191) related survivability to "meeting basic human needs"; feminist geographers such as Katz (2004) have discussed the relationship between survivability and social reproduction and define survivability as a precondition for resistance.

Chatterton and Heynen (2011) also defend a renewed focus on the everyday tactics of resistance in the face of the fact that a far-reaching revolution is rarely possible and this allows us to redefine resistance as always relational, situated in space, as a multiplicity of actions, not necessarily emancipatory or oppositional. These everyday resistances can often occur where we do not necessarily expect them to, they can be visible or invisible (we would argue that the invisible practices need much more attention), they can be intentional or nonintentional, and they are not necessarily politically conscious. We recognize, as Butler, Gambetti, and Sabsay (2016) pointed out, that vulnerability can be both a result of resistance, especially in the increasingly violent contexts in which resistance to gentrification takes place, and a precondition. As they explained, the body itself is put at risk, but vulnerability also anticipates resistance when resisting people are extremely precarious individuals. When people organize, their precarious position is exposed, politicized, and performed bodily. As collective infrastructures fail, vulnerability, and with it the possibility of resistance, emerges. Törnberg (2013) stressed the need to explore how the materiality of things influences resistance, urging us to consider survivability as a key component within processes of resistance, as access to fundamental material goods become the priority.

In thinking about survivability, we can also draw on Holloway's (2002, 2010) work (which itself draws on the struggle of the Zapatista movement of Chiapas, Mexico), which breaks with the traditional left in arguing that the possibility of revolution resides in day-to-day acts that refuse domination by capitalist society rather than seeking power through state apparatuses. His is a more optimistic view of resistance, one that the South African, antigentrification, shack-dweller movement, Abahlali baseMjondolo, lauds as "refreshing in the sense that it engenders hope," breaking with the traditions of authoritarian and vanguardist leftism in Marxist revolutionary struggles (Abahlali 2016). He is interested in the ordinary politics of ordinary people, what Abahlali baseMjondolo calls "living politics."

For Holloway, as for many fighting gentrification, resistance occurs in and through the cracks in capitalism, in interstices. He recognized that the most violent force is the force of the state (and we see this in the state's heavy involvement in planetary gentrification)

and restores human beings—individuals—in struggles. Drawing on Holloway, survivability is a moment, even explosion, of creation where the state is pushed aside but, significantly, it will not always have momentum. We must not overemphasize continuity, as survivability might not last, but this does not make it any less successful or important, because simply surviving is also a matter of dignity and self-esteem (what Holloway calls the refusal to accept humiliation and dehumanization). This does not, however, rule out the fact that context-dependent relationships can develop between everyday practices and organized struggle (DeFilippis and North 2004). The everyday building of solidarities in place is important and can at times be scaled, and individuals and neighborhoods can act as both platforms for the organization of resistance and objects of resistance (Butler et al. 2016).

Conclusion

We have begun the work here of developing an analytical framework for researching resistance to gentrification globally that is strengthened by attention to survivability in everyday practices of resistance. We argue that research on resistance to gentrification needs to extend much more toward individual, as well as collective, actions that are not organized, formal, or necessarily public or even intentionally political, actions that are linked to configurations of power in everyday life. We have found in our own scholar-activist research on resisting gentrification[3] that resistance is not always a call to arms and a storming of barricades. More often it is small-scale, haphazard, and simply reactive practices of survivability, which in some cases eventually spark collective organizing but in others do not. Resistance to gentrification can constitute a small-scale (geo)politics undertaken by rational, emotional, and embodied urban citizens; some of their acts are visible, some invisible. Through a postcolonial lens, survival per se can be seen as success in the face of brutal, hegemonic practices. Attempts to pacify, impose social and cultural norms, and evict can be met with subaltern insurrections: the ability of not conforming to imposed norms, of continuing to relate to the city as an "other" in the face of acute marginalization and indeed criminalization. Bayat's (2007) critique of Scott's (1985) "weapons of the weak" pushes the idea of the mere defensive mechanisms of the disenfranchised toward a notion of active, "offensive" mechanisms that go on to build further possibilities.

Our notion of survivability reflects the ability of threatened people to act on their agency.

Prioritizing basic material needs, like a home, is fundamental to survival. Anchoring resistance in the material, fundamental logic of survival, as Chatterton and Heynen (2011) suggested, moves us away from binary interpretations of resistance and allows us to focus on contradictions, the different identities produced, and the various scales where a reworked concept of resistance is performed. Indeed, scale is important, for although surviving and staying put are key areas for actions, at some point more organized resistance could be needed either to hold on to that survivability or to scale up the fight. Butler et al. (2016) argued that we need platforms because without them we cannot mobilize. We would argue that platforms can occur from practices of survivability, but there is no demand that they do so. Survivability gives dignity to those threatened by gentrification, but it also has the potential to be scaled up, down, and in between the individual and the collective. Scaling it up to the city level and globally in the fight against gentrification (Smith 1992) is perhaps easier than scaling down, and this could have implications for the right to the city, national, and global movements. In future research on resistance to gentrification, cases should be examined as the loci where relationships are established. In so doing, we might look again at Massey's (2005) work on the reclaiming of spaces as "the product of interrelations," where actors become entangled with each other more or less willingly and where battles of all sizes are won sometimes by the sheer ability to belong to a threatened landscape.

There needs to be a stronger and more determined international conversation on the potential of all anti-gentrification practices worldwide. We hope that this brief article goes some way toward starting such an international dialogue, the aim being to forge more successful resistances to gentrification. As part of this conversation, gentrification researchers must also ask probing questions of themselves in relation to ethics, positionality, and their working with marginalized or vulnerable groups in everyday resistance against gentrification—failure to do so will only reproduce the hegemony of gentrification itself.

Acknowledgments

We thank the referees for their very useful comments and Nik Heynen for his patience but, more importantly, we thank the brave people we have been working with fighting gentrification.

Funding

The following research grants underpin this work: EU Grant No. 625691; ESRC 2017-2020, "Gentrification, Displacement, and the Impacts of Council Estate Renewal in C21st London" (PI: Lees, L., CoIs: Hubbard, P. and Tate, N) [ES/N015053\1]; EU FP7-PEOPLE-2013 Marie Curie Action Fellowship 2014–2016, "AGAPE: Exploring Anti-gentrification Practices and Policies in Southern European Cities" (PI: Lees, L. CoI: Annunziata, S.); Antipode Activist Scholar Award 2012 "Challenging 'The New Urban Renewal': Gathering the Tools Necessary to Halt the Social Cleansing of Council Estates and Developing Community-led Alternatives for Sustaining Existing Communities" (PI: Lees, L. CoIs: London Tenants Federation, Richard Lee/Just Space and Mara Ferreri); Leicester University Impact Fund, 2015–2016.

ORCID

Loretta Lees ⓘ http://orcid.org/0000-0001-5834-8155

Sandra Annunziata ⓘ http://orcid.org/0000-0003-0353-4158

Clara Rivas-Alonso ⓘ http://orcid.org/0000-0001-7052-6575

Notes

1. On the need to strengthen the nexus between development studies and gentrification studies, see Lees (2012), although, interestingly, this has faced some kickback from some development geographers.
2. Annunziata and Rivas-Alonso (forthcoming) argued that the most useful academic writings on resisting gentrification are from scholar-activists who are involved in the fight to stay put (e.g., Hartman, mentioned earlier, and, more recently, The London Tenants Federation, Lees, Just Space, and SNAG [2014]; for the Swedish version, see Thörn, Krusell, and Widehammar (2016); see also Andrej Holm's (nd) blog. They argue that it is not academic texts but handbooks, blogs (see http://35percent.org/), mapping (antievictionmap.org), passionate writing (Colau and Alemany 2012), documentaries and movies, and artist-activist works (see lefthandrotation.com) that have the most practical value. They are, first of all, accessible and easy to read and understand; do not intellectualize the problem at stake; and go directly to possible solutions. They are written for and with communities and imply the participation of those directly affected by displacement (see also Annunziata and Lees [2016] on resistance to gentrification movements in southern European cities).

3. We have not the space here to reflect on our own personal experiences and contributions as scholar-activists involved in resisting gentrification in London, Rome, Istanbul, and elsewhere, but such reflections are important in a neoliberal academic environment that leaves little time or energy for deep community work. Any academic investigation of survivability entails working with vulnerable people (threatened with eviction or displacement) in the field, and we also need to be resilient (to survive emotionally, personally) as researchers in the face of disturbing, vicious gentrifications that disrupt and ruin people's lives. We also have an ethical responsibility not to exploit these awful stories of poverty, pain, and oppression for academic gain or intellectual ruminations.

References

Abahlali. 2016. Abahlali baseMjondolo. Accessed August 28, 2017. http://abahlali.org/node/9157/.

Annunziata, S., and L. Lees. 2016. Resisting austerity gentrification in southern European cities. *Sociological Research Online* 21 (3). Accessed August 28, 2017. http://www.socresonline.org.uk/21/3/5.html.

Annunziata, S., and C. Rivas-Alonso. Forthcoming. Resisting gentrification. In *Handbook of gentrification studies*, ed. L. Lees with M. Phillips. Cheltenham, UK: Edward Elgar.

Bayat, A. 2007. The quiet encroachment of the ordinary. *Chimurenga* 11:8–15.

Betancur, J. 2014. Gentrification in Latin America: Overview and critical analysis. *Urban Studies Research* 2014:986961. doi:10.1155/2014/986961

Blomley, N. 2004. *Unsettling the city: Urban land and the politics of property*. London and New York: Routledge.

Braidwood, E., and J. Dunton. 2016. Aylesbury Estate CPO ruling: What went wrong? *The Architect's Journal*. Accessed 28 August 2017. https://www.architectsjournal.co.uk/news/aylesbury-estate-cpo-ruling-what-went-wrong/10012171.article.

Butler, J., Z. Gambetti, and L. Sabsay. 2016. *Vulnerability in resistance*. Durham, NC: Duke University Press.

Chatterton, P., and N. Heynen. 2011. Resistance(s) and collective social action. In *A companion to social geography*, ed. V. J. Del Casino, 508–25. Chichester, UK: Wiley-Blackwell.

Chin, C., and J. Mittelman. 1997. Conceptualising resistance to globalization. *New Political Economy* 2 (1):25–37.

Cloke, P., J. May, and S. Johnsen. 2010. *Swept up: Re-envisioning the homeless city*. Chichester, UK: Wiley-Blackwell.

Colau, A., and A. Alemany. 2012. Mortgaged lives: From the housing bubble to the right to housing. *Journal of Aesthetics & Protest Press*. Accessed March 30, 2017. http://www.joaap.org/press/pah/mortgagedlives.pdf.

Crossa, V. 2013. Play for protest, protest for play: Artisan and vendors' resistance to displacement in Mexico City. *Antipode* 45 (4):826–43.

DeFilippis, J., and P. North. 2004. The emancipatory community? Place, politics and collective action in cities. In *The emancipatory city? Paradoxes and possibilities*, ed. L. Lees, 72–89. London: Sage.

Desmond, M. 2016. *Evicted: Poverty and profit in the American city*. New York: Penguin.

DeVerteuil, G. 2016. *Resilience in the post-welfare inner city: Voluntary sector geographies in London, Los Angeles and Sydney*. Bristol, UK: Policy Press.

Erie, M. 2012. Property rights, legal consciousness and the new media in China: The hard case of the "toughest nail-house in history." *China Information* 26 (1):35–59.

Gallaher, C. 2016. *The politics of staying put, condo conversion and tenants right to buy in Washington, DC*. Philadelphia: Temple University Press.

Gibson-Graham, J. K. 1996. *The end of capitalism (as we knew it): A feminist critique of political economy*. Oxford, UK: Blackwell.

———. 2006. *A postcapitalist politics*. Minneapolis: University of Minnesota Press.

Goetz, E., ed. 2016. Resistance to social housing transformation. *Cities: The International Journal of Urban Policy and Planning* 57.

González, S. 2016. Looking comparatively at displacement and resistance to gentrification in Latin American cities. *Urban Geography* 37:1245–52.

Hackworth, J., and N. Smith. 2001. The changing state of gentrification. *Tijdschrift voor Economische en Sociale Geografie* 22:464–77.

Hartman, C. 1974. *Yerba Buena: Land grab and community resistance in San Francisco*. San Francisco: Glide Publications.

———. 1984. The right to stay put. In *Land reform, American style*, ed. C. Geisler and F. Popper, 302–18. Totowa, NJ: Rowman and Allanheld.

Hartman, C., D. Keating, and R. LeGates. 1982. *Displacement: How to fight it*. Washington, DC: National Housing Law Project.

Hartman, C., and D. Robinson. 2003. Evictions: The hidden housing problem. *Housing Policy Debate* 14 (4):461–501.

Harvey, D. 1973. *Social justice and the city*. London: Edward Arnold.

———. 2013. *Rebel cities: From the right to the city to the urban revolution*. London: Verso.

Hasan, A. 2012. The gentrification of Karachi's coastline. Paper presented at the London Workshop Towards an Emerging Geography of Gentrification in the Global South, London. Accessed August 28, 2017. http://arifhasan.org/wp-content/uploads/2012/08/P16_Gentrification-Karachi-Coastline.pdf.

———. 2015. Value extraction from land and real estate in Karachi. In *Global gentrifications: Uneven development and displacement*, ed. L. Lees, H. Shin, and E. Lopez-Morales, 181–98. Bristol, UK: Policy Press.

Heynen, N. 2006. But it's alright, Ma, it's life, and life only: Radicalism as survival. *Antipode* 38 (5):916–29.

Holloway, J. 2002. *Change the world without taking power: The meaning of revolution today*. London: Pluto.

———. 2010. *Crack capitalism*. London: Pluto.

Holm, A. nd. Gentrification blog: Nachrichten zur Stärkung von Stadtteilmobilisierungen und Mieter/innenkämpfen. Accessed October 2, 2017. https://gentrificationblog.wordpress.com

Holm, A., and A. Kuhn. 2011. Squatting and urban renewal: The interaction of squatter movements and strategies of urban restructuring in Berlin. *International Journal of Urban and Regional Research* 35 (3):644–58.

International Alliance of Inhabitants. 2008. Vigilias por el derecho a vivir en el centro historico de Lima [Housing rights watch in the historical center of Lima]. Accessed August 28, 2017. http://www.habitants.org/zero_evic tions_campaign/world_zero_evictions_day_2008/ vigi lias_por_el_derecho_a_vivir_en_el_centro_historico_ de_Lima.

Janoschka, M. 2015. Politics, citizenship and disobedience in the city of crisis: A critical analysis of contemporary housing struggles in Madrid. *Die Erde: Journal of the Geographical Society of Berlin* 146:2–3.

Katz, C. 2004. *Growing up global: Economic restructuring and children's everyday lives.* Minneapolis: University of Minnesota Press.

Kolodney, L. 1991. Eviction free zones: The economics of legal bricolage in the fight against displacement. *Fordham Urban Law Journal* 18 (3):507–44.

Koopman, S. 2015. Social movements. In *The Wiley Blackwell companion to political geography*, ed. J. Agnew, V. Mamadouh, A. Secor, and J. Sharp, 339–51. Hoboken, NJ: Wiley.

Lees, L. 1999. Critical geography and the opening up of the academy: Lessons from "real life" attempts. *Area* 31 (4):377–83.

———. 2012. The geography of gentrification: Thinking through comparative urbanism. *Progress in Human Geography* 36 (2):155–71.

———. 2014. The urban injustices of New Labour's "new urban renewal": The case of the Aylesbury Estate in London. *Antipode* 46 (4):921–47.

Lees, L., and M. Ferreri. 2016. Resisting gentrification on its final frontiers: Lessons from the Heygate Estate in London (1974–2013). *Cities* 57:14–24.

Lees, L., H. Shin, and E. Lopez-Morales, eds. 2015. *Global gentrifications: Uneven development and displacement.* Bristol, UK: Policy Press.

———. 2016. *Planetary gentrification.* Cambridge, UK: Polity.

Leitner, H., K. M. Sziarto, E. Sheppard, and A. Maringanti. 2007. Contesting urban futures: Decentering neoliberalism. In *Contesting neoliberalism: Urban frontiers*, ed. H. Leitner, J. Peck, and E. Sheppard, 1–25. New York: Guilford.

The London Tenants Federation, Lees, L., Just Space, and SNAG. 2014. *Staying put: An anti-gentrification handbook for council estates in London.* Accessed March 30, 2017. https://southwarknotes.files.wordpress.com/2014/ 06/staying-put-web-version-low.pdf.

Maeckelbergh, M. 2012. Mobilizing to stay put: Housing struggles in New York City. *International Journal of Urban and Regional Research* 36 (4):655–73.

Massey, D. 2005. *For space.* London: Sage.

Mayer, M. 2009. The "right to the city" in the context of shifting mottos of urban social movements. *City* 13 (2–3):362–74.

Newman, K., and E. Wyly. 2006. The right to stay put, revisited: Gentrification and resistance to displacement in New York City. *Urban Studies* 43 (1):23–57.

Novy, J., and C. Colomb. 2013. Struggling for the right to the (creative) city in Berlin and Hamburg: New urban social movements, new "spaces of hope"? *International Journal of Urban and Regional Research* 37 (5):1816–38.

Routledge, P. 1994. Backstreets, barricades, and blackouts, urban terrains of resistance in Nepal. *Environment and Planning D: Society and Space* 12 (5):559–78.

Routledge, P. 2012. Sensuous solidarities: Emotion, politics and performance in the clandestine insurgent rebel clown army, *Antipode* 44 (2):428–52.

Samara, T., S. He, and G. Chen, eds. 2013. *Locating right to the city in the Global South.* London and New York: Routledge.

Scott, J. 1985. *Weapons of the weak: Everyday forms of peasant resistance.* New Haven, CT: Yale University Press.

Smith, N. 1992. Contours of a spatialized politics: Homeless vehicles and the production of space. *Social Text* 33:54–81.

———. 1996. *The new urban frontier: Gentrification and the revanchist city.* London and New York: Routledge.

Soymetel, E. 2014. "Belonging" in the gentrified Golden Horn/Halic neighbourhoods of Istanbul. *Urban Geography* 36 (1):1–26.

Thörn, C., M. Krusell, and M. Widehammar. 2016. Rätt Att Bokvar: en handbook I organisering mot Hyreshöjningar och gentrifiering [The right to stay put. A handbook on how to organize against rent increases and gentrification]. Accessed October 2, 2017. https:// koloni.info/Ratt_att_bo_kvar_2016.pdf

Törnberg, A. 2013. Resistance matter(s): Resistance studies and the material turn. *Resistance Studies Magazine* 1:1–15.

Velásquez-Atehortúa, J. 2014. Barrio women's invited and invented spaces against urban elitisation in Chacao, Venezuela. *Antipode* 46 (3):835–56.

Vinthagen, S., and A. Johansson. 2013. "Everyday resistance": Exploration of a concept and its theories. *Resistance Studies Magazine* 1:1–46.

———. 2016. Dimensions of everyday resistance: An analytical framework. *Critical Sociology* 42 (3):417–35.

LORETTA LEES is Professor of Human Geography at the University of Leicester, Leicester LE1 7RH, UK. E-mail: loretta.lees@le.ac.uk. She is an international expert on gentrification and her most recent book, *Planetary Gentrification* (with Hyun Bang Shin and Ernesto Lopez-Morales), is the launch text for Polity Press's new Urban Futures series. Since 2009 she has coorganized The Urban Salon: A London Forum for Architecture, Cities and International Urbanism (see http://www.theurbansalon.org/). She is also an activist-scholar who for the past decade has been involved in fighting the gentrification of council estates in London, where she lives. Her expertise has been used in two public enquiries, and she is currently continuing this work in a three-year Economic and Social Research Council project collating evidence on displacement.

SANDRA ANNUNZIATA is a lecturer in urban theory and urbanism in the Department of Architecture, University of Roma Tre, Italy, and an Honorary Research Fellow in Geography at the University of Leicester, Leicester LE1 7RH, UK. E-mail: sa644@leicester.ac.uk. She has a degree in architecture and urbanism and a PhD in urban studies, the latter of which won the 2010 Giovanni Ferraro National Award in Italy. She recently completed a European Union project with Loretta Lees on antigentrification policies and practices in three southern European cities—Rome, Madrid, and Athens. The research involved working with groups resisting gentrification in all three cities. She is currently writing up the results as an antigentrification toolkit for southern European cities and continuing her activism in Rome.

CLARA RIVAS-ALONSO is a PhD student in Geography at the University of Leicester, Leicester LE1 7RH, UK. E-mail: cra12@leicester.ac.uk. Her PhD is an investigation into everyday practices and perceptions of resistance in a neighborhood in Istanbul under threat from state-led gentrification, Okmeydani. Undertaking ethnographic research on resistance, she lived in Okmeydani during the recent Turkish government crackdowns. She is interested in the more invisible solidarities that escape institutional attempts at rent extraction. A scholar-activist, she argues that the current global urban condition calls for more innovative methods of resistance.

5 Urban Movements and the Genealogy of Urban Rights Discourses

The Case of Urban Protesters against Redevelopment and Displacement in Seoul, South Korea

Hyun Bang Shin

Despite significant contributions made to progressive urban politics, contemporary debates on cities and social justice are in need of adequately capturing the local historical and sociopolitical processes of how people have come to perceive the concept of rights in their struggles against the hegemonic establishments. These limitations act as constraints on overcoming hegemony imposed by the ruling class on subordinate classes and restrict a contextual understanding of such concepts as the right to the city in non-Western contexts, undermining the potential to produce locally tuned alternative strategies to build progressive and just cities. In this regard, this article discusses the evolving nature of urban rights discourses that were produced by urban protesters fighting redevelopment and displacement, paying particular attention to the experiences in Seoul that epitomized speculative urban accumulation under the (neoliberalizing) developmental state. Method-wise, the article makes use of archival records (protesters' pamphlets and newsletters), photographs, and field research archives. The data are supplemented by the author's in-depth interviews with former and current housing activists. The article argues that the urban poor have the capacity to challenge the state repression and hegemony of the ruling class ideology; that the urban movements such as the evictees' struggles against redevelopment are to be placed in the broader contexts of social movements; that concepts such as the right to the city are to be understood against the rich history of place-specific evolution of urban rights discourses; and that cross-class alliance is key to sustaining urban movements. *Key Words: displacement, rights discourses, Seoul, urban movements, urban protests.*

尽管当代有关城市与社会正义的辩论, 已对激进的城市政治做出显着的贡献, 但仍需充份捕捉人们在与霸权形构的斗争中, 如何理解权益的概念之在地历史与社会过程。这些限制, 成为克服统治阶级对从属阶级施加的霸权之限囿, 并限缩了对非西方城市脉络中的城市权概念的脉络性理解, 且有损生产建立激进与正义城市的在地化另类策略之潜能。因此, 本文探讨由对抗再发展和迫迁的城市抗争者转变中的城市权论述, 并特别关注首尔——一个象徵着在 (新自由主义化的) 发展形国家中的投机性城市积累之地。研究方法上, 本文运用档案纪录 (抗争者的宣传手册和通讯) 、照片与田野研究档案。这些数据, 由作者对于先前与当下的居住倡议者所进行的深度访谈补充之。本文主张, 城市中的穷人, 具有挑战国家压迫和统治阶级意识形态霸权的能力; 诸如被驱逐者反抗再发展的斗争之城市运动, 必须被置放在更广泛的社会运动脉络中; 诸如城市权的概念, 必须相对于城市权论述在特定地方的丰富演变历史进行理解; 跨阶级的结盟, 则是维系城市运动的关键。 关键词: 迫迁, 权益论述, 首尔, 城市运动, 城市抗议。

Pese a las contribuciones significativas que se aportan a la política urbana progresista, los debates contemporáneos sobre las ciudades y la justicia social claman porque se involucren también los procesos locales históricos y sociopolíticos acerca de cómo ha llegado la gente a percibir el concepto de los derechos en su lucha contra los establecimientos hegemónicos. Estas limitaciones actúan como obstáculos para vencer la hegemonía impuesta por la clase dominantes sobre las subordinadas, y restringen un entendimiento contextual de conceptos como el del derecho a la ciudad en contextos no occidentales, debilitando el potencial de producir estrategias alternativas localmente afinadas para construir ciudades progresistas y justas. A este respecto, este artículo discute la naturaleza evolutiva de los discursos sobre derechos urbanos que se originaron desde acciones de manifestantes urbanos contra el redesarrollo y el desplazamiento, prestando particular atención a las experiencias de Seúl que encarnaron la acumulación especulativa urbana bajo un estado desarrollista (neoliberalizador). En términos de método, el artículo hace uso de registros de archivo (panfletos de los manifestantes y boletines informativos), fotografías y archivos de investigación de campo. Esos datos fueron suplementados con entrevistas a profundidad del autor con activistas enfrentados al problema de vivienda, anteriores y actuales. El artículo arguye que los pobres urbanos están en capacidad de desafiar la represión estatal y la hegemonía ideológica de la clase dominante; que movimientos urbanos tales como las luchas de los desahuciados contra el redesarrollo deben ser ubicados dentro del más amplio contexto de los movimientos sociales; que conceptos por el estilo del derecho a la ciudad deben entenderse contra la rica historia de la evolución específicamente relacionada con lugar en los discursos sobe derechos urbanos; y que la alianza entre clases es clave para mantener los movimientos urbanos. *Palabras clave: desplazamiento, discursos sobre derechos, Seúl, movimientos urbanos, protestas urbanas.*

Urban built environment and social realities reflect the class interests of those who have economic and political power to produce cities in their own imagination (Lefebvre 1996; Mitchell 2003). Our highly unequal cities can therefore be regarded as the "socially just" manifestation in the eyes of the ruling class. This calls for the urgency of conferring greater power to the marginalized and disenfranchised (Marcuse 2009). All too often, however, we hear less about the voices of those who bear the brunt of profit-seeking activities of the rich and powerful. Despite significant contributions to progressive urban politics, contemporary debates on social justice are in need of adequately capturing the local historical and sociopolitical processes of how people voice out and produce their own alternative discourses against the hegemonic establishments (Gramsci 1971; Glassman 2013). These limitations undermine the production of locally tuned alternative strategies to build progressive and just cities. This is where my focus on the voices of the urban protesters against displacement comes from.

This article is an extension of ongoing efforts among critical scholars to perceive social movements and grassroots activism as "knowledge-producers in their own right" rather than objects of study (Chesters 2012, 145). By adopting a strategic–relational perspective, I examine the evolving nature of rights claims that were put forward by protesters against urban redevelopment and displacement, placing this in the context of condensed and speculative urbanization of South Korea (hereafter Korea). What the history of the evolution of rights discourses in Korea demonstrates is, I argue, how the urban poor as part of subordinate classes challenge the hegemony of private property rights and how this is made possible through the solidarity among subordinate classes and the establishment of cross-class alliance. The focus on Korea in this article is helpful for advancing the scholarship, as the emergence of urban rights discourses or Korea's "urban question" was in a political–economic context that differed from the postindustrial economies of the West. Urban movements in the West calling for strengthening urban

rights and the protection of collective consumption were in the context of eroding Keynesian welfare state, economic crisis, austerity, and neoliberalization of urban services provision (cf. Mayer 2009). Korea's experience of urban movements and the call for urban rights has been in the context of the strong authoritarian statism (in the 1960s–1980s in particular) that retained a close nexus with capital (large businesses in particular), which refrained from the provision of universal welfare and emphasized individual and family responsibility for access to collective consumption, including housing. Korea's experience also differs from the rest of Southeast and East Asian economies, because of its rich history of democracy movements that successfully challenged the state in the 1980s and 1990s, producing state–society relations that are markedly different from the era of the authoritarian state (Castells 1992; B.-G. Park 1998; Shin, Lees, and López-Morales 2016). Such changes to the state–society relations in Korea produce a space of resistance and counterhegemony, which in turn provides opportunities to collectively advance the urban rights discourses through active formation of alliance among classes and various sectors of (urban) social movements.

The study reconstructs the past trajectory of rights claims by urban protesters, focusing on the period between the 1980s and the present. Given the limitations of longitudinal qualitative research that requires real-time and recurrent engagement with events and participants (Saldaña 2003), the analysis in this article makes use of both historical data and in-depth interviews. The main historical data include (a) an archival collection of protesters' pamphlets and newsletters from the 1980s and 1990s (amounting to 143 pages); (b) photography collections (500+ images) in the Korea Democracy Foundation archive; and (c) documented materials gathered from my previous field research in the early 2000s. These data are supplemented by in-depth interviews with former and current housing activists, conducted during my field visits to Seoul between 2011 and 2015. Before presenting the key findings, the subsequent two sections present this article's theoretical framework and then the political economy of Korea's urbanization, discussing the changing state–society relations as well as socioeconomic contexts within which urban social movements by evictees and the housing poor have been embedded.

State Repression, Hegemony, and Urban Social Movements

Gramsci (1971), in his analysis of state–society relations, contended that a ruling class's overpowering of its subordinate classes is achieved through state domination in the political society and the construction of hegemony in civil society. In his words, "A social group dominates antagonistic groups, which it tends to 'liquidate,' or to subjugate perhaps even by armed force; it leads kindred and allied groups" (57). State domination largely rests on violence and coercion by mobilizing police, military, and other means of law enforcement. By contrast, hegemony is exercised through "the consent and passive compliance of subordinate classes" (Scott 1985, 316). This is where, according to Scott, Gramsci's major contribution lies. Gramsci's discussion of hegemony construction is a fine elaboration on Marx and Engels's "ruling ideas of the epoch" held by the ruling class in possession of the means of material production, an important point they raised in *The German Ideology*:

> The class which has the means of material production at its disposal, has control at the same time over the means of mental production, so that thereby, generally speaking, the ideas of those who lack the means of mental production are subject to it. (Marx and Engels 1965, 61)

Hegemony can be considered as the ruling class's imposition on subordinate classes who might internalize the ideologies of the ruling class (Gramsci 1971). The ideological hegemony of the ruling class, aided by the use of coercive state apparatuses, conditions the behavior of the subordinate classes who might be coopted, persuaded, and oppressed. If the ruling class manages to remain in power through the state domination and the construction of hegemony, the question is how the subordinate classes overthrow the ruling class.

Gramsci's concept of hegemony is often misread to explain the failure of revolutionary movements (e.g., Scott 1985; also see the critique by Hart 1991), but it would be erroneous to conclude that hegemony works to keep the subordinate classes docile and submissive to the ruling class. Rather, as Glassman (2013) asserted, Gramsci's "conception of hegemony contains a sense of the internal dynamics that can lead to hegemony's collapse" (254). In other words, the dialectical reading of hegemony, rooted in the political economy of capitalist accumulation and uneven development, allows room for the erosion of the very conditions that have given rise to the establishment of time- and place-

bounded hegemony. Such understanding of hegemony calls for attention to the accumulation of latent antiestablishment movements that challenge the state domination and the dominant ideology of the ruling class on the one hand and, on the other, changing state–society relations. These are elaborated further below.

First, although studies on (urban) social movements might often focus on major societal disruptions (e.g., Tahrir Square in 2011, Tian'anmen Square in 1989, Seoul Spring in 1980 and 1987), it would be equally crucial to understand how such major disruptions are founded on a series of quotidian and organized resistance in response to state repression and cooptation. As Chang (2015) ascertained, social movements evolve under both endogenous and exogenous pressures and, therefore, the study of social movements "need[s] a diachronic view of movement evolution that accounts for the dynamic nature of contention over time" (7). In this regard, Chang (2015) examined the buildup of antigovernmental oppositional movements by students, intellectuals, and workers during the 1970s in South Korea to understand how the major burst of democracy and labor movements in the 1980s was possible.

Large-scale mass popular movements are therefore preceded by various practices of coalition building, ideological diversification and struggles, and the framing of each contesting group's resistance during the state of latency (Johnston 2015). Such struggles involve the subordinate classes in the production of their own set of vision and political will for a just city, demonstrating a degree of organizational capacity to sustain long-term durability of their resistance to state repression (Routledge 2015a). Scott (1985) went further to argue that the subordinate classes (poor peasants in Scott's case study) have the ability to understand the structural conditions and reject the ideological imposition of the ruling class. The presence of authoritarian repressive states such as the South Korean state between the 1960s and 1980s does not necessarily equate with the absence of (urban) movements: Subordinate classes would still engage with the production of what Johnston (2015) conceptualized as "repressive repertoires," a series of "small acts of protest and opposition . . . creatively carved out of situations where social control breaks down and islands of freedom are creatively and agentically claimed by dissident actors" (628). Such capacity for subordinate classes to be able to engage with resistance and ideological struggles has been picked up by many critics (see, e.g., Parsa 2000; Schock 2005). The key to contesting the dominant hegemony and successful class struggles would eventually involve the establishment of "a series of consensual alliances with other classes and groups" (Haugaard 2006, 5) and, thus, the need for situating individual movement in a broader schema of social movements.

Second, the study of the evolutionary trajectories of urban social movements (e.g., struggles against forced eviction) requires the analysis of such struggles against the backdrop of changing state–society and sociopolitical relations, which are in turn embedded in broader socioeconomic contexts. In the context of uneven development of capitalist accumulation, "geographical variations in the relationship between states and civil society actors are important in understanding the context from which social movements emerge" (Routledge 2015b, 386). The dialectical reading of Gramsci's hegemony suggests that "economic developments are not … foundations on which politics are relatively built but rather a particularly crucial element of the entire context in which political outcomes like hegemony are generated" (Glassman 2013, 249). The geographies of (urban) social movements reflect the state–society relations of a particular time and space. In other words, the repressive capacity of the state, and by extension the hegemonic construction of ruling class ideology, enters into a contentious but constitutive relationship with movements, forming what Chang (2015) referred to as "protest dialectics."

As Boudreau (2004) summed up, the actions of the state shape the ways in which social movements are mobilized and how they develop over time. However, the relationship between state repression and social movements might not be entirely linear. In an authoritarian state context such as the one found in late twentieth-century South Korea, it is possible for the repressive state to effectively suppress, if not annihilate, dissidents or coopt them by monopolizing violence and utilizing resources for its own legitimacy gains. The opposite scenario is also possible; that is, the social movements being fueled by the atrocity of the state violence. In summarizing the complicated nonlinear relationship between state repression and social movements, Chang (2015) suggested the disentanglement of the movement, "shift[ing] our focus away from the total *quantity* of protest events to the substantive *quality* of movement characteristics" (9, italics in original), including ideological development and protester's discourses as well as the forms and strategies of protest.

This article emphasizes the significance of acknowledging ongoing ideological struggles for hegemony between ruling and subordinate classes, especially the urban poor, who have produced a series of urban rights discourses as tactical strategies to contest the state-led urban redevelopment and displacement in the midst of the state pursuit of condensed urbanization. In this article, urban protests against urban redevelopment and displacement are situated as a subcomponent of broader social movements that characterized South Korean politics since the 1980s. Taking into consideration the changing state–society relations, examination of the changing urban rights discourses also acknowledges the significance of historical conjunctures that influence the direction of urban movements: This is in recognition of the fact that the nonlinear relationship between state repression and social movements is further influenced by historic junctures or what Slater (2010) referred to as "critical antecedents." Such junctures often precipitate the disintegration of the political elite's leadership and the formation of a broader coalition of social movements (Johnston 2015). The next section examines Korea's political economy of urbanization to provide the geographical contexts within which the intensification of urban redevelopment projects came to emerge from the 1980s onward.

The Political Economy of Urbanization in Korea

Korea's urbanization can be described as condensed urbanization coupled with industrialization, a characteristic that the country shares with mainland China and other East Asian "tiger" economies such as Taiwan and Singapore (Shin 2014). Dunford and Yeung (2011) reported that East Asian economies took less than thirty years to reach a fivefold increase of their initial real gross domestic product per capita from the time of economic take-off. Conversely, other advanced economies such as those of the United Kingdom and the United States turned out to have taken more than 160 and 100 years, respectively. Among the East Asian economies, Korea's pace was the fastest, having taken only about twenty-two years to achieve this rate of development.

Nationally, the rapid economic development was achieved by the establishment of industrial estates for export-oriented manufacturing, subsidizing the costs of production for industrialists by the developmental states whose legitimacy was garnered by their ability to achieve economic developmental goals without changing the social order (Castells 1992). These industrial complexes were further supported by the construction of various infrastructure and service facilities; hence the accumulation of fixed capital in the built environment (Harvey 1978). These sites of production accompanied urbanization to accommodate workers and their families as well as other service industries. Major cities in Korea such as Ulsan and Changwon came to develop in this way. As shown in Figure 1, the 1960s and 1970s were the period of urbanization subordinated to industrialization, guided by the authoritarian and developmental state that channeled available resources (e.g., national savings, foreign loans) to subsidize the expansion of large businesses rather than expanding national welfare provision (B.-G. Park 1998; Woo-Cumings 1999; Mobrand 2008). Social welfare, including housing, was largely in the hands of individuals; hence the heavy dependence on families and social networks of individuals under the productivist welfare system (Halliday 2000).

From the mid-1980s onward, Korea entered a new era, characterized by decreasing rates of profit in the manufacturing sector, increasing costs of production, and relocation of those factories in search of low-cost labor in other countries (e.g., textile industry relocating to mainland China in the 1990s). The average net profit rate in the manufacturing industry turned out to be 16.9 percent between 1981 and 1990, whereas the figures for 1963 to 1971 and 1972 to 1980 were 39.7 percent and 27.7 percent, respectively (Jung 1995). The mid-1980s also saw the net surplus in Korea's international trade, a turning point indeed for a country that depended heavily on export-oriented industry for its economic development. The resulting overaccumulation and surplus capital as well as the accumulation of wealth by the emerging middle classes in the country were met by the surge of real estate investment and speculative urbanization (Shin and Kim 2016) on the one hand and by the labor movements calling for a fairer share of surpluses as well as the social movements demanding democracy after more than two decades of authoritarian statism on the other (Koo 2001).

The absolute amount of real estate investments also grew rapidly from the late 1980s: In comparison with the 1987 figure, the size of real estate investments in 1993 essentially quadrupled. Accordingly, whereas the share of real estate investment in gross fixed capital formation in 1987 was estimated to be 18.7 percent, this jumped to reach 30.8 percent in 1991 and 36.1 percent in 1993 (The Bank of Korea 2004). Throughout the 1990s, the figure remained at or above 30 percent. Rampant speculation ensued due to price spikes in real estate. The average price of land in Korea increased by 2,976 times between 1964 and 2013, whereas the price of daily necessities (e.g., rice) grew by only fifty to sixty times. As of 2013, real estate assets reportedly accounted for about 89 percent of national assets (N.-H. Ha 2015). In this context, with the industrial restructuring, it can be said that the post-1980s period has seen the reversal of the relationship between urbanization and industrialization (see Figure 1), whereby the highly speculative nature of urbanization (real estate investment in particular) becomes more important for asset accumulation. That is, the investment in the built environment has come to focus more on expanding speculative real estate assets than the expansion of productive investments.

The result was the surge of urban redevelopment projects from the mid-1980s onward, especially in Seoul, which has been the economic, political, and cultural center of the country. Real estate speculation to maximize profits by closing rent gaps in redevelopment neighborhoods (Shin 2009; López-Morales 2011) has become a major means for families to build up their assets, thus consolidating the hegemony of private property rights (Shin and Kim 2016). Here, I am thinking of Ley and Teo's (2014) discussion of the rise of the "cultural hegemony of property" in Hong Kong and Hsu and Hsu's (2013) proposition of "the political culture of property" in Taiwan, all of which privileged private ownership of property and supported the ascendancy of speculative real estate markets and profit-led urban redevelopment. Coupled with the aspiration of the authoritarian state to sanitize and modernize the urban landscape, especially at the time of preparing for the 1988 Seoul Olympic Games (Greene 2003), the developmental state embarked on

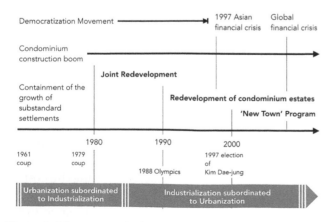

Figure 1. The process of urbanization in Korea and its key events.

a massive scale of displacement of the urban poor. For tenants in redevelopment project sites, there was initially little compensation during the early years of the program in the 1980s (S.-K. Ha 2001). An evictees' movement emerged eventually, further fueled by the democratization movement (Korea Center for City and Environment Research (KOCER) 1998). More detailed pictures of changing state–society relations are provided during the discussions of changing urban rights discourses in the next section.

Urban Protests and the Genealogy of Urban Rights Discourses

1980s: *Saengjon'gwon* or the Right to Subsistence

To understand the urban protests from the 1980s, it is necessary to understand the experience of Korean democracy movements throughout the 1970s when the country was under the dictatorship of then-President Park Chung-Hee (1961–1979). Through the use of police force, the military, the Korean Central Intelligence Agency, and emergency decrees, the authoritarian state endeavored to undermine and suppress the civil society and oppositional movements, while pursuing economic development by forming a developmental alliance with large business conglomerates known as *Chaebols* in Korean. In this context, the focus of oppositional movements was on achieving democracy, led by university students, religious groups (especially progressive Christians), and intellectuals (lawyers, journalists; Chang 2015). Labor movements were yet to be organized despite landmark, yet tragic, events such as the death of labor activist Chun Tae-il, whose self-immolation was a wake-up call for Korean intellectuals, students, and nascent labor activism. As for the protests by evictees, until the end of the 1970s, they remained isolated and sporadic, because of the high prevalence of substandard settlements and the government focus on their containment rather than unrealistic targets of complete eradication (S.-H. Kim 2011). As the alliance between the state and *Chaebols* had been at the center of economic development, the prevalence of substandard settlements was an effective means of minimizing the cost of labor reproduction for businesses (Mobrand 2008).

It was from the early 1980s that urban protests against forced eviction began to be more organized, having faced an entirely hostile set of socioeconomic and political conditions (Shim 1994; S.-H. Kim 1999). Politically, the state–business alliance was still intact despite the sudden collapse of the Park Chung-Hee

dictatorship, as the military coup in December 1979 led by General Chun Doo-Hwan kept the country more or less in the old order. The Fifth Republic headed by then-President Chun Doo-Hwan (1981–1987) continued the practices of the previous Park Chung-Hee dictatorship that resorted to the use of coercive state power to bring society under their control. Socioeconomically, the country witnessed the continued growth of middle-class populace, whose asset basis expanded substantially, thanks partly to the speculative price increases in real estate. Construction subsidiaries of *Chaebols* or large conglomerates also began to show an interest in participating in urban redevelopment projects with commercial and corporate orientation (S.-K. Ha 2001). Seoul as the national capital came to be the epicenter of commodification of space through redevelopment targeting both residential and business districts. The transformation of Seoul to host the 1986 Summer Asian Games and the 1988 Summer Olympic Games also added fuel to the proliferation of urban redevelopment (Asian Coalition for Housing Rights [ACHR] 1989).

From 1983, urban redevelopment projects targeting substandard neighborhoods in Seoul intensified with the introduction of new government policy to implement what was known as *Hapdong Jaegaebal* or joint redevelopment program, which was estimated to have affected about 10 percent of the total municipal population since implementation (Shin and Kim 2016). Facing harsh conditions of displacement and relocation, tenants' protests grew in both size and intensity. On the introduction of the joint redevelopment program, tenants were initially offered neither compensation nor any other alternative housing provision. Under the circumstances, as a former leader of a tenants' group against forced eviction in the Hawang 2–1 redevelopment district in central Seoul states, *saengjon'gwon* "came first" before any other expressions, as "resistance was to fight the exploitation of people's life spaces and the destruction of life" (Mr. Y, interview 20 August 2013). In other words, for poor tenants subject to eviction with no compensation, *saengjon'gwon* or the right to subsistence occupied center stage of their protests to survive. Such sentiment was frequently pronounced in various pamphlets and slogans throughout the 1980s (see Figure 2). Protesters' demands centered on the governmental provision of alternative relocation housing, especially in the form of public rental housing as part of addressing their immediate shelter needs. For instance, in a protest pamphlet dated 23 July 1987, tenants from Dohwa 3 district urged, "Stop the forced demolition immediately. Guarantee the *saengjon'gwon* for the urban poor" (see KOCER 1998, 332).

Figure 2. Tenants' protests in Sadang and Dongjak in January 1988, Seoul. The writing on the placard reads, "Guarantee the Right to Subsistence and Public Rental Housing with Long-Term Loan Conditions." *Source*: The Kyunghyang Shinmun (Park, Yong-Su).

1990s: *Jugeo'gwon* or the Right to Housing

The prevalence of commercialized redevelopment in the 1980s resulted in a humongous scale of brutal and forced eviction in Seoul. The ACHR reported that about 48,000 dwellings housing 720,000 urban poor people were subject to eviction between 1983 and 1988 (ACHR 1989; Greene 2003). As tenants' frustration escalated, their protests became more organized: A city-wide organization called the Seoul Council of Evictees (*Seoul Cheolgeomin Hyeob'euihui*) was formed in Seoul in 1987 at the height of the democracy movements in the 1980s, providing support for individual sites of struggle. Although the state–*chaebols* alliance was still in place after the 1987 June Democratic Uprising, as the ruling right-wing party narrowly escaped its demise by winning the 1987 December presidential election (which was largely due to the schism between opposition parties), it was under pressure to devise compensation measures to appease tenants and maintain their legitimacy. After piloting a series of incremental measures, a new policy was introduced in 1989 that included the provision of cash (living costs for three months) or in-kind (tenancy in public rental housing) compensation (K.-J. Kim et al. 1996). This arrangement subsequently remained unchanged for more than a decade. The state concession could be considered the fruits of the evictees' strenuous fights against the alliance of the state, developers, and landlords-cum-speculators, supported by other sectors of social movements.

As the new compensation measures settled in, a new language of *jugeo'gwon* or the right to housing began to emerge in the early 1990s. Rather than confining tenants' protests to obtaining *saengjon' gwon*, housing is to be seen as part of basic human rights and constitutional rights (Mr. Y, interview 20 August 2013; see Figure 3). A former student activist, who is now a district mayor in Seoul, recalled that "in the early to mid-1980s, the slogan was by and large to attain *minjung saengjon'gwon* [people's right to subsistence], and then evictees' *saengjon'gwon*. *Jugeo'gwon* came afterward. Regardless of house ownership, having a home to live was to be seen as a right" (Mr. K, interview 21 August 2013). Protest materials also reflected the changing slogan. For instance, in their pamphlet dated 18 October 1990, tenants in Nolyangjin 2–2 district

To some extent, the rise of the *saengjon'gwon* slogan could be attributed to the increasing degree of awareness of human rights concerns that emerged in the late 1970s as tactical evolution of democracy movements during times of repressive state domination that resulted in harsh physical suppression of dissidents and protesters. As Chang (2015) succinctly summarized, "Human rights became part of South Korean civil society for the first time when antigovernment dissidents made it an integral part of the larger democracy movement in the 1970s" (159). Korea's democracy movements in the 1980s culminated in the 1987 June Democratic Uprising that resulted in the authoritarian state's concession to introduce a direct presidential election. Such movements were made possible by the formation of political alliances not only among dissident communities but also among university students, progressive intellectuals, trade unions, farmers, the urban poor (e.g., informal street vendors, poor tenants in substandard settlements), and eventually white-collar workers. Each of the groups had its own movement agenda, but they came together under democratization as a shared frame for collective action. Poor evictees took part in it, too, with an understanding that a more democratic state would protect their *saengjon'gwon*, as exemplified by a statement in a pamphlet from Dohwa 3-district dated 21 July 1987: "We longed and fought for democratization, because democratization would allow a fair treatment of us who work strenuously to make the ends meet. ... [Furthermore,] there is no democratization without guaranteeing our *saengjon'gwon*" (KOCER 1998, 330).

Figure 3. Protesters in 1991 demanding the right to housing, Seoul. *Source:* The Kyunghyang Shinmun (Park, Yong-Su).

argued that "we will be fighting all the way for *saengjon'gwon* as our minimal right. ... [People of similar circumstances from] development areas should unite to be guaranteed of their *jugeo'gwon*."

The provision of public rental housing as in-kind compensation was considered by many as having met the *saengjon'gwon* of tenants experiencing forced eviction. Protests continued to emerge from a number of redevelopment project sites to address unresolved issues such as support for temporary relocation, and more violent fights broke out sporadically involving groups of ineligible tenants against displacement. The attention of activists and progressive intellectuals, however, began to steer toward improving the legal system for general housing welfare of the poor and thus the right to housing (KOCER 1998; Lee 2012). A major development was the establishment of the National Coalition for Housing Rights (NCHR) in 1990 as an umbrella organization by a number of social movement organizations, including those of evictees and housing activists and progressive religious groups: As the declaration for the NCHR establishment states, the organization aimed at the acquisition of the right to housing as its major goal, proclaiming it as a basic right of the people.

The shift toward improving the legal system and housing welfare provision throughout the 1990s can be seen as an extension of the institutionalization of social movements that constitute what Prujit and Roggerband (2014) referred to as a "dual movement structure." Autonomous and institutionalized social movements in a dual movement structure benefit from each other in the context of a more open political environment, as the former creates disruptive actions to add pressure on the state and the latter provides institutionalized support and legitimacy for social movements. The series of political changes in the first half of the 1990s in Korea enabled the transition from autonomous social movements to institutionalized social movements. The developmental state, having its legitimacy challenged by the democracy movements, made efforts to distance itself from the authoritarianism of the 1970s and 1980s. The political reform in the early 1990s also included the establishment of local assemblies from 1991 and the implementation of the direct election of mayors and provincial governors from 1995. Like many other sectors that took part in the earlier democracy movements, housing activists and supporting networks pursued the establishment of institutional arrangements to integrate housing rights and access to affordable housing as part of governmental frames. For instance, a number of Korean civil society delegates who participated in the 1996 Habitat II conference joined to establish action plans to legislate the Basic Housing Rights Act as part of advancing the right to housing (see M.-S. Park and Kim 1998; Seo 1999).

The shift from *saengjon'gwon* to *jugeo'gwon* also reflects the rapidly diminishing stocks of affordable housing for the urban poor, resulting from mounting interests in real estate investments. The developmental state still kept its close nexus with businesses: Having previously faced resistance from organized labor movements and with decreasing rates of profit in the manufacturing sector, the state–business alliance opted for *segyehwa* or globalization, involving selective overseas relocation of production bases, transnational investment, and liberalization of the financial industry. The direct election of local assembly members, mayors, and governors laid the foundation for the rise of local growth machines, further propelling investments in real estate properties and infrastructure. Large-scale urban redevelopment projects ensued, especially in Seoul, which witnessed government efforts to transform the national capital into a world city and involved active participation of construction subsidiaries of major *chaebols*. Rapidly disappearing affordable housing stocks and the sharp increase in housing rents due to megadisplacement of poor tenants led to growing awareness of housing as a basic right. For many activists working in poor neighborhoods, the major concern in the 1990s was how to ensure the housing right of poor residents who faced eviction as such neighborhoods became subject to megaredevelopment projects (Binmin Jiyeog Undongsa Balgan Wiwonhoi [BJUBW] 2017). The emphasis on housing rights continued to exert its presence, albeit with limited success, during the times of postcrisis Korean welfare statism that involved the establishment of social safety nets for the victims (including homeless people) of the economic crisis in the aftermath of the 1997 Asian financial crisis.

2000s: *Jeongju'gwon* or the Right to Settlements

Despite the efforts by civil society organizations to legislate the Basic Housing Rights Act, they faced a barrier, especially due to the severe downturn of the national economy following the Asian financial crisis. To stimulate economic recovery, the promotion of real estate development remained intact (S.-K. Ha 2010). Reformist policies such as the Basic Housing Rights Act were seen as a hindrance to real estate development, especially by "those established interests who gained much of developmental profits through redevelopment. [Private] Property rights were prioritized" (Mr. Y, interview 20 August 2013).

Investment in fixed assets, especially infrastructure and real estate, characterized the postcrisis recovery efforts, especially at the local scale. In Seoul, having experienced a brief slump after the Asian financial crisis, urban redevelopment picked up its pace again in the early 2000s, this time led by then–Seoul Mayor Lee Myung-Bak (2002–2006), whose previous position

as the CEO of Korea's largest construction firm aligned him with real estate interests (Doucette 2010). In line with his mayoral election manifesto that promised boosterish developmental projects, Myung-Bak, a member of the conservative Grand National Party, gave birth to the highly speculative megadistrict redevelopment program, euphonically coined as "new town development." Pilot projects began in northern Seoul, targeting those urban districts that escaped the fervor of urban redevelopment in the previous decade and thus witnessed widened rent gaps. Becoming a new town program site was met by an instantaneous surge in property value, thus providing opportunities for speculative gains for property owners and absentee landlords-cum-speculators (Shin and Kim 2016).

In response to the new town program as an area-wide initiative, housing activists turned their attention toward promoting *jeongju'gwon* or the right to human settlements. This shift was to acknowledge the importance of going beyond the individual housing unit and placing housing in a wider context of settlement that encompasses multiple dimensions of habitation. *Jeongju'gwon* was recognized as a concept that "encompassed *jugeo'gwon*, as well as the concept of local community [*jiyeog sahoi* in Korean]" (Mr. Y, interview 20 August 2013). In their online posting dated 8 April 2003, the National Council of Center to Victims of Forced Evictions (NCCVFE), a nongovernmental advocacy organization for the protection of people's rights against forced eviction founded in April 1993, correspondingly reframed the objective of their activities, explaining that they pursue "*jeongju'gwon* movement based on reasons and rationale. Based on *jeongju'gwon*, we do our best to prevent quality of life from degrading by redevelopment that endangers residents' *jeongju'gwon*" (NCCVFE 2003).

Post-2009: The Emergence of *Dosi'gwon* or the Right to the City

The conceptualization of *jeongju'gwon* in response to the rise of new town projects experienced further transformation after 2009. This was precipitated by the tragic conclusion of small business tenants' protests in Yongsan, Seoul, in January 2009 when six people (five protesters and one policeman) died in the midst of a police SWAT team operation carrying out military-style suppression of small business tenant protesters. This tragedy, as a key historic juncture, was a wake-up call for housing activists and critical scholars as well as civil society organizations who painfully admitted that the state violence against eviction still persisted, despite the country's nominal democratization. Another major revelation was the limitation of the 1989 compensation regime, which failed to take into account small business tenants who were left without adequate compensation. Small business tenants came to be core members of evictee organizations, as a housing activist noted in an interview (Mr. L 15 December 2011): "From 2000, more than 80 percent of the members of evictee organizations such as the NCCVFE or the Urban Poor Evictees' Union were business tenants."

Two successive national governments from 2008 were headed by presidents from the right-wing party that managed to restore its power after having lost the 1997 and 2002 presidential elections. The election of former Seoul Mayor Lee Myung-Bak as president in 2007 also signaled a major shift of economic policies toward heavier investment in the built environment, including continued expansion of real estate investments and urban redevelopment projects. This also meant that the previous efforts to institutionalize social movements and by extension to institutionalize various social rights including housing rights also faced retreat, as the state resorted to the repressive use of its power to subdue social movements and oppositional voices that were critical of the new right-wing governments. The Yongsan tragedy was seen to be an extension of such state violence.

Since the Yongsan tragedy in 2009, there has been a noticeable degree of attention to incorporating *dosi' gwon* or the right to the city concept in urban movements for social justice, influenced in part by the works of critical Korean geographers (e.g., Y.-C. Kim 2009; Hwang 2010) and human rights activists (Miryu 2010). Y.-C. Kim (2009), for instance, reflected on the tragedy of Yongsan and argued that *dosi'gwon* is to be adopted as the key slogan to fight dispossession resulting from urban redevelopment. To some extent, the attention to the right to the city was to overcome the predominant focus on residential tenants in the previous housing movements. As housing activist Mr. L pointed out (interview 15 December 2011):

> Urban researchers or those members of housing rights movement groups neglected the business tenants' problems. It was not easy for them to connect business tenants with a certain concept of right, and there was hardly any research or consideration for supporting them [the struggles of small business tenants].

Another business tenant further expressed that "to me, the struggle of commercial tenants has only just begun" (interview 20 December 2011).

Against this backdrop, human rights activists and evictee organizations have worked together in alliance to launch a campaign to legislate the Protection from Forced Eviction Act. According to a human rights activist (Ms. M, interview 15 December 2011), this was based on an increasing degree of awareness that forced eviction should be seen as a violation of basic human rights. The movement to legislate the Protection from Forced Eviction Act was to draw people's attention to the human dimension and costs associated with the demolition of building structures. In collaboration with academics such as those members of the Korean Association of Space & Environment Research and legal professions (e.g., Democratic Legal Studies Association), a draft act was motioned as a new bill by the supportive members of the National Assembly (The Kyunghyang Shinmun 2011). As of January 2017, the bill had not passed, and it was not clear how soon it would come to be fully legislated. The major barrier was thought to be the hegemony of property rights, as the act would constrain any attempt to turn properties into a "higher and better" use for speculative profit gains.

Challenging the Hegemony of Property Development and Forming Solidarity

The history of the urban poor's struggle against eviction in Korea can be understood as the history of the subordinate classes challenging the legitimacy of the capitalist accumulation regime that sought to maximize its gains from socially unjust urban transformation (cf. Weber 2002). The physical struggle accompanied an ideological struggle. The review of the archival records of pamphlets and protest materials makes it evident that there is no lack of understanding among the protesters with regard to the exploitative nature of urban redevelopment based on capturing the rent gaps. From the early days, those resisting forced eviction retained acute awareness of unequal power relations manifested in their neighborhoods, as partly noted in the previous sections. A newsletter published by the Seoul Council of Evictees (*Seoul Cheolgeomin Hyeob'euihui*) on 21 November 1987 stated on the cover page that "the urban poor has the natural obligation to fight till the end redevelopment and demolition carried out by the monopoly *chaebol* such as Hangug Geon'eob, Daelim Saneop, and Hyundai Geonseol under the auspice of military dictatorship headed by Chun and Rho [presidents]" (see KOCER 1998, 178).

Challenging the state, developers, and the hegemony of private property rights was accompanied and supported by the formation of a wide-encompassing alliance: Evictees reached out to student activists and civil society organizations, who were integral components of local activism in poor neighborhoods (BJUBW 2017). Protesters' discourses revealed their acute awareness of the importance of positioning the struggle in a broader context of fighting capitalist exploitation. This was possible, to some extent, because of the historic legacy of Korea's democracy movements (also known as *minjung* [common people's] movements) during the times of dictatorship and military regimes between the 1960s and the 1980s (see Lee [2007] for the *minjung* movement). Particularly in the 1980s, after nearly two decades of military dictatorship, Korea saw the outburst of social movements, led by intellectuals, students, farmers, the urban poor, and workers, demanding not only real democracy but also redistributive justice. In this regard, Korea was not lacking efforts to establish cross-class alliances (see also Chang 2015).

This is the environment within which housing activism, and more recently the antigentrification movement, in Korea has been embedded (see Shin 2017). Going back to the Yongsan tragedy in January 2009, on the evening of the tragedy, more than eighty civil society and political organizations held a candlelight vigil with thousands of citizens, which then led to a more violent street protest in the late evening (The Seoul Institute 2017). Overnight discussions among activists resulted in the formation of a committee that saw the participation of more than eighty civil society organizations including those working to enhance the urban poor's housing rights: They aimed at bringing justice to those who were responsible for the forced and brutal oppression of evictees (The Seoul Institute 2017).

It is interesting to note how such interaction between evictees and other social groups enabled the evictees to acquire the languages of protest and rights claims and that poor tenants' resistance to redevelopment and displacement did not emerge out of the blue. Mr. Y (interview 20 August 2013), who was the head of a tenants' group in the Hawang 2–1 redevelopment district in central Seoul in 1993, recalled that the most frequently expressed slogan was the demand for the right to housing, but this was the result of education, helping them continue their fight. J.-H. Kim's (2017) review of the history of local activism predating redevelopment in the Hawang 2–1 redevelopment district reveals how

the buildup of local activism throughout the 1980s and early 1990s enabled the effective organization of tenants' efforts to resist displacement. The tenants' organization was rooted in a children's study group organized by local activists for the poor in Hawang and adjacent neighborhoods. Local activists, who settled down in the neighborhoods from 1987, held various educational sessions to inform children's mothers about redevelopment and displacement, and the mothers also brought their husbands to be involved when the tenants' organization was to be formed. Mr. Y, quoted earlier, was one of those husbands. Local activists in the neighborhoods also came together to organize a local council of activists (1989–1994) to coordinate their activities. The key figures among the activists were a married couple, both of whom were seasoned activists for the poor. They began their activism from the early 1970s, and the husband in particular had experiences of working with tenants against displacement in the 1970s: Such experiences turned out to be beneficial for the education of local activists in the Hawang 2–1 district and adjacent neighborhoods (see J.-H. Kim 2017).

The solidarity among evictees, local activists, and other civil society organizations, as well as their efforts to pursue cross-class alliance, is quite encouraging for achieving social justice through progressive urban movements, as these initiatives allow them not to be confined to their self-interest. For a number of more persistent protesters who continue to exercise activism and engage with long-term social movement, their long-term commitments seem to develop class consciousness. The chair of the Korea Evictees Association, who has been leading the organization for more than two decades, explained how his struggle for the right to housing led to his realization of the importance of cross-class alliance: "Resolving the right to housing issue does not solve everything. We need to open our eyes to the labor movement too. Evictee's movement alone does not resolve capitalist contradictions. Workers, evictees and farmers all have to work together" (see Choi et al. 2009, 189).

Nevertheless, as discussed earlier, the fact that the efforts to legislate the Protection from Forced Eviction Act have been facing barriers suggests that the property hegemony persists. There has also been a degree of fragmentation among evictees and their organizations, resulting in the establishment of several umbrella organizations due to their different views on what would be the most effective tactics for the housing rights struggle (see S.-K. Park and Lee 2012), although they might still come together to collectively address major state oppressions like the Yongsan tragedy. Furthermore, the struggle by evictees has clear limitations of being a highly place-specific rights struggle that runs the danger of dissolution once a neighborhood disappears (Mr. Y, interview 20 August 2013). Local activists who worked hard in the 1980s and 1990s to create neighborhood-based grassroots organizations lamented that urban redevelopment projects disintegrated residents and that it was difficult to continue the organizational momentum after redevelopment and displacement. This testifies to the destructive nature of gentrification including urban redevelopment, posing serious threats to the growth of place-specific urban movements to advance the right to the city and achieve social justice.

Conclusion

Reflecting on the Korean history of urban accumulation and injustice, the production of urban space has been undeniably in the imagination of the ruling class, who imposed their own vision of an ideal city and of "a just social order" (Scott 1985, 305) on subordinate classes. The voices of the tenants facing forced eviction and increasingly unaffordable housing costs, however, have produced their own set of demands and narratives about the socially unjust nature of urban redevelopment. Their demands called for the guarantee of their *saengjon'gwon* (the right to subsistence) and *jugeo'gwon* (the right to housing), refusing to be denigrated as barriers to societal progress. The enactment of the National Basic Housing Rights Act in 2015 can be regarded as the culmination of the efforts made by the progressive urban movements. Various evictee organizations established in the early 1990s continue to operate at present, their longevity possibly helped by the ongoing injustice in the production of the built environment and also by the experience of eviction as "shared emotional connections" (Bosco 2007) that bind them together.

With the changing economic climate that questions high rates of economic development and real estate accumulation, there emerges an opportunity to think of a new way of imagining and building a new Seoul. It is perhaps time to revisit the legalist agenda put forward nearly three decades ago when the National Coalition of Housing Rights was established in 1990 and efforts were made to secure the right to housing for the general population. As the advocates of the right to the city often point out (see Mitchell 2003; Harvey 2008; Marcuse 2009), the legal provision is only one of many necessary conditions for the realization of a new alternative way of producing just cities.

Facilitated by a broader cross-class alliance, fights for the collective consumption such as housing have a direct potential to make this possible (see Harvey 2013; Merrifield 2014). It is about time to rethink seriously the ramification of speculative urbanization and gentrification and embark on producing "a genuinely humanizing urbanism" (Harvey [1976] 2009, 314) that realizes a vision that places people at the center and not profit (Brenner, Marcuse, and Mayer 2009). In this regard, the emergent discourses of the right to the city in Korea in recent years can be considered as an assuring positive shift, as such a move propels progressive Korean urban politics to go beyond the residential domain of urban social movements and to be inclusive of commercial tenants and other forms of inhabiting space.

Acknowledgments

I thank the participants and audience at the following events where various versions of this article were presented: the Centre of Korean Studies seminar, SOAS, London, February 2016; the Seoul Institute Conference on Seoul as a Model of Progressive City, Seoul, October 2015; and the 2016 Annual Conference of the American Association of Geographers, San Francisco. I would like to express my gratitude to Tim Butler, Paul Waley, Jaeho Kang, Chai-Kwan Lee, Soo-Hyun Kim, Jesook Song, Nik Heynen, my interviewees who kindly shared their valuable time with me, and the journal's anonymous reviewers for their encouragement, constructive comments, and helpful suggestions. I take sole responsibility for any possible errors in this article.

Funding

The author acknowledges financial support from the National Research Foundation of Korea Grant funded by the Korean Government (NRF-2017S1A3A2066514).

References

Asian Coalition for Housing Rights (ACHR). 1989. *Battle for housing rights in Korea: Report of the South Korea Project of* the Asian Coalition for Housing Rights. Bangkok: Asia Coalition for Housing Rights [and] Third World Network.

The Bank of Korea. 2004. *National accounts 2004*. Seoul: The Bank of Korea.

Binmin Jiyeog Undongsa Balgan Wiwonhoi (BJUBW) [Publication Committee for Movements in Poor Neighborhoods]. 2017. *Searching for the origin of neighborhood community movements*. Seoul: Han'ul.

Bosco, F. J. 2007. Emotions that build networks: Geographies of human rights movements in Argentina and beyond. *Tijdschrift poor Economische en Sociale Geografie* 98 (5):545–63.

Boudreau, V. 2004. *Resisting dictatorship: Repression and protest in Southeast Asia*. New York: Cambridge University Press.

Brenner, N., P. Marcuse, and M. Mayer. 2009. Introduction: Cities for people, not for profit. *City* 13 (2&3):176–84.

Castells, M. 1992. Four Asian tigers with a dragon head: A comparative analysis of the state, economy, and society in the Asian Pacific rim. In *States and development in the Asian Pacific Rim*, ed. R. Appelbaum and J. Henderson, 33–70. Newbury Park, CA: Sage.

Chang, P. Y. 2015. *Protest dialectics: State repression and South Korea's democracy movement, 1970–1979*. Stanford, CA: Stanford University Press.

Chesters, G. 2012. Social movements and the ethics of knowledge production. *Social Movement Studies* 11 (2):145–60.

Choi, H.-W., M.-S. Ahn, I.-S. Kim, S.-C. Jageuni, H.-S. Kim, R. Kim, H.-S. Park, M.-J. Yeonjeong, H.-Y. Kim, S.-O. Lee, G. Lee, D. Kang, and I.-H. Jang. 2009. *There are people here*. Seoul: Sam'i Boineun Chang.

Doucette, J. 2010. The terminal crisis of the "participatory government" and the election of Lee Myung Bak. *Journal of Contemporary Asia* 40 (1):22–43.

Dunford, M., and G. Yeung. 2011. Towards global convergence: Emerging economies, the rise of China and western sunset. *European Urban and Regional Studies* 18 (1):22–46.

Glassman, J. 2013. Cracking hegemony: Gramsci and the dialectics of rebellion. In *Gramsci: Space, nature, politics*, ed. M. Ekers, H. Gillian, S. Kipfer, and A. Loftus, 241–57. Chichester, UK: Wiley-Blackwell.

Gramsci, A. 1971. *Selections from the prison notebooks*, trans. G. Nowell Smith and Q. Hoare. London: Lawrence & Wishart.

Greene, S. J. 2003. Staged cities: Mega-events, slum clearance, and global capital. *Yale Human Rights & Development Law Journal* 6:161–87.

Ha, N.-H. 2015. Land price increased by 3000 times, whereas the price of rice and oil increased by 50 and 77 times respectively. *ChungAng Daily* November 17. Accessed December 1, 2016. http://news.joins.com/article/19080234.

Ha, S.-K. 2001. Substandard settlements and joint redevelopment projects in Seoul. *Habitat International* 25: 385–97.

———. 2010. Housing crises and policy transformations in South Korea. *International Journal of Housing Policy* 10 (3):255–72.

Halliday, I. 2000. Productivist welfare capitalism: Social policy in East Asia. *Political Studies* 48:706–23.

Hart, G. 1991. Engendering everyday resistance: Gender, patronage and production politics in rural Malaysia. *The Journal of Peasant Studies* 19 (1):93–121.

Harvey, D. [1976] 2009. *Social justice and the city*. Rev. ed. Athens: University of Georgia Press.

———. 1978. The urban process under capitalism: A framework for analysis. *International Journal of Urban and Regional Research* 2 (1–4):101–31.

———. 2008. The right to the city. *New Left Review* 53:23–40.

———. 2013. *Rebel cities: From the right to the city to the urban revolution*. London: Verso.

Haugaard, M. 2006. Conceptual confrontation. In *Hegemony and power: Consensus and coercion in contemporary politics*, ed. M. Haugaard and H. Lentner, 3–19. Oxford, UK: Lexington.

Hsu, J.-Y., and Y.-H. Hsu. 2013. State transformation, policy learning, and exclusive displacement in the process of urban redevelopment in Taiwan. *Urban Geography* 34 (5):677–98.

Hwang, J.-T. 2010. *Doo-ri-ban and the right to the city*. Accessed January 10, 2016. http://sarangbang.or.kr/bbs/view.php?board=hrweekly&id=1591.

Johnston, H. 2015. "The Games's afoot": Social movements in authoritarian states. In *The Oxford handbook of social movements*, ed. D. D. Port and M. Diani, 619–33. Oxford, UK: Oxford University Press.

Jung, H.-N. 1998. Land prices and land markets in Korea, 1963–1996: Explanations from political economy perspectives. *The Korea Spatial Planning Review* 27:127–46.

Kim, J.-H. 2017. Unrealised dream: Moving towards a collective again. In *Searching for the origin of neighborhood community movements*, ed. Binmin Jiyeog Undongsa Balgan Wiwonhoi (BJUBW), 381–427. Seoul: Han'ul.

Kim, K.-J., I.-J. Lee, S.-H. Jeong, and Y.-G. Jeon. 1996. *Substandard housing redevelopment in Seoul*. Seoul: Seoul Development Institute.

Kim, S.-H. 1999. The history of evictees' movement in Seoul. *Urbanity and Poverty* 36:51–77.

———. 2011. Squatter settlement policies and the role of the state. *Housing Studies Review* 19 (1):35–61.

Kim, Y.-C. 2009. From physical redevelopment of cities to the right to the city. *Changjaggwa Bipyeong* 144:339–53.

Koo, H. 2001. *Korean workers: The culture and politics of class formation*. Ithaca, NY: Cornell University Press.

Korea Center for City and Environment Research (KOCER). 1998. The history of housing rights movement in Seoul. In *Eviction in the eyes of evictees: The history of evictees' movement in Seoul*, ed. KOCER, 85–111. Seoul: KOCER.

The Kyunghyang Shinmun. 2011. Human rights organizations to produce "Protection from Forced Eviction Act." 11 July:12.

Lee, N. 2007. *The making of Minjung: Democracy and the politics of representation in South Korea*. Ithaca, NY: Cornell University Press.

Lefebvre, H. 1996. *Writings on cities*. Oxford, UK: Blackwell.

López-Morales, E. 2011. Gentrification by ground rent dispossession: The shadows cast by large scale urban renewal in Santiago de Chile. *International Journal of Urban and Regional Research* 35 (2):330–57.

Marcuse, P. 2009. From critical urban theory to the right to the city. *City* 13:185–97.

Marx, K., and F. Engels 1965. *The German ideology*. London: Lawrence & Wishart.

Mayer, M. 2009. The "right to the city" in the context of shifting mottos of urban social movements. *City* 13 (2–3):362–74.

Merrifield, A. 2014. *The new urban question*. London: Pluto.

Miryu. 2010. City and human rights should meet. Accessed January 20, 2017. http://sarangbang.or.kr/bbs/view.php?board=hrweekly&id=1628.

Mitchell, D. 2003. *The right to the city: Social justice and the fight for public space*. New York: Guilford.

Mobrand, E. 2008. Struggles over unlicensed housing in Seoul, 1960–80. *Urban Studies* 45 (2):367–89.

National Council of Center to Victims of Forced Evictions (NCCVFE). 2003. What is the right to settlements. Accessed January 20, 2017. http://www.nccmc.org/bbs/board.php?bo_table=promotion&wr_id=1230&page=83.

Park, B.-G. 1998. Where do tigers sleep at night? The state's role in housing policy in South Korea and Singapore. *Political Geography* 74 (3):272–88.

Park, I.-K., and S.-Y. Lee. 2012. Analyzing changes in space and movements for opposition and alternatives in Seoul. *Space and Society* 22 (4):5–50.

Park, M.-S., and E.-H. Kim. 1998. Adaptation of the Habitat II agenda to housing rights organizations in Korea. *Dosi Yeongu* 4:217–41.

Parsa, M. 2000. *States, ideologies, and social revolutions: A comparative analysis of Iran, Nicaragua, and the Philippines*. New York: Cambridge University Press.

Pile, S. 1996. Introduction: Opposition, political identities and spaces of resistance. In *Geographies of resistance*, ed. M. Keith and S. Pile, 1–32. London and New York: Routledge.

Prujit, H., and C. Roggeband. 2014. Autonomous and/or institutionalized social movements? Conceptual clarification and illustrative cases. *International Journal of Comparative Sociology* 55 (2):144–65.

Routledge, P. 2015a. Engendering Gramsci: Gender, the philosophy of praxis, and spaces of encounter in the Climate Caravan, Bangladesh. *Antipode* 47 (5):321–45.

———. 2015b. Geography and social movements. In *The Oxford handbook of social movements*, ed. D. D. Port and M. Diani, 383–93. Oxford, UK: Oxford University Press.

Saldaña, J. 2003. *Longitudinal qualitative research: Analyzing change through time*. Lanham, MD: Rowman & Littlefield.

Schock, K. 2005. *Unarmed insurrections*. Minneapolis: University of Minnesota Press.

Scott, J. C. 1985. *Weapons of the weak: Everyday forms of peasant resistance*. New Haven, CT: Yale University Press.

Seo, J.-G. 1999. A movement to legislate National Housing Basic Rights for the attainment of the right to housing. Accessed December 30, 2016. http://www.peoplepower21.org/Welfare/643157.

The Seoul Institute. 2017. Yongsan tragedy white paper. Seoul Metropolitan Government, Seoul, Korea.

Shim, S.-G. 1994. The evaluation of the progress of the struggle against eviction. *Urbanity and Poverty* 6:2–17.

Shin, H. B. 2009. Property-based redevelopment and gentrification: The case of Seoul, South Korea. *Geoforum* 40 (5):906–17.

———. 2014. Contesting speculative urbanisation and strategising discontents. *CITY: Analysis of Urban Trends, Culture, Theory, Policy, Action* 18 (4–5):509–16.

———. ed. 2017. *Anti-gentrification: What is to be done?* Paju: Dongnyok.

Shin, H. B., and S.-H. Kim. 2016. The developmental state, speculative urbanisation and the politics of displacement in gentrifying Seoul. *Urban Studies* 53 (3):540–59.

Shin, H. B., L. Lees, and E. López-Morales. 2016. Introduction: Locating gentrification in the global East. *Urban Studies* 53 (3):455–70.

Weber, R. 2002. Extracting value from the city: Neoliberalism and urban redevelopment. *Antipode* 34 (3):519–40.

Woo-Cumings, M. ed. 1999. *The developmental state*. Ithaca, NY: Cornell University Press.

HYUN BANG SHIN is Associate Professor of Geography and Urban Studies at the London School of Economics and Political Science, London WC2A 2AE, UK. E-mail: h.b.shin@lse.ac.uk. He is also Eminent Scholar at Kyung Hee University, Seoul, South Korea. His research centers on the critical analysis of the political economic dynamics of urbanization, the politics of redevelopment and displacement, gentrification, housing, the right to the city, and megaevents as urban spectacles, with particular attention to Asian cities.

6   Urban Precarity and Home

There Is No "Right to the City"

Solange Muñoz

Employing qualitative and ethnographic data from field research in 2009 and 2017 in Buenos Aires, Argentina, this article seeks to position urban housing and the home at the center of discussions about social and spatial justice and the right to the city. A focus on housing and home is largely absent in the literature on the right to the city. This article contributes to the literature on the right to the city through an analysis of the struggle for the right to housing and to the city by social organizations and residents who live in informal housing in Buenos Aires, Argentina. I argue that precarious housing—through the threat of eviction and lack of affordable options—affects poor urban residents' right to housing and right to the city and hinders the struggle for social justice. In this study, *home* is theorized as a central space from which urban dwellers are able to create stable spaces, access urban resources, and contribute to the social fabric and development of the city. As such, precarious housing is understood not simply as experienced at the scale of the home but rather as a primary space from which the right to the city is endeavored, challenged, and denied. I examine how urban injustice and precarity are routinely produced and experienced and argue that without access to stable housing and social change, there is no right to the city. *Key Words: home, Latin America, precarious housing, right to the city, urban geography.*

本文运用 2009 年和 2017 年在阿根廷布宜诺斯艾利斯的田野工作取得的质性与民族志数据, 企图将城市居住与住家, 置于有关社会及空间正义与城市权讨论的核心。城市权的文献, 大幅缺乏对于居住与住家的关注。本文透过分析居住于阿根廷布宜诺斯艾利斯的非正式住宅中的社会组织及居民争取居住权和城市权的斗争, 对城市权的文献做出贡献。我主张, 不安定的居住——透过威胁进行驱逐并缺乏可负担的选择——影响了贫穷城市居民的居住权与城市权, 并阻碍了社会正义的斗争。本文将住家理论化为城市居民能够创造安定空间、取得城市资源、并对城市的社会纹理和发展做出贡献的核心空间。于此, 不安定的居住, 并非被单纯理解为在家的尺度上经历之事, 而是城市权努力耕耘、挑战和受到拒绝的主要空间。我检视城市不正义和不安定如何持续被生产与被经验, 主张若缺乏取得稳定的居住与社会变迁的管道, 城市权将不复存在。 关键词: 住家, 拉丁美洲, 不安定的居住, 城市权, 城市地理学。

Con datos cualitativos y etnográficos obtenidos por investigación de campo en 2009 y 2017 en Buenos Aires, Argentina, en este artículo se busca posicionar la vivienda urbana y la casa u hogar en el centro de las discusiones sobre justicia social y espacial, y sobre el derecho a la ciudad. En gran medida, el enfoque centrado en la vivienda y la casa no aparece en la literatura sobre el derecho a la ciudad. Este artículo contribuye a esta literatura especializada a través de un análisis de la lucha por el derecho a la vivienda y a la ciudad adelantada por organizaciones sociales y por residentes que habitan viviendas informales en Buenos Aires, Argentina. Yo sostengo que la vivienda precaria—por la amenaza de desahucio y la carencia de opciones costeables—afecta el derecho a vivienda y el derecho a la ciudad de los residentes urbanos pobres y dificulta la lucha por la justicia social. Es este estudio, la *casa* se teoriza como un espacio central a partir del cual los habitantes urbanos pueden crear espacios estables, tener acceso a los recursos urbanos y contribuir a la fábrica social y desarrollo de la ciudad. Por definición, la vivienda precaria es entendida no simplemente como se experimenta a la escala de casa, sino mejor como un espacio primario a partir del cual el derecho a la ciudad es disputado, retado y denegado. Examino el modo como la injusticia y la precariedad urbanas son producidas y experimentadas rutinariamente y arguyo que sin acceso a una vivienda estable y sin cambio social, no existe derecho a la ciudad. *Palabras clave: casa, América Latina, vivienda precaria, derecho a la ciudad, geografía urbana.*

La ciudad es de los vecinos! La ciudad es de los vecinos! No del gobierno, ni de los inmobiliarios.

Somos nosotros que tenemos que definir lo que queremos!

[The city belongs to the residents! The city belongs to the residents!

Not to the government, nor to the real estate developers. We are the ones who have to decide what we want!]

For many urban residents, Buenos Aires, Argentina, offers economic opportunities and a relatively higher quality of life than many other cities throughout the region, including access to health care, education, various social programs, and economic support. Renting an apartment, however, is practically impossible without the proper economic or social capital, even with some source of stable income. Real estate prices and rents have risen steadily since 2001, making it increasingly difficult for even the middle and lower classes to rent or buy an apartment or house in the city (Baer 2008; Ciccolella and Baer 2008; Ostuni and Van Gelder 2015; Rodríguez, Rodríguez, and Zapata 2015; Van Gelder, Cravino, and Ostuni 2016). Furthermore, urban migrants and residents must furnish landlords with a garantía,[1] paystubs, and additional costly up-front fees, to which the poor or even middle-income inhabitants do not easily have access. These requirements, together with policies and practices that have historically neglected the issue of affordable housing in the city, as well as more recent neoliberal urban initiatives, routinely exclude poor urban residents from the city's formal housing market (Oszlak 1991; Pastrana et al. 1995; Centner 2012; Rodríguez, Rodríguez, and Zapata 2015; Janoschka and Sequera 2016; Van Gelder, Cravino, and Ostuni 2016).

Under these conditions, poor urban residents of Buenos Aires are forced to live with others, rent bedrooms in boarding houses, or live in one of many villas miserias (informal settlements) on the borders of the city. For those who look for housing outside the city but who continue to work and study in Buenos Aires, this can mean an expensive daily commute of two or three hours or more on multiple modes of transportation. Others find housing in the city in casas tomadas—empty buildings or informal hotels—that have been taken over by individuals who then rent out or sell rooms to (often unsuspecting) and desperate families.

Many of these informal housing options occupy a peculiar space in the city. Often located in the center of the city, residents are able to live close to jobs, education, health care, and other services and resources, albeit in highly precarious conditions in which they have few, if any, protections or security of tenure. Presented with a notice of eviction, residents might wait months or even years, amidst crowded and deteriorating conditions, until they are finally evicted. Once evicted, most families enter into another informal arrangement, continuing a cycle of precarious housing in the city.

This article examines precarious housing and notions of house and home through the optic of the right to the city and the struggle for social and spatial justice (Lefebvre 1991; Purcell 2002; Fenster 2005; Harvey 2008; Marcuse 2009; Soja 2009; Aalbers and Gibb 2014; Rolnik 2014). Through a review of the multiple meanings and uses of house and home and their intrinsic social, political, and spatial relationship with the city and its resources and services, I argue that a conversation about the right to the city must begin at the scales of the house and home. In this research, both *house* and *home* are conceptualized as central spaces from which individuals employ place-making strategies to appropriate urban spaces, access urban resources, and participate in urban constructions and governance (Turner 1968; Coolen 2006; Simone 2008; Brickell 2012; Rolnik 2014), what Lefebvre broadly referred to as practices of *autogestion* (in Purcell 2013; see also Rolnik 2014). Although I employ the concepts of house and home together, my intention is not to conflate them but rather to acknowledge their multiple and intersecting significance as a place, site, and location; through the material and practical meanings and uses of house and home; and also as a place of complex emotions, cultural meanings, and identity, all within the context of the right to the city. In their seminal work *Home*, Blunt and Dowling (2006) employed a critical geography approach through an analysis of home as both material and imaginative; as tied to power and identity; and as multiscalar (see also Brickell 2012). These approaches emphasize the multiple meanings and uses of house and home as not exclusive of one another but rather as fluid and overlapping.

Precarious homes and housing—through the material threat of eviction, displacement, and lack of affordable alternatives in the city, as well as the emotional toll that precarity takes on families, individuals, and their future prospects—affect residents' abilities to claim urban spaces, limiting and disabling opportunities for social justice and the right to the city (Brickell 2012; Desmond 2016). These processes are effectively part of broader urban processes and directly bound to urban residents' (in)abilities to prosper and partake of and contribute to collective constructions and decision making (Brickell 2012; Vasudevan 2015; Fernández-Arrigoitia 2017). Without access to stable, affordable housing, from which urban residents are able to engage in long-term homemaking practices, access urban resources, and actively and publicly engage in urban life, there is no right to the city.

Employing an empirical analysis of poor urban residents struggling to remain in Buenos Aires amidst highly precarious housing and home conditions and the threat of eviction, this research offers insight into the spatial, material, social, and political uses and meanings of housing and home within the broader context of the struggle for the right to the city. Specifically, this article contemplates the following research questions:

1. What is the place of housing and home in the struggle for the right to the city?
2. What does the struggle for the right to the city entail in the daily lives of some of the city's most vulnerable inhabitants?

The objective of this approach is to frame and examine how those who have few rights to the city struggle to secure a place for themselves and their families amidst great instability. Paraphrasing Marcuse (2009), it is the right to the city of those who do not have that right with which this research is concerned.

Methodological Considerations

This article is based on field research conducted in 2009 and 2017 in Buenos Aires, Argentina. In 2009, I spent ten months working with the housing organization *Coordinador de Inquilinos de Buenos Aires*[2] (CIBA), a social organization that fights for the rights of residents waiting for eviction and struggling to remain in the city. Employing ethnographic methods such as participant observation, informal and semiformal interviews, and extensive field notes, I documented the daily conditions and residents' practices inside informal hotels and casas tomadas, their relationships with and demands of CIBA, and their contact with the city government. More recently in 2017, I spent five weeks in Buenos Aires, again working with CIBA, visiting casas tomadas and observing how the housing situation had changed since 2009. Drawing on findings from my fieldwork and on the literature on home, this article argues that despite important research and theoretical discussions that continue to advance a right to the city framework, the significance of housing and home as a material and imaginative place from which urban residents can demand and participate in the right to the city has not been adequately considered. At the same time, I argue that as access to housing becomes increasingly difficult in cities around the world, it is

necessary to bridge complex theoretical considerations with more empirical analyses of routine resistance and struggle for the right to the city that occur in often paradoxical, contradictory, and contentious ways.

The Right to Housing, Home, and the City

There is a rich body of work on the right to the city that encompasses a broad series of concerns from urban governance and citizenship to social movements and theoretical explorations of the very meaning of rights (Isin 2000; Purcell 2002, 2013; Harvey 2008; Brenner, Marcuse, and Mayer 2009; Attoh 2011). At the heart of much of this research is a reaction to the increasing effects of the neoliberal project on the global urban order (Purcell 2002; Harvey 2008; Marcuse 2009) and an appeal for "the urgent political priority of constructing cities [and societies] that correspond to human social needs rather than to the capitalist imperative of profit-making" (Brenner, Marcuse, and Mayer 2009, 2). In this sense, for many scholars and activists, the right to the city is a morally urgent claim, one that moves beyond the legal realm and that promotes a social order that incorporates the individual and material needs of a broader collective whole (Purcell 2002, 2013; Harvey 2008; Brenner, Marcuse, and Mayer 2009).

The right to the city literature is grounded in the notion that to inhabit the city is explicitly tied to one's ability to actively determine how they use it in ways that allow them to contribute to their spatial and social mobility, stability, quality of life and to the city itself (Lefebvre 1991; Purcell 2002; Fenster 2005; Harvey 2008; Brenner, Marcuse, and Mayer 2009; Rolnik 2014). The ability to access housing and to construct stable home spaces through homemaking practices and relationships provides a reliable and consistent entrance or right to the city and its many resources. As such, without the right to housing or the right to engage in stable homemaking practices inside the city, other forms of social and spatial mobility are severely restricted or nonexistent. Simply put, the right to the city begins at the home.

Notwithstanding, much of the academic literature on the right to the city has paid scant attention to the importance of housing and home as central to stable, long-term, placemaking practices in urban areas (Fenster 2005; Aalbers and Gibb 2014; Kant and Ronald 2014; Rolnik 2014). This might partly be due to the stigma that continues to surround research on housing and particularly on the home. Duncan and Lambert's (2004) comment, "There still appears to be a lingering sense that home … is trivial compared with the public worlds of business, politics or even public pleasures" (382), seems relevant today. Similarly, many scholars continue to view home as relegated to a private, personal sphere, reinforcing a false binary that contrasts the collective and public nature of social life and city living with home as a separate, private realm (Mallett 2004; Blunt and Dowling 2006). Marxist scholars and approaches have also largely disregarded the multiple social and political meanings of house and home, limiting their analyses and understanding of these spaces as simply for the social reproduction of labor (Blunt and Dowling 2006). Scholars of home, however, have argued against this binary and instead highlight the multiscale and fluid meanings and uses of home and the multiple ways in which home lives and experiences are bound to broader collective, public, and political scales and processes (Massey 1994; Wardhaugh 1999; Mallett 2004; Fenster 2005; Blunt and Dowling 2006; Brickell 2012; Aalbers and Gibb 2014; Rolnik 2014; Desmond 2016; Brickell, Fernández-Arrigoitia, and Vasudevan 2017). These scholars have argued and demonstrated that home and its multiple meanings and uses are much more complex and multiscalar than is often recognized. Among these approaches, feminist and critical scholarship have both challenged and explored the tension between definitions that position home as an intimate, safe, and private space, and more dynamic frameworks that imagine home as more ambiguous, complex, and interconnected with other spaces and processes (Turner 1968; Fenster 2005; Massey 2005; Blunt and Dowling 2006; Coolen 2006; Brickell 2012; Rolnik 2014; Brickell, Fernández-Arrigoitia, and Vasudevan 2017). They offer a framework for analysis of the ways in which the poor and other marginalized urban actors both experience and employ urban spaces in often dynamic, creative, and innovative ways to negotiate their presence, access basic resources and services in the city, and make demands (Simone 2004; Davis 2006; Roy 2011; Vasudevan 2015). Similarly, they address the ways in which home spaces and experiences both represent and influence the construction of individual, political, and social identities that stretch beyond the home (Massey 1994).

Research on home, in the context of the right to the city, can offer a broad conceptualization of the multiple meanings and practices of home spaces and the many ways in which individuals and communities are able to draw from house and home to resist gentrification, claim urban spaces, access resources, and contribute to the urban landscape. In their most basic sense, house and home provide a location and a place from which to access urban resources and services and to appropriate the city in individual and collective ways. They also provide a context from which individuals and communities are able to organize and imagine their identity and positionality in relationship to the city. For many urban residents, proximity to the city center offers them more possibilities and opportunities in the form of social, spatial, and material networks (Turner 1968; Fernández-Arrigoitia 2017). As a result, poor residents will sacrifice certain comforts and securities to remain close to resources and other opportunities (Turner 1968; Simone 2008). In Turner's (1968) discussion on the functionality of urban housing, he emphasized the importance of location, explaining that for residents to "maximize their opportunities," they "must live near the source of those jobs where subsistence goods and available housing are cheap and transport costs and times are negligible" (356).

Many of the residents of informal hotels and casas tomadas who I interviewed said that they stay in the city because of the numerous resources and benefits available to them. The political, economic, and geographic distinction between the formally titled Autonomous City of Buenos Aires (CABA) and the areas and neighborhoods that make up the twenty-four *partidos* (counties) that surround the city is stark. For those living in the wealthy counties and towns that are well connected through public transportation or direct routes, there is little difference. For the vast majority, however, living outside of Buenos Aires means poorer access to resources and services, traveling long distances, and high transportation costs. Residents also explained that services such as health care clinics, education, and after-school programs were closer and much better quality than outside of the city. The women I interviewed, whose work entails cleaning houses and businesses, told me that they were able to take on multiple jobs in the city, allowing them to make more money. Living in the city also meant that parents had more control over their time with and commitments to their children. Finally, it is much safer to live in the city, in part because of the population density, use, and organization of the streets and neighborhoods.

A focus on house and home as central to any right to the city project must also acknowledge the significance of gender and other identities in distinguishing the different ways in which people and collectives access, use, and draw meaning from the city (Fenster 2005). House and home as primary places from which to access the right to the city can highlight how particular spaces produce gender identities and roles that are performed and experienced differently at multiple scales throughout the city. For example, in my research, everyday uses and claims to the city, as well as access to housing and the struggle to remain in Buenos Aires, were largely undertaken by women, whereas men often remained on the margins both inside casas tomadas and in the more public struggle for the city.

Home, as an emotional, imagined, and idealized place that is created and maintained through intimate, familial, and collective placemaking practices, responds to a personal as well as political experience from where identity, meaning, and belonging are produced (Mallet 2004; Fenster 2005; Blunt and Dowling 2006; hooks 2009; Brickell 2012). Within this framework, stable home spaces and experiences can empower urban residents to both use and appropriate the city and to make claims to resources and services through a sense of belonging and entitlement (Lefebvre 1991; Rolnik 2014). Residents in stable dwellings have a degree of power and agency that is experienced both inside and outside the home. Residents without secure home dwellings, however, often live in a marginalized state of continuous waiting and uncertainty, experiencing power, time, and space distinctly from those living in stable conditions (Bourdieu 2000; Simone 2004; Wingate-Lewinson, Hopps, and Reeves 2010; Auyero 2012; Harms 2013).

Yet, as Wingate-Lewinson, Hopps, and Reeves (2010) suggested, even in precarious living arrangements, people engage in homemaking and placemaking practices for themselves and their families. Citing Wright (1997), they claimed, "As active agents it is clear that poor people, like all people attempt to reassert their 'place' in society, to establish a 'homeplace' in the midst of deprivation, humiliation, and degradation" (5). Thus, despite the hardships that many poor and marginalized communities endure, they are continually working to create opportunities, access resources, and carve out spaces to build some form of sustainable livelihood for themselves and their families (Simone 2004, 2008; Desmond 2012; Procupez 2015; Vasudevan 2015). Although precarious home spaces still offer residents the ability to construct alternative spaces, networks, and opportunities for themselves and their families in the city, they are often unable to resolve the precarity that characterizes their conditions. When housing and

the ability to create a stable home are denied or limited, residents' abilities to actively participate in the construction and development of the city are also obstructed. In this context, urban residents' access to housing and home and their use of the city is not guaranteed but rather determined by (often precarious) relationships and social networks grounded in power, structural inequalities, and, to some degree, sheer luck (see Ribot and Peluso 2003).

Gentrification and Displacement in Latin America

In stark contrast to Lefebvre's conceptualization of the right to the city, the ability to access stable housing, to engage in homemaking practices, and in the right to the city have become increasingly challenging, if not impossible. Rising rent costs, lack of affordable housing options, threat of eviction, and displacement from the city, fueled in part by neoliberal political and economic structures that continue to privilege investments, wealth, and power over the needs and demands of those who live, dwell, and use the city, all characterize the reality of many urban dwellers around the world (Brickell, Fernández-Arrigoitia, and Vasudevan 2017). Although a global phenomenon, urban housing crises and the ways in which they are experienced and resisted are embedded within the local and regional contexts in which they take place (Kadi and Ronald 2014; Brickell, Fernández-Arrigoitia, and Vasudevan 2017). Research on urban development, gentrification, and displacement emphasizes the multiple historical, structural, and contingent forces and scales that lead to and produce particular conditions and experiences within specific geographical contexts (Betancur 2014; Kadi and Ronald 2014; Brickell, Fernández-Arrigoitia, and Vasudevan 2017).

These processes of urban revitalization and their impacts have not been lost on Latin American cities. Betancur (2014) argued that like in other areas of the world, these urban processes were originally promoted by international institutions pushing neoliberal restructuring and urban competitiveness in the Global South. In many Latin American cities, this has led to the dismantling of urban neighborhoods and the dispossession of poor and working-class families to advance state-supported redevelopment by private interests with projects geared toward upper-class consumers and international tourists (Janoschka and Sequera 2016; Rodríguez and Di Virgilio 2016). In

Buenos Aires, urban renewal through heritage tourism has seen the transformation of key neighborhoods that for decades were havens for poor and working-class families (Janoschka and Sequera 2016). This has meant not only the displacement of poor and working-class families from these neighborhoods but also their further marginalization and dispossession of the city through revanchist rhetoric of "taking back the city" and the material "loss of *use of value* of land, environmental key resources and access to mobility and public services" (López-Morales 2015; see also Rodríguez and DiVirgilio 2016). For a region in which an estimated 20 to 25 percent of the population live in informal settlements (Fernandes 2011, cited in Janoschka and Sequera 2016), urban renewal and displacement are only exacerbating the spatial and socioeconomic inequality and segregation that have historically characterized the region.

The recent literature on gentrification and urban displacement in Latin America has highlighted not only the different actors and processes involved but also the multiple ways in which different communities challenge and resist them (Betancur 2014; López-Morales 2015; González 2016; Janoschka and Sequera 2016). Employing a comparative perspective, these scholars identify the different ways in which urban processes of gentrification, displacement, and resistance occur in a regional context. Through a multicity comparative perspective, Janoschka and Sequera (2016) identified distinctive processes of gentrification that have been carried out in some of the most prominent cities of Latin America, specifically Buenos Aires, Rio de Janeiro, Mexico City, and Santiago. González (2016) argued that in Latin America, processes of gentrification take a "more violent and aggressive form" but that they are accompanied by more determined, creative, and radical forms of resistance and contestation than in the Global North (1247). González was not suggesting that these processes are simply more confrontational; rather, she identified the multiple and nuanced "speeds, rhythms and intensities" in which processes of gentrification and resistance are experienced and carried out in Latin America (Delgadillo, cited in González 2016, 1248). In Buenos Aires, gentrification and urban development have exacerbated the housing crisis for the urban poor in the city, who have historically been marginalized from the formal housing market (Centner 2012). More recently, neighborhoods like La Boca, San Telmo, and Almagro, where some of city's poor found refuge in the past, are now touted as cultural centers geared toward international tourists and students (Centner 2012; Janoschka and Sequera 2016). This has led to an increase in evictions and displacement of some of the city's most vulnerable residents.

As urban development and gentrification lead to an increase in property taxes and rent prices and eviction and urban displacement become more common, discussions about the right to the city must begin with the right to stable housing and to create stable home spaces. The following case study discusses how poor residents who stay in Buenos Aires by living in casas tomadas experience housing precarity and struggle to make a home space for themselves and their family. Through this study, I focus on the importance of house and home in the larger struggle for social justice and the city.

The Struggle to Make a Home

The residents included in my research were all living in casas tomadas, large informal hotels where entire families live in a small bedroom and share kitchens and bathrooms with many other families, as they wait to be evicted. With few housing options and the possibility of receiving a city-sponsored temporary housing subsidy of a few hundred dollars a month for up to ten months, families remain inside casas tomadas for weeks and years, waiting for the day of eviction.

The informal, crowded character of casas tomadas, along with the imminent threat of eviction, make them highly unstable and precarious spaces for poor residents trying to make a home for themselves and their family in the city. At the same time, residents live in these spaces because they allow them to remain in Buenos Aires, close to economic opportunities, and other resources and services. Furthermore, as one woman explained when I asked her why she lived in a casa tomada:

> I live here because it is close to my job and to my son's school. [Also] because it is cheaper for me and because in other places they told me they wouldn't rent to families with kids. It doesn't matter if they are younger or older, they don't want them. They only want single people. Single.

Many of the residents included in my research were women migrants who had originally traveled alone from Peru, Paraguay, or the northern provinces of Argentina to Buenos Aires to find work and were later followed by other family members. Arriving in the city alone, migrants are often able to find employment and to rent a room in a hotel with relative ease. Reunited with loved ones, they are often denied access to stable housing and the ability to create a home space, told by hotel owners that they will not rent to families with children. The women I interviewed discussed their early experiences of migrating to Buenos Aires through a narrative that highlighted their pride in making enough money to send to their family and their ability to construct independent lives and access opportunities unavailable to them before they migrated. Natalia, who had arrived to be with her older sister and to study at the university, explained, "I am proud of my work, and what I have, and to be doing the things I am doing in the short time I have been here."

Natalia was young and single at the time I interviewed her. Other women who had started families in Buenos Aires or who had brought their children and spouses from abroad described a somewhat different scenario, highlighting the challenges of finding a place to live and making a home for their families. All of a sudden, with children and family, they are barred from the few housing options available, forcing them to live in highly precarious conditions and highlighting how through the "'private' their right to use [the city] is denied" (Fenster, 2005, 220). In this context, poor urban residents are refused access to housing and to the city because of their identity as mothers, fathers, partners, caregivers, and children, at precisely the moment when they are beginning to settle down and make a home in the city. As one seven-year-old boy cried to his mother after a long day of looking for a place to live, "Oh, Mom! Just give me away because I am the problem. It's my fault that no one wants to rent you an apartment!"

For those families who are repeatedly denied a room by hotel managers and owners once they reveal that they have children, access to casas tomadas through the purchase, renting, or occupying of a room allows them to remain inside Buenos Aires's city limits. Staying in the city is important for poor families for a number of reasons, as Sonia's account highlights here:

> There is also a lot less security [outside the city]. Here I can walk around, but in Lomas de Zamora[3] I can't. A lot of things happen, robberies, gunshots, kids smoking. ... For me the city is safer. Even living in [a casa tomada], I feel safer here. I feel comfortable here, because the hospital is close, the shopping mall is close, [my kids] can go outside in the street and play and run and it isn't dangerous.

Sonia lived in a casa tomada in the city during the week so that her children could go to school and be closer to some of the many low-cost or free services and resources the city offered. In the neighborhood, there were places where she and her children could go

to sculpt, along with a workshop for games and gymnastics.

> My daughter goes to the library, I go to workshops for women on nutrition and other things. I feel good! I am finishing a knitting class. They teach you how to knit while the kids play and learn things.

Many of the residents I interviewed took advantage of the resources and services that existed around them, including Abasto Shopping, the large, upscale shopping mall that has a children's interactive museum on the top floor.[4]

Sonia was also able to save time and money. On the days when she traveled to the city from her home in Lomas de Zamora, she would spend on average about 10 pesos a day, or about 300 pesos a month for the three of them to travel. Using public transportation, she would travel about forty minutes to drop her son off at school and then take the other one to day care. Afterward, she would buy another ticket to return to Lomas de Zamora, arriving around 11:00 a.m., returning to the city around 3:30 p.m. to pick up her kids and take them back home. Staying in the city was a common concern among all residents I interviewed. Like Sonia, they all worked, had families, and sent their children to schools and clinics and afterschool programs in the city. They explained that the resources available and the conditions in general were much better inside the city limits of Buenos Aires. Despite the precarious housing situation in which all of the residents found themselves, their lives were in the city. Once they were evicted, moving out of the city often meant that they were separated from the networks, jobs, opportunities, and resources that made up their daily lives and gave them some sense of security and stability. This type of routine hypermobility keeps urban residents on the move in their everyday lives, limiting their ability to create stable home spaces and community.

Although the city provides many opportunities and resources, the inability to access stable housing creates a highly precarious and disruptive situation of long-term marginalization and uncertainty for many urban residents. Living in highly precarious conditions and waiting for eviction, poor urban residents are denied the ability to fully create stable home spaces that they and their family members can properly benefit from that access and also contribute to the city. Nevertheless, residents of casas tomadas work to create a sense of home for their children and families despite precarious, crowded, and temporary conditions.

The struggle to create and provide a stable home space is closely associated with women's identities, their abilities to care for their families, their daily responsibilities, and the routine demands of living in such precarious quarters. Inside casas tomadas, women assume most of the tasks of taking care of their families through cooking, cleaning, and caring for children. They must also spend time and energy resolving social and spatial conflicts and are constantly negotiating daily use of spaces, resources, and time with many other residents.

Women also represent the public face of the struggle for housing in Buenos Aires. For CIBA and other housing organizations, the visible presence of women and children during street protests and resistance to evictions offers a powerful message that is used to publicly illustrate the violence and injustice of the housing crisis in the city. At the same time, once families are evicted, it is usually women who must sit and wait, sometimes for hours, at the welfare office each month to receive the city government–sponsored housing subsidy (Auyero 2012). In this sense, gender roles and identities, often associated with, produced, and reproduced in home spaces, are not confined to the home but rather are performed and reinforced in multiple spaces and places, by residents, activists, and the state alike in the struggle for housing and the right to the city.

Inside casas tomadas there is always an uneasy tension that is part of residents' daily encounters, demands, and negotiations. Residents know that they must rely on and work with other residents to get things done, yet precarious conditions make this extremely difficult and outcomes are always uncertain. As a result, along with constant and long-term feelings of fatigue, stress, and fear about the future, residents both resent and often mistrust one another, a condition that makes it difficult to collectively organize inside casas tomadas and in the broader context of the city and the struggle for the right to the city.

A focus on the precarious conditions of poor residents living in casas tomadas in Buenos Aires, Argentina, illustrates the importance of housing and home spaces in the context of urban residents' access to and use of the city, its resources, and opportunities. Despite conditions of extreme precarity, urban residents will continue to struggle to make claims to the city and to challenge neoliberal notions of who deserves the city and who does not, but this is not enough. Without guaranteed access to stable housing that offers opportunities for homemaking, community, and collective action, there is no social justice, nor is there a chance for any right to the city.

Conclusion

Urban residents' struggle for the right to the city begins at the house and home. Opportunities and the ability to participate and contribute to a broader public and collective notion of the city begins by being able to live in the city. Despite the large body of literature that continues to promote the right to the city, we are witnessing a very different trend on the ground, in which cities are increasingly becoming exclusive spaces from which the urban poor, communities of color, and other groups and individuals are being marginalized and displaced. The term housing crisis is now used to refer to processes happening in most large cities around the world, and the few real initiatives in place to alleviate the hardships rarely address the broader underlying causes behind these crises. Focus on house and home provides a repositioning of the literature on the right to the city at a time when affordable housing crises and the struggle of many urban residents to stay in the city have become central issues and concerns in many of the debates now happening in cities around the world.

Among the literature on the right to the city, housing and home have largely remained outside the purview of many urban justice and right to the city scholars. Yet, as I have argued throughout this article, housing and home must take a more central role in the scholarship on the right to the city, not simply because of their material significance as space and place from which urban residents can access resources but also because of home's much broader emotional, imaginative, multiscalar, and collective significance as a foundation from where networks, individual and collective identities, and opportunities originate and are reinforced. Like the right to the city, a discussion about the right to house and home is part of a call for an end to capitalist forms of wealth accumulation exacerbated by the global economy and neoliberalism. It also provides a much more expansive conversation about what else the right to the city can entail and what has historically been left out of the literature. Centering house and home within the broader scholarship on the right to the city can provide more radical, explicit, multiscalar, and fluid frameworks and approaches that acknowledge the multiple scales, spaces, places, identities, and encounters in which urban residents engage and that are representative of the routine and everyday ways that urban residents make claims to the city.

ORCID

Solange Muñoz ⓘ http://orcid.org/0000-0001-5353-184X

Notes

1. The *garantía* consists of owning (or knowing someone who owns) property in Buenos Aires. By requiring a garantía, the owner can draw from the value of the property acting as the garantía. Without one, an individual or family cannot rent an apartment inside the city. In addition, other economic requirements, discriminatory attitudes, and discriminatory practices exclude the poor from accessing any options inside the formal housing market.
2. Translation: Coordinator of Inhabitants of Buenos Aires.
3. Lomas de Zamora is a county on the southern border of the City of Buenos Aires.
4. Abasto Shopping was part of a larger development project funded by George Soros and geared toward the renovation and development of the Abasto neighborhood located in the center of the city.

References

Aalbers, M. B., and K. Gibb. 2014. Housing and the right to the city: Introduction to the special issue. *International Journal of Housing Policy* 14 (3):207–13.

Attoh, K. 2011. What kind of right is the right to the city? *Progress in Human Geography* 35 (5):669–85.

Auyero, J. 2012. *Patients of the state: The politics of waiting in Argentina.* Durham, NC: Duke University Press.

Baer, L. 2008. Precio del suelo, actividad inmobiliaria y acceso a la vivienda: El caso de la ciudad de Buenos Aires luego de la crisis de 2001/2002 [Land prices, real estate activity and access to housing: The case of Buenos Aires after the 2001/2002 crisis]. *Ciudad y Territorio Estudios Territoriales* XL (156):345–59.

Betancur, J. J. 2014. Gentrification in Latin America: Overview and critical analysis. *Urban Studies Research.* Accessed December 1, 2017. https://doi.org/10.1155/2014/986961

Blunt, A., and R. Dowling. 2006. *Home.* London and New York: Routledge.

Bourdieu, P. 2000. *Pascalian meditations.* Cambridge, UK: Polity.

Brenner, N., P. Marcuse, and M. Mayer, eds. 2009. *Cities for people, not for profit: Critical urban theory and the right to the city.* London and New York: Routledge.

Brickell, K. 2012. "Mapping" and "doing" critical geographies of home. *Progress in Human Geography* 36 (2):225–44.

Brickell, K., M. Fernandez Arrigoitia, and A. Vasudevan, eds. 2017. *Geographies of forced eviction: Dispossession, violence, resistance.* London: Palgrave Macmillan.

Centner, R. 2012. Moving away, moving onward: Displacement pressures and divergent neighborhood politics in Buenos Aires. *Environment and Planning A* 44 (11):2555–73.

Ciccolella, P., and L. Baer. 2008. Buenos Aires tras la crisis: ¿Hacia una metrópolis más integradora o más excluyente? [Buenos Aires after the crisis: Is it a more integrated or exclusive metropolis?] *Ciudad y Territorio Estudios Territoriales* XL (158):641–60.

Coolen, H. 2006. The meaning of dwellings: An ecological perspective. *Housing, Theory and Society* 23 (4):185–201.

Davis, M. 2006. *Planet of slums.* London: Verso.

Delgadillo, V. 2016. Selective modernization of Mexico City and its historic center. Gentrification without displacement? *Urban Geography* 37 (8):1154–74.

Desmond, M. 2012. Disposable ties and the urban poor. *American Journal of Sociology* 117 (5):1295–1335.

———. 2016. *Evicted: Poverty and profit in the American city.* New York: Crown.

Duncan, J., and D. Lambert. 2004. *Landscapes of home: A companion to cultural geography,* ed. J. Duncan, N. Johnson, and R. Schein, 382–403. London: Blackwell.

Fenster, T. 2005. The right to the gendered city: Different formations of belonging in everyday life. *Journal of Gender Studies* 14 (3):217–31.

Fernandes, E. 2011. *Regularization of informal settlements in Latin America.* Cambridge, MA: Lincoln Institute of Land Policy.

Fernández-Arrigoitia, M. 2017. Unsettling resettlements: Community, belonging and livelihood in Rio de Janeiro's Minha Casa Minha Vida. In *Geographies of forced eviction: Dispossession, violence, resistance,* ed. K. Brickell, M. Fernandez Arrigoitia, and A. Vasudevan, 71–96. London: Palgrave Macmillan.

González, S. 2016. Looking comparatively at displacement and resistance to gentrification in Latin American cities. *Urban Geography* 37 (8):1245–52.

Harms, E. 2013. Eviction time in the New Saigon: Temporalities of displacement in the rubble of development. *Cultural Anthropology* 28 (2):344–68.

Harvey, D. 2008. The right to the city. *New Left Review* 53 (September–October):23–40.

hooks, b. 2009. *Belonging: A culture of place.* London and New York: Routledge.

Isin, E. 2000. Introduction: Democracy, citizenship and the city. In *Democracy, citizenship and the global city,* ed. E. Isin, 1–21. London and New York: Routledge.

Janoschka, M., and J. Sequera. 2016. Gentrification in Latin America: Addressing the politics and geographies of displacement. *Urban Geography* 37 (8):1175–94.

Kadi, J., and R. Ronald. 2014. Market based housing reforms and the "right to the city": The variegated experiences of New York, Amsterdam, and Tokyo. *International Journal of Housing Policy* 14 (3):268–92.

Lefebvre, H. 1991. *The production of space.* Malden, MA: Blackwell.

López-Morales, E. 2015. Gentrification in the Global South. *City* 19 (4):564–73.

Mallet, S. 2004. Understanding home: A critical review of the literature. *The Sociological Review* 52 (1):62–89.

Marcuse, P. 2009. From critical urban theory to the right to the city. *City* 13 (2–3):185–87.

Massey, D. 1994. *Space, place and gender.* Minneapolis: University of Minnesota Press.

——— 2005. *For space.* Thousand Oaks, CA: Sage.

Ostuni, F., and J. L. Van Gelder. 2015. Housing policy in the city of Buenos Aires. In *Housing and belonging in Latin America,* ed. C. Klaufus and A. Ouweneel, 149–63. New York: Berghahn.

Oszlak, O. 1991. *Merecer la Ciudad: Los Pobres y el Derecho al Espacio Urbano* [To deserve the city: The poor and the right to urban space]. Buenos Aires: Colección CEDES—Humanitas.

Pastrana, E., M. Bellardi, S. Agostinis, and R. Gazzoli. 1995. Vivir en un cuarto: Inquilinatos y hoteles en el Buenos Aires actual [To live in a room: Tenement houses and hotels in present-day Buenos Aires]. *Revista Medio Ambiente y Urbanización* 14 (50–51):1–143.

Procupez, V. 2015. The politics of housing emergency in Buenos Aires. *Current Anthropology* 56 (11):55–65.

Purcell, M. 2002. Excavating Lefebvre: The right to the city and its urban politics of the inhabitant. *Geojournal* 58:99–108.

———. 2013. The right to the city: The struggle for democracy in the urban public realm. *Policy & Politics* 41 (3):311–27.

Ribot, J., and N. Peluso. 2003. A theory of access. *Rural Sociology* 68 (2):153–81.

Rodríguez, M. C., and M. M. DiVirgilio. 2016. A city for all? Public policy and resistance to gentrification in the southern neighborhoods of Buenos Aires. *Urban Geography* 37 (8):1215–34.

Rodríguez, M. C., M. Di Virgilio, S. Arqueros Mejica, M. F. Rodríguez, and M. Zapata. 2015. Contradiciendo la Constitución de la Ciudad: Un análisis de los programas habitaciónales en la ciudad de Buenos Aires en el período 2003–2013 [Contradicting the city's constitution: An analysis of the housing programs in Buenos Aires from 2003–2013]. *Instituto de Investigaciones Gino Germani: Documentos de Trabajo* 72:1–55.

Rodríguez, M. C., M. F. Rodríguez, and M. C. Zapata. 2015. La casa propia, un fenómeno en extinción. La "inquilinización" en la ciudad de Buenos Aires [One's own house, a phenomenon in extinction. The "rentification" of the City of Buenos Aires]. *Cuadernos de Vivienda y Urbanismo* 8 (15):68–85.

Rolnik, R. 2014. Afterword: Place, inhabitance and citizenship. Right to housing and the right to the city in contemporary urban world. *International Journal of Housing Policy* 14 (3):293–300.

Roy, A. 2011. Slumdog cities: Rethinking subaltern urbanism. *International Journal of Urban and Regional Research* 35 (2):223–38.

Simone, A. 2004. *For the city yet to come: Changing African life in four cities.* Durham, NC: Duke University Press.

———. 2008. Remaking urban socialities: The intersection of the virtual and the vulnerable in inner-city Johannesburg. In *Gendering urban space in the Middle East, South Asia and Africa,* ed. M. Reiker and A. K. Asdar, 135–68. New York: Palgrave Macmillan.

Soja, E. 2009. The city and spatial justice. *Justice Spatiale/ Spatial Justice* 1:1–5.

Turner, J. C. 1968. Housing priorities, settlement patterns, and urban development in modernizing countries. *Journal of the American Institute of Planners* 34 (6):354–63.

Van Gelder, J. L., M. C. Cravino, and F. Ostuni. 2016. Housing informality in Buenos Aires: Past, present and future? *Urban Studies* 53 (9):1958–75.

Vasudevan, A. 2015. The makeshift city: Towards a global geography of squatting. *Progress in Human Geography* 39 (3):338–59.

Wardhaugh, J. 1999. The unaccomodated woman: Home, homelessness and identity. *The Editorial Board of the Sociological Review* 47 (1):91–109.

Wingate-Lewinson, T., J. G. Hopps, and P. Reeves. 2010. Liminal living at an extended stay hotel: Feeling "stuck" in a housing solution. *Journal of Sociology and Social Welfare* 37 (2):9–34.

Wright, T. 1997. *Out of place: Homeless mobilizations, subcities, and contested landscapes.* New York: SUNY Press.

SOLANGE MUÑOZ is an Assistant Professor in the Department of Geography at the University of Tennessee, Knoxville, TN 37996-0925. E-mail: imunoz@utk.edu. Her research interests include spatial inequality in urban centers with emphasis on housing and home precarity, evictions, and urban displacement.

7 The Anti-Eviction Mapping Project

Counter Mapping and Oral History toward Bay Area Housing Justice

Manissa M. Maharawal and Erin McElroy

The Anti-Eviction Mapping Project is a data visualization, data analysis, and oral history collective documenting gentrification and resistance in the San Francisco Bay Area. In this article, we discuss the history and methodology of our narrative mapmaking, situating our work in the tradition of critical geography, critical race studies, as well as feminist and decolonial science studies. Aligned with activist work that is fighting for a future beyond the current tech-dominated political economy of speculative real estate and venture capital, our project maps sites of resistance, while remembering spaces lost and struggled for. In this article, we highlight the connections between countermapping, oral history, and housing justice work. *Key Words: countermapping, eviction gentrification, oral history, San Francisco, social justice.*

反对驱逐的製图计画，是记录三藩市湾区的贵族化及反抗的资料可视化、资料分析和口述历史的集体。我们于本文中探讨自身的叙事性製图历史及方法，并将我们的研究置于批判地理学、批判种族研究和女权主义与去殖民科学研究的传统之中。与社会运动者奋力追求超越当前受科技支配的房地产投机与创投资本的政治经济之未来的努力一致，我们的计画绘製反抗的场域，同时追忆失落与奋斗过的空间。我们于本文中强调反抗製图、口述历史与居住正义工作之间的关联性。 关键词： 反抗製图, 驱逐式贵族化, 口述历史, 三藩市, 社会正义。

El Proyecto Cartográfico contra el Desalojo es un colectivo de visualización de datos, análisis de datos e historia oral que documenta el aburguesamiento y la resistencia en el Área de la Bahía de San Francisco. En este artículo discutimos la historia y la metodología de nuestra narrativa de la elaboración de mapas, situando nuestro trabajo en la tradición de la geografía crítica, los estudios críticos de raza, así como los estudios de ciencia feminista y descolonial. En línea con el activismo que propende por un futuro alejado de la actual economía política de especulación inmobiliaria y capital de riesgo, dominada por la técnica, nuestro proyecto cartografía los sitios de resistencia, mientras recuerda los espacios perdidos que fueron objeto de disputa. En este artículo, destacamos las conexiones que existen entre contra-mapeo, historia oral y el trabajo sobre justicia en vivienda. *Palabras clave: contra-mapeo, aburguesamiento por desalojo, historia oral, San Francisco, justicia social.*

In 2013, at the height of the San Francisco Bay Area's most recent eviction crisis, the Anti-Eviction Mapping Project (AEMP) began publishing digital maps and analyzing eviction data. The project was formed in response to the devastating impacts of venture capital, urban neoliberal politics, and real estate speculation in the Bay Area that we were witnessing in our everyday lives. Thus, the AEMP was formed as a data visualization, data analysis, and digital storytelling collective with the aim of documenting dispossession to make visible and actionable the terrain of gentrification and resistance in the city. To date the AEMP has produced more than 100 maps and data visualizations, community power maps, and a narrative cartographic project called Narratives of Displacement and Resistance (NDR). This NDR project embeds oral history and video work in a digital geospatial interface with the intent of making tangible the life stories and community experiences of people at the forefront of the Bay Area's eviction epidemic—not just as victims but as actors and activists in the process of urban transformation.

As founder of the AEMP (McElroy) and cofounders of NDR (Maharawal and McElroy), in this article we describe the process, tactics, and ideas behind our work; justify how and why we lattice countermapping and oral history practices; and set forth an analysis of how these can be used in the fight for housing justice.

We offer the following analysis as a contribution to both recent studies on Bay Area gentrification (Mirabal 2009; Walker and Schafran 2015; McNeill 2016; Stehlin 2016; Werth and Merianthal 2016; Maharawal 2017a, 2017b; McElroy 2017) and the growing scholarship on AEMP itself (Brahinsky 2014; Opillard 2015; McElroy and Opillard 2016; Shaffer 2016; Maharawal and McElroy 2017). Although both AEMP and NDR currently operate in both San Francisco and Alameda Counties (with AEMP further working in San Mateo and Los Angeles Counties), for the sake of this article, we focus on San Francisco, where both AEMP and the NDR project first emerged. This focus on San Francisco is not meant to reify the city as the center of the Bay Area gentrification crisis; rather, it is to highlight how and why AEMP and the NDR project came to be. We also want to acknowledge that we are just two of the project's numerous members. During AEMP's weekly meetings, which occur in both San Francisco and Oakland, sometimes dozens and other times just a handful of volunteers show up—volunteers consisting of activist scholars, oral historians, cartographers, disgruntled tech employees, youth, people new to the area, and those who have lived in the region for decades. These volunteers, along with the numerous community partners we have worked with, contribute vital labor to the project, shaping the direction it has taken and building its future.

Countermapping the Eviction Epidemic

The AEMP has by no means been the only group mapping San Francisco's contemporary tech boom. The boom, at times referred to as the Tech Boom 2.0, the Dot-Com Boom 2.0, or the App Boom, is thought to have emerged roughly in 2011, following the 2008 foreclosure crisis and in the long shadow of the first dot-com boom and bust of the late 1990s and early 2000s. Along with this tech boom and its attendant new waves of gentrification came a plethora of maps, each with its own novel geographic imaginary of the Bay Area. For example, in 2014, the luxury apartment complex NEMA—located in the "Twitter Tax Break Zone," a 2011 tax break for tech companies in the mid-Market area of San Francisco that led to increased property values throughout downtown (Lang 2015)—released its own marketing map of San Francisco neighborhoods. This map erased Chinatown and the largely working-class southern neighborhoods of the

city and renamed the Castro, a historically gay neighborhood, as Eureka Valley/Dolores Heights. The list of such real estate–driven neoliberal fantasy maps goes on (McElroy 2016) and is situated within a deep history of capitalist cartography privileging racialized, classed, and gendered geographic perceptions. As Wood and Krygier (2009) argued, deciding what to include in a map "surfaces the problem of knowledge in an inescapable fashion, as do symbolization, generalization and classification" (10). We developed our maps to counter such speculative real estate imaginaries, methodologically aligned with Kwan's (2002) conception of feminist visualization or a mode in which geographic information systems (GIS) can be used through feminist analytics and praxis. Ultimately for the AEMP, how the Bay Area is visualized is itself a terrain of struggle. Countermapping is for us a political act and one that (we insist) should also be accompanied by political action.

By countermapping, we refer to a set of critical cartographic and feminist data visualization practices that seek to render visible the landscapes, lives, and sites of resistance and dispossession elided in capitalist, colonial, and liberal topographies (Kwan 2002; Wood and Krygier 2009; Kurgan 2013; Ignazio 2015; Voyles 2015; Van der Vlist 2017). Methodologically, countermapping questions how, why, and with whom maps are made. In our work, the narrative and countermapping practices of AEMP and NDR seek not only to create a geospatial archive of loss but also to intentionally build solidarity and political collectivity among the projects' participants. That is, beyond mapping for mapping's sake, the project's goal in countermapping has always also been to assist and contribute to the rich terrain and history of activism throughout the region. Further, drawing on our experiences of creating collective and public community power maps, we argue for engaged methodological approaches that not only perform "autopsies" of evicted communities or produce what Woods (2002) described as "social death" for research subjects. Instead, whereas real estate speculators map investment opportunities, we map loss, dispossession, resistance, and struggle. Whereas real estate maps work hand in hand with neoliberal urban policymakers and property developers, our maps are produced through collaborations with activists and tenants fighting their evictions. Whereas their maps seek to produce an urban future of speculative capital accumulation, ours are designed to strengthen intersectional approaches theorizing risk, displacement, and resistance. In doing so, they

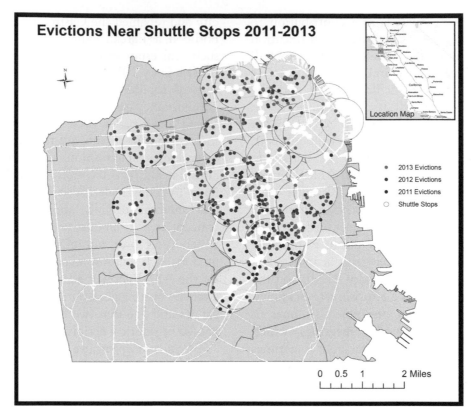

Figure 1. Tech Bus Stop Eviction Map, showing proximity of evictions to tech bus infrastructure. As we found, 69 percent of no-fault evictions between 2011 and 2013 occurred within four blocks of tech bus stops (Anti-Eviction Mapping Project 2014).

produce an alternative "geographic imagination" (Katz 2011, 58; Harvey 1990), elucidating new possibilities and modes of analysis. The AEMP's cartographic practice is thus in the critical geographic tradition of the Detroit Geographic Institute and Expedition (Barnes and Heynen 2011; Bunge 2011).

In this vein, AEMP pushes an engaged and activist geographical work through explicitly feminist, decolonial, antiracist cartographic practices in tandem with everyday political struggles. For instance, we have taken up an intersectional mapping approach in our partnership with the Eviction Defense Collaborative, a San Francisco–based nonprofit legal clinic that provides legal representation to tenants who have received eviction notices. Working with them to analyze and map their eviction and relocation data, we produced data visualizations showing that San Francisco's poor and working-class black and Latino residents are more likely to be displaced than white residents (Anti-Eviction Mapping Project and Eviction Defense Collaborative 2016). Interestingly, these eviction numbers inversely correlate with the hiring statistics of leading Bay Area tech companies (McElroy 2016). By combining these data sets, we have

contributed to an intersectional analysis of gentrification in the city, pointing to the racialized and classed nature of "evictability" (Van Baar 2016). Thus, our visualizations produce something akin to what Shabazz (2015) described as "ghost mapping," a conjuring of that which causes disappearance—in this case, white male tech capitalist geographies.

We countered such tech geographies early on in our project when we produced our Tech Bus Stop Eviction Map (Figure 1), responding to public outrage and protest regarding private luxury shuttles, colloquially known as the "Google buses" used by tech companies. These buses illegally (at the time) used the city's public bus stops to pick up their employees who commuted to and from Silicon Valley–based campuses. Not only were many San Francisco residents angered that private tech companies were taking over the city's public transportation infrastructure by using public bus stops for free, often delaying public buses in the process, but, further, as many tenants suspected, the new luxury bus lines were also causing property speculation and thus inciting evictions (Maharawal 2014; McElroy 2017).

We substantiated this suspicion through cartographic data analysis, finding that 69 percent of San

Francisco's "no-fault" evictions between 2011 and 2013 occurred within four blocks of private tech bus stops (Anti-Eviction Mapping Project 2014). In San Francisco, no-fault evictions are issued to tenants who have not violated their leases, whereas "fault" evictions are issued due to lease violations (nevertheless, fault evictions are often given for benign offenses). No-fault evictions are often used by real estate speculators to evict tenants and, as we found, the proximity of tech bus stops causes further speculation. Another study conducted by Dai and Weinzimmer (2014) concluded that up to 40 percent of those riding buses would not live in San Francisco if the buses did not exist. Our mapping was conducted in tandem with the Google bus blockades, which were direct actions that we also took part in organizing. These blockades drew attention to the connections between the private tech transportation infrastructure and evictions in the city, something that we made visually accessible through mapping. Further, our maps were used by activists in City Hall hearings on regulating the buses, demonstrating its public utility. Tech itself was not the problem, we argued, but rather real estate speculators were being given license to prey on the new geographies and wealth that tech generated.

Our Tech Bus eviction map built on the first map the AEMP produced, our Ellis Act Eviction Map, which depicted the accumulation of Ellis Act evictions in San Francisco since 1994. This map visualizes the alarming growth of Ellis Act evictions, a type of no-fault eviction prevalent in rent-controlled California cities. The Ellis Act is a California state law that permits landlords to "exit" the rental market, evict tenants due to no fault of their own, and change the "use" of the building—most often into ownership units (e.g., condos)—effectively destroying affordable rental housing (San Francisco Tenants Union 2016). Utilizing a JavaScript data visualization library, D3, to create a time-lapse map, we depicted Ellis Act evictions through a series of "explosions" in which red dots erupt across the city, corresponding to the number of units evicted (as filed with the San Francisco Rent Board). The map provided a quantitative yet visceral geographic representation of displacement in the city, the red eviction dots leaving the city pockmarked and blemished by the end of the time lapse.

Analyzing the data, one eviction at a time, and cross-referencing with Planning Department data and recorded real estate transactions, we calculated that Ellis Act evictions were increasingly being used by speculators to evict rent-controlled tenants and flip buildings. In fact, in San Francisco this seemed to be the Ellis Act's primary use, rather than its use by long-time landlords to exit the rental market, the latter being a myth promulgated by the real estate industry. As we discovered, 60 percent of Ellis Act evictions transpired within the first year of ownership and 79 percent within the first five years (Anti-Eviction Mapping Project and Tenants Together 2014). Collaborating with the statewide tenant rights organization Tenants Together, as well as the San Francisco housing rights coalition the Anti-Displacement Coalition, these data were used in political campaigns designed to curb Ellis Act eviction–induced real estate speculation. Unfortunately, both citywide and statewide measures failed, due in large part due to the immense lobbying and financing of counter campaigns by the real estate industry.

Our early maps and data visualizations drew much attention and were picked up by news outlets and politicians, as well as housing organizations. To many, they offered a conceptual foothold for grasping the seemingly amorphous and ubiquitous process of gentrification and social transformation occurring in the city. Our work pointed to the contours of processes of enclosure currently taking place in San Francisco, in which public goods (bus stops, parks, and rent-stabilized housing) were being undermined and enclosed by techno-capital (McNeill 2016; Maharawal 2017a; McElroy 2017). Inspired by the response, particularly that from activists and tenant organizers, we amplified our work, producing more cartographic experiments and partnering with housing organizations, activist collectives, and arts groups, from the San Francisco Tenants Union to the Unsettlers Project.

Yet, as activists, organizers, and academics, something was bothering us about our cartographic creations. Our everyday lives were surrounded by the experiences and stories of eviction, loss, and refusal, yet these rich social worlds were not being represented on our maps. We realized that our data-driven cartographic activism, vital as it was, was also reducing complex lifeworlds to dots on a map.

The idea that maps can be reductionist is, of course, not new (see Pickles [1995] as a seminal critique of GIS as positivist), and there have been many debates about GIS (e.g., the so-called GIS wars) that have led to the emergence of participatory GIS (PGIS) and public participatory GIS (PPGIS; Weiner and Harris 2003; Rouse, Bergeron, and Harris 2009). Ultimately, however, our realization that AEMP's maps needed to be accompanied by life stories came from our experiences organizing with people being evicted rather than the academic debates about the use and power of GIS.

Oral History toward Collective Resistance

Oral history, as a coproduced archival practice, inspired the AEMP to generate the NDR project, formed as a collective practice of recording community and life histories. Methodologically, the NDR project uses the format of oral history to produce what Frisch (1990) termed "shared authority," allowing "a more profound sharing of knowledges, an implicit and sometimes explicit dialogue from very different vantages about the shape, meaning and implications of history" (xxi–xxii). For instance, some of our interviewees, especially those publicly and actively fighting their evictions, had been interviewed by journalists, and their stories had helped to create media narratives about Bay Area gentrification. In their extractive format, though, media interviews tended to reduce these stories to simple narratives about victimhood and loss, producing tenants as subjects of processes happening to them, rather than as actors who are intentionally contesting, resisting, and thereby also shaping such processes. Thus, these journalistic interviews were antithetical to an ethos of "shared authority." They often produced an image of passivity and docility in the face of displacement and eviction. For us, oral history works to counter such representations, cocreating and fomenting collective political analyses while building resistance (cf. Kerr 2008). As such, the NDR project seeks to "share" analytic authority both within and beyond the moment of the interview. This approach values nuance and does not shy away from the messiness of politics or the complexity of personal histories.

From the outset, though, we remained concerned with how to map and represent life histories, neighborhood stories, and complex social worlds while utilizing oral history for *housing justice* (Kerr 2008). We wanted to take on representational critiques emergent from decolonial and postcolonial scholarship and critical race studies (Spivak 1988; Povinelli 2002; Stoler 2002; Woods 2002; Simpson 2014), as well as feminist and critical geography (Harvey 1984; Massey 1994; Kwan 2002; Wood and Krygier 2009; Kurgan 2013; Chambers et al. 2014). Specifically, we endeavored to record the complex social and political worlds that were being disappeared *nonreductively*, empowering

those involved. We were wary of documentary projects that viewed recording dispossession stories as an end in itself. We did not want merely to record "eviction stories" and risk reducing tenants' lives solely to their eviction nor merely bear witness to stories of suffering. Although we do record sorrowful stories of loss and pain, the oral history format ensures that these stories are not *all* the project records; rather, they are entangled with stories of joy, resistance, laughter, and contradiction.

Critical of both quantitative and qualitative liberal epistemic traditions, we thus embedded our work within the engaged practice of political organizing and social movement building. In constructing a qualitative GIS project (Elwood and Cope 2009) that understands GIS fundamentally as a power relation (Pavlovskaya 2009), our project has had to interface with diverse social and political worlds—from the activists and tenants we were organizing with to policymakers and media outlets who were sometimes the targets of our campaigns. As we further discuss later, the project often has oscillated at the blurred boundaries between analytical, authorial, affective, and political labor. As participants and scholar-activists (Cope 2008; Mitchell 2008; Autonomous Geographies Collective 2010), we found ourselves collaborators at the muddy crossroads of oral history, countermapping, academic knowledge production, direct action, and community organizing.

Ultimately, the goals of the NDR project are threefold: to (1) create an archive and historical record of the eviction epidemic through the stories of communities under threat of displacement and cultural erasure; (2) generate stories and data useful to activists and tenants in their campaigns; and (3) build solidarity and collectivity among the project's participants who could help one another in fighting evictions and collectively combat the alienation that eviction produces. These aspirations led to the creation of a participatory ethnographic oral history format that privileges stories of how people forge resistance around the concept of home through the intimate everyday politics of place. The NDR is an ongoing experiment in political community making, crafted through collective labor, mapping quantitative and qualitative data alongside direct action, and housing a diverse range of revolutionary aspirations. As many of the interviewees became interviewers and vice versa, the project is a collectively produced archive of community history, loss, and resistance, as well as an important historical document of San

Figure 2. Narratives of Displacement and Resistance mural, painted in the Mission's Clarion Alley in collaboration with Clarion Alley Mural Project. More photos and videos of the mural live can be found online at https://antievictionmap.squarespace.com/mural-in-clarion-alley, and the map's digital version can be found at http://www.antievictionmappingproject.net/narratives.html. (Color figure available online.)

Francisco's massive political and economic transformations. Moreover, as many of these narrators and archivists are also involved in various forms of activism and organizing together, the NDR project itself functions as just one facet of a broader effort to build a community of resistance. Thus, the tactical work of archiving publicly accessible stories works alongside other strategies—such as campaigns to fight specific types of eviction, as well as legal and direct action campaigns against various engines of gentrification. These strategies, tactics, and forms of resistance all amount to a process of place making in the Bay Area.

Place Making and Power Mapping

Place making is a vital component of social justice work. As Gilmore (2002) wrote, "The violence of abstraction produces all kinds of fetishes: states, races, normative views of how people fit into and make places in the world. A geographical imperative lies at the heart of every struggle for social justice; if justice is embodied, it is then therefore always spatial, which is to say part of the process of making a place" (16). The place-making practices of the NDR project are multiple and embodied in various ways. Although the digital archive and maps primarily live online, it takes various embodied and material forms offline as well.

For instance, when we released our Narratives Map in the spring of 2015, after having recorded our first thirty oral history interviews, we partnered with the Clarion Alley Mural Project and a team of muralists to paint an image of the oral history map on the Mission's Clarion Alley wall (Figure 2). In addition to depicting evictions across the city, the mural features nine portraits, each paired with a five-minute oral history clip that passersby could access and listen to through a "call-the-wall" function that operated by visitors simply calling a number painted on the wall. In this way, visitors could hear from tenants themselves, learning from stories of loss as well as resistance as they move through the spaces of the city.

The mural also featured a portrait of Alex Nieto, a Latino man who was killed by the San Francisco police in 2014 in Bernal Heights Park. Born and raised in the Mission, Nieto was killed while eating a burrito in the park before he went to his job as a security guard, after Justin Fritz, a white man who was new to the neighborhood, called 911 reporting Nieto as a person acting "suspiciously." When the police arrived, they shot and killed Nieto on the spot, allegedly mistaking his security guard's Taser for a gun. Many activists and residents in the Mission subsequently connected his death to the violence of gentrification in San Francisco (Solnit 2016). Our mural was positioned directly across from the Mission police station, where the police who killed him were stationed—and on it

Alex's portrait (partly painted by his parents) was accompanied by his parents' narrative of the aftermath of his death on their lives. By including his portrait, we sought to honor not only Alex but also the work of the Justice for Alex Nieto Coalition and to contribute to a political analysis that connects gentrification, racialized surveillance, and police brutality.

Leading up to the release of our oral history map and mural, AEMP volunteers also simultaneously produced a zine titled *We Are Here*, featuring transcriptions of some our oral histories, as well as photographs of antieviction actions, poems, artwork, essays on the idea of displacement written by activists, and page-by-page collages. In the back is a "know-your-rights" information section, explaining how to fight an eviction and find support in the region, followed by a list of wins achieved through direct action by groups in which members of AEMP were involved. We distributed the first print edition of *We Are Here* in Clarion Alley during the mural's unveiling, where we also invited each of the tenants featured on the wall, as well as Alex Nieto's parents, to be part of a dedication ceremony. At the time of the ceremony, all of the tenants featured were still in their homes, each of them choosing to

fight his or her eviction through a variety of direct action tactics including street protests, call-in campaigns, and a refusal to leave. Thus, the dedication ceremony was also a celebration of the power of protest and resistance, a sign of refusal to simply becoming a statistic or a docile dot on a map of loss.

Building on these experiences in San Francisco, we worked toward a regional analysis of gentrification and resistance struggles by starting a collaborative mapping project with the statewide group Tenants Together. Using eviction data from Alameda County (focusing on the cites of Oakland, Fremont, and Alameda), as well as data from the Oakland Rent Board, the project began mapping evictions in these cities, in tandem with producing oral histories and video work. In collaboration with groups that included the Oakland Creative Neighborhood Coalition, the Alameda Renters Coalition, Fremont RISE, Filipino Advocates for Justice, Causa Justa/Just Cause, and more, this cartographic and narrative work coalesced into report entitled *Counterpoints: Data and Stories for Resisting Displacement* (Graziani et al. 2016). As part of our partnership with the Oakland Creative Neighborhood Coalition, we also created a Community Power Map

Figure 3. Oakland Community Power Map, made in collaboration with the Betti Ono Gallery and the Oakland Creative Neighborhood Coalition. A digital crowdsource-able version of the map is available at http://arcg.is/2bnNUMa. (Color figure available online.)

(Figure 3) in the Betti Ono Gallery in downtown Oakland. This collaborative map aimed to reframe representations of the spaces of gentrification and struggle in Oakland by overlaying existing geographies with images of community power. The base layer for the map was collectively drawn by AEMP and Betti Ono members on two walls, representing Oakland's geography. Subsequently, gallery visitors could add what they considered assets and markers of community power on the map. As we wrote on the wall:

> In a city that has historically faced disinvestment by the powers that be, the current tide of changes and development in Oakland does not take into account what the heart and soul of Oakland want. What is valuable to our cultural identities, and what threatens our very place here. It is crucial, at this time, that we let the city know what we have, what we value, and what we want. This is a community power map. Your offerings to the map will live beyond this installation in an online map made with Anti-Eviction Mapping Project in collaboration with The Oakland Creative Neighborhoods Coalition. (Graziani et al. 2016, 17)

Before taking down the Community Power Map, we digitized its contents, so that it now lives online and is crowdsource-able, continuing to reframe the ways in which Oakland is imagined (Anti-Eviction Mapping Project 2016). As such, we hope that it will feed political imaginations antithetical to real estate development and foment imaginaries of everyday means of resisting hypergentrification.

Since producing this first community power map, we have produced several more, aligned with traditions of participatory GIS (Weiner and Harris 2003; Rouse, Bergeron, and Harris 2009). For instance, in partnership with the Bay Area Video Collaborative and Seven Tepees, we coproduced maps of youth power assets. Further, with University of California Berkeley students, we are currently mapping resources for undocumented students on campus. In these ways, through collaborative mural painting, organizing community events, and producing power maps alongside more traditional mapping, archiving, and storytelling, we have sought to build collective resistance to regional processes of dispossession.

Conclusion

Both the AEMP and the NDR project are multifaceted: at once an archive, a coproduced digital ethnographic object, a mural, a zine, and a collective political project of community and place making. In building AEMP and NDR, we have had to ask ourselves this: How can producing narrative and cartographic work online help foment embodied and material sociopolitical change? How can such a methodological approach work with and not merely for impacted communities (Tallbear 2014)? We have found numerous answers to these questions, answers that point to the ways in which archiving, mapmaking, storytelling, and political organizing can be intertwined and symbiotic, treated as important tools in an arsenal of tactics and strategies for resistance, place making, and political community building in the Bay Area.

One participant in the NDR project, Claudia Tirado, an elementary school teacher in San Francisco who fought her Ellis Act eviction, described how through the process of contesting her eviction, she found a political home: "I talk to Patricia, I talk to Benito, I talk to people who have been evicted before and who are there to fight evictions. They understand, they understand what it feels like. They understand what it is about and I feel a little more at home there" (this interview is available in full at http://www.antievictionmappingproject.net/narratives.html). Embedded within the larger AEMP, the NDR project demonstrates how the landscapes of property speculation are not an abstract terrain but rather an intimate topography composed of the clatter and clang of objects moved, lives and homes disrupted. Mapping this intimate terrain points to ways that collective resistance can be waged, and in doing so we must keep asking how cartographic activism and storytelling can support those waging rebellion and how such projects can provide a political "home" for people as they fight to save their physical homes.

Acknowledgments

Both authors contributed equally to this article and are listed in alphabetical order. We thank the collective members (past and present) of the AEMP for their vital work on the project. This project would not be possible without the many narrators who contributed, and we are grateful for their words and stories. Two anonymous reviewers and Zoltan Gluck provided exceedingly helpful comments and we thank them for this.

Funding

Manissa M. Maharawal was provided funding for research and writing by the Wenner-Gren Foundation for Anthropological Research, the American Council for Learned Societies/Mellon, The New York Council of Humanities, and a Digital Innovation Grant from The Graduate Center, CUNY. Erin McElroy received funding for this research from the Creative Work Fund, UCSC's Blum Center on Poverty, and UCSC's Eugene V. Cota-Robles Fellowship.

References

Anti-Eviction Mapping Project. 2014. Tech bus evictions. Accessed November 10, 2016. http://www.anteviction mappingproject.net/techbusevictions.html.

———. 2016. Oakland community power map. Accessed November 10, 2016. http://arcg.is/2bnNUMa.

Anti-Eviction Mapping Project and Eviction Defense Collaborative. 2016. *Eviction report, 2015: A report by the eviction defense collaborative and the anti-eviction mapping project.* San Francisco: Anti-Eviction Mapping Project.

Anti-Eviction Mapping Project and Tenants Together. 2014. The speculator loophole: Ellis Act evictions in San Francisco. Accessed March 20, 2017. http://antie victionmappingproject.net/speculatorloophole.html.

Autonomous Geographies Collective. 2010. Beyond scholar activism: Making strategic interventions inside and outside the neoliberal university. *ACME: An International E-Journal for Critical Geographies* 9 (2):245–75.

Barnes, T., and N. Heynen. 2011. A classic in human geography: William Bunge's (1971) Fitzgerald: Geography of a revolution. *Progress in Human Geography* 35 (5):712–20.

Brahinsky, R. 2014. The death of the city? *Boom: A Journal of California* 4 (2):43–54.

Bunge, W. 2011. *Fitzgerald: Geography of a revolution.* Athens: University of Georgia Press.

Chambers, K., J. Corbett, C. Keller, and C. Wood. 2014. Indigenous knowledge, mapping, and GIS: A diffusion of innovation perspective. *Cartographica: The International Journal for Geographic Information and Geovisualization* 39 (3):19–31.

Cope, M. 2008. Becoming a scholar-advocate: Participatory research with children. *Antipode* 40:428–35.

Dai, D., and D. Weinzimmer 2014. *Riding first class: Impacts of Silicon Valley shuttles on commute & residential location choice.* Berkeley: University of California, Berkeley, Institute of Transportation Studies.

Elwood, S., and M. Cope. 2009. Introduction: Qualitative GIS: Forging mixed methods through representations, analytical innovations, and conceptual engagements. In *Qualitative GIS: A mixed methods approach*, ed. M. Cope and S. Elwood, 1–12. Los Angeles: Sage.

Frisch, M. 1990. *Shared authority: Essays on the craft and meaning of oral and public history.* Albany: State University of New York Press.

Gilmore, R. W. 2002. Fatal couplings of power and difference: Notes on racism and geography. *The Professional Geographer* 54 (1):15–24.

Graziani, T., E. McElroy, M. Shi, and L. Simon-Weisberg. 2016. *Counterpoints: Data and stories for resisting displacement.* San Francisco: Anti-Eviction Mapping Project.

Harvey, D. 1984. On the history and present condition of geography: An historical materialist manifesto. *The Professional Geographer* 36 (1):1–11.

———. 1990. Between space and time: Reflections on the geographical imagination. *Annals of the Association of American Geographers* 80 (3):418–34.

Ignazio, C. 2015. *What would feminist data visualization look like?* MIT Blog. Accessed February 10, 2017. https://civic.mit.edu/feminist-data-visualization.

Katz, C. 2011. Accumulation, excess, childhood: Toward a countertopography of risk and waste. *Documents D'anàlisi Geogràfica* 57 (1):47–60.

Kerr, D. 2008. Countering corporate narratives from the street: The Cleveland Homeless oral history project. In *Oral history and public memories*, ed. P. Hamilton and L. Shopes, 231–51. Philadelphia: Temple University Press.

Kurgan, L. 2013. *Close up at a distance: Mapping, technology, and politics.* Cambridge, MA: MIT Press.

Kwan, M. 2002. Feminist visualization: Re-envisioning GIS as a method in feminist geographic research. *Annals of the Association of American Geographers* 92 (4):645–61.

Lang, M. 2015. Companies avoid $34M in city taxes thanks to "Twitter tax break." *SF Gate* October 29. Accessed August 8, 2016. http://www.sfgate.com/busi ness/article/Companies-avoid-34M-in-city-taxes-thanks-to-6578396.php.

Maharawal, M. 2014. Protest of gentrification and eviction technologies in San Francisco. *Progressive Planning* 199:20–24.

———. 2017a. San Francisco's tech-led gentrification: Public space, protest, and the urban commons. In *City unsilenced: Urban resistance and public space in the age of shrinking democracy*, ed. J. Hou and S. Knierbein, 30–43. London and New York: Routledge.

———. 2017b. Black Lives Matter, gentrification and the security state in the San Francisco Bay area. *Anthropological Theory.* Advance online publication.

Maharawal, M., and E. McElroy. 2017. In the time of trump: Housing, whiteness, and abolition. In *Contested property claims: What disagreement tells us about ownership*, ed. M. Hojer Bruun, P. Cockburn, B. Skærlund Risager, and M. Thorup. Abingdon and New York: Routledge.

Massey, D. 1994. *Space, place and gender.* Minneapolis: University of Minnesota Press.

McElroy, E. 2016. Intersectional contours of loss and resistance: Mapping Bay Area gentrification. Accessed March 29, 2017. https://antipodefounda tion.org/2016/08/18/intersectional-contours-of-loss-and-resistance.

———. 2017. Postsocialism and the Tech Boom 2.0: Techno-utopics of racial/spatial dispossession. *Social Identities* 23:1–16.

McElroy, E., and F. Opillard. 2016. Les Villes Américaines/Objectivité dans l'action et Cartographie collective dans le san francisco Néolibéral: Du Travail du Collectif anti-eviction mapping project [Objectivity in action and collective mapping in neoliberal San Francisco. On the work of the anti-eviction mapping project collective]. Accessed March 29, 2017. http://www.revue-urbanites.fr/les-villes-americaines-objectivite-dans-laction-et-cartog

raphie-collective-dans-le-san-francisco-neoliberal-du-travail-du-collectif-anti-eviction-mapping-project/.

McNeill, D. 2016. Governing a city of unicorns: Technology capital and the urban politics of San Francisco. *Urban Geography* 374:494–513.

Mirabal, N. R. 2009. Geographies of displacement: Latina/os, oral history, and the politics of gentrification in San Francisco's Mission District. *The Public Historian* 31 (2):7–31.

Mitchell, K., ed. 2008. *Being and practicing public scholar.* London: Blackwell.

Pavlovskaya, M. 2009. Non-quantitative GIS. In *Qualitative GIS: A mixed methods approach,* ed. M. Cope and S. Elwood, 13–37. Los Angeles: Sage.

Pickles, J. 1995. *Ground truth: The social implications of geographic information systems.* New York: Guilford.

Povinelli, E. A. 2002. *The cunning of recognition: Indigenous alterities and the making of Australian multiculturalism.* Durham, NC: Duke University Press.

Rouse, L. J., S. J. Bergeron, and T. M. Harris. 2009. Participating in the geospatial web: Collaborative mapping, social networks and participatory GIS. In *The geospatial web,* ed. A. Scharl and K. Tochtermann, 153–58. London: Springer.

San Francisco Tenants Union. 2016. Ellis Act evictions. Accessed November 30, 2016. https://www.sftu.org/ellis/.

Shabazz, R. 2015. *Spatializing blackness: Architectures of confinement and black masculinity in Chicago.* Chicago: University of Illinois Press.

Shaffer, A. 2016. West side stories. *Oral History Review* 43 (2):427–30.

Simpson, A. 2014. *Mohawk interruptus: Political life across the borders of settler states.* Durham, NC: Duke University Press.

Solnit, R. 2016. Death by gentrification: The killing that shamed San Francisco. *The Guardian* March 21. Accessed February 10, 2017. https://www.theguardian.com/us-news/2016/mar/21/death-by-gentrification-the-killing-that-shamed-san-francisco.

Spivak, G. C. 1988. Can the subaltern speak? In *Colonial discourse and post-colonial theory: A reader,* ed. P. Williams and L. Chrisman, 66–111. New York: Columbia University Press.

Stehlin, J. 2016. The post-industrial "shop floor": Emerging forms of gentrification in San Francisco's innovation economy. *Antipode* 48 (2):474–93.

Stoler, A. L. 2002. Colonial archives and the arts of governance. *Archival Science* 2 (1–2):87–109.

Tallbear, K. 2014. Standing with and speaking as faith: A feminist-indigenous approach to inquiry. *Journal of Research Practice* 10 (2):17. Accessed September 12, 2017. http://jrp.icaap.org/index.php/jrp/article/view/405/371

Van Baar, H. 2016. Evictability and the biopolitical bordering of Europe. *Antipode* 49:1–19.

Van der Vlist, F. 2017. Counter-mapping surveillance: A critical cartography of mass surveillance technology after Snowden. *Surveillance & Society* 15 (1):137–57.

Voyles, T. 2015. *Wastelanding: Legacies of uranium mining in Navajo Country.* Minneapolis: University of Minnesota Press.

Walker, R., and A. Schafran. 2015. The strange case of the Bay Area. *Environment and Planning A* 47 (1):10–29.

Weiner, D., and T. Harris. 2003. Community-integrated GIS for land reform in South Africa. *URISA Journal* 15 (2):61–73.

Werth, A., and E. Merienthal. 2016. "Gentrification" as a grid of meaning: On bounding the deserving public of Oakland First Fridays. *City* 20 (5):719–36.

Wood, D., and J. Krygier. 2009. Critical cartography. In *The international encyclopedia of human geography,* ed. R. Kitchin and N. Thrift, 340–44. New York and London: Elsevier.

Woods, C. 2002. Life after death. *The Professional Geographer* 54 (1):62–66.

MANISSA M. MAHARAWAL is Assistant Professor in the Department of Anthropology at American University, Washington, DC 20016. E-mail: manissa@gmail.com. She is the cofounder of the Narratives of Displacement and Resistance Oral History project as part of the Anti-Eviction Mapping Project. Her research interests include urban transformation, social movements and resistance, race, oral history, and political subjectivity.

ERIN McELROY is a Doctoral Candidate in Feminist Studies at University of California Santa Cruz, Santa Cruz, CA 95064. E-mail: emmcelro@ucsc.edu. She is also cofounder of the Anti-Eviction Mapping Project based out of San Francisco. Her research interests focus on techno-utopics, racialized dispossession, feminist technology studies, and postsocialism.

8　From New York to Ecuador and Back Again

Transnational Journeys of Policies and People

Kate Swanson

In this article, I explore the surprising and unexpected turns that have developed since zero tolerance policing was exported from New York to Ecuador at the turn of the new millennium. Drawing from fifteen years of ethnographic research with young indigenous Ecuadorians, I demonstrate how the impacts of displacement can extend far beyond the local scale. Street work has long been a key survival strategy for the indigenous Kisapincha. Yet, as growing poverty forced rising numbers onto the streets, cities in Ecuador responded by importing punitive neoliberal urban policies to cleanse and sanitize the streets. Deprived of critical income, many Kisapincha turned to transnational migration to seek better opportunities in the United States. Since then, young Kisapincha men and women have endured brutal 9,000-km journeys through South America, Central America, and Mexico to work in garment sweatshops and as day laborers in the United States. This research reveals how existing inequalities are reproduced and exacerbated in the drive to gentrify and modernize cities. I argue that zero tolerance policing in Ecuador pushed many former street vendors to migrate to New York City. These transnational displacements and scalar disruptions have led to profound injustices and intergenerational trauma for the Kisapincha. To untangle the hidden geographies of urban change, I suggest that scholars adopt ethnographic and longitudinal approaches to expose the long-term and unforeseen ramifications of policy mobilities over time and space. *Key Words: Ecuador, indigenous, migration, policy mobilities, zero tolerance.*

我于本文中，探讨自新世纪以来，零容忍的警备从纽约引入厄瓜多尔后，令人惊讶且超乎预期的发展转折。我运用对年轻的厄瓜多尔原住民进行为期十五年的民族志研究，显示迫迁的影响如何远远超出地方尺度。对于基萨平查 (Kisapincha) 的原住民而言，街头工作一向是主要的生存策略。但随着渐增的贫穷迫使越来越多人走上街头，厄瓜多尔的城市，开始透过引入惩罚性的新自由主义城市政策来清洗并淨化街道。被剥夺为生的关键收入后，许多基萨平查人转成跨国移民，到美国追求更好的机会。自此，基萨平查的年轻男女便开始忍受穿越南美洲、中美洲与墨西哥的九千公里之严酷旅程，以在美国的血汗成衣工厂工作，以及从事按日计酬的临时工。本研究揭露，既存的不均，如何在贵族化与现代化城市的驱力中再生产并加剧。我主张，厄瓜多尔的零容忍警备，将诸多过往的街头小贩推向移民纽约市之路。这些跨国迫迁与尺度化的扰动，已对基萨平查人造成深刻的不正义和跨世代的创伤。为了解开城市变迁的隐藏地理，我建议学者可採取民族志和长程的研究法，以揭露政策移动跨越时间与地理的长期且不被看见的后果。*关键词：厄瓜多尔, 原住民, 移民, 政策移动, 零容忍。*

En este artículo exploro los giros sorprendentes e inesperados que han surgido desde que el ejercicio policial de cero tolerancia se exportó desde Nueva York a Ecuador a la vuelta del nuevo milenio. Basándome en quince años de investigación etnográfica entre indígenas ecuatorianos jóvenes, demuestro cómo el impacto del desplazamiento puede llegar mucho más allá de la escala local. Desde hace mucho tiempo el trabajo callejero ha sido una estrategia de supervivencia clave para los indígenas kisapincha. Sin embargo, a medida que la expansión de la pobreza empujó a un número creciente de familias hacia las calles, las ciudades ecuatorianas respondieron importando políticas urbanas neoliberales punitivas para limpiar y sanear el espacio público. Privados de este crucial ingreso, muchos kisapinchas apelaron a la migración transnacional en los Estados Unidos, buscando mejores oportunidades de vida. Desde entonces, jóvenes kisapinchas de ambos sexos han soportado brutales viajes de 9.000 km a través de América del Sur, América Central y México, para terminar trabajando en talleres clandestinos de ropa y como jornaleros en los Estados Unidos. Esta investigación revela el modo como las desigualdades existentes son reproducidas y exacerbadas en el intento por aburguesar y modernizar las ciudades. Sostengo que las prácticas policiales de cero tolerancia en Ecuador impulsaron a muchos comerciantes callejeros a migrar hacia la Ciudad de Nueva York. Estos desplazamientos y perturbaciones transnacionales han conducido a profundas injusticias y trauma intergeneracional para los kisapinchas. Para desenredar las geografías ocultas del cambio urbano, sugiero que los eruditos adopten enfoques etnográficos y longitudinales para exponer las ramificaciones imprevistas y de largo plazo de las políticas de movilidades a través del tiempo y el espacio. *Palabras clave: Ecuador, indígenas, migración, política de movilidades, tolerancia cero.*

Following policy over extended periods of time can uncover surprising interconnections between seemingly unrelated and distant phenomena. In this article, I explore the "unexpected turns" (Peck and Theodore 2012, 29) that have developed since zero tolerance policing was exported from New York to Ecuador at the turn of the new millennium. As is now well known, zero tolerance policing was developed under Mayor Rudy Giuliani and Police Commissioner William Bratton in New York City in the early 1990s. Since then, this "fast policy model" (Peck and Theodore 2001) has circulated around the world. It has gained particular sway in Latin America in the guise of *mano dura*—or iron fist policing—as local politicians attempt to showcase their hard-line approaches to crime. Of course, like many mobile policies, zero tolerance has mutated and transformed in its implementation. Elsewhere I have argued that zero tolerance in Latin America represents an especially punitive policing model that explicitly targets the region's marginalized and racialized poor (Swanson 2013).

Here I am concerned with the long-term consequences of zero tolerance's transfer to Ecuador. By "following the policy" (Peck and Theodore 2012) using a broad topographical approach (Robinson 2011), we can trace its surprising impacts across borders, states, and regions. In effect, we can trace how the transnational flow of zero tolerance policy from New York to Ecuador has had a direct impact on the transnational flow of undocumented people from Ecuador to New York. The connection between Ecuador and New York might seem counterintuitive; yet, it transpires from "studying through," or following policy across space and time to explore the lives of those most affected (McCann and Ward 2012; Peck and Theodore 2015). In the Ecuadorian context, those arguably most affected by the importation of Ecuador's zero tolerance policing strategies are those who earned their livings in the urban informal sector, particularly indigenous street vendors and beggars. I focus on the case of the Kisapincha, a high Andean Quichua migrant community that relied on street vending and begging as a significant source of family income (Swanson 2010). After the implementation of zero tolerance—which paved the way for gentrification and urban restructuring in key tourist districts

of Ecuador's largest cities—it became increasingly difficult for the Kisapincha to earn a living on the streets. As a result, many decided to seek their fortunes on the streets of New York City instead.

Most often, policy outcomes are examined at the local scale and over short periods of time (Cochrane 2011). I have been working with Kisapincha rural–urban migrants in Ecuador for the last fifteen years and with Kisapincha transnational migrants in New York City for the last nine years. In studying connections and relations between sites within geographies of policy, McCann (2011) and Peck and Theodore (2015) suggested adopting a global ethnographic approach, often referred to as the extended case method. As stated by McCann (2011), this approach is valuable because it "focuses simultaneously on specific sites and on global forces, connections and imaginaries [and] reflects a concern with how to theorize the relationships between fixity and mobility, or territoriality and relationality, in the context of geographies of policy" (123). In this article, I examine unexpected connections between mobile policies and mobile people to uncover how zero tolerance has affected the lives of the Kisapincha over more than a decade. I argue that doing so uncovers profound injustices and intergenerational trauma, which can be directly linked to quick fix applications of zero tolerance policing. To do so, I draw from interviews, focus groups, informal conversations, surveys, and artwork, all of which were collected from indigenous children and adults living in Ecuador and New York between 2002 and 2017.

The rest of this article is structured as follows. I begin with a brief overview of the literature on policy mobilities and outline how my research furthers this field of study. I then dive into the specifics of the case study on the Kisapincha. I describe their motivations, their migration journeys, and their lives in New York City. I also examine the aspirations of children and young people who remain in Ecuador and explore their imaginaries of transnational life and migration. Finally, I discuss another unexpected twist in this tale of policy mobility: the long-term impacts of zero tolerance policing in New York and skyrocketing housing values in previously disinvested neighborhoods. For the Kisapincha, rapid gentrification in New York is leading to further displacement.

Zero Tolerance: Following the Policy

At a global scale, urban policies are on the move. As cities become more interconnected, politicians are increasingly looking abroad for examples of successful urban policies. Models that work are being replicated at rapid speeds, so much so that some scholars have taken to calling them fast policy (Peck and Theodore 2001). The speed and scale of connectivity have intensified as urban practitioners replicate models pertaining to creative cities (Florida 2002), urban drug policy (McCann and Temenos 2015), conditional cash transfers (Peck and Theodore 2015), governmentality (Robinson 2011), business improvement districts (Ward 2011), and zero tolerance (Swanson 2013), among others. Fast policy evangelists, such as Richard Florida and Rudy Giuliani, praise the triumphs of their models and are frequently contracted to replicate their policies around the world, garnering substantial consulting fees in the process.

Yet, context is everything; inevitably, conditions on the ground shape policy implementation. Local realities are often messy and unpredictable, which can lead to unanticipated outcomes. The area of policy mobilities is an emerging field in geography, and much of the research to date has explored immediate and short-term effects, partly due to methodological constraints (Cochrane and Ward 2012). Others have emphasized the role of policy actors and how key individuals circulate policies among cities (McCann and Ward 2012). My research shifts this approach to explore not only how a particular policy has circulated across cities but also how this policy has contributed to circulating *people* across cities. By following zero tolerance policy over an extended time period and an expansive geographical space, my research uncovers the unexpected ways punitive policing policies affect the lives of residents at the margins.

Zero tolerance policing first came to Ecuador in 2002. The mayor of Ecuador's largest coastal city, Guayaquil, contracted renowned "super cop" William Bratton to help shape the city's anticrime strategy (El Universo 2004). Although Bratton put forth an expansive list of reforms, the city lacked the resources to fully comply with what became known colloquially as Plan Bratton. As a result, the implementation was superficial and focused more on ridding the streets of undesirables than on tackling endemic police corruption and other policing problems. Reports of police violence and abuse became numerous as the city's most marginal residents were imprisoned, fined, and kicked off the streets. Key tourist districts were especially targeted as the city cleansed the streets and plazas of street children, vendors, beggars, and others who offended newly developed municipal codes regarding "urban norms" and "citizen image" (Municipalidad de Guayaquil 2004).

The Andean city of Quito followed suit, although its program was based more on heightened police surveillance and active antistreet commerce campaigns. The end goal was the same: Push informal street vendors and beggars out of the key tourist districts. By 2003, the municipality had succeeded in removing 6,900 informal workers from the streets of the city's historical center. Thereafter, heavily armed police roamed the streets to keep them free of informal vendors. The physical presence of the police was also accompanied by a new high-tech video surveillance system dubbed "Eyes of the Eagle" (Swanson 2007). By 2006, Quito put forth an anti–street commerce campaign, with prominent signs posted around the city to discourage buying on the streets. The signs declared that by buying from street vendors, not only were you "putting your safety in danger," you were also encouraging delinquency and filth in the city. In 2007, President Correa's new administration also turned its attention to begging. Shortly thereafter, they launched a prominent diverted giving and media campaign to discourage begging across the nation. Although the Give Dignity for an Ecuador without Begging (*Da Dignidad por un Ecuador sin Mendicidad*) campaign does purport to invest in rural communities, some dispute the depth of this investment. As stated by Malena,[1] a twenty-seven-year-old Kisapincha woman who began working on the streets as a child, "How does it help us if they give us an apple and a bag of candy once a year? What's the point?"

These policy shifts were taking place in the larger context of rapid neoliberalization, increasing debt, political instability, economic crisis, declining rural incomes, and rising urban migration. With poverty rates at 60 percent nationally and almost 90 percent among indigenous peoples, many Ecuadorian families were struggling to make ends meet in the early 2000s (Sistema Integrado de Indicadores Sociales del Ecuador 2003). The situation was especially difficult for the indigenous Kisapincha, who live in the high Andean region of Quisapincha Alto in the province of Tungurahua. For hundreds of years, the Kisapincha have lived on steep mountain slopes overlooking Mama Tungurahua and Taita Chimborazo, two massive volcanoes that dominate the region's mountainous landscape. Until the 1990s, subsistence agriculture was the economic mainstay for the Kisapincha. During the same period, limited access to education meant that illiteracy rates were above 80 percent. Moreover, poverty and a diet based largely on root crops meant that nine out of ten children were malnourished (Centro Ecuatoriana de Servicios Agrícolas 1992; Cruz et al. 1994).

Despite their remote location, mobility always remained a part of Kisapincha lives. Some migrated to the local markets in the city of Ambato to work as *cargadores*—heavy lifters for potatoes and grains, among other bulky items. In the 1970s, others migrated even further to work on coastal banana plantations to support Ecuador's growing banana economy. Life really began to change for Quisapincha Alto's approximately 6,000 residents with the construction of their first road in 1992. Up until that point, the only way in or out of the region was along steep and circuitous mountain paths, which took approximately two hours to traverse up or down. Although the new road did not necessarily decrease the time of travel—as the rainy season often made the road a slick river of mud—it did make it easier to leave the high Andean region, particularly for women with young children—and leave they did, en masse.

By the early 2000s, street vending, begging, and shoe shining in Quito and Guayaquil had become valuable sources of income for Kisapincha families. In the city, they targeted key tourist districts, which they dubbed *Gringopampa* (Kichwa for the "Field of Gringos"), as prime earning locations. Eventually, the Kisapincha gained a niche on the streets as beggars and vendors, much to the municipalities' chagrin. Given that they were no longer able to subsist from agriculture alone and they desperately needed money to pay for their children's educations, the money they earned on the streets went a long way toward improving their life conditions. Beyond education, they used their earnings to buy animals, such as sheep and cattle; replace mud and thatch homes with cinder block homes; open small community stores; purchase trucks to use as community transport vehicles; and buy commodities such as radios, televisions, and bicycles. Zero tolerance and punitive policing strategies eventually put an end to this. With rising harassment on the streets of Quito and Guayaquil, the Kisapincha were forced to look elsewhere for earnings. By 2005, a handful of Kisapincha women and children began migrating to Colombia's biggest cities to see whether begging and vending could be lucrative there. After the Colombian police temporarily seized a few of their children on the premise of child endangerment, though, they realized that they needed a better long-term strategy (Swanson 2007).

Ecuadorians have a long history of migration to New York (Jokisch 2002; Miles 2004; Pribilsky 2007). In fact, New York is often cited as Ecuador's third largest city (Weismantel 2003). Yet, much of this migration has been from southern Ecuador in the Cañar and Azuay regions. The first Kisapinchas to turn to the dangerous transnational migration route through the Americas left in 2007. Since then, many young Kisapincha men and women have endured brutal 9,000-km-long journeys through South America, Central America, and Mexico to work in New York City. Teenage girls I had previously encountered begging on the streets of Quito and Guayaquil now labor in sweatshops in Brooklyn; young shoe shiners I met on busy Ecuadorian streets now earn their livings as day laborers outside of New York City's big box construction stores. I cannot imagine that Bratton or Giuliani saw this coming.

As much as I would like to tell a neatly packaged story about linear cause and effect, I recognize that there is more nuance involved here. Although I argue that zero tolerance and punitive policing were instrumental in pushing indigenous Kisapincha off the streets and into transnational migration, I cannot discount the roles of ongoing poverty, marginalization, the search for better opportunities, and perhaps even a thirst for adventure. As the first Kisapincha migrants left and recounted their experiences and—more importantly—began to send remittances to their families back home, more and more young people decided to leave. What began with a few individuals triggered a chain migration; as of 2017, there were hundreds of indigenous Kisapincha living in Brooklyn's Bushwick neighborhood. Nevertheless, there is no doubt that most Kisapincha, particularly those with young children, would much rather remain at home than embark on such high risk and costly journeys. As stated by Pedro in New York, "Here, you're far from family, to fight and work and dedicate every day to work. . . . I know we make money, a bit of money here. But for me, I'd much rather live in my own country." Yet, once zero tolerance policing made it increasingly difficult to earn a living on the streets, the Kisapincha were left with few other viable options.

Ecuador–New York–Ecuador

Tomás and his nephew were the first Kisapincha to migrate to New York City. They arranged the trip with a *coyote* over the phone and given that they were migrant pioneers from the region, the coyote took advantage of their inexperience to charge US$18,500 per person. After securing a loan from a *chulquero*—a high-interest informal money lender—they began their 9,000-km trek. Tomás thought the trip would be easy because the coyote reassured them that they would have plenty of food and shelter along the way. "*Pero no fue así*" (it didn't turn out that way), he said. In retrospect, he added, he had no idea what he was getting into. They began by taking a four-hour bus ride to Quito and then traveled many more hours to the coast. From there they boarded a small fishing boat made for fifteen to twenty people. The boat was much fuller than that, though: It overflowed with more than 130 people. They were on the boat for ten days until they reached the coast of Guatemala. For most of the trip they were forced to stay in the hold so that they would not be seen by border patrol or customs agents. Tomás described his trip as "*entre vida y muerte*" (between life and death). Once they made it to Guatemala, they traveled by car, truck, or foot. He said that at times, they went days without food and suffered ter-

rible treatment at the hands of their coyotes. In total, his journey to the United States lasted seventy days.

When I spoke to Beatrisa and Natalia about their journeys, they both told me, "*El camino es feo*" (the journey is ugly). Like many of the more recent migrants, Beatrisa did not travel by boat. Instead, she flew to Honduras. From there, she traveled north in trucks and cars. At times, they were packed into the back of trucks for days "like sheep," "*sin aire para respirar*" (with barely enough air to breathe). At the U.S.–Mexico border, they crossed the Rio Grande on rafts. After that, their guide abandoned them in the desert. Her husband, Carlos, had a similar experience a few years earlier. In 2008, Carlos recounted how he spent four days wandering through the Arizona desert without food or water. With temperatures that can reach 50°C during the summer months and descend to freezing during the winter months, the risks of hypo- and hyperthermia are high. This is especially the case for migrants, many of whom arrive wearing only the clothes on their backs (Swanson and Torres 2016).

Figure 1. The prompt reads, "Please draw a picture of a migrant's journey to the United States." In this drawing, a sad migrant with his arms outstretched is about to be caught by the Border Patrol. The Border Patrol says, "I'm going to catch you." The migrant says, "I regret coming." His shirt says, "agua," expressing his need for water in the hot desert.

Fortunately, Carlos and Beatrisa were found before their situations became dire.

Not everyone has made it across the U.S.–Mexico border. Manuel described how immigration authorities caught him in Sonora, a Mexican state adjacent to Arizona. This was after he had already traveled nearly 5,000 km. Isabel, a former beggar and one of Kisapincha's first high school graduates, explained how her husband tried to migrate to the United States twice and failed both times. He was detained once in Mexico and once in Arizona. Now they owe over US$5,000 to the *chulqueros* and she has no idea how they will pay. Beyond debt and detention, there has also been death. In 2010, the Zeta cartel kidnapped and detained seventy-two Latin American migrants in the border state of Tamaulipas. After refusing to do the cartel's bidding, the migrants were lined up and massacred. Two of those killed were young Ecuadorian indigenous women from Quisapincha Alto. Elvira was

eighteen years old, and Magdalena was twenty-one. Both left young children behind as they journeyed to the United States in search of better opportunities (La Hora 2010). In 2013, another young Kisapincha man died on his way to the United States. Juan Toala Guamán, twenty-two years old, drowned on the last leg of his journey while crossing the Rio Grande (El Universo 2013).

Despite significant risks, many Kisapincha remain motivated to leave. During a focus group discussion I conducted with eight Kisapincha men in Brooklyn in 2012, Kléber summarized their reasons for leaving as follows: "Like everyone has said, the truth is that we migrated for necessity, for greater capacity, and due to political problems. The truth is that we want to give our children better educations and things we weren't able to have." All of them spoke about the intense poverty and lack of opportunity in Ecuador. They spoke about racism in Ecuador and difficulties for

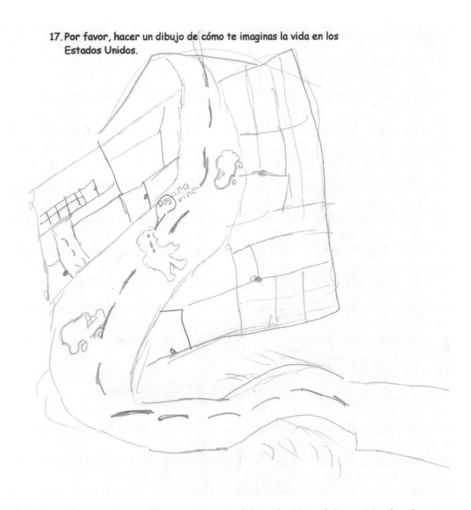

Figure 2. The prompt reads, "Please draw a picture of how you imagine life in the United States." In this drawing, a crying migrant walks along a busy city street and says, "I didn't want to, but I came."

indigenous peoples. They spoke about how they risked their lives to get to the United States, all so that they could improve the lives of their families and community back home.

Their lives in New York have not been easy, however, especially for the earliest migrants. Carlos explained how he arrived in January 2008 with no money, no work, and no warm clothes. He said, "For three months I went without work, and I struggled, struggled, struggled." Kléber added, "Now that I've arrived, I understand what life is like here. . . . If you work, you can eat what you like. But if you don't work, you can die of hunger." Another couple explained the precariousness of their lives in New York City. Natalia, who had worked on the streets of Quito and Guayaquil for years, migrated to New York in 2008 when she was nineteen years old. In the United States, she worked twelve-hour shifts, six days a week in a clothing sweatshop. Her husband, Nelson, worked as a day laborer in construction. Both of them supplemented their daytime work with nighttime flyer deliveries until 1 a.m. In general, they were lucky to get four hours of sleep per night. Meanwhile, they left their three- and one-year-old daughters in Ecuador in the care of their grandmother. Tomás explained how depression is now a part of his life in New York City; he struggles without his wife and three children and he is plagued by the traumas of his journey.

Children in Ecuador are also increasingly aware of the dangers of transnational migration. In a survey I conducted with sixty schoolchildren in 2015, I asked them to draw pictures of migrant journeys to the United States. Many drew grim scenes of thirsty migrants in the desert or of U.S. Border Patrol agents in pursuit (Figure 1). I also asked them to draw pictures of migrant life in New York. Although some drew pictures of happy migrants, others depicted loneliness and hardship (Figure 2). Despite this, many young people still wish to migrate, particularly among sixth-graders, which is often a terminal grade in Ecuador. In fact, 63 percent of sixth-graders said they hope to migrate to New York when they get older. Among ninth-graders, only 30 percent expressed a desire to migrate, perhaps because being slightly older, they experienced recent migrant deaths more viscerally. One ninth-grader explained how her parents do not want her to migrate because, "To migrate to the U.S. is to suffer. Sometimes they can't walk and they die. It's suffering." When I asked if she thought she would migrate regardless, she said, "Yes." Why? "Because it's beautiful and I want to be close to my family." Two of her siblings had migrated to New York a few months prior to my survey.

The costs of zero tolerance policing in Ecuador have been significant for the Kisapincha. Deprived of key income on the streets of Quito and Guayaquil, many have put their lives at risk to endure great hardships as undocumented migrant workers in New York City. Their children back home have suffered, too. Although youth in Ecuador might benefit from remittances and a newfound ability to purchase coveted goods and gadgets, the emotional costs of growing up with parents in absentia are profound (Pribilsky 2007).

There is one more twist to this tale on the unforeseen consequences of policy mobilities. When the Kisapincha moved to New York, they settled in the Bushwick area of Brooklyn. Bushwick has long been a disinvested immigrant neighborhood known for high crime and poverty. When Natalia and Nelson relocated to Bushwick, they were able to rent a three-bedroom apartment for $900 per month. They shared their apartment with several other Kisapincha to keep costs down. In recent years, however, Bushwick has been gentrifying rapidly (New York City Comptroller 2017). In fact, the long-term impacts of zero tolerance policing in New York City have led to rising rents across the city, displacing lower income residents into the region's outer boroughs. William Bratton, who once again became New York City's Police Commissioner under Mayor Bill de Blasio, takes full credit for dropping crime rates alongside escalating property values (Kelling and Bratton 2015). Yet, as homeowners benefit, renters suffer. In the search for affordable rents, Bushwick has become the latest urban mecca for hipsters, artists, and young urban professionals. Data demonstrating this rapid urban change are striking: Between 2000 and 2015, Bushwick's white population increased by 610 percent, whereas its Latino population decreased by 13 percent. Moreover, the U.S.-born population increased by 22 percent and the foreign-born population decreased by 14 percent (Small 2017). For Natalia and Nelson, this meant that by 2015 their rent had risen to more than $3,000 per month, a price they could no longer afford to pay. Whereas some Kisapincha are being displaced further east to Canarsie—the last stop on the L train—Nelson and Natalia decided to pack their bags and move back to Ecuador instead. Ironically, then, Bratton's efforts to export zero tolerance policing to Ecuador not only pushed Kisapincha street workers from Ecuador to New York, but his ongoing efforts to restore "order and civility across the five boroughs" (Kelling and Bratton 2015) served to push them back to Ecuador again.

Conclusion

Only by following policy over an extended period of time can we untangle the "surprising encounters, unexpected turns, and unforeseen conclusions" of policy mobilities (Peck and Theodore 2012, 29). In this article, I have traced the long-term impacts of zero tolerance policing across time and space to explore how it has shaped the lives of those at the margins. In doing so, I have demonstrated how policy mobilities can lead to surprising outcomes with far-reaching consequences. Zero tolerance policing has served as an impetus for the transnational dislocations and scalar disruptions experienced by the Kisapincha. They have been displaced across cities, states, and borders in their quest to improve the material quality of their families' lives. Yet, although families might be materially better off, there are lasting emotional, embodied, and intergenerational consequences. For instance, when Natalia and Nelson moved back to Ecuador, their eight-year-old daughter rejected them outright. Despite her parents' deep sacrifices, she refuses to live with them, preferring to stay with her grandmother instead.

By taking a global ethnographic approach and "studying through," I have demonstrated the topographies of relationality between seemingly disparate and distant spaces (Jacobs 2012). I have traced zero tolerance policy in multiple directions—from New York to Ecuador and Ecuador to New York and back again. In doing so, I heed Peck and Theodore's (2012) call to provide an ethnographic account that troubles and disturbs conventional wisdom. This research demonstrates that the impacts of displacement can extend far beyond the local scale. By focusing on those at the margins—in this case, indigenous peoples—it reveals how existing inequalities are reproduced and exacerbated in the drive to gentrify and modernize cities. The transnational displacements of the Kisapincha provides a cautionary tale for scholars. Space and scale are critically important if we are to untangle the hidden geographies of gentrification. As stated by Massey (2013), space is "like a pincushion of a million stories." Whose stories are absent in top-down, short-term urban research? Although more methodologically challenging, urban scholars must embrace ethnographic and longitudinal research to uncover multiple stories over time and space. This is especially critical if

we hope to challenge unforeseen injustices, particularly for the most marginalized urban residents.

Acknowledgments

Many thanks to Anne-Marie Debbané, William Brewer, Sara Koopman, Nik Heynen, and anonymous reviewers for help with various aspects of this article. I also thank Dimitri Jones, my now eight-year-old research assistant, for helping me with data collection in Ecuador and New York City. Of course, I offer my heartfelt thanks to the Kisapincha of Quisapincha Alto for welcoming me into their lives.

Note

1. All names are pseudonyms.

References

Centro Ecuatoriana de Servicios Agrícolas (CESA). 1992. *Prediagnostico Participativo de las Organizaciones Campesinas de las Parroquias de Quisapincha, Pasa, y San Fernando* [Participatory prediagnostic of Quisapincha, Pasa and San Fernando's Peasant Organizations]. Quito, Ecuador: CESA.

Cochrane, A. 2011. Foreword. In *Mobile urbanisms*, ed. E. McCann and K. Ward, ix–xi. Minneapolis: University of Minnesota Press.

Cochrane, A., and K. Ward. 2012. Researching the geographies of policy mobility: Confronting the methodological challenges. *Environment and Planning A* 44:5–12.

Cruz, A., V. H. Fiallo, E. Hinojoza, R. Moncayo, B. Rendon, and J. Sola. 1994. *Participación y Desarrollo Campesino: El Caso de San Fernando-Pasa-Quisapincha (Síntesis)* [Peasant participation and development: The case of San Fernando-Pasa-Quisapincha (synthesis)]. Quito, Ecuador: Centro Ecuatoriana de Servicios Agrícolas.

El Universo. 2004. William Bratton: Se necesita mejorar calidad de policías [William Bratton: The quality of the policy must be improved]. *El Universo* March 2. Accessed December 29, 2016. http://www.eluniverso.com/2004/03/02/0001/10/A7B9E2AB15DA47E5806BB6734A5996D9.html

———. 2013. Comunidad hizo sepelio simbólico de desaparecido [Community holds symbolic funeral for the disappeared]. *El Universo* April 9. Accessed December 29 2016. http://www.eluniverso.com/2013/04/09/1/1422/comunidad-hizo-sepelio-simbolico-desaparecido.html

Florida, R. 2002. *The rise of the creative class: And how it's transforming work, leisure, community and everyday life.* New York: Basic Books.

Jacobs, J. M. 2012. Urban geographies I: Still thinking cities relationally. *Progress in Human Geography* 36:412–22.

Jokisch, B. D. 2002. Migration and agricultural change: The case of smallholder agriculture in highland Ecuador. *Human Ecology* 30:523–50.

Kelling, G. L., and W. Bratton. 2015. Why we need broken windows policing. *City Journal* Winter. Accessed December 29, 2016. http://www.city-journal.org/html/why-we-need-broken-windows-policing-13696.html

La Hora. 2010. Masacre México tiñe de dolor a Quisapincha [Mexican massacre pains Quisapincha]. *La Hora* November 13. Accessed December 29, 2016. http://lahora.com.ec/index.php/noticias/show/1101047812/-1/Masacre_M%C3%A9xico_ti%C3%B1e_de_dolor_a_Quisapincha.html#.WGWOo7YrJo4

Massey, D. 2013. Doreen Massey on space. *Social Science Bites*. Accessed May 29, 2017. http://www.socialsciencespace.com/wp-content/uploads/DoreenMassey.pdf

McCann, E. 2011. Urban policy mobilities and global circuits of knowledge: Toward a research agenda. *Annals of the Association of American Geographers* 10:107–30.

McCann, E., and C. Temenos. 2015. Mobilizing drug consumption rooms: Inter-place networks and harm reduction drug policy. *Health & Place* 31:216–23.

McCann, E., and K. Ward. 2012. Assembling urbanism: Following policies and "studying through" the sites and situations of policy making. *Environment and Planning A* 44:42–51.

Miles, A. 2004. *From Cuenca to Queens: An anthropological story of transnational migration*. Austin, TX: University of Texas Press.

Municipalidad de Guayaquil. 2004. Art. 13-De la Imagen Ciudadana y Normas de Urbanidad in *Ordenanza Reglamentaria de la Zona de Regeneración Urbana del Centro de la Ciudad* [Article 13 - Citizen Image and Urban Norms in the *Regulatory Ordinance for the City Center's Urban Regeneration Zone*]. January 14. Concejo Municipal de Guayaquil: Guayaquil. Accessed September 29, 2017. http://www.guayaquil.gob.ec/Ordenanzas/Regeneraci%C3%B3n%20Urbana/14-01-2004%20Ordenanza%20reglamentaria%20de%20la%20zona%20de%20regeneraci%C3%B3n%20urbana%20del%20centro%20de%20la%20ciudad.pdf

New York City Comptroller. 2017. *The new geography of jobs: A blueprint for strengthening NYC neighborhoods*. New York: Bureau of Policy and Research.

Peck, J., and N. Theodore. 2001. Exporting workfare/importing welfare-to-work: Exploring the politics of Third Way policy transfer. *Political Geography* 20:427–60.

———. 2012. Follow the policy: A distended case approach. *Environment and Planning A* 44:21–30.

———. 2015. *Fast policy: Experimental statecraft at the threshold of neoliberalism*. Minneapolis: University of Minnesota Press.

Pribilsky, J. 2007. *La chulla vida: Gender, migration and the family in Andean Ecuador and New York City*. Syracuse, NY: Syracuse University Press.

Robinson, J. 2011. The spaces of circulating knowledge: City strategies and global urban governmentality. In *Mobile urbanisms*, ed. E. McCann and K. Ward, 15–40. Minneapolis, MN: University of Minnesota Press.

Sistema Integrado de Indicadores Sociales del Ecuador (SIISE). 2003. *La Década de los 90s en Cifras: La Pobreza y la Extrema Pobreza de Consumo* [The 1990s in numbers: Poverty and extreme poverty]. Quito, Ecuador: SIISE.

Small, A. 2017. The gentrification of Gotham. *CityLab*. Accessed May 2, 2017. https://www.citylab.com/life/2017/04/the-gentrification-of-gotham/524694/

Swanson, K. 2007. Revanchist urbanism heads south: The regulation of Indigenous beggars and street vendors in Ecuador. *Antipode* 39:708–28.

———. 2010. *Begging as a path to progress: Indigenous women and children and the struggle for Ecuador's urban spaces*. Athens: University of Georgia Press.

———. 2013. Zero tolerance in Latin America: Punitive paradox in urban policy mobilities. *Urban Geography* 34:972–88.

Swanson, K., and R. M. Torres. 2016. Child migration and transnationalized violence in Central and North America. *Journal of Latin American Geography* 15:23–48.

Ward, K. 2011. Policies in motion and in place: The case of business improvement districts. In *Mobile urbanisms*, ed. E. McCann and K. Ward, 71–95. Minneapolis: University of Minnesota Press.

Weismantel, M. J. 2003. Mothers of the patria: La Chola Cuencana and La Mama Negra. In *Millennial Ecuador: Critical essays on cultural transformations and social dynamics*, ed. N. E. Whitten, Jr., 325–54. Iowa City: University of Iowa Press.

KATE SWANSON is an Associate Professor in the Department of Geography at San Diego State University, San Diego, CA 92182. E-mail: kswanson@mail.sdsu.edu. Her research interests include migration, inequality and exclusion in Latin America, and the U.S.–Mexico border region.

9 Police Torture in Chicago

Theorizing Violence and Social Justice in a Racialized City

Aretina R. Hamilton and Kenneth Foote

Harvey's *Social Justice and the City* was published just as a major case of social injustice was unfolding on Chicago's South Side. Starting in the early 1970s and continuing for almost twenty years, police officers under Detective Jon Burge tortured confessions from as many as 200 black men. We use the contrast between *Social Justice and the City* and the Chicago police torture cases to emphasize the work that geographers have accomplished since the 1970s in theorizing race, space, and place. The torture cases have led to a decades-long struggle for justice and reparations waged by survivors, families, and activists. Here we examine the spatiality of the torture and how racialized practices of policing, housing, and employment operate across scales, sometimes amplifying the effects of each practice. Looking backward from the 1970s it is possible to see the torture cases as an extension of a long history of racial violence rooted deeply in U.S. history. Looking forward, these cases have clear parallels with contemporary events, including recent "blue-on-black" police killings and the rise of the Black Lives Matter movement. Our aim, more broadly, is to begin theorizing violence within the larger debate over social justice in the contemporary U.S. city and to examine more closely how violence has been used repeatedly in Chicago and other U.S. cities to enforce social and spatial definitions of "race" and racial boundaries. *Key Words: Chicago, police violence, race, torture, violence.*

大卫．哈维的着作《社会正义与城市》，正好发表于芝加哥南部爆发一个重大的社会不正义案件之际。从 1970 年代早期以降，警探琼．伯格手下的警察，便对高达两百位的黑人进行刑求逼供，历时将近二十年。我们运用《社会正义与城市》和芝加哥警方的刑求案例之间的对比，强调地理学者自 1970 年代以降理论化种族、空间与地方的成就。刑求案例引发了生还者、家属以及行动者数十年来对于正义与补偿修复的斗争。我们于此检视刑求的空间性，以及警备、居住和就业的种族化实践，如何跨越尺度进行操作，有时放大各种操作的效应。从 1970 年代开始回顾，得以将刑求案例视为深植于美国历史的种族暴力之长期历史的延伸。展望未来，这些案例和当前的事件之间具有清晰的相似处，包括晚近 "执法人员针对黑人" 的警察杀戮，以及 "黑人的命也是命" 的运动之兴起。我们更进一步的目标在于，在有关美国当代城市中的社会正义之广泛辩论中着手理论化暴力，并更仔细地检视芝加哥与其他美国城市，如何反覆运用暴力来执行"种族"与种族疆界的社会与空间定义。 关键词： 芝加哥, 警察暴力,种族,刑求,暴力。

El libro de Harvey *Social Justice and the City* [Justicia social y ciudad] fue publicado precisamente cuando se ventilaba un caso importante de injusticia social en el South Side de Chicago. Empezando desde principios de los 1970 y continuando durante casi veinte años más, los agentes de la policía dirigidos por el detective Jon Burge obtuvieron confesiones bajo tortura de no menos de 200 hombres negros. Usamos el contraste entre *Social Justicie and the City* y los casos de tortura de la policía de Chicago para destacar el trabajo que los geógrafos han cumplido desde los años 1970 para teorizar raza, espacio y lugar. Los casos de tortura han llevado a las luchas que durante décadas por justicia y reparación han librado sobrevivientes, familiares y activistas. En este artículo examinamos la espacialidad de la tortura y cómo las prácticas racializadas de la acción policial, la vivienda y el empleo operan a través de diferentes escalas, algunas veces ampliando los efectos de cada práctica. Mirando hacia atrás desde los años 1970 es posible ver los casos de tortura como extensión de un largo acontecer de violencia racial profundamente arraigado en la historia de los EE. UU. Mirando hacia adelante, estos casos representan un claro paralelo de eventos contemporáneos, incluyendo los recientes asesinatos policiales de "azul-contra-negro" y la aparición del movimiento Las Vidas Negras Importan. Nuestro propósito, en términos más generales, es empezar a teorizar la violencia dentro del más amplio debate de la justicia social en la ciudad norteamericana contemporánea y examinar a mayor profundidad cómo se ha utilizado repetidamente la violencia en Chicago y otras ciudades de los EE. UU. para ejecutar las definiciones sociales y espaciales de "raza" y las fronteras raciales. *Palabras clave: Chicago, violencia policial, raza, tortura, violencia.*

Thus grew up a double system of justice, which erred on the white side by undue leniency and the practical immunity of red-handed criminals, and erred on the black side by undue severity, injustice, and lack of discrimination.

—Du Bois ([1903] 2003, 127)

David Harvey's (1973) *Social Justice and the City* was published just as a major case of social injustice was unfolding on Chicago's South Side. Starting in the early 1970s and continuing for almost twenty years, police officers under the direction of Jon Burge tortured confessions from as many as 200 black men. Burge was fired from the Police Department in 1993 after an internal review of the torture cases. Convicted later, in 2010, of obstruction of justice and perjury relating to a civil suit stemming from the torture, Burge eventually served just over three and half years in prison. Few of the other officers involved in the torture were punished in any substantial way, whereas the men they tortured lost years and decades of their lives. Some torture victims were eventually released, but others died in prison, and a few remain incarcerated even today.

We use this episode to focus on how far geography has come in theorizing race and violence since *Social Justice and the City* first appeared. In retrospect, it is notable that the book's index lists no entries under "violence" and only four under "race relationships." Furthermore, there is no mention of earlier writers such as Ida Wells and W. E. B. Du Bois, whose works called out racial violence as a key element of U.S. life, as emphasized in our epigraph from Du Bois's *The Souls of Black Folk.*

Our intention is not, however, to fault Harvey for this oversight. Classic works such as Drake and Cayton's ([1945] 2015) *Black Metropolis* and other earlier publications of the Chicago School of urban studies paid little attention to racial violence. Harvey's research was focused instead on different issues. In *Social Justice and the City* (1973), he was beginning to detail his now widely influential Marxist critique of capitalist urbanization. This broader theoretical frame had the potential to encompass race and violence, even if these topics were not addressed directly in the book. Yet Harvey's push for an encompassing Marxist theory has been questioned by even some of his closest students and collaborators, such as Smith (2008):

> But what is the response to critics who are simultaneously wedded to the centrality of capitalism, and the

Marxist critique, and to ... insistence that the oppression of women and of people of color is thoroughly integral to what makes capitalism in the first place and to political strategies for its overthrow? The left since the 1980s has voiced a desire for its own "unified theory" that will happily resolve in one big conceptual tent all our dilemmas of race, class, gender, sexuality and other social differences. But that has not happened, and it is not going to happen. (152)

At the same time, advances have been made since the 1970s in understanding the spatialization of race and violence. Subsequent research has explored how social, economic, political, and cultural processes, including violence, interact across a range of scales to reinforce racial boundaries. The Chicago police torture cases highlight a wide gap between what we know about the dynamics of racial violence and what can be done to stop such violence. Contemporary events including the recent killings of black men and women, the rise of the Black Lives Matter movement, and instances of violence directed toward many other groups only emphasize the importance of this work in terms of both theory and practice.

Theorizing Race, Place, and the Production of Black Geographies

The contrast between *Social Justice and the City* and the Chicago police torture cases is important theoretically. It allows us to emphasize the work that geographers have accomplished since the 1970s in theorizing race and urban landscapes. Harvey's book might have sidestepped these issues, but such omissions are not surprising given Kobayashi's (2014) observation that "the concept of race has a benighted biography among those who have created our discipline" (1101). Her point is that geographers have too often accepted the idea of race uncritically but, more important, have in some cases been complicit in its propagation.

Since the release of *Social Justice and the City*, geographers have tried to correct this lacuna (Kobayashi and Peake 1994). A great deal has been published on race, place, and geography on a variety of topics. These include nuanced examinations of landscapes (J. B. Jackson 1970; K. T. Jackson 1985), belonging and contestation (Sibley 1995), gender and space (Longhurst 1995), whiteness (Bonnett 1997; Shaw 2007), queer geographies (Bell and Valentine 1995), and state-sanctioned violence (Tyner 2012; Tyner and Inwood

2014). In addition to these works, scholars began to explicitly address the absence of race in geography (Kobayashi and Peake 2000; Peake and Schein 2000).

Of these areas of theoretical innovation most related to our topic, the first is the emergence of a literature on black geographies beginning in the post–civil rights era of the late 1960s. Bunge's (1965) Detroit Geographic Expedition was one of the first steps in this direction in its exploration of the lived experience of blackness in Detroit. Rose's address to the Association of American Geographers in 1978 as past president, "The Geography of Despair," was another key step forward. Rose (1978) focused on the high incidence of homicides among black males but also revealed the systemic roots of inequality that existed in black communities. This work served as an entry point for scholarship specifically focused on blackness. Two decades later, Woods's (1998) Development Arrested generated another shift toward the establishment of black geographies as a subarea. Woods argued that the black working class developed a "blues epistimology" that contested conventional assumptions regarding social relations in the South.

McKittrick's (2006) Demonic Grounds introduced an intersectional approach to the study of black women's cartographies, that blackness and gender are not binary categories but, instead, involve fluid blendings of identities. Another innovation in this area was Wilson's (2000) use of the theme of apartheid to explore racial segregation in Birmingham, Alabama. Recent research on black geographies has expanded to include black queer sexualities (Walcott 2007; Eaves 2017) and how blackness is spatialized outside of an American context (Fanon 1963; Gilroy 1993; Wynter 2003; Nimako and Small 2009) has helped broaden this earlier work. These geographies are not only local, they are transnational and part of a larger diaspora that is "underscored by a long history of racial domination" (McKittrick and Woods 2007, 8).

Among the most important developments since the 1970s are works on carceral geographies and policing. Carceral powers discipline space, a fact that geographers began to examine through the work of Goffman (1961) and Foucault (1979). Philo (2001) was among the first to build on the work of Foucault by exploring spaces of confinement. Gilmore's (1999) work alerted geographers to the ways in which state building was often complicit in the creation of a prison state. Her Golden Gulag (Gilmore 2007) argued that politics and race are intrinsic to understanding the dynamics and expansion of the prison and the criminal justice system in California. Interest in carceral geographies also created an entry point for scholars concerned with the policing of blackness.

Herbert (1997) wrote about the territorial control used by police in Los Angeles to regulate the personal geographies of black citizens. These earlier works began to look at the systemic causes of violence and police prejudice. Other scholars viewed policing as tools of neoliberal policies and sites of antiblack racism (Herbert and Brown 2006; Beckett and Murakawa 2012). Within these black communities, all citizens are marked as suspects and put under surveillance. This point was underscored in Shabazz's (2015) Spatializing Blackness. For Shabazz, carceral power and the racialization of space moved from the public spaces of the streets into the home, as well as into practices of masculinity. Writing about kitchenettes, one-bedroom apartments, he noted, "Racialized practices expressed via geography became normalized, rationalized, and even (for some segments of the society) desirable" (Shabazz 2015, 71).

The upshot of these theoretical insights is to recognize that for these citizens, the concept of race made their communities, their homes, and their bodies contested sites where the state could employ many tools of discipline (including the police) to subdue or remove them from the landscape. These points are essential to understanding the geographical spaces that made it possible for the victims of the Chicago police torture to be brutalized. Separated from the rest of the city with clear material boundaries, the people who were tortured were selected because they had little or no societal or spatial autonomy. Race operates across various scales (public and private spaces, home, streets, neighborhoods) and largely defines how African American residents in Districts 2 and 3 move throughout them (Boddie 2010; Hague 2010; Lipsitz 2011).

Within the neighborhoods that Burge and his officers occupied, racialized violence became a practice of everyday life and a means of social control (Blomley 2003). Over a period of two decades, members of the Chicago Police Department used violence and terror to ensure that racial boundaries were enforced. At the same time, the 1970s and 1980s were decades of tremendous economic and social change in Chicago and elsewhere—a period of weakened social control—so Burge's torture can be seen in a different light. In both cases, these points complicate Harvey's (1973) argument in

Social Justice and the City. His articulation of social justice ignores how institutional structures have resulted in increased violence for both reasons.

The Case Study: Historical and Political Contexts

The Chicago police torture cases provide a window on how violence continues to be used to enforce spatial racial, ethnic, social, and economic divides. Other U.S. cities offer somewhat parallel examples, but Chicago is a particularly good case. It has played a role in some of the most important social and economic transformations in U.S. society from the mid-nineteenth century onward. As the birthplace of the Chicago School of sociology, it has also served as a focus of social research for a century (Park, Burgess, and McKenzie 1925; Suttles 1968; Hunter 1974; Reed 2011; Drake and Cayton [1945] 2015). A further reason for focusing on Chicago is that violence has been part of its history from the destruction of the first white settlement, Fort Dearborn, in 1812, right up to the present. Violence has erupted in Chicago repeatedly to mark and enforce social, racial, and ethnic boundaries, and the Burge torture cases can be viewed within this broader context (Table 1).

Up to World War II, Chicago was a battleground of labor rights, but after the war, violence focused increasingly on issues of race and the desegregation of the city's housing. As Rothstein (2017) noted, in the first five years after the war there 357 acts of violence against blacks attempting to buy or rent homes in white neighborhoods and such attacks continued to occur for decades. The Burge case also arose during a period when a sense of siege enveloped many major U.S. cities as confrontations erupted not only over race but also related to the Vietnam War and other issues. During this period, some major cities were placed under de facto military occupation by the National Guard. Some police and civil authorities came to see "violence against violence" as a suitable response to the civil unrest of the 1960s and 1970s. In Chicago, the rise of powerful street gangs, the growth of the Nation of Islam, and the emergence of the black and Puerto Rican nationalist movements created additional tension and produced a situation primed for violence.

The 1970s were also a period of political transition in Chicago. Richard J. Daley, Chicago's mayor from 1955 until his death in 1976, built one of the strongest

Table 1. A selection of major riots and events of mass violence in Chicago and neighboring communities up to the start of the Burge torture cases in the 1970s

1855, Lager Beer Riot. Enforcement of liquor laws targets German and Irish immigrants.

1886, Haymarket Affair. A protest favoring an eight-hour day develops into one of the watershed events in U.S. labor and legal history.

1894, Pullman Strike. A nationwide railroad strike leads to violence over reduced wages in a company town established by George Pullman.

1919, Chicago Race Riot. One of the largest or most violent riots in the United States during the "Great Migration" of black Americans from the South as these new migrants began to compete for jobs and housing with German, Irish, and other white immigrants.

1931, Chicago Rent Strike Riot. Violence during the Great Depression resulting from high levels of unemployment.

1937, Republic Steel Strike/Memorial Day Massacre. Shooting of demonstrators during Little Steel Strike.

1946, Airport Homes Race Riots. White violence against blacks moving into Chicago Housing Authority homes on southwest side.

1947, Fernwood Park Race Riot. White violence against blacks moving into Chicago Housing Authority homes on South Side.

1949, Englewood Race Riot. White violence against Blacks over integration of housing.

1951, Cicero Riot. White mob violence over integration of housing.

1950s, Founding of Major Chicago Street Gangs. These include, among others, the Black P. Stone Nation/Blackstone Rangers/El Rukn; Almighty Vice Lord Nation; and Latin Kings.

1964, Dixmoor Riot. Arrest results in violence in black community.

1966, Division Street/West Town/Humboldt Park Riots. Riots in Puerto Rican neighborhoods after police shooting.

1966, Chicago West Side Riots. Riots following arrest of man for opening fire hydrant during heat wave.

1968, Chicago Riots/West Side Riots. Violence following the assassination of Martin Luther King, Jr.

1968, Democratic National Convention Protests. Antiwar demonstrations to disrupt convention.

1969, Days of Rage Protests. Political actions organized by the Weatherman faction of the Students for a Democratic Society.

1969, Chicago Police and FBI Attack on Black Panthers. Assassination of Fred Hampton and Mark Clark.

1972, Start of Burge Torture Cases. These continue until 1989.

1977, Humboldt Park Riot. Riot in Puerto Rican neighborhoods after police shooting.

and most influential Democratic political machines in the twentieth century. His death led to a period of leadership changeovers until his son, Richard M. Daley, was elected mayor in 1989, eventually serving until 2011. It was during the period between 1976 and 1989 that the police torture cases peaked. One irony is that Harold Washington, Chicago's first black mayor

(1983–1987), served during this period until his sudden death in 1987. Eugene Sawyer, at that time Chicago's longest serving black alderman, was elected by the city council to complete the remainder of Washington's second term (1987–1989). Among the issues that remain difficult to answer about the torture cases are those of "who knew what and when"—because Washington and Sawyer represented constituencies on Chicago's South Side. Although these questions about the political context of the torture cases go beyond the scope of this article, it is important to note that future mayor Richard M. Daley served as Cook County State's Attorney from 1981 to 1989. A prison doctor's report of torture was forwarded to his office in 1982 (as noted later) but received no response, although Daley would eventually be involved in resolving the torture cases as mayor.

Many details about the torture cases and Jon Burge's career are documented in a variety of sources (Conroy 2000; Van Cleve 2016; Chicago Torture Justice Memorials 2017). Burge, an Army veteran, joined the Chicago Police in 1970 after serving in South Korea and Vietnam. He rose through the ranks and received numerous commendations and promotions along the way. Beginning around 1972, Burge and other officers began to torture suspects into making confessions. They tended to use painful or stressful techniques that left few physical marks on their victims. Some of the most painful devices were cattle prods, stun guns, an old-fashioned hand-cranked telephone, modified hair dryers, and a "violet wand," an electrically charged sex toy. All of these devices were used to deliver painful electrical shocks to the suspects' faces, gums, genitals, anuses and were even inserted into their rectums. Mock executions were also performed, including pointing guns at suspects and putting plastic bags over their heads.

A turning point in the case occurred in 1982 when two police officers were killed on 9 February. The hunt for the killers—perhaps spurred by the killing of two other officers earlier in the year—ended in the arrest of brothers Andrew and Jackie Wilson on 14 February. The police had Andrew Wilson's confession by the end of the day, at which time he was taken to a local hospital with cuts on his head and face, chest bruises, and burns on his thighs (Conroy 2007). Dr. John Raba, who examined Wilson at the hospital, as well as another doctor at Cook County Prison, noted their suspicions of torture to the Police Superintendent as well as

to State's Attorney Richard M. Daley, as noted earlier. In 1983, Andrew Wilson was sentenced to death, his brother to life imprisonment. Eventually, Jackie Wilson's conviction was overturned and Andrew Wilson was granted a second trial. Sentenced to life imprisonment, he died in prison in 2007.

The most important aspect of Andrew Wilson's story is that he never gave up in pressing his claim of torture. He began in 1986 by filing, on his own, a civil suit in federal court claiming torture, a suit that was tried three times: the first ending in a mistrial; the second deciding that Wilson's rights had been violated but that he had not been subjected to excessive force; and the third in 1996 with a verdict in Wilson's favor. Wilson's case is important not only because he won but because the information that surfaced between 1986 and 1996 (including anonymous tips coming from within the police department) led to the exposure of Burge and his colleagues. Since the case began, more than 110 torture victims have been identified. Many of their verdicts have been overturned or reversed. Some of the defendants have been pardoned or had their sentences commuted, but others remain incarcerated.

The Spatiality of Torture

It is difficult to imagine that such widespread and long-term violence could continue for so long unless it was isolated spatially in ways that allowed the Chicago police to conduct their activities unchecked. This involved not one but many ways in which racial exclusion operated across many scales, from the local to the national. At each level—whether of policing, housing, or economic opportunity—the overlap of racist attitudes and actions helped amplify the effects of the others, allowing officials to deny that such policies were aimed at blacks or other racial and ethnic groups. Perhaps the most obvious spatial dimension was where the torture took place. Chicago's police department is organized hierarchically into three areas (north, central, south), twenty-five districts, sixty-nine sectors, and finally 277 beats (Figure 1).

For most of his career, Burge's commands were in Districts 2 and 3 on Chicago's South Side, an area that was home to one of the largest black communities in the city. Through time, these districts had been isolated spatially by the construction and

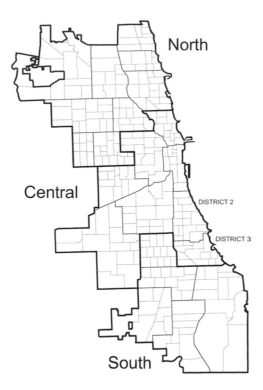

Figure 1. The spatial organization of Chicago's police areas, districts, and beats, excluding O'Hare International Airport to the northwest of the map. *Source*: Map by Zachary Taylor based on http://gis.chicagopolice.org/pdfs/district_beat.pdf.

expansion of the Dan Ryan Expressway (I-90/I-94) running north to south along the western edge of the district, the Stevenson Expressway (I-55) cutting the area off from Chicago's central business district to the north, and Lakeshore Drive and Lake Michigan bordering the districts to the east (Figure 2). Some of the largest public housing projects in the city were built there—the Robert Taylor Homes, Harold Ickes Homes, Stateway Gardens, Dearborn Homes, and Hilliard Homes—although some of these have since been demolished. At the scale of the city, these districts served to enforce what Massey and Denton (1995) termed "American apartheid," segregated enclaves defined socially, economically, and spatially. As Shabazz (2015) wrote:

> Continued economic and spatial disenfranchisement of Black residents also contributed to the failure of the housing projects and produced, in many instances, intergenerational poverty based on geography. Additionally, the projects failed because of architecture. Postwar public housing in Chicago relied almost exclusively on high-rise models, which packed large number of people in less space, but such design isolated them from the rest of the city. ... A

consequence of using high-rise housing structures to enable racial segregation was that they made it possible keep Blacks in certain parts of the city, effectively cutting them off from the resources that would enable social and economic growth. (56, 63)

When these new residential projects did not work as expected, efforts were made to police them more rigorously:

> The abundance of security measures that shaped the project—chain-link fencing that covered the balconies, policing, video surveillance, perimeter patrols, controlled visitation, curfews, apartment sweeps, metal detectors, stop-and-frisk techniques, turnstiles, onsite [police] courts and the resident identification program—radically transformed the built environment of the project. It also changed its function, moving it from a home space to a liminal place between home and prison. (Shabazz 2015, 68)

Given this situation, Districts 2 and 3 were large areas within which the police could move with little fear of consequence. As Chicago journalist John Conroy (2000, 24) wrote about the search for the Wilson brothers:

> Every young black male in sight was being stopped and questioned, and ... the Reverend Jesse Jackson proclaimed that the black community was living under martial law, in "a war zone ... under economic, political, and military occupation," that the police department was holding "the entire black community hostage for the crimes of two. ..."

Police kicked down doors, invaded and threatened people in their homes, and detained or arrested people on the slightest pretext. When Andrew Wilson was finally arrested, he was tortured at the district police headquarters. Its interrogation rooms provided confined, publicly inaccessible places where officers could do almost anything to prisoners while, at the same time, deny everything they did. In some cases, the police took men to distant isolated locations where no one could see, hear, or document their torture.

Many first-person accounts of these episodes of torture have been documented in legal affidavits and documents. One important example comes from the 1985 testimony of Darrell Cannon. He described in detail how he was tortured at an isolated industrial area on Chicago's South Side (Figure 3; Cannon 1985). The officers resorted to physical and psychological torture—a cattle prod pressed to Cannon's testicles, a shotgun pushed into his mouth, a flashlight used as a club—until he confessed (Figure 4). In this

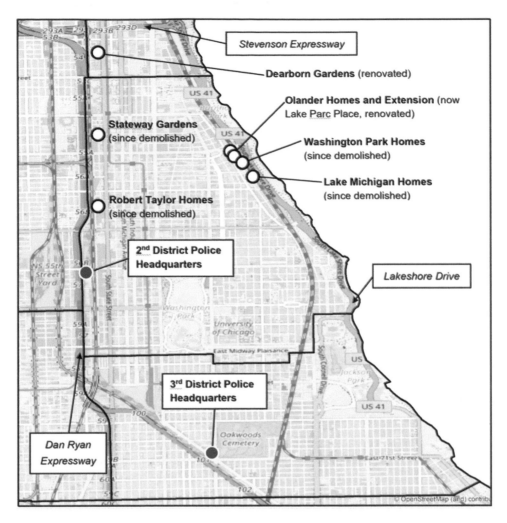

Figure 2. Police Districts 2 and 3 on Chicago's South Side including major expressways and public housing projects. *Source:* Base map from OpenStreetMap. Map by Ken Foote and Rich Mrozinski. (Color figure available online.)

case, the torture moved from the scale of the city and neighborhood to the personal, intimate spaces of the body.

Given the racial, socioeconomic, and professional disparities in status between the police and their suspects, claims by the torture victims were easy to ignore or deny. Cannon's wife filed a complaint with the Police Department's Office of Professional Standards (OPS) but the complaint was dismissed.

Discussion: Deny, Defend, and Deflect

That Burge and his colleagues were able to continue torturing prisoners for so long speaks to the layers of protection that buffered them from repercussions. Space was one dimension of this protection insofar as Districts 2 and 3 were more occupied than protected by the police. By labeling and isolating these districts

as crime-ridden and lawless, the police created a layer of deniability around their own lawless activities. The hierarchical organization of police jurisdictions and the autonomy offered officers working their beats allowed them to keep activities hidden from public view. Space alone might be an insufficient explanation for what happened in Chicago, but it helped to confine and hide the violence for many years because racialized policing, housing, and employment seemed to amplify their effects in space.

Equally important were the tremendous social, economic, and professional disparities between the police officers and the men they tortured. Who would believe claims of police violence made by criminals and gangsters from the South Side? To Burge, his officers, and much of the world outside of Districts 2 and 3, the suspects were "human vermin" and "guilty vicious criminals" (Spielman 2015). Their status made a difference whenever a complaint was filed about police brutality.

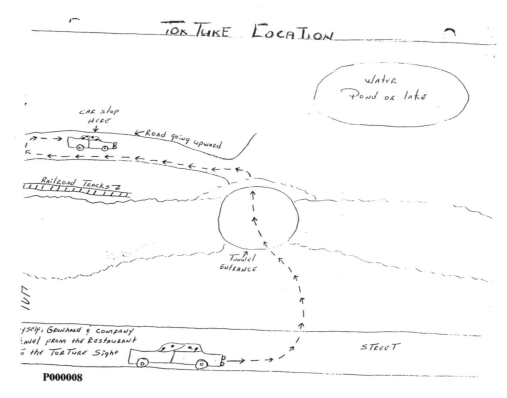

Figure 3. Darrell Cannon's sketch of torture location. *Source:* Cannon (1985). Used with permission.

Studies of complaints made against the Chicago police indicate that overall only about 2.6 percent were sustained in recent years, but only about 1.6 percent of those filed by black residents were (Shifflett et al. 2015; Invisible Institute 2016). As the Shifflett et al. (2015) study bluntly noted, "Most unsustained

Figure 4. Darrell Cannon's sketch and description of one episode in his torture. *Source:* Cannon (1985). Used with permission.

allegations had black complainants. Most sustained allegations had white complainants." A *Chicago Tribune* study yielded slightly different percentages—but comparably low figures, also noting that most sustained complaints resulted in, at best, minor sanctions (Gorner and Hing 2015). These findings and other complaints have spurred a U.S. Department of Justice investigation, but this is only in its earliest stages (U.S. Department of Justice 2015).

Indeed, making a complaint about the Chicago police can be a complicated process that is stacked against the complainant. Investigations and punishments involve three separate agencies—the Independent Police Review Authority (IPRA); the police department itself, including its Bureau of Internal Affairs; and the Chicago Police Board (City of Chicago, Chicago Police Board 2016). Officers against whom complaints are filed are provided with legal support and advice during the review process by their union, the Fraternal Order of Police, Chicago Lodge 7, whereas the complainants must cover their legal fees themselves. Moreover, by state law, complaints will only be investigated in Illinois if they are accompanied by "a sworn affidavit that certifies that the allegation is true and correct. If the person making the complaint did not witness the alleged conduct, they must certify that they believe that the facts in the allegation are true" (City of Chicago, IPRA 2016). People arrested or harassed are very hesitant to file such paperwork because details of their complaints can be used against them in court.

Together these bureaucratic barriers helped to reinforce what has been termed the "blue wall of silence," "blue code," or "blue shield," the unwritten conventions among police officers in Chicago and other departments not to report mistakes, misconduct, crimes, or violence. This means that, if questioned about such incidents, police officers deny or claim that they were unaware of any wrongdoing. Breaking this wall of silence has proven very difficult, even today when the use of personal digital media and body cameras makes it increasingly hard for police officers to engage in unlawful behavior without fear of repercussions.

Finally, even when the wall of silence is breached, a final defense is the "official inquiry." The Chicago Police Department has repeatedly faced cases of internal misconduct, crime, and corruption that have been publicly exposed (Table 2; Lindberg 1998). Each time, a commission is formed,

Table 2. A selection of major events and investigations of the Chicago Police Department relating to violence and race

1919, *Chicago Defender* publishes story detailing police brutality during race riot.

1922, Chicago Commission on Race Relations issues report on 1919 riots, *The Negro in Chicago*.

1928, Chicago Crime Commission established as an independent, nonpartisan civic organization aimed at coordinating work of the business community, government agencies, and law enforcement to address organized crime.

1945-1946, Commission on Human Relations established.

1966, Chicago Citizens Committee to Study Police Community Relations created following 1966 riots.

1967, National Advisory Commission on Civil Disorders, the Kerner Commission, focuses on race riots in Chicago and other cities.

1968, National Commission on the Causes and Prevention of Violence, Chicago Study Team investigates violence during Democratic National Convention, issues final report, *Rights in Conflict* (the Walker Report).

1972, Chicago Human Relations Commission reviews police files on police brutality.

1976, U.S. Senate Select Committee to Study Governmental Operations with Respect to Intelligence Activities (Church Committee) issues report on FBI's actions to destroy Black Panther Party as part of COINTELPRO program including complicity in raid that killed Fred Hampton and Mark Clark.

1972, Metcalfe Report released: *The Misuse of Police Authority in Chicago*.

1974, Chicago Law Enforcement Study Group issues report: *The Question of Police Discipline in Chicago*.

1991, Amnesty International calls for an investigation into allegations of police torture.

1992–1993, Burge faces internal hearing and is fired from Department.

2008, Burge charged with obstruction of justice and perjury.

2011–2014, Burge serves jail sentence for obstruction of justice and perjury, not torture or other charges.

2015, Task Force on Police Accountability formed to review police system of accountability, oversight, and training.

2015, Justice Department opens investigation into the Chicago Police Department.

an investigation is convened, and a report is issued—with little long-term follow-up. Certainly, it is possible to point to some long-standing changes based on investigations of police corruption. Indeed, the Chicago Crime Commission, created as an independent, nonpartisan oversight organization in 1919, was based on efforts to fight organized crime and the police corruption it spawned. In so many other cases, however, committees, commissions, and investigations seem to be mainly public relations efforts that deflect attention away from police (Platt 1971). The general pattern of these public relations responses is to deny any allegation, defend any officer, and deflect attention away from the department. In the end, then, space is not the only barrier to police accountability but is interwoven with these other means of oppression. Police officers know that, even if charged with violence toward suspects, they are likely to avoid punishment or at least delay sanctions for many years.

It is highly ironic that when redress did occur in this case, it was because local organizations "jumped scale" and brought the claims of police torture to the attention of the United Nations. In December 2007, Chicago's People's Law Office presented a report to the UN Committee on the Elimination of Racial Discrimination, and in November 2014, the group We Charge Genocide submitted a report to the United Nations Committee Against Torture. Both reports generated a larger conversation within the international realm, resulting in calls for investigation. The international spotlight was what led eventually to the prosecution of Burge and his officers and to the award of reparations to the victims of his torture. In turn, the city developed subcommittees on police brutality, and investigations were begun by the Civil Rights Division of the U.S. Department of Justice. At the same time, these actions might spur only temporary correctives. According to the project Mapping Police Violence (2016), the most recent data indicate that a "majority (51 percent) of unarmed people who were killed were black," making these rates four times the rate of white victims. Despite the seemingly bittersweet ending of the Chicago police torture cases, the prevalence of police brutality continues.

Conclusion: The More Theory Advances, the More Reality Remains Unchanged

Between 1972 and 1991, Chicago Police Districts 2 and 3 were transformed into "laboratories for organized violence, where new forms of suppression, punishment, and political control were practiced and refined" (Thomas 2011, xxii). Within this space, Burge and his officers could suspend the law and hide their violence. The aftermath of their work is sobering. A reparation fund of $5.5 million was approved for victims, but this sum pales in comparison with the funds the City of Chicago has spent defending itself and its police force in

legal actions stemming from the torture cases. Nevertheless, the broader question is whether the litigation of the torture cases has reduced the use of racially directed violence in Chicago or elsewhere around the country. Clearly, this has not happened. Reparations, no matter how large, do not remove the traumas associated with the torture cases. As one of the victims, Anthony Holmes (2012) noted:

> There is always this inner fear that I will get tied into something I didn't do, and they will tie me up with something. You can never describe that first feeling when they call you or see you. There is nothing I can do. That is why I no longer live in the City. I always have the fear with police—oh boy here they come. I am just a little or a lot paranoid.

Carceral power was ingrained in the geographies of black Chicago, making police violence a common feature of African American communities and black occupied space. Hence, decades of police violence and surveillance unchecked removed any sense of safety that might have existed for Holmes.

Even as we write this article, continuing urban violence has claimed the lives of African Americans such as Tamar Rice, Sandra Bland, Michael Brown, Eric Garner, Keith Scott, and Anthony Lamar Smith. The police officers in these cases were white and exonerated of all charges. In many respects, the situation is hardly different now from that in 1903, when W. E. B. Du Bois wrote ([1903] 2003):

> It was not then a question of crime, but rather one of color, that settled a man's conviction on almost any charge. Thus Negroes came to look upon courts as instruments of injustice and oppression, and upon those convicted in them as martyrs and victims. (127)

In 1973, Harvey's solution to the spatialization of race was to suggest moving away from the von Thünen bid-rent model and toward "a socially controlled urban land market and socialized control of the housing sector" (Harvey 1973, 137). But isn't this social (and racialized) control of space, land, and housing a factor that led to police violence in Chicago?

Our point is that this theorization of race, place, social justice, and the city has advanced considerably since the 1970s, but our despair is that the more theory advances, the more the reality of racialized violence remains unchanged. We have no ready answer to this paradox but can suggest three issues for further research. These are investigations into the long-term use of violence as a means of maintaining or policing social, racial, and ethnic boundaries; the increasing reliance on covert, hidden, or deniable violence such as those employed by Burge and his officers to enforce racial and social order; and the use of official investigations to deflect attention away from change and reform. These are long-term efforts, but such long-term efforts are perhaps critical to changing the patterns of spatialized and racialized violence that are so much a part of American life.

ORCID

Aretina R. Hamilton ⓘ http://orcid.org/0000-0002-9246-8617

References

Beckett, K., and N. Murakawa. 2012. Mapping the shadow carceral state: Toward an institutionally capacious approach to punishment. *Theoretical Criminology* 16 (2):221–44.

Bell, D., and G. Valentine. 1995. *Mapping desire: Geographies of desire*. London and New York: Routledge.

Blomley, N. 2003. Law, property, and the geography of violence: The frontier, the survey, and the grid. *Annals of the Association of American Geographers* 93 (1):121–41.

Boddie, E. C. 2010. Racial territoriality. *UCLA Law Review* 58:401–63.

Bonnett, A. 1997. Geography, "race" and whiteness: Invisible traditions and current challenges. *Area* 29 (3):193–99.

Bunge, W. 1965. Racism in geography. *The Crisis* 73 (8):494–97.

Cannon, D. 1985. Affidavit of defendant Darrell Cannon. Accessed October 7, 2017. http://chicagotorture.org/files/2012/03/17/Darrell_Cannon_Affidavit_and_Drawings.pdf.

Chicago Torture Justice Memorials. 2017. Home page. Accessed December 10, 2017. http://chicagotorture.org/.

City of Chicago, Chicago Police Board. 2016. Police discipline. Accessed October 7, 2017. https://www.cityofchicago.org/city/en/depts/cpb/provdrs/police_discipline.html.

City of Chicago, Independent Police Review Authority. 2016. Complaint against a Chicago police officer. Accessed October 7, 2017. https://www.cityofchicago.org/city/en/depts/copa.html.

Conroy, J. 2000. *Unspeakable acts, ordinary people: The dynamics of torture*. New York: Knopf.

———. 2007. The persistence of Andrew Wilson. *The Chicago Reader* November 29. Accessed October 7, 2017. http://www.chicagoreader.com/chicago/the-persistence-of-andrew-wilson/Content?oid=999832.

Drake, S., and H. R. Cayton. [1945] 2015. *Black metropolis: A study of Negro life in a northern city*. Chicago: University of Chicago Press.

Du Bois, W. E. B. [1903] 2003. *The souls of black folk*. New York: Barnes and Noble.

Eaves, L. 2017. Black geographic possibilities: On a queer black south. *Southeastern Geographer* 57 (1):80–95.

Fanon, F. 1963. *The wretched of the Earth*, trans. C. Farrington. New York: Grove.

Foucault, M. 1979. *Discipline and punish: The birth of the prison*. New York: Vintage.

Gilroy, P. 1993. *The black Atlantic*. Cambridge, MA: Harvard University Press.

Gilmore, R. W. 1999. You have dislodged a boulder: Mothers and prisoners in the post-Keynesian California landscape. *Transforming Anthropology* 8(1–2):12–38.

———. 2007. *Golden gulag: Prisons, surplus, crisis, and opposition in globalizing California*. Berkeley: University of California Press.

Goffman, E. 1961. *Asylums: Essays on the social situation of mental patients and other inmates*. Garden City, NY: Anchor.

Gorner, J., and G. Hing. 2015. Tribune analysis: Cops who pile up complaints routinely escape discipline. *Chicago Tribune* June 13. Accessed October 7, 2017. http://www.chicagotribune.com/news/ct-chicago-police-citizen-complaints-met-20150613-story.html.

Hague, E. 2010. "The right to enter every other state"—The Supreme Court and African American mobility in the United States. *Mobilities* 5 (3):331–47.

Harvey, D. 1973. *Social justice and the city*. London: Edward Arnold.

Herbert, S. K. 1997. *Policing space: Territoriality and the Los Angeles Police Department*. Minneapolis: University of Minnesota Press.

Herbert, S., and E. Brown. 2006. Conceptions of space and crime in the punitive neoliberal city. *Antipode* 38 (2):755–77.

Holmes, A. 2012. Violent victim impact statement. Defendant's name: Jon Burge, Case number: 2007R00712. Accessed October 8, 2017. https://chicagotorture.org/files/2012/03/17/Holmes_sentencing_statement.pdf.

Hunter, A. 1974. *Symbolic communities: The persistence and change of Chicago's local communities*. Chicago: University of Chicago Press.

Invisible Institute. 2016. Projects. Accessed October 7, 2017. http://invisible.institute/police-data/.

Jackson, J. B. 1970. *Landscapes: Selected writings of J. B. Jackson*, ed. E. H. Zube. Amherst: University of Massachusetts Press.

Jackson, K. T. 1985. *Crabgrass frontier: The suburbanization of the United States*. New York: Oxford University Press.

Kobayashi, A. 2014. The dialectic of race and the discipline of geography. *Annals of the American Association of Geographers* 104 (6):1101–15.

Kobayashi, A., and L. Peake. 1994. Unnatural discourse: "Race" and gender in geography. *Gender, Place & Culture* 1 (2):225–43.

———. 2000. Racism out of place: Thoughts on whiteness and an antiracist geography in the new millennium. *Annals of the American Association of Geographers* 90 (2):392–403.

Lindberg, R. 1998. *To serve and collect: Chicago politics and police corruption from the Lager Beer Riot to the Summerdale Scandal 1855–1960*. Carbondale: Southern Illinois University Press.

Lipsitz, G. 2011. *How racism takes place*. Philadelphia: Temple University Press.

Longhurst, R. 1995: The body and geography. *Gender, Place and Culture: A Journal of Feminist Geography* 2 (1):97–105.

Mapping Police Violence. 2016. Home page. Accessed October 7, 2017. http://mappingpoliceviolence.org/.

Massey, D. A., and A. Denton. 1995. *American apartheid: Segregation and the making of the underclass*. Cambridge, MA: Harvard University Press.

McKittrick, K. 2006. *Demonic grounds: Black women and the cartographies of struggle*. Minneapolis: University of Minnesota Press.

McKittrick, K., and C. Woods, eds. 2007. *Black geographies and the politics of place*. Cambridge, MA: South End Press.

Nimako, K., and S. Small. 2009. Theorizing black Europe and African diaspora: Implications for citizenship, nativism and xenophobia. In *Black Europe and the African diaspora: Blackness in Europe*, ed. D. C. Hine, T. D. Keaton, and S. Small, 212–37. Urbana: University of Illinois Press.

Park, R. E., E. W. Burgess, and R. D. McKenzie. 1925. *The city*. Chicago: University of Chicago Press.

Peake, L., and R. Schein. 2000. Racing geography into the new millennium: Studies of "race" and North American geographies. *Social and Cultural Geography* 1 (2):133–42.

Philo, C. 2001. Accumulating populations: Bodies, institutions and space. *International Journal of Population Geography* 7 (6):473–90.

Platt, A. M. 1971. *The politics of riot commissions, 1917–1970: A collection of official reports and critical essays*. New York: Collier.

Reed, C. R. 2011. *The rise of Chicago's black metropolis, 1920–1929*. Urbana: University of Illinois Press.

Rose, H. M. 1978. The geography of despair. *Annals of the American Association of Geographers* 68 (4):453–64.

Rothstein, R. 2017. *The color of law: A forgotten history of how our government segregated America*. New York: Liveright.

Shabazz, R. 2015. *Spatializing blackness: Architectures of confinement and black masculinity in Chicago*. Urbana: University of Illinois Press.

Shaw, W. 2007. *Cities of whiteness*. Malden, MA: Blackwell.

Shifflett, S., A. Scheller, S. Alecci, and N. Forster. 2015. Police abuse complaints by black Chicagoans dismissed nearly 99 percent of the time: Investigators rarely sustain allegations of any kind. *The Huffington Post* December 7. Accessed October 7, 2017. http://data.huffingtonpost.com/2015/12/chicago-officer-misconduct-allegations.

Sibley, D. 1995. *Geographies of exclusion: Society and difference in the West*. London and New York: Routledge.

Smith, N. 2008. Book review essay: Castree, N. and Gregory, D., *David Harvey: A critical reader*. *Progress in Human Geography* 32:147–55.

Spielman, F. 2015. Disgraced Chicago cop Jon Burge breaks silence, condemns $5.5 million reparations fund. *Chicago Sun-Times* April 17. Accessed October 7, 2017. http://chicago.suntimes.com/politics/disgraced-chicago-cop-jon-burge-breaks-silence-condemns-5-5-million-reparations-fund/.

Suttles, G. D. 1968. *The social order of the slum: Ethnicity and territory in the inner city*. Chicago: University of Chicago Press.

Thomas, M. 2011. Introduction: Mapping violence onto the French colonial mind. In *The French colonial mind: Vol. 2. Violence, military encounters, and colonialism*, ed. M. Thomas, xi–1. Lincoln: University of Nebraska Press.

Tyner, J. 2012. *Space, place, and violence: Violence and the embodied geographies of race, sex, and gender*. London and New York: Routledge.

Tyner, J., and J. Inwood. 2014. Violence as fetish: Geography, Marxism and dialectics. *Progress in Human Geography* 38 (6):771–84.

U.S. Department of Justice, Office of Public Affairs. 2015. Justice Department opens pattern of practice investigation into the Chicago Police Department. Accessed October 7, 2017. https://www.justice.gov/opa/pr/justice-department-opens-pattern-or-practice-investigation-chicago-police-department.

Van Cleve, N. G. 2016. *Crook County: Racism and injustice in America's largest criminal court*. Stanford, CA: Stanford University Press.

Walcott, R. 2007. *Homopoetics: Queer space and the black queer diaspora*. Cambridge, MA: South End Press.

Wilson, B. M. 2000. *America's Johannesburg: Industrialization and racial transformation in Birmingham*. Lanham, MD: Rowman & Littlefield.

Woods, C. 1998. *Development arrested*. New York: Verso.

———. 2007. No one knows the mysteries at the bottom of the ocean. In *Black geographies and the politics of place*, ed. K. McKittrick and C. Woods, 1–13. Cambridge, MA: South End Press.

Wynter, S. 2003. Unsettling the coloniality of being/power/truth/freedom: Towards the human, after man, its overrepresentation—An argument. *CR: The New Centennial Review* 3 (3):257–337.

ARETINA R. HAMILTON is an Adjunct Professor at Georgia Gwinnett University, Lawrenceville, GA 30043. E-mail: ahamilton11@ggc.edu. Her research focuses on anti-black violence, race and trauma in urban spaces, as well as black queer geographies in the U.S. South.

KENNETH FOOTE is Professor and Head of the Department of Geography at the University of Connecticut, Storrs, CT 06269-4148. E-mail: ken.foote@uconn.edu. His research focuses on violence in U.S. society as well as issues of public memory and commemoration.

10 The Uneven Geographies of America's Hidden Rape Crisis

A District-Level Analysis of Underpolicing in St. Louis

Alec Brownlow

The article uses ratios of rape and homicide to explore the underpolicing of rape in U.S. cities. In doing so, I build on Yung's (2014) rate-based model and identify a statistic (the *c*-value) that can be used to rapidly assess or rank the policing behaviors of different metropolitan departments. I apply this method to a finer scale analysis of district-level crime in the city of St. Louis, Missouri. When combined with demographic data in a geographic information system, results suggest that police in precincts serving majority black constituencies are more likely to undercount rape than their peers attached to precincts that serve constituencies with fewer blacks. *Key Words: policing, race, rape, statistics, St. Louis.*

本文运用强暴与谋杀率，探讨美国城市对强暴的警备不足。我藉此根据荣 (Yung 2014) 建立在比率上的模型，并指认能够用来快速取得或排名大都会区不同部门的警备行为之统计 (c值)。我将此一方法应用至密苏里圣路易市较微观的街区层级犯罪尺度。当与地理信息系统中的人口数据结合时，研究结果显示，在主要为黑人选区服务的警方，较服务于较少黑人的选区之同僚，更倾向低报强暴案件。 *关键词： 警备，种族，强暴，统计，圣路易。*

Este artículo usa ratios de violación y homicidio para explorar el déficit de atención policial a la violación en las ciudades norteamericanas. Para hacer esto, me apoyo en el modelo de Yung basado en tasa (2014) e identifico una estadística (el valor *c*) que puede usarse para evaluar rápidamente o categorizar los comportamientos policiales de diferentes departamentos metropolitanos. Aplico este método en un análisis a escala fina del crimen a nivel de distrito en la ciudad de San Louis, Missouri. Cuando se le combina con datos demográficos en un sistema de información geográfica, los resultados sugieren que la policía de los precintos que sirven electorados negros estará más inclinada a subvalorar la violación que sus colegas adscritos a precintos que sirven a electorados con menos negros. *Palabras clave: vigilancia policial, raza, violación, estadística, San Luis.*

Policing and race in the United States have captured the attention of the U.S. public and media as perhaps never before (Weitzer 2015). An ever-diversifying landscape of video and social media technologies and accessibilities has eliminated once-resilient spatial, temporal, and political hurdles to information access and availability (Meek 2012). As a result, with growing frequency the public is eyewitness to a distressingly routine pattern of police brutalities toward black males especially and can see and judge for itself the circumstances involved in these often deadly encounters. As a result, this emerging era of hashtag activism and journalism (Yang 2016) is challenging dominant police narratives of the "justified shooting" and hegemonic representations of crime, space, and black male bodies by the police and in the media (Kristoff 2014; Bonilla and Rosa 2015; Chaney and Robertson 2015; Derickson 2016). To this end, the new hypervisibilities of racialized police injustice are a welcome, powerful, and validating instrument among social justice activists, police reform advocates, and scholars of police–race relationships (e.g., Smith, Visher, and Davidson 1984; Weitzer and Tuch 2006; Brunson and Miller 2006a, 2006b; Stewart 2007; Rice and White 2010). Across the social sciences, the new visibilities underscoring this age of Ferguson (Bernard 2015) are catalyzing new inquiries and critiques into the production and political economies of urban space that approach these "geographically diffuse events [as] somehow part of a deeper culture of policing, racism, and the security state" (Derickson 2016, 2; see also Brownlow 2009).

Buried among the visibilities and media spectacles signifying the age of Ferguson, and unavailable to bystander filming or hashtag journalism, however, are urban geographies of racialized policing that remain doggedly hidden and spatially discreet, in particular the underpolicing of female bodies of color that are the victims of rape in the private sphere. Despite regular investigative reports by local news outlets detailing the inventive measures taken by police to suppress or eliminate rape from official crime rates across urban America—from Philadelphia (Fazlollah, Matza, and McCoy 1999) to St. Louis (Kohler 2004, 2005), from Atlanta (Bruner 2003) to New York (Rayman 2010), from Baltimore (Hermann 2010) to New Orleans (Maggi 2009; Daley and Martin 2014)—and regardless of the steady drum of tell-all accounts and admissions to these very practices by former police officials from across the United States (Bouza 1990; Burnham 1996; Stamper 2006; Eterno and Silverman 2012), the underpolicing of rape has been strikingly less successful in capturing the attention of or conjuring cries of injustice or demands for police reform among the U.S. public or mainstream media.

My goal in this article is to help make visible the hidden geographies of rape-hiding by police in the urban United States and to demonstrate where in U.S. cities police pursue this practice and who is at risk. To achieve this, I adopt Yung's (2014) model of identifying and discerning questionable rape policing practices and geographies. I introduce Yung's rate-based model and suggest two shortcomings, one statistical and the other spatial. In response, I advance Yung's model by identifying a descriptive statistic, the *c-value*, that corresponds to and reflects accurately Yung's (2014) results. I use *c*-values to demonstrate the accuracy and argue for the utility of a quickly derived, ratio-based model of policing. With this alternative approach, I extend Yung's metropolitan scale of analysis to a finer and, arguably, more explanatory spatial scale—the police district, a geography of sociospatial control that geographers (e.g., Herbert 1996, 1997) and criminologists (e.g., Klinger 1997; Kane 2002) alike consider central to police culture, identity, and behavior. I pursue this district-level analysis in the City of St. Louis, a historically segregated and economically struggling U.S. metropolis where undercounting rape and sexual violence by city police has been revealed in the local media (Kohler 2004, 2005). My findings, when combined with demographic data in a geographic information system (GIS), suggest that reporting victims of rape who reside in police districts dominated by communities of color are especially vulnerable to police malfeasance or neglect.

Yung and How to Lie with Rape Statistics

Responding to the growing number of media investigations that collectively suggest the systemic mishandling of rape victims and their reports by urban police departments across the United States, Yung (2014) proposed a method to discern those cities whose police effectively hide rape from official crime rates from those whose police appear to pursue the offense more robustly. By standardizing and tracking across time rates of rape (a statistic notoriously vulnerable to police fudging and manipulation) against standardized rates of homicide (a crime more difficult to recast and a statistic more difficult to massage), Yung distinguished *typical* (i.e., no evidence of data tampering) from *undercounting* (i.e., evidence of data tampering) police departments in hundreds of U.S. cities. He concluded that no fewer than 796,000 forcible rapes were deliberately hidden by police from official crime rates between 1995 and 2012 (i.e., 44,000–63,000 per year), a national epidemic he called "America's hidden rape crisis" (see Chemaly 2014; also McCoy, Matza, and Fazlollah 1998; Hermann 2010).

For his model, Yung (2014) measured rates of rape vis-à-vis rates of homicide for all U.S. cities with populations greater than or equal to 100,000 between the years of 1995 and 2012. His identification of homicide data as a baseline against which to assess the reliability of rape data is predicated on three key assumptions. First, over any given period of time (measured in years) and regardless of analytical scale (e.g., police district, city, region, state), the frequency of murder per population unit is several times rarer and more unevenly distributed temporally (e.g., crime waves, seasonal spikes) and spatially (e.g., hot spots) than is the more spatially and temporally persistent crime of rape (Brownmiller 1993; Hester, Kelly, and Radford 1996). According to the FBI's annual Uniform Crime Report (UCR), the annual national rape:homicide (R:H) ratio throughout the eighteen-year period of Yung's study averaged 5.5:1 (never dipping below 4.5:1). Other instruments, such as the Bureau of Justice's National Crime Victimization Survey, place this ratio even higher (see National Research Council 2014). Second, relative to rape, homicide data are generally accurate; that is, homicide is not a crime that easily avails itself to statistical manipulation or recasting by police. Although recent reports suggest the cunning used by police in Chicago (Bernstein and Isackson, 2014) and Detroit (LeDuff and Santiago 2009) to keep some homicides (e.g., of homeless victims) off the official crime sheets, hiding bodies or labeling homicide as anything but homicide is a decision fraught with political risk and thus tends to be vigorously policed as homicide (Klinger 1997; Zimring 2007; cf. Chambliss 2000). This is not the case with rape, which, for a variety of reasons and by a variety of means and instruments, is widely and historically undercounted (i.e., downgraded, hidden) by police and throughout the criminal justice system (Hanmer, Radford, and Stanko 1989; Carrington and Watson 1996; Hester, Kelly, and Radford 1996; Kelly and Radford 1996; Cahill 2001; Leuders 2006; Brownlow 2009). Third, when plotted over time, the incidental relationship between the two crime types, rape and homicide, is strongly correspondent; that is, over a given period of time (measured in years), their relative upward and downward trends generally track one another closely albeit at different frequencies. This can be seen, for example, in the relative stability of the national R:H ratio over Yung's eighteen-year study period as measured by the UCR.

Yung (2014) used data from the FBI's UCR, an annual compilation of crime data solicited from jurisdictions across the nation. Significantly, jurisdictions are not obligated to participate in the UCR process either entirely or in part. Further, only data "known to police" are reported to the FBI by local jurisdictions; crimes "unknown to police" are not included. Thus, the data compiled in and distributed through the annual UCR are only as accurate as the data gathering instruments, policies, personnel, and institutions that record and report them, a fact that has brought the UCR under intense scrutiny and criticism for its inaccuracies and exploitable methodology (Levitt 1998; Chambliss 2000). Still, the UCR is widely used for analyses and assessments of crime rates, patterns, and trends both in the media and academia. Moreover, insofar as crime data are locally generated and reported, the UCR avails itself to analyses of questionable bookkeeping and statistical maneuvers through anomalies in the crime frequencies, ratios, and relationships discussed earlier.

With these assumptions in mind, and using annual UCR data for the eighteen-year period between 1995 and 2012, Yung examined the relationship between rates of rape (RR) and rates of homicide (HR) for all U.S. cities with populations of 100,000 or greater

$(n = 210)$. Rates of each crime type are calculated at the per 100,000 population level for each metropolitan area and for each year. To account for national trends (see Cohen and Felson 1979; Blumstein and Wallman 2006), annual RR and HR for each city are standardized by, respectively, dividing each by national rape (RR_n) and homicide (HR_n) rate for that same year. So, for any given year in City A:

$$RR_s = RR_A/RR_n \quad \text{and} \quad HR_s = HR_A/HR_n/$$

When plotted over the eighteen-year period, the relationship of these standardized rates vis-à-vis one another demonstrate, according to Yung, a given police department's commitment toward policing rape. For Yung, a *typical* city (i.e., one whose police force does not [noticeably] fudge rape numbers) demonstrates standardized rates of rape and homicide that track one another closely over the eighteen-year period (Assumption 3 earlier). Alternatively, a city identified by Yung as an *undercounter* demonstrates standardized homicide rates several times greater than standardized rates of rape. As suggested earlier (Assumption 1), given the time span of Yung's study, this should never be the case. Yung's findings suggest that police in New Orleans (Maggi 2009; Daley and Martin 2014), Atlanta (Bruner 2003), Baltimore (Hermann 2010), Philadelphia (McCoy, Matza, and Fazlollah 1998), and St. Louis (Kohler 2004, 2005), among others, all chronically undercount rape. Independently, local media investigations in each of these cities have reached the same conclusion.

Although an intuitive and powerful method for identifying questionable policing practices in particular cities, Yung's method contains two shortcomings. The first is the absence of a meaningful statistic that can rapidly identify and accurately and reliably distinguish typical from undercounting city police departments. Second is his emphasis on the scale of the city, a unit of analysis that fails to distinguish varying policing patterns potentially identifiable at finer scales. Identifying different cities as either typical or undercounting is a helpful first step in discerning and drawing attention to potentially questionable policing geographies; however, by failing to resolve patterns at more local scales, Yung's analysis neglects the history, spatial variability, and place-based contingencies of police practice and, in doing so, runs the risk of inaccurately branding entire police departments and their personnel as either malfeasant or truthful, when in reality neither is ever entirely the case.

c-Values and the R:H Ratio

To address the first shortfall, I replicated Yung's method of standardizing city rape and homicide rates at the per 100,000 population level for all U.S. cities with populations of 250,000 or more ($n = 61$) between the years 1995 and 2012. Next, for each city, I determined the annual relationship between the two standardized rates by creating a ratio (RR:HR) between the two.

$$RR_s: HR_s \text{ for year } x = RR_s/HR_s$$

For each city, ratios were calculated for each year over the eighteen-year period. Eighteen-year averages were calculated to provide a numeric expression of a given city's RR_s:HR_s relationship. I call these *correspondence values*, or *c-values*. Unsurprisingly, cities identified by Yung (2014) as typical have *c*-values approximating 1.0. That is, over the eighteen-year period covered, standardized rates of rape and homicide maintained an average tracking correspondence approximating 1:1, thus reflecting the third assumption listed earlier. Examples include Seattle (0.94), San Antonio (1.03), Omaha (1.03), and Boston (1.04), among others. Similarly, cities identified by Yung as undercounters had significantly smaller *c*-values, indicating rates of rape considerably lower than both national trends and the city's homicide rate. Examples include New Orleans (0.16), Washington, DC (0.19), Baltimore (0.21), Miami (0.25), and St. Louis (0.26), among others.[1]

Significantly, *c*-values can be accurately calculated without the additional step of standardization. I achieved this by, first, determining the R:H ratio for a given city in a given year, where R:H = #rapes/#homicides. For example, in 2012 the City of Boston reported to the UCR 249 rapes and 57 homicides, for an R:H of 4.4 (i.e., 249/57). Meanwhile, also in 2012, the City of New Orleans reported to the UCR 136 rapes and 193 homicides, for an R:H of 0.7 (i.e., 136/193). Using UCR data I calculated annual R:H ratios for all cities with populations of 250,000 or greater for the years 1995 through 2012. Second, national R:H ratios were determined for each year in the eighteen-year period; that is:

$$R:H_n = \text{\#rapes in the U.S.}/ \text{\#homicides in the U.S.}$$

Third, following Yung, to account for national trends, annual R:H values for any given city were standardized

by division with annual R:H_n values. For example, the R:H for Boston in 2012 was 4.4 (see earlier). The R:H_n that same year was 5.7. Thus, R:H_s for Boston in 2012 = 4.4/5.7 = 0.77.

Finally, for any given city, I averaged the annual R:H_s values over the eighteen-year time period. For all cities, the resulting R:H_s figures accurately replicated those *c*-values calculated using homicide and rape rates (mean difference = 0.007), the implication being that unstandardized R:H ratios are, in and of themselves, statistically accurate and intuitively meaningful measures of police behavior in cases of rape. For example, cities identified by Yung as typical (i.e., eighteen-year *c*-values approximating 1.0) revealed average, unstandardized R:H ratios at or near the national R:H average (5.5) over this period. Similarly, cities that, according to their *c*-values, undercount have corresponding R:H values that are significantly below the national mean.

Policing and the Police District

Influenced by and stemming from ecological theories of crime and criminality (e.g., Shaw and McKay 1942; Stark 1987; Reiss and Tonry 1988; Bursik and Grasmick 1993; Sampson and Wilson 1995), theorists of police and policing increasingly consider the significance of place to police behavior (Smith 1986; Stark 1987; Herbert 1997; Klinger 1997; Kane 2002; Terrill and Reisig 2003). From this perspective, police culture—that is, the rules, values, and norms that define and structure normative police work; that guide behaviors, decision making, and social relations; and that function to ensure group coherence and security in uncertain and hazardous environments (see Herbert 1998; Paoline 2003)—is an emergent and responsive phenomenon, neither rigid nor homogeneous across the police landscape. Rather, it is necessarily adaptive, developing in place and over time in response to local, place-specific contexts within which individual officers and units operate and function daily.

For many of these researchers, the police district (i.e., the precinct) is the organizational and administrative spatial unit that (1) is most entrenched within the urban milieu; (2) is most responsive to and reflective of its local context; (3) informs and shapes how law and the norms of policing are interpreted and pursued; and (4) receives, embeds, and socializes new recruits within this bounded, place-contingent organizational and occupational culture, thus ensuring its reproduction. For example, Herbert's (1996, 1997) ethnography of officers and police work in the Los Angeles Police Department's Wilshire Division identifies the district as the operational geography of police territoriality and spatial control. How these spatial pursuits unfold and are mobilized on the ground and the meanings and interpretive values assigned to them by individual officers can, according to Herbert, be interpreted within a more general framework of "normative orders" that, respectively and in combination, provide structure, meaning, value, and coherence to the occupation of policing. For Herbert, the six normative orders of law, bureaucratic control, machismo, safety, competence, and morality constitute the basis for police action and decision making and the prism through which situations and people are judged, assessed, categorized, and responded to (or not; Herbert 1998); they are, according to Herbert, the building blocks of police culture.

Herbert's is a compelling illustration of the formalized and institutionalized boundedness of police perceptions of, movement through, and control over urban space, one where both police culture and "the district" are performed, made meaningful, and dialectically reproduced through place-based policing (Herbert 1997, 1998). What remains unclear and underaddressed in Herbert's writing on the subject, however, is how the normative orders of policing that he identified as central to police culture and identity differentially develop and operate, both independently and vis-à-vis one another, in response to district-specific conditions. What accounts for the spatial variability of police behaviors and actions across any city's district landscape, where some districts might be notoriously corrupt or abusive and others are not? What, in other words, is the role of place in the development of the normative orders of police culture (see Marston 1997)?

The ecological theories of Klinger (1997) and Kane (2002) emphasize the place-based characteristics, patterns, and longevities of the police district, adding additional insight to the developmental geographies of police culture. Klinger (1997), building on Stark's (1987) thirty-point framework of place-based deviance, argued that the general resiliency of any given district's crime profile (i.e., continually high or violent crime vs. continually low or nonviolent crime) is foundational to the formation, development, and persistence of distinct, district-level police cultures. Here, district officers, as "boundary personnel ... utterly immersed in the environment of the districts they

patrol" (Klinger 1997, 287), assume the norms of local (i.e., district) police culture, including attitudes and behaviors toward everything from how crime is perceived and responded to (e.g., what counts as "normal" crime; leniency vs. vigor in their response to particular crimes) to how victims are perceived and responded to (e.g., "deservedness" of victimization or of police follow-through). Generally speaking, high-crime districts with high police workloads result in high thresholds of what constitutes normal crime, police leniency toward (or neglect of) crimes perceived to be "low priority" (including rape), and increased cynicism toward victims (including victims of rape; Hanmer, Radford, and Stanko 1989; Carrington and Watson 1996; Kelly and Radford 1996), whereas officers assigned to low-crime districts pursue even low-priority crimes with vigor and are more sympathetic toward crime victims.

Following Klinger, Kane (2002) demonstrated the significance of district-level police culture to the racialized geography and nature of police misconduct in New York City, where patterns of aggressive overpolicing (e.g., of black men) or neglectful underpolicing, of victim neglect or victim sympathy, reflect and are differentiated by the demographic profile of any given police district. Kane found that police in districts populated primarily by communities of color (i.e., historically segregated, economically marginalized, perceived as criminal; more likely to concentrate violent crime; etc.) are more likely to host local police cultures that engage in, turn a blind eye toward, or even encourage police misconduct than are officers assigned to districts whose whiter or more affluent constituents enjoy an altogether different policing experience (Smith, Visher, and Davidson 1984; Brunson and Williams 2006a, 2006b; Weitzer and Tuch 2006; Rice and White 2010).

In what follows, I apply Yung's (2014) quantitative model, albeit modified as described earlier, to an ecologically informed, district-level analysis of potential rape-hiding in the city of St. Louis, Missouri.

Method

As demonstrated earlier, R:H ratios accurately replicate c-values, are easily calculated and, when graphed longitudinally, offer an intuitive and insightful glimpse into the place-specific and temporal machinations and politics of rape policing in the urban United States. In this section, I apply this statistic toward a finer scale

Table 1. Mean demographic characteristics and R:H data for police districts in St. Louis, Missouri (1995–2012)[a]

District	1	2	3	4	5	6	7	8	9
Percentage Black[b]	17.9	10.1	35.6	64.7	91.1	91.6	82.1	92.4	63.7
Percentage FamPov	16.2	7.6	24.4	33.6	33.7	26.1	26.5	30.5	25.1
R:H									
Range[c]	1.8–4.2	1.7–9.5	1.4–5.7	1.4–5.0	0.4–1.4	0.5–1.7	0.7–2.2	0.3–2.4	0.7–4.4
Mean R:H	3.0	5.8	2.4	3.2	0.8	1.0	1.3	1.2	2.7
# Years ≤ 1.0	0	0	0	1	13	9	6	8	2
Percentage of years	0	0	0	5.8	76.5	53.0	35.3	47.0	11.8
# Years ≤ 2.0	3	2	7	6	17	17	16	15	7
Percentage of years	17.6	11.8	41.2	35.3	100	100	94.1	88.2	41.2

Note: Data from U.S. Census Bureau (1990–2010) and St. Louis Metropolitan Police Department (1995–2012).
[a]Does not include district-level data from 2001 ($n = 17$ years).
[b]Demographic statistics were calculated by averaging U.S. census data from 1990, 2000, and 2010.
[c]High and low R:H values were removed as outliers.

analysis of policing in and across the City of St. Louis, Missouri, between 1995 and 2012. I first acquired district-level annual rape and homicide data.[2] For each police district ($n = 9$), I calculated annual R:H ratios. These annual ratios, when averaged over the seventeen-year period (district-level data were not available for 2001), provide a numerical indicator of district-level rape policing profiles. I next gathered tract-level demographic data for census years 1990, 2000, and 2010. For the purposes of this study, tract-level data for percentage black and percentage families in poverty are emphasized. In a GIS, and using police district shapefiles provided by the St. Louis Metropolitan Police Department (SLMPD), I calculated percentage black and percentage families in poverty for each

police district for each decennial census year ($n = 3$; 1990–2010). When averaged, these provide demographic profiles of each police district over the time span covered, against which district-level R:H data can be compared over the same period (Table 1). The results are discussed later.

St. Louis, Missouri

Compared with all U.S. cities with populations of at least 250,000 (mean R:H = 5.5), city-wide R:H values in St. Louis were abysmal, remaining below 2.0 through the first ten years of the study period and dipping below or hovering near the 1.0 mark between 1999 and 2004 (Figure 1). Following a two-year

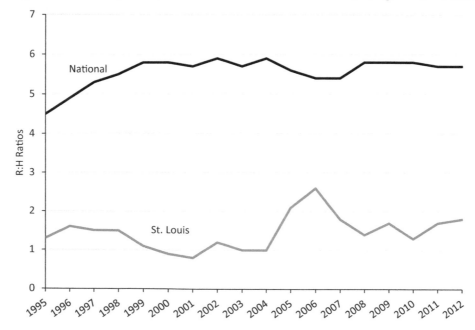

Figure 1. St. Louis, Missouri, R:H ratios (city and national; 1995–2012. Data from FBI Uniform Crime Reports [1995–2012]).

investigative exposé in the *St. Louis Dispatch* that accused the SLMPD of deliberately undercounting rape (Kohler 2004, 2005), the city-wide R:H rose noticeably, if unimpressively (never breaching the 3.0 mark), over the next two years, only to drop below 2.0 again for the remainder of the study period.

The SLMPD is divided into nine police districts (Figure 2).[3] As Figure 3 demonstrates, on average, over a quarter (2.3 out of 9) of all districts reported R:H ratios less than or equal to 1.0 (i.e., disturbingly low) for each year of the study period. This number rises to over 50 percent (average = five districts) at R:H less than or equal to 2.0 (i.e., questionably low). In only one year (2006) do no districts report disturbingly

low R:H ratios; in no year do fewer than four districts report questionably low R:H ratios.

District comparisons reveal distinct profiles that suggest local disparities in how rape is policed. Seventeen-year district R:H averages range from a low of 0.8 (District 5) to a high of 5.5 (District 2), with only the latter approaching the national R:H mean during this period (Table 1). Out of nine police districts, four (44.4 percent) have mean R:H ratios less than 2.0 (Districts 5–8); only three (33.3 percent) have mean ratios above or equal to 3.0 (Districts 1, 2, and 4). Police Districts 5 through 8 consistently occupy the questionably low (R:H ≤ 2.0) or disturbingly low (R:H ≤ 1.0) range of R:H values, whereas other districts are only occasional visitors to these lower R:H regions

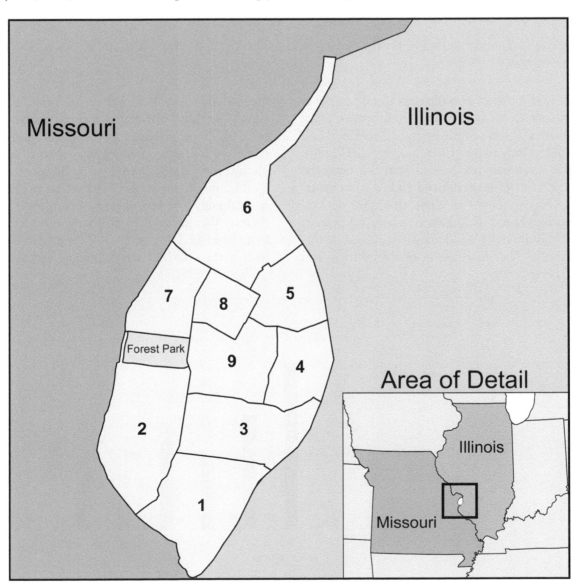

Figure 2. Districts, City of St. Louis Police Department (1995–2012). Data from St. Louis Metropolitan Police Department (1995–2012). Map created by Cassie Follett.

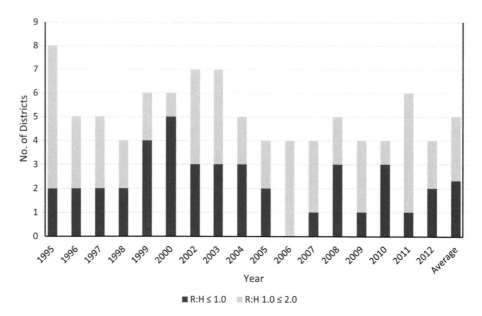

Figure 3. Annual incidences of disturbingly low (≤1.0) and questionably low (≤2.0) R:H ratios (1995–2012). Data from St. Louis Metropolitan Police Department (1995–2012).

(Figure 4). R:H ratios in two districts (5 and 6) never eclipsed the questionably low mark in seventeen years; a third (District 7) did so only once (2.2 in 1995) and District 8 did so only twice (1998 [3.0] and 2012 [2.4]).

Over the seventeen-year period analyzed here, the nine SLMPD districts experienced 153 "district years" (9 × 17). Over a quarter of these ($n = 39$) demonstrate disturbingly low R:H ratios; nearly 60 percent ($n = 90$) of all district years demonstrate questionably low R:H ratios. The distribution of disturbingly and

questionably low R:H values is, however, far from equal. Districts 5 through 8 account for nearly 95 percent of all occurrences in the disturbingly low R:H range, with District 5 alone accounting for a third of all disturbingly low district years. By contrast, Districts 1 and 2 rarely display R:H ratios that might qualify as questionably low, and never as disturbingly low. Nonetheless, the mean R:H of District 2 (5.8) is nearly twice that of District 1 (3.0); it is the only district within the SLMPD whose mean R:H approximates

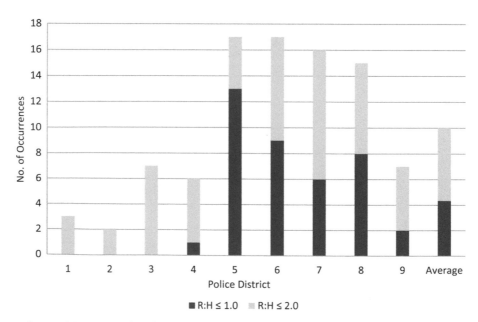

Figure 4. Per district incidence of disturbingly low (≤1.0) and questionably low (≤2.0) R:H ratios (1995–2012). Data from St. Louis Metropolitan Police Department (1995–2012).

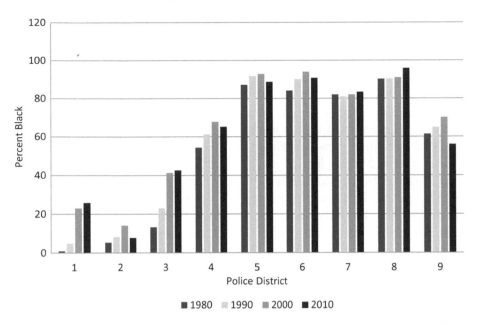

Figure 5. Percentage black population in St. Louis police districts (1990–2010). Data from U.S. Census Bureau (1990–2010).

that of the rest of the urban United States over this time period.

In short, although city-wide data suggest that rape-hiding by the SLMPD is a pervasive, persistent, and systemic problem, side-by-side comparisons of district R:H profiles suggest that the City of St. Louis's poor track record is driven by persistently abysmal R:H values in four out of the SLMPD's nine police districts. This district group (Districts 5–8) occupies the north-central and northern portions of the City of St. Louis (Figure 2). These low values are potentially indicative of malfeasant practices and behaviors among police assigned to these district areas (Kohler 2004, 2005). Moreover, their place-based longevity suggests the entrenchment of norms and standard operating procedures toward the policing of rape that could very well be described as cultural (see Stark 1987; Herbert 1997; Klinger 1997; Kane 2002). Victims of sexual violence

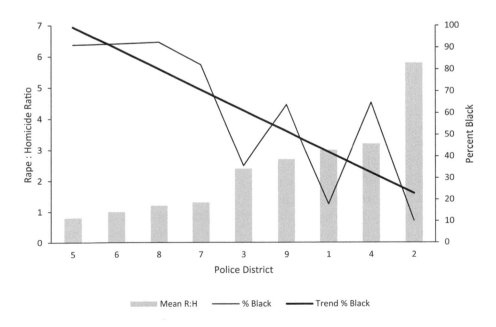

Figure 6. Mean district R:H ratios as a function of racial constituencies. Data from St. Louis Metropolitan Police Department (1995–2010) and U.S. Census Bureau (1990–2010).

who live in communities served by these districts appear to be at particular risk of having their complaint downgraded, dismissed, or otherwise wiped from the official crime rate. I now turn to the question of who these victims are.

To identify the social inequities of spatially uneven policing, I analyzed tract-level demographic data from the 1990, 2000, and 2010 decennial censuses within the spatial framework of the SLMPD's district map, whose boundaries and configuration remained stable over this time period (Figure 2). This was performed using ArcInfo GIS (Version 10.1, Esri, Redlands, CA, USA). As Table 1 demonstrates, six police districts (4–9) serve populations in central and north St. Louis that are majority African American. Constituencies in Districts 5 through 8 are almost entirely African American and have remained so for at least the past three decades (Figure 5; Tighe and Ganning 2015). Districts 4 and 9, although majority black, are just so and maintain sizable and, in the case of District 9, growing white populations. Since 1990, black populations have grown in those parts of the city served by Districts 1 and 3 but do not constitute a majority in either. District 2 serves a constituency that is overwhelmingly and historically white (Tighe and Ganning 2015).

District-level R:H ratios correlate convincingly with the racial constituency served within each police district. Specifically, police districts in St. Louis with disturbingly and questionably low mean R:H values serve constituencies that are, and have historically been, overwhelmingly African American (Figure 6). Generally speaking, when moving across the SLMPD landscape, as district constituencies become "whiter" (i.e., proportionately fewer African Americans), R:H ratios rise accordingly.

The trenchantly nonblack District 2 maintains the highest mean R:H of all nine SLMPD districts; it also is least likely to dip into disturbingly or questionably low R:H ratios in any given year. The opposite is true of the trenchantly black districts of the city's north. Meanwhile, Districts 4 and 9, where African Americans maintain only slight majorities of the district constituency, display intermediary mean R:H values and fewer visits than their northerly neighbors to the low end of the R:H range in any given district year, suggesting the possible tensions, conflicts, and contradictions that influence normative police behavior and practice (i.e., police culture) when confronted with the shifting political and economic pressures and power dynamics that accompany demographic change.

Conclusion

Feminist scholars and activists have long decried the patriarchal politics of police and rape, often with little political effect (Hanmer, Radford, and Stanko 1989; Brownmiller 1993; Hester, Kelly, and Radford 1996; Cahill 2001; Leuders 2006). Scholars and activists of race and policing have raised similar complaints of racialized police abuses and, until recently, faced similar skepticism and inattention among the U.S. public and media. The age of Ferguson, by making hypervisible the seemingly routine and systemic nature of racialized police injustices, has effectively, if not entirely, challenged the hegemonic narrative of the latter (Derickson 2016). In doing so, however, the invisibilities of racialized, patriarchal policing have potentially become even more tenacious and entrenched.

Within this context, Yung's (2014) rate-based model is a compelling first attempt to glean and distinguish what has remained hidden and invisible from data that are widely available and under our noses (see Hiemstra 2017). His model, however, is not without its shortcomings. On the one hand, I have demonstrated in this article that Yung's rate-based model can be easily and accurately simplified by exchanging his rate-based analysis with a quickly computed R:H ratio. I identify this correspondence through their mutual relationship with what I have called a c-value. Simplifying Yung's method makes it accessible to scholars and activists alike in their efforts to make rape more visible, demand police reform, and pursue justice for victims of rape and sexual violence. On the other, I have advanced Yung's metropolitan-based analysis to a study of police districts, those bounded, political and social geographies within the context of which, it is reasoned, police culture assumes distinct form and character (Herbert 1997; Klinger 1997; Kane 2002). The focus on the scale of the district, rather than the city, brings greater resolution to those geographies that are the drivers of any given city's rape-policing profile. For example, in the case of St. Louis—a city identified by Yung and a police department accused by local media as a chronic undercounter of rape—a district-by-district analysis clearly demonstrates that that city's poor performance is the result of disturbingly low and questionably low R:H numbers in a handful of police districts on the city's historically black north side; districts with majority white constituencies on the city's south side maintained R:H ratios closer to the national mean, but their values and frequencies were both indiscernible and negligible when combined in city-wide analyses with the lower and more frequent lows of the north side. Three important points emerge from this finding. First, it appears that the racialized underpolicing, or neglect, of female rape victims of color is as much a part of racialized policing and is as entrenched within police culture as is the racialized overpolicing of black men. Although the age of Ferguson has clearly drawn our attention to the latter, the complexities and injustices of gender, race, and policing go well beyond tragic spectacles of police beatings and shootings (Brunson and Miller 2006a). Second, by failing to look deeper into the spatial machinations of underpolicing, and by identifying cities as either typical or undercounting, Yung inaccurately, unnecessarily, and unfairly runs the risk of branding entire police departments and their personnel as either wholly corrupt or not corrupt. A growing body of police scholarship and memoirs consistently indicates that neither is ever completely the case (Bouza 1990; Burnham 1996; Herbert 1997; Stamper 2006; Eterno and Silverman 2012). Third, the notion of district drivers is as applicable to typical cities, like Ft. Worth or Omaha, as it is for undercounters, like St. Louis or New Orleans. It is entirely possible, and indeed likely, that the typicality of a city like, for example, Seattle (c-value close to 1.0) is a product of that city's majority white and wealthy population overshadowing and effectively hiding the undercounting of rape victimization that likely is practiced by police assigned to those fewer districts containing that city's relatively scant minority population. Typicality does not, in all likelihood, preclude injustice as much as it masks it. Each of these points suggests that district-level applications of Yung's (2014) model offer greater clarity of the history, machinations, and relationships among race, gender, and policing in the urban United States.

This study was compelled by academic and investigative accounts of malfeasance toward rape and rape victims by police in the City of St. Louis. Caution is emphasized, however, in any more generalized application to or interpretation of police or policing in other cities. The model proposed here is case specific, and it is my intention to draw attention to low R:H ratios as generally problematic and demanding greater clarity more than to suggest police as generally corrupt. Low R:H ratios are not inevitably indicative of (solely) malfeasant policing. Drivers of low R:H rates might include inter alia (1) greater police attention to rape in "whiter" districts; (2) negligence toward rape and rape victims by police assigned to

"less white" districts; (3) greater trust in and reporting of rape to the police by rape victims in "whiter" districts; (4) diminished trust in and, therefore, diminished reporting of rape to police by rape victims in "less white" districts; (5) downgrading of rape or dismissal of rape reports at other levels of the criminal justice system (e.g., investigators, prosecutors, judges, juries, etc.; see Carrington and Watson 1996; Kelly and Radford 1996); or (6) any combination of these drivers. Combined with corresponding accusations by the local media, disturbingly low (R:H $\leq$ 1.0) rates of the kind persistently measured in St. Louis suggest the likelihood of malfeasant police methods and bookkeeping as, at least, one part of what might be a more complex and complicated SLMPD puzzle. Whatever their explanation, low R:H ratios and their unequal distribution across the urban landscape are worrisome, requiring greater investigation and clarity on a city-by-city basis.

Acknowledgments

The author is grateful to Allison Williams, Nikki Chaffin, Cassandra Follett, Patrick McHaffie, and the Social Science Research Center at DePaul University for their assistance and suggestions in the development and production of this research. Thanks are also due to the helpful comments and suggestions of two anonymous reviewers.

Notes

1. Cities with c-values significantly greater than 1.0 (e.g., Colorado Springs, Colorado [3.18], St. Paul, Minnesota [2.28], Corpus Christi, Texas [2.24]) suggest a third profile type neither identified nor discussed by Yung.
2. These data are publicly available in the form of annual police reports published in either hard copy or online by respective city police departments.
3. In 2014, the number of St. Louis police districts was reduced to six.

References

Bernard, E. 2015. Between the world and me: Black American motherhood. *The Atlantic* July 28. Accessed April 30, 2017. http://www.theatlantic.com/national/archive/2015/07/between-the-world-and-me-black-motherhood/399668/.

Bernstein, D., and N. Isackson. 2014. Are crime stats getting washed? Special report, Part 2. *Chicago Magazine* May 18:70–75, 136, 141–44.

Blumstein, A., and J. Wallman, eds. 2006. *The crime drop in America.* New York: Cambridge University Press.

Bonilla, Y., and J. Rosa. 2015. #Ferguson: Digital protest, hashtag ethnography, and the racial politics of social media in the United States. *American Ethnologist* 42:4–17.

Bouza, A. 1990. *The police mystique.* New York: Plenum.

Brownlow, A. 2009. Keeping up appearances: Profiting from patriarchy in the nation's "safest" city. *Urban Studies* 46:1680–1701.

Brownmiller, S. 1993. *Against our will: Men, women, and rape.* New York: Ballantine.

Bruner, T. K. 2003. Discipline recommended in handling of rape files. *The Atlanta Journal-Constitution* September 5: 3B.

Brunson, R. K., and J. Miller. 2006a. Gender, race, and urban policing: The experience of African American youths. *Gender & Society* 20:531–52.

———. 2006b. Young black men and urban policing in the United States. *British Journal of Criminology* 46:613–40.

Burnham, D. 1996. *Above the law: Secret deals, political fixes, and other misadventures of the U.S. Department of Justice.* New York: Scribner.

Bursik, R. J., and H. G. Grasmick. 1993. *Neighborhoods and crime.* New York: Lexington.

Cahill, A. J. 2001. *Rethinking rape.* Ithaca, NY: Cornell University Press.

Carrington, K., and P. Watson. 1996. Policing sexual violence: Feminism, criminal justice, and governmentality. *International Journal of the Sociology of Law* 24:253–72.

Chambliss, W. J. 2000. *Power, politics, and crime.* New York: Westview.

Chaney, C., and R. V. Robertson. 2015. Armed and dangerous? An examination of fatal shootings of unarmed black people by police. *Journal of Pan African Studies* 8:45–78.

Chemaly, S. 2014. How did the FBI miss over 1 million rapes? *The Nation* June 27. Accessed April 30, 2017. https://www.thenation.com/article/how-did-fbi-miss-over-1-million-rapes/.

Cohen, L. E., and M. Felson. 1979. Social change and crime rate trends: A routine activities approach. *American Sociological Review* 44:588–608.

Daley, K., and N. Martin. 2014. NOPD's statistics on rape questioned. *The Times-Picayune* May 14:A01.

Derickson, K. D. 2016. Urban geography II: Urban geography in the age of Ferguson. *Progress in Human Geography* 40:1–15.

Eterno, J. A., and E. B. Silverman. 2012. *The crime numbers game: Management by manipulation.* Boca Raton, FL: CRC.

Fazlollah, M., M. Matza, and C. R. McCoy. 1999. After FBI questioned one tactic, another was found. *Philadelphia Inquirer* October 18:A8.

Federal Bureau of Investigation (FBI). 1995–2012. *Crime in the U.S.* Accessed September 28, 2017. https://ucr.fbi.gov/crime-in-the-u.s/.

Hanmer, J., J. Radford, and E. A. Stanko, eds. 1989. *Women, policing, and male violence: International perspectives.* London and New York: Routledge.

Herbert, S. 1996. The normative ordering of police territoriality: Making and marking space with the Los Angeles Police Department. *Annals of the Association of American Geographers* 86:567–82.

———. 1997. *Policing space: Territoriality and the Los Angeles Police Department.* Minneapolis, MN: University of Minnesota Press.

———. 1998. Police subculture reconsidered. *Criminology* 36:343–69.

Hermann, P. 2010. Downgrading rape cases not a new problem. *Baltimore Sun* June 30:A6.

Hester, M., L. Kelly, and J. Radford, eds. 1996. *Women, violence, and male power.* Buckingham, UK: Open University Press.

Hiemstra, N. 2017. Periscoping as a feminist methodological approach for researching the seemingly hidden. *The Professional Geographer* 69:329–36.

Kane, R. J. 2002. The social ecology of police misconduct. *Criminology* 40:867–96.

Kelly, L., and J. Radford. 1996. "Nothing really happened": The invalidation of women's experiences of sexual violence. In *Women, violence, and male power,* ed. M. Hester, L. Kelly, and J. Radford, 99–115. Buckingham, UK: Open University Press.

Klinger, D. 1997. Negotiating order in patrol work: An ecological theory of police response to deviance. *Criminology* 35:277–306.

Kohler, J. 2004. Rosy crime numbers were wrong. *St. Louis Post Dispatch* November 18:A1.

———. 2005. What rape? *St. Louis Post Dispatch* January 17:A1.

Kristoff, N. 2014. When whites just don't get it: After Ferguson, race needs more attention, not less. *New York Times* August 30. Accessed April 30, 2017. https://www.nytimes.com/2014/08/31/opinion/sunday/nicholas-kristof-after-ferguson-race-deserves-more-attention-not-less.html.

LeDuff, C., and E. Santiago. 2009. Detroit police routinely underreport homicides. *The Detroit News* June 18:A1.

Leuders, B. 2006. *Cry rape.* Madison: University of Wisconsin.

Levitt, S. D. 1998. The relationship between crime reporting and police: Implications for the use of Uniform Crime Reports. *Journal of Quantitative Criminology* 14:61–81.

Maggi, L. 2009. Decline in rape reports raises questions. *The Times-Picayune* July 12:1.

Marston, S. 1997. Who's policing what space? Critical silences in Steve Herbert's *Policing Space. Urban Geography* 18:385–88.

McCoy, C. R., M. Matza, and M. Fazlollah. 1998. Statistical manipulation by police goes back decades. *The Philadelphia Inquirer* November 1:A27.

Meek, D. 2012. YouTube and social movements: A phenomenological analysis of participation, events, and cyberspace. *Antipode* 44:1429–48.

National Research Council. 2014. *Estimating the incidence of rape and sexual assault.* Washington, DC: National Academies Press.

Paoline, E. A. 2003. Taking stock: Toward a richer understanding of police culture. *Journal of Criminal Justice* 31:199–214.

Rayman, G. 2010. NYPD Tapes 3: A detective comes forward about downgraded sexual assaults. *The Village Voice* June 8:1.

Reiss, A. J., and M. Tonry, eds. 1988. *Communities and crime*. Chicago: University of Chicago.

Rice, S., and M. White, eds. 2010. *Race, ethnicity, and policing*. New York: NYU Press.

Sampson, R. J., and W. J. Wilson. 1995. Toward a theory of race, crime, and urban inequality. In *Crime and inequality*, ed. J. Hagan and R. D. Peterson, 37–51. Palo Alto, CA: Stanford University Press.

Shaw, C. R., and H. D. McKay. 1942. *Juvenile delinquency and urban areas*. Chicago: University of Chicago Press.

Smith, D. A. 1986. The neighborhood context of police behavior. In *Communities and crime*, ed. A. J. Reiss and M. Tonry, 313–41. Chicago: University of Chicago Press.

Smith, D. A., C. A. Visher, and L. A. Davidson. 1984. Equity and discretionary justice: The influence of race on police arrest decisions. *Journal of Criminal Law & Criminology* 74:234–49.

Stamper, N. 2006. *Breaking rank: A top cop's exposé of the dark side of American policing*. New York: Nation Books.

Stark, R. 1987. Deviant places: A theory of the ecology of crime. *Criminology* 25:893–99.

Stewart, E. A. 2007. "Either they don't know or they don't care": Black males and negative police experiences. *Criminology & Public Policy* 6:123–30.

St. Louis Metropolitan Police Department. 1995–2012. Annual report to the community. St. Louis, MO: Metropolitan Police Department.

Terrill, W., and M. D. Reisig. 2003. Neighborhood context and police use of force. *Journal of Research in Crime & Delinquency* 40:291–321.

Tighe, J. R., and J. P. Ganning. 2015. The divergent city: Unequal and uneven development in St. Louis. *Urban Geography* 36:654–73.

United States Census Bureau/American FactFinder. 1990–2010. *Summary File 1: P3, Race and Hispanic or Latino origin*. Accessed September 28, 2017. https://factfinder.census.gov/faces/nav/jsf/pages/searchresults.xhtml?refresh=t

Weitzer, R. 2015. American policing under fire: Misconduct and reform. *Social Science and Public Policy* 52:475–80.

Weitzer, R., and S. A. Tuch. 2006. *Race and policing in America*. New York: Cambridge University Press.

Yang, G. 2016. Narrative agency in hashtag activism: The case of #BlackLivesMatter. *Media & Communication* 4:13–17.

Yung, C. R. 2014. How to lie with rape statistics: America's hidden rape crisis. *Iowa Law Review* 99:1197–1256.

Zimring, F. E. 2007. *The great American crime decline*. New York: Oxford University Press.

ALEC BROWNLOW is an Associate Professor in the Department of Geography at DePaul University, Chicago, IL 60614. E-mail: alec.brownlow@depaul.edu. His research explores geographies of violence and sacrifice and their relationship to development processes and politics.

11 Building Relationships within Difference

An Anarcha-Feminist Approach to the Micropolitics of Solidarity

Carrie Mott

Nuanced understandings of praxes of solidarity are critical for grassroots political activists from diverse backgrounds to be able to work together. Since the early 2000s, the primary concern of grassroots political activism in Tucson has been migrant justice and opposition to the militarization of the U.S.–Mexico border. In the aftermath of Arizona's notorious 2010 racial profiling legislation, SB 1070, The Protection Network Action Fund (ProNet) was founded as a collaboration between migrant activist members of The Protection Networks and their allies, with the expressed goal of fundraising to support migrant-led activism in Tucson. ProNet's strategy is rooted in long-term relationship building between migrant activists and predominantly white allies and a commitment to address micropolitical challenges within Tucson, where white-led humanitarian aid groups often remain unaware of activism in Spanish-speaking Chican@ and Latin@ communities. This article examines ProNet as an example of anarcha-feminist solidarity work rooted in a praxis of autonomous horizontal organizing that also takes deliberate steps to negotiate the differences in embodied social privilege that accompany race, class, gender, language, and documentation status. *Key Words: activism, anarcha-feminism, micropolitics, social justice, solidarity.*

细緻地理解团结实践, 对于来自不同背景的草根政治行动者得以共同合作来说至关重要。自 2000 年代早期以降, 图森的草根政治行动主义之主要考量, 便是移民正义与反对美墨边境的军事化。自 2010 年亚历桑那州通过恶名昭彰的 SB 1070 种族定性立法之后, 便成立了保护网络行动基金 (ProNet), 作为保护网络的移民社运成员及其联盟之间的合作, 而其目标在于募集资金来支持图森中以移民为首的行动主义。ProNet 的策略, 植基于移民社运者和主要由白人所构成的同盟之间长期建立的关系, 以及应对图森中的微政治之承诺, 其中以白人为首的人道救援团体, 经常未能察觉说西语的在美墨西哥人和拉丁社群中的社会运动。本篇文章检视 ProNet 作为植基于自发性水平组织实践的无政府女权主义团结行动之案例, 该组织方式同时採用谨慎的步骤, 协商随着种族、阶级、性别、语言与身份状态而来的社会优势中的差异。 关键词: 行动主义, 无政府女权主义, 微政治, 社会正义, 团结。

La flexibilidad en las maneras de entender las prácticas de solidaridad es crucial para que las bases del activismo político con diferentes antecedentes puedan trabajar juntas. Desde principios de los 2000, las principales preocupaciones del activismo político de base en Tucson han sido la justicia para el migrante y la oposición a la militarización de la línea fronteriza EE.UU.–México. Inmediatamente después de la legislación de Arizona en 2010 con sesgo racial, la SB 1070, se fundó el Fondo de Acción de la Red de Protección (ProNet), a modo de colaboración entre los miembros migrantes activistas de las Redes de Protección y sus aliados, con la finalidad expresa de conseguir fondos para ayudar al activismo orientado por migrantes en Tucson. La estrategia de ProNet se fundamenta en la tarea de construir una relación duradera entre los migrantes activistas con aliados predominantemente blancos y con el compromiso de enfrentar los retos micropolíticos dentro de Tucson, donde grupos de ayuda humanitaria dirigidos por blancos con frecuencia siguen sin saber del activismo que opera en comunidades hispanohablantes chican@s y latin@s. Este artículo estudia el ProNet como un ejemplo de trabajo solidario anarco-feminista arraigado en una praxis organizadora horizontal autónoma que también se involucra con pasos deliberados en negociar las diferencias en privilegios sociales enconchados que van de la mano con estatus de raza, clase, género, idioma y documentación. *Palabras clave: activismo, anarco-feminismo, micropolítica, justicia social, solidaridad.*

This land was Mexican once,
Was Indian always
and is.
And will be again.

—Gloria Anzaldúa (1987, 3)

In southern Arizona, the question of how solidarity is materially enacted is crucial for understanding the ways in which grassroots political work can be relevant and useful for the people rendered most vulnerable by anti-immigrant legislation and militarized bordering practices. Activists in the region come from diverse backgrounds, including people from directly affected communities as well as allied supporters. All must negotiate interpersonal tensions that arise within activist networks. The micropolitics of small-scale grassroots ventures present significant challenges that stem from the positionalities of people relative to the struggle, whether one's identity means that one is directly targeted by state violence or whether one occupies a position of relative immunity in the eyes of the law. "Every politics," write Deleuze and Guattari (1987), is "simultaneously a *macropolitics* and a *micropolitics*" (213, italics in original). Tucson's migrant justice movement broadly articulates shared political goals. Simultaneous moments of privilege and marginalization within the activist networks, however, reflect micropolitical tensions that arise from participants' differences in race, class, gender, language, and documentation status.

A perennial tension for grassroots activism is the capacity for people to continue their work together in the long term. Walia (2013) reflects on the challenges that come up when activist organizers are not from directly affected communities:

> A future that continues to be led by students and professionals, those who speak in a glossary of activist terms played on repeat, will not be led anywhere, nor arrive anywhere. Every time an undocumented mother walks into a school to enroll her child, it is an act of resistance and defiance. ... Simply staring down the bared face of violence and continuing to breathe is incredible resistance. Linking our political organizing to this chain of freedom is critical and one of our most urgent concerns. (281)

The stakes are high for undocumented migrant activists and their resistance occurs not only in overt acts of protest but in the spaces of the everyday. To work in solidarity requires mindfulness of the profound differences between where white allies and migrant activists are coming from and ongoing collaboration and relationship building that leaves power and leadership in the hands of the communities directly affected by unjust immigration and bordering practices.

The Protection Network Action Fund (ProNet) is a group based in Tucson, Arizona, dedicated to fundraising in support of migrant-led organizing. Through the example of ProNet, I address the micropolitical aspects of anarcha-feminist solidarity work within networks composed primarily of Latin American migrants, longtime Chican@ and Latin@ residents, and white allies. Through looking at the internal dynamics of grassroots social justice work based on my participant observation with ProNet, I examine solidarity as it functions within small-scale, locally rooted activist collaborations. In particular, I discuss Rooting for Change, a 2015 ProNet campaign that was at once a way to provide training for grassroots organizers and fundraisers while fostering solidarity among Tucson activists from different backgrounds.

ProNet does not identify itself as explicitly feminist or anarchist, but I contend that it is, in fact, both. In doing so, I agree with Wright (2009) and others (Barker and Pickerill 2012) who show that labels such as feminist or anarchist might not be strategically mobilized by grassroots activist groups themselves but that the presence or absence of such labels does not alter the character of the work undertaken. Regardless of the explicit political labels a group uses to describe itself, academics and activists alike have much to gain from understandings of the material praxes of horizontal organizing and solidarity work. There is a danger that debate about appropriate labels ends up obscuring the actual strategies for building solidarity that grassroots groups mobilize—strategies that offer useful tools for anyone concerned with the ways in which we might materially challenge oppressive social structures. In considering ProNet as an example of anarcha-feminist praxis, I contribute to conversations about solidarity occurring in both anarchist (Graeber 2009; Lagalisse 2011; Barker and Pickerill 2012; Chatterton, Featherstone, and Routledge 2013; Routledge and Derickson 2015) and feminist geographical circles (Mohanty 2003; Koopman 2008, 2014; Nagar 2008; Wright 2010), which often discuss solidarity and nonhierarchical organizing in similar ways, despite the fact that they remain disparate discourses. Further, by utilizing an explicitly anarcha-feminist analysis, I disrupt the trend within anarchist scholarship to privilege white male self-described anarchists as constituting the theoretical foundations of anarchism (e.g., Nettlau 1996; Eltzbacher 2004; Ward 2004; Guérin 2005; Springer 2016).

The story of ProNet constitutes a countertopography (Katz 2004; Mountz 2011b) of bordering processes that speaks to calls for heightened engagement with conditions produced by nation-state boundaries (Johnson et al. 2011). Like all international borderlands, the U.S.–Mexico border is a pool of emotion, fear, and memory (Paasi 2011) expanding beyond the actual boundary, "a dividing line, a narrow strip along a steep edge ... a vague and undetermined place created by the emotional residue of an unnatural boundary" (Anzaldúa 1987, 3). In Arizona, on the front lines of border militarization, this complexity warrants "creative ways of mapping borders," as Mountz (2011a, 65) argues, to contribute new understandings of how militarization permeates everyday life and the dimensions of resistance that are possible.

In what follows, I situate anarcha-feminism amid understandings of anarchist and feminist solidarities. I consider first how anarchism, an approach rooted in autonomous nonhierarchical organizing, can be merged with feminism, a politics mindful of the role of difference in constituting experiences of the everyday, and how this connection can enrich our understandings of the micropolitics of sociospatial phenomena. Second, I discuss ProNet's anarcha-feminist praxis of solidarity work amid the larger context of the Arizona borderlands to show the group's efforts to facilitate interpersonal connection across the raced, classed, and language-based divisions in Tucson's activist networks. Through doing so, I highlight the practical strategies employed by ProNet to challenge problematic power dynamics by deliberately and conscientiously enacting solidarity and supporting the people rendered most vulnerable by militarization and surveillance in the U.S.–Mexico borderlands.

Anarcha-Feminist Solidarities

Anarcha-feminism brings together anarchism, rooted in anticapitalism, antistatism, and horizontal approaches to social organization, with feminism's emphasis on the significance of intersectional difference in shaping everyday relations of power. A number of works since the 1970s have expressly focused on anarcha-feminism as a nonhierarchical approach to social organization that takes into account identity-based difference (Dark Star Collective 2012). Early works in anarcha-feminism were rooted in a white feminist approach emphasizing gendered injustice (Kornegger 1975); however, the work of Black and other feminists of color pushed feminism more broadly to incorporate intersectional analyses (hooks 1981; Combahee River Collective 1983; Crenshaw 1991). In this section, I discuss various approaches that show the compatibilities of anarchism and feminism alongside some of the challenges that often prevent the two from functioning well together.

In Mohanty's articulation, the core of antiracist feminist solidarity is the understanding that solidarity across difference should be "a political as well as ethical goal" in the practice of decolonization and anticapitalist struggle (Mohanty 2003, 3). Pulido's (2006) analysis of activism in the 1960s and 1970s in Los Angeles explores solidarity work between Black, Chican@, and Japanese activists, showing how "movements are more than the sum of their parts. Their character, size, and shape are also determined by their interactions with other organizations and individuals" (Pulido 2006, 153). The micropolitical dynamics of difference within activist networks include some of the "parts" that Pulido described, and the negotiation of solidarity activism across difference plays a significant role in whether or not activist projects are able to continue their collaborative efforts in the long term.

Within the U.S. anarchist movement, a lack of acknowledgment of the significance of difference has been a persistent problem. Anarchist author Landstreicher (2001), for example, argues that identity-based concerns are detrimental to a larger anarchist movement: "The so-called privileges enumerated in the mea culpas of guilt ridden radicals are really nothing more than means for constructing social identities that serve the ruling class by producing artificial divisions among those they exploit." This illuminates an attitude commonly found in masculinized anarchist spaces (see also Heartfield 2013) that the "real" enemy is the capitalist state. Concerns about the ways in which difference and intersectionality as they shape experiences of oppression are treated as insignificant in comparison to the injustices of capitalism and, further, are often portrayed as divisive to anarchist movements. As others have argued, however, any challenge to the hierarchical oppression of the capitalist state requires that anarchism interrogate white supremacy and patriarchy (Olson 2009). For example, Rogue and Willis (2012) show how intersectionality works with anticapitalist analyses, advocating for an anarchism that reflects "how the daily lives of people can be used to talk about the ways in which structures and institutions intersect and interact" (43).

An intersectional anarcha-feminist approach highlights how we are all affected uniquely by forces of oppression, depending on our proximity to what Lorde (1984) termed the "*mythical norm*," who is someone "white, thin, male, young, heterosexual, Christian, and financially secure" (116, italics in original). Samudzi (2017) recently discussed why she deliberately avoids "canonical anarchism," which is typically European-based, white, and male, explaining that her "most important and foundational left politics have been derived from observing organizations and political formations themselves" such as the Black Panthers, the Zapatistas, and African revolutionary figures such as Amilcar Cabral and Thomas Sankara. Intersectional anarcha-feminism takes into account the meanings of anarchism in different contexts and the role of positionality in shaping those understandings. Hall (2016), for example, argues that "Indigenist feminist and anarchist intersections are vital spaces," through which struggles to reclaim lands and restore Indigenous systems of governance occur alongside "the centrality of women's leadership" and "the fluidity of gender and choices about sexual freedom [that are] rooted in Indigenous worldviews" (82–83).

The complexities of solidarity activism and racialized difference have been articulated in multiple ways (Nopper 2003; Benally 2013; Kendall 2013; Walia 2013; Garza 2014; Goggans 2014; Woods 2014). Although each speaks to specific place-based struggles, a recurring topic is the challenging nature of activist collaboration between whites and people of color. The urgency of grassroots political work often demands a swift process that moves from the birth of an idea to the execution of an action. Consequently, there is not always time for critical reflection on the interpersonal dynamics at play or to confront problematic power relations that exist within activist networks. "Movement building requires reflexivity," as Walia (2013, 173) shows, and although there should be systematic collective reflection on organizing processes, such conversations frequently take place in informal spaces, reflecting the complex micropolitics at play within activist networks (Barker and Pickerill 2012; Rouhani 2012; Mott 2016, 2017). Movements

dedicated to challenging oppressive structures contain their own problematic power relations through which microaggressions occur, often in ways that serve to replicate white supremacist and patriarchal norms prevalent throughout larger society (Koopman 2007; Warburton 2016), a phenomenon reflected in how internal conflicts are (or are not) processed. In what follows, I consider the empirical example of ProNet and the ways in which this group negotiates some of the micropolitical tensions within Tucson's activist networks and how the group's approach constitutes an anarcha-feminist praxis of solidarity.

The Case of ProNet

The existence of the militarized border and anti-immigrant policing strategies make everyday life precarious for many in Tucson, and grassroots activist groups have emerged within communities of migrants and longtime Chican@ and Latin@ residents. The Protection Networks are a coalition of six migrant-led organizations based in Tucson: The Southside Worker Center, Derechos Humanos, Tierra y Libertad Organization, Fortin de las Flores, Corazón de Tucson, and Mariposas sin Fronteras (Protection Network Action Fund n.d.). In 2012, the majority of these groups formalized their connections with one another as *Las Redes de Protección*, a coalition of grassroots community groups dedicated to social change and the belief "in the power of community organizing led by the people most affected by unjust policing and inhumane policies" (Protection Network Action Fund n.d.). The Protection Networks focus on different aspects of the experience of Latin@ and Chican@ residents in Tucson, providing services such as offering legal aid and advice, supporting queer and transgendered migrants who have been detained, or working for migrant labor rights.

In the early 1980s, Tucson moved to prominence as a center for social justice activism through the Sanctuary Movement, where faith leaders provided sanctuary to Central American refugees, openly defying federal immigration laws as increasing numbers of Salvadoran and Guatemalan migrants crossed the Arizona border to escape violence in their home countries. By the early 2000s, the U.S.–Mexico border's Tucson sector saw increasing numbers of migrant deaths as militarization strategies tightened around urban centers such as El Paso/Ciudad Juárez and San Diego/Tijuana and migration routes moved into more remote and

127

dangerous desert terrain. In 2010, numbers of migrant deaths in the Tucson sector peaked, with 225 bodies reported recovered, a number likely much higher, based on research revealing widespread underreporting of deaths by the Border Patrol (Rubio-Goldsmith et al. 2007; Trevizo 2015). To address the humanitarian crisis of migrant deaths in the Sonoran Desert, several grassroots organizations emerged throughout the early 2000s, among them Humane Borders (2000), The Tucson Samaritans (2002), and No More Deaths (2004). Although these groups occupy an important position within the landscape of activism in Tucson, there is a disconnect that exists between, on the one hand, the important work done by these humanitarian aid groups and, on the other, the organizing done by migrants themselves within their own communities (Mott 2017). There have been some attempts to bridge the disjunctures within Tucson's activist communities, however. The We Reject Racism campaign was one such example aimed to repeal unjust anti-immigrant legislation and to facilitate long-term collaboration between various autonomous networks of activists (Loyd 2012). A collaboration between members of No More Deaths and Tierra y Libertad Organization, We Reject Racism operated through outreach to neighborhoods and businesses in Tucson as a conscious effort to generate wider awareness about the impacts of anti-immigrant legislation in the city.

After the 2010 passage of Arizona's notorious racial profiling legislation, Senate Bill 1070 (SB 1070), also known as the Support our Law Enforcement and Safe Neighborhoods Act, the political climate in Arizona was decidedly hostile for Latin@ and Chican@ residents of Tucson. SB 1070 was intended to induce trauma into the daily lives of migrants, setting "a national precedent for restrictive immigration legislation that aims to disrupt the everyday lives of undocumented immigrants to such a degree that they 'self deport'" (Williams and Boyce 2013, 896). Through bringing municipal police into closer collaboration with Border Patrol, SB 1070 meant that, for anyone who appeared to be Latin@, interaction with police would require proof of citizenship (Menjívar 2014).

SB 1070's aims to induce fear and trauma into the daily lives of migrants were certainly effective, although this was experienced very differently throughout the city. "Residents in some parts of Tucson," as Loyd (2012, 138) explains, "often do not even know that migration sweeps occur in the city, nor how a simple traffic stop can lead to deportation." The focus of migrant policing is typically on South

Tucson and the surrounding neighborhoods,[1] whereas residents in predominantly white parts of the city might never see evidence of the militarized border. This "low grade state terrorism" (Loyd 2012, 138) appears very differently depending on where one is in the city, such that people in certain parts of Tucson regularly experience the power of the state to disrupt daily life through terror, and others remain unaware that such practices are even happening.

For many migrant activist projects, the urgent threat posed by SB 1070 created a situation where organizations shifted their energies toward fundraising to bond community members out of the detention system and publicly protesting detention and deportation.[2] The Southside Worker Center,[3] in particular, was hit hard by immigration enforcement after SB 1070, and the severe toll that detentions had on their community was the source of many discussions about how to bond members out without getting into an endless cycle of fundraising that left them unable to pursue the real aims of the organization, supporting migrants' rights to work.

In late 2012, migrant and allied activists began to sketch out plans for ProNet. The project began as a collaboration between white and migrant activists who had already established relationships with one another through the groups that comprise The Protection Networks. ProNet's stated goals were to fundraise bond money for the migrant activist community and to alleviate some of the strain brought about by SB 1070. Collectively, ProNet is composed of activist allies, most of whom were white at the time of my involvement.[4] The group prioritizes relationship building and horizontal organizing within multiracial coalitions, and many of those involved in fundraising for ProNet have also been active with migrant justice work with The Protection Networks and other groups in Tucson, as well as with humanitarian aid in the desert. When ProNet started in 2012, their primary focus was fundraising bond money. Since 2015, however, the group has been able to set aside an organizing fund for general use in support of each of The Protection Networks' individual campaigns (Protection Network Action Fund n.d.).

I volunteered with ProNet for about a year in 2014 and 2015. During my time participating with the project, the group consisted of anywhere from five to ten core activists who met regularly to strategize fundraising campaigns, keep track of donations, and discuss ways to build the group's capacity. Much of my own participation was through the routine business of the group—attending meetings, helping to draft documents and promotional materials, and assisting with event planning. During my participant observation, I recorded interviews with core members of ProNet and engaged in conversation and reflection with others, as well as with members of The Protection Networks.[5]

Part of the ethic driving ProNet's solidarity work involves a clear distinction between the work of ProNet's fundraisers, as allies, and the rest of The Protection Networks, who are the communities directly targeted by anti-immigrant legislation and border militarization. This distance emphasizes that ProNet's goals are to support work for migrant justice, in line with the desires of directly affected communities in Tucson. Decisions about how money should be spent and determinations about who should be bonded out of detention with ProNet funds lie with members of The Protection Networks rather than with ProNet itself. Each of The Protection Networks has its own process for deciding whether or not someone who has been detained should be a candidate for ProNet funds. Since the inception of ProNet, there has been deliberate distance between the group's fundraising efforts and the decisions made by the migrant activists of The Protection Networks about how those funds should be allocated. Paige, one of the core members of ProNet who has also been involved directly with The Protection Networks for several years, explained the dynamics of this relationship:

> There's a set of internal conversations to when [The Protection Networks] approach the action fund. But we've been pretty explicit that we want those conversations to happen in house, in organization so that we're not put into a position of making decisions. I couldn't really even tell you with the exception of a couple organizations what those stipulations currently are. So it's something that ... those organizations take on that responsibility within their own groups to determine that, and I'm just going to trust that process and when they come to us, you know, there's a trust that the conversation has been had. (Interview 2016)

A significant aspect of the work that ProNet does is rooted in relationship building and attempting to bridge gaps between the predominantly white and English-speaking humanitarian aid organizations and the work done in Spanish-speaking Latin@ and Chican@ communities through The Protection Networks. As Paige explained, ProNet worked to counter the barriers between networks of activists in Tucson who were often split by racialized and linguistic divides, noting that she hoped to invite "a sense of

'Well, we can do both of these things,' because both of these things are approaching the issue in distinct ways, but let's focus on our affinity" (interview 2016). Beyond simply raising money, much of ProNet's work aims to "focus on our affinity," as she said, to facilitate connection between the work of the different communities in the city dedicated to migrant justice.

Rooting for Change was a ProNet campaign that spanned several months and included a number of community gatherings in South Tucson, held in both English and Spanish. These gatherings had several aims. First and foremost was relationship building between among communities of activists of different backgrounds who all represented different projects throughout the city. ProNet saw these gatherings as a way to address the distance between predominantly white humanitarian aid groups and grassroots organizing done by migrants and other directly affected communities in Tucson. Second, these gatherings sought to communicate tools and strategies for fundraising, something that was hoped would be useful for participants' individual organizational affiliations as well as for ProNet.

The Rooting for Change campaign ran from January through May 2015. In late 2014, ProNet announced that the campaign was on the horizon and that the group sought to organize a group of community members

> interested in improving their skills as grassroots organizers and fundraisers, deepening their analyses of how race and class impact our organizing work, building community across race and class, and making a short term commitment to do personal fundraising for the Protection Network Action Fund. (Protection Network Action Fund 2014)

The events associated with Rooting for Change included potluck meals, workshops on race and class, trainings on how to ask for money as fundraisers, and tours of several of The Protection Networks' bases of operation. The events associated with the campaign were all conducted in a mix of Spanish and English, and attendees included Chican@ and Latin@ members of The Protection Networks, university students, and activists from various backgrounds who worked with humanitarian aid projects in the desert.

The primary goal of Rooting for Change was to build ProNet's capacity through relationship building. Certainly, fundraising was an important aspect of the campaign, but it was considered secondary to the conversations and connections that occurred through the spaces of communal meals and interactive workshops. The organization of the Rooting for Change campaign was, like all of ProNet's work, connected to ongoing discussions with members of The Protection Networks themselves about what would be most meaningful for them in terms of workshop topics and trainings, as well as in educating Tucson's activist networks about the work happening within the communities directly affected by the militarized border and anti-immigrant legislation. ProNet's Rooting for Change campaign exemplifies a praxis of solidarity that is based in nonhierarchical and horizontal organizing strategies and that takes into account the significance of racial and other differences amid a critique of capital and oppressive state policies. Through taking seriously what Mohanty (2002) described as "the micropolitics of content, subjectivity, and struggle" (501), ProNet does its work to raise funds in support of directly affected communities through a slow process of relationship building directed by the six organizations within The Protection Networks. Solidarity is contextually situated, as Routledge and Derickson (2015) show, and in this case the knowledge of how to proceed is directly related to the ongoing micropolitical tensions within Tucson's activist networks.

The Rooting for Change campaign reveals both the anarchist and feminist character of ProNet's approach to solidarity work. Throughout the campaign, organizers worked with The Protection Networks to involve their communities directly in the various workshops and gatherings, while also ensuring that the concerns of The Protection Networks were represented because some who attended Rooting for Change events had little knowledge of the activism occurring in the Chican@ and Latin@ communities in Tucson. Activism undertaken by self-proclaimed allies has rightly been critiqued for tendencies to shift power away from directly affected communities (Nopper 2003; Benally 2013; Kendall 2013; Walia 2013; Garza 2014; Goggans 2014; Woods 2014). The activists involved with ProNet were aware of such critiques and took pains not to replicate problematic dynamics to the greatest extent possible, although this was also negotiated with their knowledge that embodied privilege remains attached to bodies despite one's attempts to distance oneself from it (Mott 2017).

ProNet does not describe itself as an anarcha-feminist project, but it is nonetheless rooted in anarchist organizing strategies and feminist ethics of solidarity. Often when anarchism and feminism come together, a deliberate distance is maintained from the "canonical anarchism" that Samudzi (2017) describes. Rather than focusing on anarchist theory, such activism is rooted in a praxis of horizontalism and autonomous organizing that supersedes affiliation with specific Leftist labels. The Protection Networks and the work of ProNet exemplify anarcha-feminism in the way in which the group's structure deliberately leaves decision-making power in the hands of The Protection Networks' member organizations, each of which functions autonomously. At the same time, ProNet works on fundraising to support The Protection Networks while intentionally decentering the social privileges of many of the activists involved with ProNet. The group operates with a consciousness of micropolitical concerns that appear throughout Tucson's activist networks—issues of privilege that stem from race, class, gender, and documentation status; the challenges of organizing in both Spanish and English; and the ways in which border militarization is lived differently throughout the spaces of the city. The negotiation of these concerns reflects a larger feminist ethic driving ProNet's work. As Mohanty (2003) points out, "Diversity and difference are central values" in relationships of solidarity (7), suggesting that meaningful feminist work requires a recognition of difference within communities of activists such that embodied social privilege can be ethically and productively negotiated.

Conclusion

Anarcha-feminism is often articulated in terms of a praxis that supersedes alignment with the lineage of predominantly European white male theorists who are typically called on as foundational anarchist voices (Hall 2016; Warburton 2016; Samudzi 2017). Many activist ventures operate with deliberate anti-authoritarian and horizontal organizing strategies but without necessarily calling themselves anarchist because of racialized and gendered associations carried by the term. Feminism is a similarly loaded concept. The contentious history of white feminism often means that activists of color do not see space for themselves within a larger feminist movement and thus might not mobilize the term in their own work (hooks 1981; Kendall 2013). An intersectional anarcha-feminism has the potential to rise above some of these challenges through foregrounding the unique constellations of difference that constitute identity amid the specific contexts of horizontally organized activism.

An anarcha-feminist approach to grassroots political action takes into account the micropolitics of identity-based inequalities that themselves perpetuate unethical hierarchies within social justice movements. Knowledge of the nuances of place is critical to any understanding of how people from privileged backgrounds can meaningfully practice solidarity with directly affected communities, a dynamic significant for activists as well as academics. There is no singular formula for how to go about doing this, as every struggle is specific to its own context (Mohanty 2003; Pulido 2006; Routledge and Derickson 2015). ProNet mobilizes strategically to respond to the demands of the micropolitics of Tucson's migrant justice movement but also as a deliberate practice of horizontal organizing rooted in solidarity work with directly affected communities. A significant aspect of ProNet's work is to engage with The Protection Networks through relationship building and continual dialogue, simultaneously reaching out to the larger Tucson activist networks to facilitate bridges across differential divides. Ultimately, this type of work constitutes an anarcha-feminist practice of solidarity that is ongoing and constantly evolving alongside the needs of the communities most targeted by militarization and surveillance in the U.S.–Mexico borderlands.

Notes

1. South Tucson is a distinct municipality within the larger city of Tucson and is predominantly Spanish speaking.
2. There are many reasons to bond someone out of immigrant detention, including the obvious drive to keep families and communities together. Those facing deportation proceedings, however, have a much better chance of fighting deportation from outside of detention facilities. Deportation proceedings happen much more quickly when one is physically in detention, as opposed to having been bonded out.
3. The Southside Worker Center originated in 2006, as the collaboration between Southside Presbyterian Church and the migrant labor community in Tucson. As they explain on their Web site, they typically see "approximately 50 men daily, who, in spite of the hostile climate created by employer sanction laws and anti-immigrant legislation such as Arizona Senate Bill 1070, continue to maintain that they too have a right to work by gathering at the Center" (http://www.southsidecentro.org/about-us.html).
4. The question of race here is complicated, and the specific dynamics of whiteness in Tucson's activist networks is a topic I have discussed elsewhere (Mott 2016, 2017). During my involvement with ProNet in 2014 and 2015, the group was composed of people who understood themselves to be privileged in many ways relative to the migrant activists in The Protection Networks, something articulated variously in terms of race, class, education, and documentation status. Although ProNet described themselves to be a group of predominantly white allies, not all members were white.
5. An in-depth discussion of my own positionality relative to this work is outside the scope of this article, but I have discussed this more substantively elsewhere (Mott 2015, 2016, 2017).

References

Anzaldúa, G. 1987. *Borderlands/La frontera: The new mestiza.* San Francisco: Aunt Lute.

Barker, A. J., and J. Pickerill. 2012. Radicalizing relationships to and through shared geographies: Why anarchists need to understand indigenous connections to land and place. *Antipode* 44 (5):1705–25.

Benally, K. 2013. Klee Benally on decolonialization. *Deep Green Philly.* Accessed November 1, 2017. http://www.deepgreenphilly.com/?cat=9&paged=3.

Chatterton, P., D. Featherstone, and P. Routledge. 2013. Articulating climate justice in Copenhagen: Antagonism, the commons, and solidarity. *Antipode* 45 (3):602–20.

Combahee River Collective. 1983. The Combahee River Collective statement. In *Home girls: A black feminist anthology,* ed. B. Smith, 264–74. New York: Kitchen Table: Women of Color Press.

Crenshaw, K. 1991. Mapping the margins: Intersectionality, identity politics, and violence against women of color. *Stanford Law Review* 43 (6):1241–99.

Dark Star Collective. 2012. *Quiet rumors: An anarcha-feminist reader.* Oakland, CA: AK Press.

Deleuze, G., and F. Guattari. 1987. *A thousand plateaus: Capitalism and schizophrenia.* Minneapolis: University of Minnesota Press.

Eltzbacher, P. 2004. *The great anarchists: Ideas and teachings of seven major thinkers.* New York: Dover.

Garza, A. 2014. A herstory of the #BlackLivesMatter movement. *The Feminist Wire.* Accessed November 1, 2017. http://www.thefeministwire.com/2014/10/blacklivesmatter-2/.

Goggans, A. 2014. Dear white people: Ferguson protests are a wake not a pep rally. *The Well Examined Life.* Accessed November 1, 2017. http://wellexaminedlife.com/2014/11/26/dear-white-people-ferguson-protests-are-a-wake-not-a-pep-rally/.

Graeber, D. 2009. *Direct action: An ethnography.* Oakland, CA: AK Press.

Guérin, D. 2005. *No gods, no masters.* Oakland: AK Press.

Hall, L. 2016. Indigenist intersectionality: Decolonizing and reweaving an indigenous eco-queer feminism and anarchism. *Perspectives on Anarchist Theory* 29:81–93.

Heartfield, J. 2013. Intersectional? Or sectarian? *Mute.* Accessed May 20, 2017. http://www.metamute.org/community/your-posts/intersectional-or-sectarian.

hooks, b. 1981. *Ain't I a woman?* Boston: South End Press.

Johnson, C., R. Jones, A. Paasi, L. Amoore, A. Mountz, M. Salter, and C. Rumford. 2011. Interventions on rethinking "the border" in border studies. *Political Geography* 30 (2):61–69.

Katz, C. 2004. *Growing up global: Economic restructuring and children's everyday lives*. Minneapolis: University of Minnesota Press.

Kendall, M. 2013. #SolidarityIsForWhiteWomen: Women of color's issue with digital feminism. *The Guardian*, August 14. Accessed May 24, 2017. https://www.theguardian.com/commentisfree/2013/aug/14/solidarityisforwhitewomen-hashtag-feminism.

Koopman, S. 2007. A liberatory space? Rumors of rapes at the 5th World Social Forum. *Journal of International Women's Studies* 8 (3):149–63.

———. 2008. Imperialism within: Can the master's tools bring down empire? *ACME: An International E-Journal for Critical Geographies* 7 (2):283–307.

———. 2014. Making space for peace: International protective accompaniment in Columbia. In *The geographies of peace*, ed. F. McConnell, N. Megoran, and P. Williams, 109–30. London: I. B. Tauris.

Kornegger, P. 1975. Anarchism: The feminist connection. *The Anarchist Library*. Accessed November 1, 2017. https://theanarchistlibrary.org/library/peggy-kornegger-anarchism-the-feminist-connection.

Lagalisse, E. 2011. "Marginalizing Magdalena": Intersections of gender and the secular in anarchoindigenist solidarity activism. *Signs* 36 (3):653–78.

Landstreicher, W. 2001. A question of privilege. *Willful Disobedience* 2. Accessed May 20, 2017. http://theanarchistlibrary.org/library/various-authors-willful-disobedience-volume-2-number-8#toc4.

Lorde, A. 1984. *Sister outsider: Essays and speeches by Audre Lorde*. Berkeley, CA: Crossing Press.

Loyd, J. M. 2012. Human rights zone: Building an antiracist city in Tucson, Arizona. *ACME: An International E-Journal for Critical Geographies* 11 (1):133–44.

Menjívar, C. 2014. The "Poli-Migra": Multilayered legislation, enforcement practices, and what we can learn about and from today's approaches. *American Behavioral Scientist* 58 (13):1805–19.

Mohanty, C. T. 2002. Revisiting "Under Western Eyes": Decolonizing feminist scholarship: 1986. *Signs* 28 (2):499–535.

———. 2003. *Feminism without borders: Decolonizing theory, practicing solidarity*. Durham, NC: Duke University Press.

Mott, C. 2015. Notes from the field: Re-living Tucson—Geographic fieldwork as an activist-academic. *Arizona Anthropologist* 24:33–41.

———. 2016. The activist polis: Topologies of conflict in indigenous solidarity activism. *Antipode* 48 (1):193–211.

———. 2017. Precious work: White anti-racist pedagogies in southern Arizona. *Social & Cultural Geography*. Advance online publication. http://dx.doi.org/10.1080/14649365.2017.1355067.

Mountz, A. 2011a. Border politics: Spatial provision and geographical precision. *Political Geography* 30 (1):65–66.

———. 2011b. Where asylum-seekers wait: Feminist counter-topographies of sites between states. *Gender, Place & Culture* 18 (3):381–99.

Nagar, R. 2008. Languages of collaboration. In *Feminisms in geography: Rethinking space, place, and knowledges*, ed. P. Moss and K. F. Al-Hindi, 120–29. Lanham, MD: Rowman and Littlefield.

Nettlau, M. 1996. *A short history of anarchism*. London: Freedom Press.

Nopper, T. 2003. The white anti-racist is an oxymoron. *Race Traitor*. Accessed November 1, 2017. http://racetraitor.org/nopper.html.

Olsen, J. 2009. The problem with infoshops and insurrection: U.S. anarchism, movement building, and the racial order. In *Contemporary anarchist studies*, ed. R. Amster, A. Deleon, L. Fernandez, A. Nocella II, and D. Shannon, 35–44. London and New York: Routledge.

Paasi, A. 2011. Borders, theory and the challenge of relational thinking. *Political Geography* 30 (1):62–63.

The Protection Network Action Fund. 2014. *Protection Network Action Fund, end of year report*.

———. n.d. ProNet. Accessed January 5, 2017. https://pronetaction.wordpress.com/.

Pulido, L. 2006. *Black, brown, yellow, and left: Radical activism in Los Angeles*. Berkeley: University of California Press.

Rogue, J., and A. V. Willis. 2012. Insurrection at the intersections: Feminism, intersectionality, and anarchism. In *Quiet rumors: An anarcha-feminist reader*, 43–46. The Dark Star Collective. Oakland, CA: AK Press.

Rouhani, F. 2012. Anarchism, geography, and queer space-making: Building bridges over chasms we create. *ACME: An International E-Journal for Critical Geographies* 11 (3):373–92.

Routledge, P., and K. D. Derickson. 2015. Situated solidarities and the practice of scholar activism. *Environment and Planning D* 33:391–407.

Rubio-Goldsmith, R. M., M. McCormick, D. Martinez, and I. M. Duarte. 2007. A humanitarian crisis at the border: New estimates of deaths among unauthorized immigrants. Washington, DC: Immigration Policy Center. Accessed November 1, 2017. http://www.immigrationpolicy.org.

Samudzi, Z. 2017. On a black feminist anarchism. Presentation at the Orange County Anarchist Book Fair. Accessed May 19, 2017. https://www.youtube.com/watch?v=F09BowIVEQo].

Springer, S. 2016. *The anarchist roots of geography: Toward spatial emancipation*. Minneapolis: University of Minnesota Press.

Trevizo, P. 2015. UA researchers seek standard in handling of border deaths. *Arizona Daily Star*. Accessed November 1, 2017. http://tucson.com/news/ua-researchers-seek-standard-in-handling-of-border-deaths/article_df35a0af-fe2a-50f7-9873-9b6f157e9b63.html.

Walia, H. 2013. *Undoing border imperialism*. Oakland, CA: AK Press.

Warburton, T. 2016. Coming to terms: Rethinking popular approaches to feminism and anarchism. *Perspectives on Anarchist Theory* 29:66–78.

Ward, C. 2004. *Anarchism: A very short introduction*. New York: Oxford University Press.

Williams, J., and G. A. Boyce. 2013. Fear, loathing and the everyday geopolitics of encounter in the Arizona borderlands. *Geopolitics* 18 (4):895–916.

Woods, J. 2014. Becoming a white ally to black people in the aftermath of the Michael Brown murder. What matters. Accessed November 1, 2017. http://janee woods.com/2014/08/14/becoming-a-white-ally-to-black-people-in-the-aftermath-of-the-michael-brown-murder/.

Wright, M. W. 2010. Geography and gender: Feminism and a feeling of justice. *Progress in Human Geography* 34 (6):818–27.

———. Gender and geography: Knowledge and activism across the intimately global. *Progress in Human Geography* 33 (3):379–86.

CARRIE MOTT is an Instructor in the Department of Geography at Rutgers, The State University of New Jersey, New Brunswick, NJ 08901. E-mail: carrie.mott@rutgers.edu. Her research interests center on the dynamics of racialized difference in the United States, particularly in relation to whiteness and settler colonialism.

12 Praxis in the City

Care and (Re)Injury in Belfast and Orumiyeh

Lorraine Dowler and A. Marie Ranjbar

This article builds on the geographic literature of nonviolence with the feminist literature of care ethics and positive security to explore the potential for a praxis that promotes relational urban social justice. We examine two cities—Belfast, Northern Ireland, and Orumiyeh, Iran—that have historically endured political struggles that continue to undermine the quality of urban life. We analyze vulnerability to political, environmental, and infrastructural violence in these two urban landscapes with an eye toward "just praxis" and "positive security," as we outline the ways in which Belfast and Orumiyeh are reinjured by institutional practices that purportedly seek urban social justice. First, we argue for the importance of care praxis in the light of the entanglement of a murder investigation with the Boston College oral history program "The Belfast Project," which recorded testimony from former and current members of paramilitary groups. Second, we examine an environmental justice movement in Orumiyeh, where activists navigate a contested political terrain shaped by state violence toward ethnic minorities and punitive economic sanctions from the international community. From this perspective, a just praxis acknowledges the ubiquity of violent conflict while it distinguishes global readings that occur from a distance to the intimate and interminable experiences of violence that take place in urban places. We argue that a more critical engagement with the relationship between care and vulnerability reveals the enormous potential of imagining geographies of existing and evolving relationalities of care rather than global assumptions from afar about vulnerable communities. *Key Words: care, feminist geopolitics, (re)injury.*

本文根据非暴力的地理文献以及照护伦理和正面安全的女权主义文献, 探讨提倡关系性城市社会正义的实践潜能。我们检视北爱尔兰的贝尔法斯特与伊朗的乌尔米耶这两座城市, 它们历史上经历了持续损害城市生活素质的政治斗争。我们在描绘贝尔法斯特和乌尔米耶被本应寻求城市社会正义的制度实践再度伤害的方式时, 分析这两座城市地景对于政治、环境和结构性暴力的脆弱性, 并关照"正义实践"与"正向安全"。首先, 我们主张依照一启谋杀案调查和波士顿学院口述历史计画"贝尔法斯特计画"的纠葛中, 强调照护实践的重要性, 该计画纪录准军事团体的前任与现任成员的证词。再者, 我们检视乌尔米耶的环境正义运动, 其中行动者航行于国家针对少数族裔的暴力和国际团体的惩罚性经济制裁所形塑的争夺政治领域之中。从上述视角看来, 正义的实践认知到暴力冲突无所不在, 同时区别从远处进行的全球性阅读到在城市空间中发生的亲密且无止境的暴力经验。我们主张, 更为批判性地涉入照护与脆弱性之间的关系, 揭露出想像既有与变化中的照护关系性地理之强大潜能, 而非对于脆弱社群的远距全球预设。 *关键词: 照护, 女权主义地缘政治, (再度) 伤害。*

Con la literatura feminista de la ética del cuidado y la seguridad positiva, este artículo contribuye a la literatura geográfica de la no violencia para explorar el potencial de una praxis que promueve la justicia social urbana relacional. Examinamos dos ciudades—Belfast, en Irlanda del Norte, y Orumiyeh, Irán—centros urbanos que a través de la historia han tenido que soportar luchas políticas que siguen afectando la calidad de la vida urbana. Analizamos la vulnerabilidad a la violencia política, ambiental e infraestructural de estos dos paisajes urbanos, con la mirada puesta en la "praxis justa" y la "seguridad positiva," al tiempo que esbozamos las maneras como Belfast y Orumiyeh son lesionadas de nuevo por prácticas institucionales que supuestamente propenden por la justicia social urbana. Primero, discutimos sobre la importancia de la praxis del cuidado a la luz del lío mayor de una investigación de asesinato, con el programa de historia oral del Boston College titulado "El Proyecto Belfast," que registró el testimonio de miembros pasados y actuales de grupos paramilitares. Segundo, examinamos un movimiento de justicia ambiental en Orumiyeh, donde los activistas navegan un terreno político disputado, al que han dado forma la violencia del estado contra minorías étnicas y las sanciones económicas punitivas de la comunidad internacional. Desde esta perspectiva, una praxis justa reconoce la ubicuidad del conflicto violento al tiempo que distingue las lecturas globales que se dan a distancia de las experiencias íntimas e interminables de la violencia que ocurre en los lugares urbanos. Sostenemos que un compromiso más crítico con las relaciones entre el cuidado y la vulnerabilidad revela el enorme potencial de imaginar las geografías de las relacionalidades del cuidado, existentes y en evolución, más que los supuestos globales vistos desde lejos acerca de comunidades vulnerables. *Palabras clave: cuidado, geopolítica feminista, (re)lesión.*

That we can be injured, that others can be injured, that we are subject to death at the whim of another, are all reasons for both fear and grief. What is less certain, however, is whether the experiences of vulnerability and loss have to lead straightaway to military violence and retribution. There are other passages. If we are interested in arresting cycles of violence to produce less violent outcomes, it is no doubt important to ask what, politically, might be made of grief besides a cry for war.

—Butler (2004, XII)

Looking for violence and calling it critique will not reveal peace.

—Bregazzi and Jackson (2016, 3)

In understanding the scars of unremitting acts of urban violence, it is critical to understand the interdependent nature of care and vulnerability to find—as suggested in the opening quotes by Butler and Bregazzi and Jackson—paths other than violence. Unquestionably, cities have been sites of vulnerabilities, such as conflict and war, environmental degradation, overbuilding, collapsing infrastructures, and "natural" disasters (Schneider and Susser 2003; Till 2012). Vulnerability is, however, a multidimensional condition that varies across different forms of urban political life, such as gender, ethnicity, race, religion, geopolitical location, and competing nationalisms.

In this article, we consider the ways in which vulnerability intersects with violence in urban landscapes in a way that takes us beyond thinking of urban wounds as binary linear processes or as incidents that have a clear "before or after" (Fluri 2009; Koybayashi 2009; S. Smith 2009; Inwood and Tyner 2011; Megoran 2011, 2013; Tyner and Inwood 2011; Williams and McConnell 2011; Fregonese 2012; Loyd 2012; Till 2012; Bjorkdahl 2013; McConnell, Megoran, and Williams 2014; Springer 2014; Williams 2015; Laliberte 2016). Instead, we suggest that cities should be viewed as places where palimpsests of past injuries imbricate with current vulnerabilities in a blended pattern of vulnerabilities, where individual significance and group interaction produce new forms of grieving as layers of injury coalesce. Through a comparison of a Global North and a Global South city, we examine the reinjuring of Belfast, Northern Ireland, and Orumiyeh, Iran, when outsider institutions center their interventions on hypervisible political events that eclipse less visible forms of urban vulnerabilities.

At first glance, these blended layers of vulnerability might not be obvious to practitioners responsible for policy, political intervention, or research exploration, which can result in a care praxis that unthinkingly eradicates that which it seeks: social justice (Atkins, Hassan, and Dunn 2006; Bosco 2007; Barnett 2010). In the case of Northern Ireland, we examine how Boston College implemented an oral history project with an aim of bringing together former combatants from both sides of the conflict to explain their motivations and actions during the conflict. Given the obstacles to Northern Ireland instituting restorative justice initiatives, this oral history project was viewed by the originators as more than the cataloging of individual experiences during the conflict. Rather, they intended this project to operate as a lens into the mindset for violence by political actors to advance the peace process not only in Northern Ireland but in other postconflict sites (McMurtrie 2014). In this case, we examine how instead of advancing the peace process, this project destabilized the Belfast community by rendering them more vulnerable to political violence. For Iran, we examine how the United Nations' (UN) primary focus on Iran's nuclear program to promote regional security ignored local calls for environmental security. The failure to recognize vulnerability to environmental degradation, while ignoring hydrodevelopment projects that undermine the UN's work on environmentally sustainable development in Iran, results in the erasure of violences stemming from environmental injustice. As a result, we argue that these outsider organizations that, broadly defined, are meant to support care instead compounded the gendered and raced violence of the state.

Admittedly, a comparison of a postconflict Western European city to a postcolonial Middle Eastern city appears incongruous. Although Belfast and Orumiyeh might have different histories of occupation, international political visibility, gender politics, and methods of state resistance, both places are still shaped by a geographical imagination of violence, hostility, and international intervention. As J. Robinson (2016) argued, "Theorizing cities can benefit from a comparative imagination" as long as such theorization is accompanied by a "reconfiguring of the ontological foundations of comparison" (5). Ho, Boyle, and Yeoh (2015) agreed that care is rooted in a social ontology of connection across space. J. Robinson (2016) warned against a "territorializing trap" of comparing relatively similar cities. Rather, she suggested examining the "circulation and connections" that shape cities, allowing for comparisons of a much wider range of contexts (J. Robinson 2016), whereas Ho, Boyle, and Yeoh (2015) stated that any understanding of care must include the "power geometries of relational space" (209). Using these insights, we argue that urban justice conceived through an analysis of care and vulnerability provides an ontological comparison that exposes a commonality of injuries that are otherwise obscured by their place-specific stories of violence, disruption, and grief.

Importantly, our case studies are connected through invisible forms of infrastructural injustice that are often eclipsed by spectacular physical statements of state sovereignty (W. Brown 2010; Bjorkdahl 2013; Dowler 2013a, 2013b; Till 2013). Infrastructural forms of violence, such as physical structures and state policies, are common to cities and might provide short-term physical security but symbolically situate people within a social hierarchy at a variety of scales (W. Brown 2010; Mattar 2012). Any resistance to this social hierarchy is often viewed by the state as a threat to sovereignty. As Björkdahl and Selimovic (2016) argued, the appearance of material spaces across post-conflict sites contributes to an agonistic peace as "the potential for peace is tied to many concrete, material places; often mutually incompatible concepts of peace will remain in and among different places and communities" (2).

Like many urban landscapes, the geographies of Belfast and Orumiyeh are shaped through the presence of constructed physical barriers, such as peace walls in Belfast and dams in Orumiyeh, that mark social difference (Nevins 2012). Despite how infrastructure delineates identity, urban residents from different places, races, and ethnicities share the impact of more invisible "unjust barriers or practices" (Rodgers and O'Neill 2012, 405). This emphasis on infrastructural disconnection underpins a geography of "social exclusion that fundamentally questions notions of citizenship, rights and membership claims by the poor and otherwise vulnerable" (Rodgers and O'Neill 2012, 407; Ong 2000).

Through a comparison of two seemingly disparate urban landscapes marked by state violence, we attend to the power geometries of vulnerability to understand these injuries of disconnection. We propose that a just praxis approach that complements the geographic literatures of nonviolence by what Gibson and Reardon (2007) referred to as "positive security" (63) and what Miller (2011) suggested as a "global duty to care" (392) to maximize care and minimize vulnerability in these urban contexts.

Who Cares?

A just praxis approach broadens and subscribes to burgeoning geographic scholarship that is rooted in a positive politics of nonviolence (McConnell, Megoran, and Williams 2016). Megoran's call for political geographers to actively build peace rather than passively study violence (Megoran 2011) prompted debates questioning whether it is possible to study one without the other (Koopman 2011a, 2011b; Ross 2011; Laliberte 2016). As feminist scholars, we are sympathetic to Megoran's challenge and also mindful of the limitations of theorizing about peace and violence through a bifurcated logic. Notions of "grounding geopolitics" to link international representation to the geographies of everyday life is not a new argument, and feminist geopolitical scholars in particular have questioned how we can make everyday forms of violence visible without rendering the commonality of transnational suffering invisible (Dowler and Sharp 2001, 172; Hyndman 2001). There is a long tradition of dismantling binaries in feminist geography, such as the intimate–global, war–peace, and hot–banal, that suggest that everyday violences need to be understood as a single complex (Wright 2005; Katz 2007; Kearns 2008; Loyd 2011, 2012; Pratt 2012; Pain and Staeheli 2014; Brickell 2015; Christian, Dowler, and Cuomo 2015; Pain 2015). More recently, when considering possibilities for positive peace in geographies marked by violence, a single complex agonistic approach further destabilizes understandings of peace that are based on supposed consensus (Shinko 2008; Ross 2011; Bjorkdahl 2012; Bregazzi and Jackson 2016). In this vein, concepts and practices that might ordinarily be associated with peace rather than violence are implicated in furthering suffering, such as peace treaties and development projects (Bregazzi and Jackson 2016). A just praxis recognizes destructive ideologies while engaging in a more positive politics (Sharp 2009,

2011) and places an emphasis on precarity to make visible the fragility of nonviolence (Butler 2004, 2009; Woon 2011, 2014). This emphasis allows for a more diverse geographic practice that can be repositioned for peace or, as Nagar (2014) suggested, "to both separate and intimately link the question of scholarship with that of political action" (2).

As previously stated, our call for a just praxis is anchored in this critical scholarship of nonviolence, while being attentive to the multidimensional nature of vulnerability (Tronto 1987; M. P. Brown 2003; Staeheli and Brown 2003; Susser and Schneider 2003; Young 2006; Lawson 2007; Milligan et al. 2007; Butler 2009; Till 2012a, 2012b; Nagar 2014; Dowler, Christian, and Ranjbar 2015; Laliberte 2016). We acknowledge that contemporary care theories can run the risk of supporting exclusion rather than contesting power imbalances, especially those imbalances that are inscribed by race, culture, and geopolitics (F. Robinson 2011). As part of any discussion of the politics of care, we—like other feminist scholars—reject exclusionary concepts of care, which Narayan (2003) argued are embedded in the discourses of colonialism.

We envision a just praxis as the single complex of security and justice enacted across relational spaces (Ho, Boyle, and Yeoh 2015). We conceptualize positive security as an ethical *duty* that obliges institutions to actively care across distant spaces for those made vulnerable by conflict in spatially proximate ways (Miller 2011). Whereas security interventions, particularly in conflict regions, are largely reactive, positive security is preventative and identifies how vulnerability itself is a potential conflict driver. As opposed to meeting the need for security after violence has already occurred, the duty to care is proactive, requiring institutions tasked with protecting vulnerable populations to eliminate the very circumstances and structures that generate violence. From this perspective, a just praxis acknowledges the ubiquity of violent conflict and at the same time distinguishes global readings that occur from a distance to the intimate and interminable experiences of violence that take place in urban places (F. Robinson 2011). We argue that a more critical engagement with the relationship between care and vulnerability reveals the enormous potential of imagining geographies of existing and evolving relationalities of care rather than global assumptions from afar about vulnerable communities (Mohanty 1991; Raghuram, Madge, and Noxolo 2009; Naxolo, Raghuram, and Madge 2011).

We consider a just praxis as extending beyond the protocols of institutions tasked with care and security, such as university institutional review boards and interventions by international organizations like the United Nations. Binary distinctions between institutional forms of care (e.g., humanitarian care) implemented from a distance and intimate care thought to be spatially proximate are not useful and often result in "disembodied, cool and detached" interventions that can deepen vulnerability to violence (Ho, Boyle, and Yeoh 2015, 209).

In the cases that follow, we hope to demonstrate that geography is an important influence to consider in terms of care and responsibility by demonstrating the relationality of urban places that connects individuals to each other over time, across spaces, and across scale (D. M. Smith 1998, 2000; Parr and Philo 2003; Popke 2007; Spark 2007a, 2007b; McEwan and Goodman 2010; Milligan and Wiles 2010; Sin 2010; F. Robinson 2011; Till 2012). We first write to the importance of a just praxis in the light of the entanglement of a Northern Irish murder investigation with the Boston College oral history program, the Belfast Project, which recorded testimony from current and former members of paramilitary groups. We then examine an Iranian environmental justice movement in Orumiyeh, where activists navigate a contested political terrain shaped by state violence toward ethnic minorities and punitive multilateral sanctions from the international community. In both of these cases, we suggest that institutional practices indeed cared about urban social justice yet paradoxically rendered these urban dwellers more vulnerable.

Disturbing the Peace

> Belfast, more than any other European city, has been stereotyped to death; its complex history in permafrost; its geo-cultural life as a port, haven, hell-hole, spectacle, dumbed down before the term was invented. Dawe (2004, 207)

Belfast, at one time considered on the fringes of a global economy, is now home to new development projects, such as the Odyssey Arena, the Obel Tower, and the Titanic Center, that perform as odes to the power of capitalism in branding Northern Ireland as a global destination forged out of the ashes of a war zone. Northern Ireland remains, however, as a place still stuck—as the quote from Dawe suggests—in the permafrost or binaries of fringe and center, peace and conflict, developed and developing. Despite its rebranding as a global city, notions of community in the neighborhoods most injured by the conflict in Belfast are synonymous with violence and geographic segregation.

Contemporary Belfast still is plagued by an infrastructural array of social patterns of widespread residential segregation along ethnic lines. A telling example is that 97 percent of children in Belfast still attend segregated schools, which are accompanied by sporadic low-level violence, especially in what are called interface areas (Shirlow and Murtagh 2006). Many parts of the city are either Protestant or Catholic, demarcated by nationalist flags, graffiti, murals, and other infrastructural markers of territory as the number of peace walls increases. Since the cessation of armed conflict, the two communities have remained largely peaceful but in conditions that might be described as a tense coexistence (Shirlow and Murtagh 2006).

One recent challenge to this fragile, or negative, peace did not come from a rival paramilitary group or at the hands of the state, although their historic violent actions certainly created an infrastructure of distrust that is intimately tied to this new threat. Instead, in 2001, an oral history project was launched in Belfast out of Boston College, with the objective of cataloging interviews with political actors from both sides of the conflict, some of whom had committed gruesome acts in the name of nationalism. The informants were promised confidentiality until their death, at which time their recordings would be released (Cullen 2014).

The idea of the project was admirable but naïve given a continued culture of suspicion between interface areas and within ethnic enclaves, such as the Irish Catholic area of West Belfast. This was further compounded by the British government's continued investigation of crimes that occurred during the conflict, despite the signing of the 1995 Good Friday Peace Agreement that resulted in the release of political prisoners from jail. This refusal to grant universal amnesty has undermined efforts to establish more restorative forms of justice, such as truth and reconciliation commissions, in Northern Ireland.

The vulnerability of peace was evident when two subjects in the Belfast Project, both former Irish Republican Army (IRA) members, indicated that Gerry Adams (the former Sinn Fein leader) was one of the planners of the 1972 murder of Jean McConville. McConville was a widowed mother of ten children who was accused of being a British operative by the IRA. In 2014, the U.S. Department of Justice received numerous subpoenas requesting that the Boston College interview materials be submitted to the United Kingdom. Subsequently, eleven[1] of the original confidential tapes were turned over to the Police Service of Northern Ireland (PSNI) by the United States.[2] As a result, Gerry Adams was detained and interrogated by the PSNI. Adams was released after four days of police questioning while crowds of supporters and detractors protested outside of police headquarters. Although the cross-community violence might have ended, Boston College project managers did not understand that tensions remained within enclaves. Individuals who are viewed as informants, as was the accusation against Jean McConville, remain potential targets of paramilitary forms of justice (McIntyre 2011; Moloney 2011; O'Neill 2011). One local historian, who was a former combatant and a primary investigator for the project, received death threats as local rivalries in the former IRA were ignited and he was forced to relocate outside of the country.

The release of the Belfast Project's first interview with Irish Nationalist Brendan Hughes in 2010 increased the vulnerability of the Belfast community in several ways that were not anticipated by Boston College's administrators. First, renewed discussions over Jean McConville's death resurrected old political binaries over whether this was a legitimate military execution of a British operative or a war crime that should be prosecuted. Second, the interview reignited disputes within the Irish Nationalist community that challenged the outcomes of the 1995 peace process, which resulted in Northern Ireland remaining within the United Kingdom and relinquishing any notions of a United Ireland. Third, calls for the prosecution of this particular crime renewed suspicions that the PSNI was targeting the Irish Catholic community,[3] as there was no request for the Belfast Project interviews with Protestant paramilitaries.[4]

Attitudes about violence were also bordered in masculinized understandings of conflict as death. The Belfast Project researchers did not take into account the relationality of certain vulnerabilities, such as the insecurity that one death produces for a larger community. For instance, the impact of intense interest in the 1972 McConville murder effectively rendered invisible the more than 3,000 unresolved open murders from the period of the conflict (Dickson 2010). Recently, fifty-six killings were marked for "legacy inquests," including the Ballymurphy 11, who died at the hands of the British Military. As opposed to the renewed attention on the McConville case, the

Northern Ireland Executive has not approved funding of inquests for the other 3,000 unresolved murder cases (Kearney 2016). We suggest that the hypervisibility of the McConville case has less to do with caring and death as much as the reemergence of a tired trope of the appropriate actions of women in conflict. Was McConville an innocent mother whose life was taken in a heinous act of murder? Or was she a British operative who could now be stripped of the protection of her gender and simultaneously rendered a bad mother (Dowler 1998, 2002)?

The reemergence of the McConville case also reinforced the borders around veneration of certain lives and not others (Butler 2004). Despite the disproportionate attention paid to the McConville case, we continue to see the privileging of paramilitary voices as opposed to the stories of McConville's children who witnessed their mother's removal from their home at the hands of the IRA. This is a striking example of the uneven relationship between care and vulnerability.

We now expand on our insights from the Belfast Project's focus on the agents of violence at the expense of nonviolent actors with the case of an environmental justice movement in Orumiyeh, Iran. Through Orumiyeh, we examine how the UN and other Western nations, notably the United States, regard the Iranian state as a violent actor and global threat for not disbanding the infrastructure around its uranium enrichment program. The hypervisibility around the nuclear discussions led to the erasure of other infrastructural violences, specifically hydroeconomic development projects in Iran that resulted in the silencing of local needs at the expense of global security.

Securing Environmental Rights

"We should not give much more time to the Iranians, and we should not waste time," Ban said. "We have seen what happened with the DPRK. . . . It ended up that they [were] secretly, quietly, without any obligations, without any pressure, making progress." The U.N. Security Council must "show a firm, decisive and effective, quick response" (interview with Ban Ki-moon; Gearan 2013). The overarching rationale for the UNDP Country Programme in the Islamic Republic of Iran is to support the Government in achieving its sustainable and inclusive development objectives. This rationale [is] aimed at "leaving no one behind" (overview, United Nations Development Programme [UNDP] 2017).

In late August 2011, images posted to Facebook from a protest in Orumiyeh circulated throughout the Iranian diaspora. In one of the photographs, a group of demonstrators stands on the western shore of Lake Orumiyeh, holding a large green banner addressed to the Secretary General of the UN: "Mr. Ban Ki-moon: Isn't the plan of desiccating Urmia [Orumiyeh] Lake as disastrous as Iran's nuclear program?" A jug of water lies near their feet, a familiar fixture of protests in Orumiyeh that often accompanies the chant, "Let's cry and fill Lake Urmia with our tears." The faces of demonstrators are hidden under white masks that have the shape of the lake outlined in blue, rendering them anonymous to the Iranian state and the global theater.

This particular protest was in response to the Iranian Parliament voting down yet another funding proposal to help restore what was once the largest saltwater lake in the Middle East. Lake Orumiyeh has lost approximately 80 percent of its water area and volume in recent years due to inadequate environmental management, the construction of dozens of dams along the lake's tributary rivers, and a prolonged regional drought. As the lake continues to lose surface area, exposed salt flats have left the adjacent environment vulnerable to salt storms, affecting agriculture and increasing rates of asthma and cancer in the region (Najafi 2012; Pengra 2012). Rather than addressing the concerns of those affected by the rapid desiccation of the lake, several members of Parliament instead proposed that Orumiyeh's residents be relocated. This sparked outrage and fear of displacement, particularly among Orumiyeh's Azeri and Kurdish communities, ethno-linguistic minorities that have been historically marginalized by the Iranian state.

Orumiyeh activists initially focused demonstrations on their vulnerability to environmental change, with demands centered on the right to live and work in a healthy environment. They called on the government to save the lake by halting dam construction, monitoring illegal well digging, and restoring the lake's water levels by diverting water from nearby rivers. In response to protests, the government acknowledged that the lake is in critical condition. The government, however, remained silent on damming and irrigation projects, instead attributing the drying of the lake to climate change. Despite international protections afforded to Orumiyeh as both a UNESCO Biosphere Reserve and Ramsar site, the UN also failed to substantively intervene.[5]

The security interests of both the Iranian state and the UN have eclipsed the Orumiyeh community's calls for environmental justice. Within Iran, the failure of the

Iranian government to restore the Orumiyeh ecosystem while pursuing an aggressive hydrodevelopment agenda in Iranian Azerbaijan has led to accusations of environmental racism against local Azeri and Kurdish populations. The Western Azerbaijan province is composed of 76 percent Azeris and 22 percent Kurds, compared with 16 percent and 7 percent of Iran's national population, respectively (Hassan 2008). These populations have been historically marginalized by the Iranian state, and the government considers regional ethnonationalism a significant national security threat given Lake Orumiyeh's geostrategic location bordering the Kurdish regions of Iraq and Turkey.

Despite UN development and environmental programming in Iran, state violence enacted through unjust environmental policies has been largely ignored by the international community, which is primarily focused on the containment of Iran's nuclear program to advance geostrategic goals of the stabilization in the Middle East.[6] Security for Orumiyeh residents, however, is the protection of the environment. During the August 2011 protest referenced earlier, demonstrators pointedly asked the UN—as the primary governing institution responsible for the equal protection of human rights—why it continues to prioritize the securitization of Iran's nuclear program over the care and protection of Iranian citizens. This particular protest signals the failure of the international community to recognize environmental rights as universal rights, as well as its uneven enforcement of security. Instead, security for the international community comes at the cost of fundamental human rights of Iranian citizens.

Similar to our critique of masculinized understandings of conflict and vulnerability in Belfast, the UN's myopic conceptualization of security compounds everyday violences in Iran. This neglect of Iranian citizens' needs is twofold. First, the failure to recognize environmental rights as critical to regional security undermines local needs, thereby exacerbating long-standing political tensions between the Iranian Azerbaijan region and the central government. A positive reconceptualization of security would protect vulnerable populations from environmental injustices to prevent future instability and conflict over critical resources. Second, the UN's primary focus on—and response to—Iran's nuclear ambitions both eclipses and deepens the insecurity of Orumiyeh's citizens. During a heightened period of protests in Orumiyeh between 2009 and 2012, the UN facilitated multilateral sanctions that have had grave consequences for Iranians, including limited mobility, economic stagnation, hyperinflation, and restricted access to certain goods and medicines. Nuclear sanctions have been partially justified through Iran's dismal human rights record, including the violent repression of protests. Through international containment policies that meld security with human rights, however, the international community has harmed the very people that it seeks to protect.

Iranian citizens, therefore, face a complicated double bind. Protests intended to draw attention to political and environmental injustices result in state violence through the repression of protests, which is then compounded by punitive international interventions. Far from enabling Iranians to make environmental or broader human rights claims, this uneven enforcement of rights and security instead reifies Global North–South power structures, where Iranians are further marginalized through international geopolitical tensions. Through the neglect of Orumiyeh's needs, coupled with multilateral sanctions over Iran's nuclear program, the international community has reinjured already vulnerable Iranian citizens through economic and political violence. The failure to recognize the multifaceted and multiscalar vulnerabilities of ethnic minority communities in northwestern Iran, particularly in terms of urban residents' relationality to Lake Orumiyeh, is, in large part, due to a one-dimensional understanding and top-down enforcement of security.

A Just Praxis

A just praxis recognizes the universal need for security while also acknowledging that place-based security needs differ. In these case studies, we have demonstrated how both cities and their residents have been subject to certain forms of injury—political, environmental, and otherwise—and how institutions tasked with providing security based on certain understandings of vulnerability have unwittingly reinjured urban populations. Although we have written with the understanding that vulnerability is a universal condition and security is a universal need, these cases demonstrate how vulnerability is differentially experienced. A just praxis, therefore, warns against universalizing when and how needs are met.

The first lesson of a just praxis is that we must take relationality seriously. In both Belfast and Orumiyeh, residents' lives are inextricably tied to the infrastructure of the city. In Belfast, masculine notions of security are manifested through physical structures that are intended to promote peace through segregation or economic development. Yet, infrastructure, such as peace walls, demarcates social difference and facilitates social exclusion, ignoring important relationships within

Belfast communities, thereby deepening vulnerabilities within the Irish Catholic community. In Orumiyeh, there is an important nature–society relationality where the impending death of a lake not only poses risks to health and livelihoods but signifies grievous injury to the social identity of a community made vulnerable to environmental and political violence. Under the auspices of hydroeconomic development, the Orumiyeh community's most basic needs are ignored, leading to heightened insecurity in the region.

Second, to revisit our earlier reference to the palimpsest—where past injuries shape contemporary vulnerabilities—institutions must also recognize the multifaceted and multidimensional conditions of vulnerability. In both of these cases, there is only one vulnerability that is recognized by institutions: conflict. Ignoring relationality and shared vulnerability, these institutions focus solely on conflict "actors," neglecting both the environment and the wider social networks of those affected by conflict. In Northern Ireland, the retelling of The Troubles through the Belfast Project is hypermasculinized, narrated by former paramilitary members, and excludes the stories of nonviolent actors injured in the protracted conflict. The only acknowledgment of the conflict's impact on local communities is told through the sensationalist story of Jean McConville. Although the case of McConville is indeed tragic, the hypervisibility of this story effectively erases the myriad impacts of The Troubles on everyday life.

The grievances of Orumiyeh residents are similarly eclipsed by an ongoing militarized discourse between Iran and the United States that is reminiscent of Cold War dynamics. Through the sole focus on state actors, what becomes lost is how environmental degradation is the main conflict driver within this region. State and international indifference to the fate of Lake Orumiyeh is interpreted by the Orumiyeh community as disregard for Azeri and Kurdish lives. A potential conflict in northwestern Iran does little to threaten the interests of the Global North, however. The refusal of the international community to listen to the needs of Iranian minority communities—while also using their narratives selectively to justify international intervention against the Iranian state—means that promises to uphold human rights ring hollow for many Iranians.

Further, these institutions wrongly assume that death signifies an end to relationality and vulnerability, in terms of relationships between members of the community and urban residents with their environment. The Belfast Project's conceptualization of confidentiality and death betrays a linear view of vulnerability. This one-dimensional understanding of vulnerability begins and ends with the death of one person, as though he or she did not have deep connections to Belfast and that this passing has a negligible impact on her or his respective communities. Similarly, the state and the UN do not recognize the critical relationship between Lake Orumiyeh and the local population. In Orumiyeh, the death of a lake signifies the neglect of an entire urban community. In both cases, institutional interventions were reactive and always too late. A just praxis in these cities would instead center on a caring and positive security that protects against injury.

Notes

1. Over the course of the project (2001–2006), hundreds of hours of testimonies were collected, totaling twenty-six interviews with former members of the IRA and twenty interviews with former Ulster Volunteer Force (UVF) members. After the deaths of former IRA member Brendan Hughes (2006) and former UVF volunteer and Unionist politician David Ervine (2007), excerpts from their transcripts were detailed in the book *Voices from the Grave* written by Irish journalist and project administrator for the Belfast Project, Ed Moloney (2011).
2. Pursuant to the Mutual Legal Assistance Treaty between the United States and the United Kingdom, British authorities sought to obtain these interviews. The United Kingdom asked the U.S. Department of Justice to subpoena Boston College for all materials relating to the Hughes and Price interviews to assist British officials in a prosecution for McConville's murder.
3. The PSNI argued that they only needed the tapes with IRA paramilitaries, as they were investigating a specific murder case.
4. Gerry Adams has long denied that he was a member of the IRA and even if the tape did not offer enough evidence to prosecute him for the death of Jean McConville, it certainly could discredit his political leadership.
5. Lake Orumiyeh is designated a national park and is a protected Ramsar site and a UNESCO Biosphere Reserve (see http://www.ramsar.org/about-the-ramsar-convention).
6. UN development projects in Iran, Conservation of Iranian Wetlands Project.

Acknowledgments

We express our sincere gratitude to the two anonymous reviewers whose thoughtful comments made this a much stronger article. We also thank the graduate students in the Spring 2016 Critical Geopolitics seminar for their encouraging feedback. Finally, we thank Emma Velez for her meticulous copy editing and Sarah

Clark Miller for her counsel vis-à-vis the philosophies of care.

References

Atkins, P. J., M. M. Hassan, and E. E. Dunn. 2006. Toxic torts: Arsenic poisoning in Bangladesh and the legal geographies of responsibility. *Transactions of the Institute of British Geographers* 31 (3):272–85.

Barnett, C. 2010. Geography and ethics: justice unbound. *Progress in Human Geography* 35:236–55.

Björkdahl, A. 2012. A gender-just peace: Exploring the post-Dayton peace process. *Journal of Peace and Change* 37 (2):286–317.

Björkdahl, A., and K. Höglund. 2013. Precarious peace-building: Friction in global local encounters. *Peacebuilding* 1 (3):289–99.

Björkdahl, A., and M. Selimovic. 2016. A tale of three bridges: Agency and agonism in peacebuilding. *Third World Quarterly* 37 (2):321–35.

Bosco, F. J. 2007. Hungry children and networks of aid in Argentina: Thinking about geographies of responsibility and care. *Children's Geographies* 5:55–76.

Bregazzi, H., and M. Jackson. 2016. Agonism, critical political geography, and the new geographies of peace. *Progress in Human Geography.* Advance online publication. doi:10.1177/0309132516666687.

Brickell, K. 2015. Towards intimate geographies of peace? Local reconciliation of domestic violence in Cambodia. *Transactions of the Institute of British Geographers* 40 (3):321–33.

Brown, M. P. 2003. Hospice and the spatial paradoxes of terminal care. *Environment and Planning A* 35 (5):833–51.

Brown, W. 2010. *Walled states, waning sovereignty.* Brooklyn, NY: Zone Books.

Butler, J. 2004. *Precarious life: The powers of mourning and violence.* New York: Verso.

———. 2009. *Frames of war: When life is grieveable.* New York: Verso.

Christian, J., L. Dowler, and D. Cuomo. 2015. Fear, feminist geopolitics and the hot and the banal. *Political Geography* 54:64–72.

Cullen, K. 2014. In Belfast, the gunmen, the shadows, the dame done: BC exercise in idealism reopened old wounds: With a promise of secrecy, Boston College recorded for history the voices of The Troubles in Ireland. But, the promise now broken, the aftershocks in Belfast are testing a fragile peace. *Boston Globe* July 6. 2014. Accessed December 21, 2016. https://www.bostonglobe.com/news/world/2014/07/05/belfast-the-shadows-and-gunmen/D5yv4DdNIxaBXMl2Tlr6PL/story.html.

Dawe, G. 2004. The revenges of the heart: Belfast and the poetics of space. In *The cities of Belfast,* ed. N. Allen and A. Kelly. 199–210. Dublin: Four Courts Press.

Dickson, B. 2010. *The European Convention on Human Rights and the conflict in Northern Ireland.* Oxford, UK: Oxford University Press.

Dowler, L. 1998. And they think I'm just a nice old lady: Gender identities and war, in Belfast, Northern Ireland. *Gender, Place and Culture* 2 (5):159–76.

———. 2002. Till death due us part: Masculinity, friendship and nationalism in Belfast Northern Ireland. *Environment and Planning D: Society and Space* 20:53–71.

———. 2013a. Cracks in the wall: The peace-lines of Belfast, Northern Ireland. *Political Geography* 30:9–10.

———. 2013b. Waging hospitality: Feminist geopolitics and tourism in West Belfast, Northern Ireland. *GeoPolitics* 18:779–99.

Dowler, L., J. Christian, and A. Ranjbar. 2015. A feminist visualization of intimate spaces of security. *Area* 46 (4):347–49.

Dowler, L., and J. Sharp. 2001. Feminist geopolitics. *Space and Polity* 5 (3):165–76.

Fluri, J. 2009. Geopolitics of gender and violence "from below." *Political Geography* 28:259–65.

Fregonese, S. 2012. Urban geopolitics 8 years on: Hybrid sovereignties, the everyday, and geographies of peace. *Geography Compass* 6 (5):290–303.

Gearan, A. 2013. Iran could use U.N. talks as cover to build bomb, Ban Ki-moon says. *The Washington Post* February 14. Accessed September 29, 2017. https://www.washingtonpost.com/world/national-security/iran-could-use-un-talks-as-cover-to-build-bomb-ban-ki-moon-says/2013/02/14/4260e7ea-76d0-11e2-95e4-6148e45d7adb_story.html?utm_term=.499d7f9d6822.

Gibson, I., and B. Reardon. 2007. Human security: Toward gender inclusion. In *Protecting human security in a post 9/11 world: Critical and global insights,* ed. G. Shani, M. Sato, and K. P. Mustapha, 50–63. Hampshire, UK: Palgrave Macmillan.

Hassan, H. 2008. *Iran: Ethnic and religious minorities.* Washington, DC: CRS Report for Congress.

Ho, E., M. Boyle, and B. Yeoh. 2015. Recasting diaspora strategies through feminist care ethics. *Geoforum* 59:206–14.

Hyndman, J. 2001. Towards a feminist geopolitics. *Canadian Geographer* 45 (2):210–22.

Inwood, J., and J. Tyner. 2011. Geography's pro-peace agenda: An unfinished project. *ACME: An International E-Journal for Critical Geographies* 10 (3):442–57.

Katz, C. 2007. Banal terrorism, spatial fetishism and everyday terrorism. In *Violent geographies: Fear, terror and political violence,* ed. D. Gregory and A. Pred, 349–62. London and New York: Routledge.

Kearney, V. 2016. The troubles: Judge begins review of inquests. Accessed May 2, 2017. http://www.bbc.com/news/uk-northern-ireland-35327073.

Kearns, G. 2008. Progressive geopolitics. *Geography Compass* 2 (5):1599–1620.

Kobayashi, A. 2009. Geographies of peace and armed conflict: Introduction. *Annals of the Association of American Geographers* 99 (5):819.

Koopman, S. 2011a. Alter-geopolitics: Other securities are happening. *Geoforum* 42 (3):274–84.

———. 2011b. Let's take peace to pieces. *Political Geography* 30:193–94.

Laliberte, N. 2016. Geographies of human rights: Mapping responsibility. *Geography Compass* 9 (2):57–67.

Lawson, V. 2007. Geographies of care and responsibility. *Annals of the Association of American Geographers* 97:1–11.

Loyd, J. 2011. War is not healthy for children and other living things. *Environment and Planning D: Society and Space* 27 (3):403–24.

———. 2012. Geographies of peace and antiviolence: Peace and antiviolence. *Geography Compass* 6 (8):477–89.

Mattar, D. 2012. Did walls really come down? Contemporary b/ordering walls in Europe. In *Walls, border and boundaries: Spatial and cultural practices in Europe*, ed. M. Silberman, K. Till, and J. Ward, 77–93. Oxford, UK: Berghan.

McConnell, F., N. Megoran, and P. Williams. 2014. Introduction: Geographical approaches to peace. In *Geographies of peace: New approaches to boundaries, diplomacy and conflict resolution*, ed. F. McConnell, N. Megoran, and P. Williams, 1–28. London: I. B. Tauris.

———. 2016. Geography and peace. In *Palgrave handbook of disciplinary and regional approaches to peace*, ed. O. Richmond, S. Pogodda, and J. Ramovic, 123–38. London: Palgrave Macmillan.

McEwan, C., and M. Goodman. 2010. Place geography and the ethics of care: Introductory remarks on the geographies of ethics, responsibility and care. *Ethics, Place and Environment* 13 (2):103–12.

McIntyre, A. 2011. Affidavit, United States District Court, District of Massachusetts. June 7, M.B.D No. 11-MC-91078.

McMurtrie, B. 2014. Secrets from Belfast: How Boston College's oral history of the Troubles fell victim to an international murder investigation. *Chronicle of Higher Education* January 27. Accessed December 21, 2016. http://www.chronicle.com/interactives/belfast.

Megoran, N. 2011. War and peace: An agenda for peace research and practice in geography. *Political Geography* 30 (4):193–94.

———. 2013. Violence and peace. In *The Ashgate research companion to critical geography*, ed. K. Dodds, M. Kuus, and J. Sharp, 189–212. Aldershot, UK: Ashgate.

Miller, S. C. 2011. A feminist account of global responsibility. *Social Theory and Practice* 37 (3):391–412.

Milligan, C., S. Atkinson, M. Skinner, and J. Wiles. 2007. Geographies of care: A commentary. *New Zealand Geographer* 63:135–40.

Milligan, C., and J. Wiles. 2010. Landscapes of care. *Progress in Human Geography* 34 (6):736–54.

Mohanty, C. 1991. Under Western eyes: Feminist scholarship and colonial discourses In *Third world women and the politics of feminisms*, ed. C. Mohanty, A. Russo, and L. Torres, 51–80. Bloomington: Indiana University Press.

Moloney, E. 2011. Affidavit, United States District Court, District of Massachusetts. June 7, M.B.D No. 11-MC-91078.

Nagar, R. 2014. *Muddying the water: Coauthoring feminism across scholarship and activism*. Chicago: University of Illinois Press.

Najafi, A. 2012. Socio-environmental impacts of Iran's disappearing Lake Urmia. Accessed October 1, 2012. http://climateandsecurity.org/2012/05/18/socio-environmental-impacts-of-irans-disappearing-lake-urmia.

Narayan, U. 2003. *Feminism without borders, decolonizing theory*. Durham, NC: Duke University Press.

Nevins, J. 2012. Police mobility, maintaining global apartheid from South Africa to the United States. In *Beyond walls and cages: Prisons, borders, and global crisis*, ed. J. Loyd, M. Mitchelson, and A. Burridge, 19–26.

Noxolo, P., R. Raghuram, and C. Madge. 2011. Unsettling responsibility: Postcolonial interventions. *Transactions of the Institute of British Geographers* 37 (3):418–29.

O'Neill, R. K. 2011. Affidavit of Robert K. O'Neill, United States District Court, District of Massachusetts. June 7, M.B.D No. 11-MC-91078.

Ong, A. 2000. Graduated sovereignty in South-East Asia. *Theory, Culture and Society* 17 (4):55–75.

Pain, R. 2015. Intimate war. *Political Geography* 44:64–73.

Pain, R., and L. Staeheli. 2014. Introduction: Intimacy-geopolitics and violence. *Area* 46 (4):344–47.

Parr, H., and C. Philo. 2003. Rural mental health and social geographies of care. *Social and Cultural Geography* 4 (4):471–88.

Pengra, B. 2012. *The drying of Iran's Lake Urmia and its environmental consequences*. Sioux Falls, SD: United Nations Environmental Programme.

Popke, J. 2007. Geography and ethics: Spaces of cosmopolitan responsibility. *Progress in Human Geography* 31:509–18.

Pratt, G., and V. Rosner. 2012. *The global and the intimate: Feminism in our time*. New York: Columbia University Press.

Raghuram, P., C. Madge, and P. Noxolo. 2009. Rethinking responsibility and care for a postcolonial world. *Geoforum* 40:5–13.

Robinson, F. 2011. *The ethics of care: A feminist approach to human security*. Philadelphia: Temple University Press.

Robinson, J. 2016. Thinking cities through elsewhere: Comparative tactics for a more global urban studies. *Progress in Human Geography* 40:3–39.

Rodgers, D., and B. O'Neill. 2012. Infrastructural violence: Introduction to the special issue. *Ethnography* 13 (4):401–12.

Ross, A. 2011. Geographies and war and the putative peace. *Political Geography* 30 (4):197–99.

Schneider, J., and J. Susser. 2003. *Wounded cities: Destruction and reconstruction in a globalized world*. New York: Bloomsbury Academic.

Sharp, J. 2009. *Geographies of postcolonialism: Spaces of power and representation*. London: Sage.

———. 2011. Subaltern critical geopolitics of the war on terror: Postcolonial security in Tanzania. *Geoforum* 42 (3):297–305.

Shinko, R. 2008. Agonistic peace: A postmodern reading. *Millennium: Journal of International Studies* 36:473–91.

Shirlow, P., and B. Murtagh. 2006. *Belfast, segregation, violence and the city*. London: Pluto.

Silberman, M. K., and J. W. Till. 2012. *Walls, borders, boundaries: Spatial and cultural practices in Europe*. Oxford, UK: Berghan.

Sin, H. L. 2010. Who are we responsible to? Locals tales of volunteer tourism. *Geofourm* 41:983–92.

Smith, D. M. 1998. How far should we care? On the spatial scope of beneficence. *Progress in Human Geography* 22:15–38.

———. 2000. *Moral geographies: Ethics in the world of difference*. London: Edinburgh University Press.

Smith, S. 2009. The domestication of geopolitics: Buddhist-Muslim conflict and the policing of marriage and the body in Ladakh, India. *Geopolitics* 14 (2):197–218.

Spark, M. 2007a. Acknowledging responsibility for space. *Progress in Human Geography* 31:395–403.

———. 2007b. Geopolitical fears, geoeconomic hopes and the responsibilities of geography. *Annals of the Association of American Geographers* 97:338–49.

Springer, S. 2014. War and pieces. *Space and Polity* 18 (91):85–96.

Staeheli, L., and M. Brown. 2003. Where has welfare gone? Introductory remarks on the geographies of care and welfare. *Environment and Planning A* 35:771–77.

Susser, I., and J. Schneider. 2003. *Wounded cities: Destruction and reconstruction in a globalized world.* Oxford, UK: Berg.

Till, K. 2012a. Resilient politics and a place-based ethics of care: Rethinking the city through the District Six Museum in Cape Town, South Africa. In *Collaborative resilience: Moving through the crisis to opportunity*, ed. B. Goldstein, 283–307. Cambridge, MA: MIT Press.

———. 2012b. Wounded cities: Memory-work and a place-based ethics of care. *Political Geography* 31:3–14.

———. 2013. Walls, resurgent sovereignty and infrastructures of peace. *Political Geography* 33:52–53.

Tronto, J. 1987. Beyond gender difference to theory of care. *Signs: Journal of Women in Culture and Society* 12:644–63.

Tyner, J., and J. Inwood. 2011. *Nonkilling geography.* Honolulu, HI: Center for Global Nonkilling.

United Nations Development Programme (UNDP). 2017. Overview. Accessed May 2, 2017. http://www.ir.undp.org/content/iran/en/home/ourwork/overview.html.

Williams, P. 2015. *Everyday peace politics, citizenship and Muslim lives in India.* Oxford, UK: Wiley.

Williams, P., and F. McConnell. 2011. Critical geographies of peace. *Antipode* 4 (43):927–33.

Woon, C. Y. 2011. *Undoing violence, unbounding precarity: Beyond the frames of terror in the Philippines* 42:285–96.

———. 2014. Precarious geopolitics and the possibilities of nonviolence. *Progress in Human Geography* 38 (5):654–70.

Wright, M. W. 2005. Paradoxes, protests and the mujeres de negro of northern Mexico. *Gender, Place and Culture* 12 (3):277–92.

Young, I. M. 2006. Responsibility and global justice: A social connection model. *Social Philosophy & Policy* 23 (1):102–30.

LORRAINE DOWLER is an Associate Professor in the departments of Geography and Women Gender and Sexuality Studies at the Pennsylvania State University, University Park, PA 16802. E-mail: lxd17@psu.edu. Her research interests include feminist geopolitics, feminist care ethics, and critical militarization studies.

A. MARIE RANJBAR is an Assistant Professor in the Department of Women, Gender and Sexuality Studies at Ohio State University, Columbus, OH 43210. E-mail: ranjbar.3@osu.edu. Her research interests include feminist political geography, with a concentration on human rights, environmental justice, and social movements in Iran.

13 Without Space

The Politics of Precarity and Dispossession in Postsocialist Bucharest

Jasmine Arpagian and Stuart C. Aitken

The eviction of families from historically nationalized and recently restituted houses in Romania is tied in complicated ways to postsocialist transitional justice policies. Delayed enactment of restitution legislation and inconsistent application leave families, and neglected houses, in a precarious state. As families remain in place, they create a politics that pushes against dispossession. Evidence of this push comes from a study of Roma families, who are arguably the most marginalized of Romania's low-income peoples. Theoretically, we draw on Butler and Athanasiou's understanding of precarity and dispossession and Askins's emotional citizenry, from which we find a glimmer of hope in the everyday performance of the political among threatened families. *Key Words: Bucharest, dispossession, emotional citizenry, precarity, postsocialism.*

在罗马尼亚, 将家户驱逐出历史上属于国家但晚近获得补偿的房舍, 与后社会主义的转型正义政策有着复杂的连结。补偿立法的拖延与不一贯的实施, 导致许多家户和被疏忽的房舍处于不稳定的境况。随着这些家户续留在原地, 他们创造出抵抗驱逐的政治。此一推力之证据, 来自于对罗马家庭的研究, 他们可说是罗马尼亚低收入者中最为边缘化者。我们运用巴特勒和阿萨纳修对于不稳定和驱逐的理解, 以及阿斯金的感性公民之理论, 从而在受到威胁的家庭的每日生活政治展演中寻得一丝希望。 关键词: 布加勒斯特, 驱逐, 感性公民, 不稳定性, 后社会主义。

El desalojo de familias de residencias nacionalizadas históricamente y recientemente restituidas está ligado de manera complicada con las políticas de la justicia transicional postsocialista de Rumania. La demora en implementar la legislación relacionada con la restitución y su aplicación inconsistente deja a las familias y a las casas desatendidas en un estado precario. En tanto las familias permanezcan en el lugar, crean una situación política que presiona contra la desposesión. La evidencia de esta presión proviene de un estudio de familias de Roma, consideradas por algunos como lo más marginado de la gente de bajos ingresos de Rumania. Dese el punto de vista teórico, nos apoyamos en la forma como entienden Butler y Athanasiou la precariedad y la desposesión, lo mismo que en la ciudadanía emocional de Askin, desde donde hallamos un atisbo de esperanza en el desempeño cotidiano de lo político entre las familias amenazadas. *Palabras clave: Bucarest, desposesión, ciudadanía emocional, postsocialismo.*

Bodies on the street are precarious—exposed to police force, they are standing for, and opposing, their dispossession. These bodies insist upon their collective standing, organize themselves without and against hierarchy, and refuse to become disposable: they demand regard.

—Butler and Athanasiou (2013, i)

Ana-Maria watched as bulldozers demolished her home of twenty years. It was an old building subdivided during Bucharest's socialist era to provide housing for working-class families. We met Ana-Maria living on the street as she rinsed her pots and hung her freshly cleaned clothes. Romania's 1989 December Revolution created tensions between new market-driven forms of governance and economics and the legacies of communism (Ramet 2010), tensions that made an appearance as the new government tried to redress past human rights violations such as the nationalization of houses and businesses. Transitional justice slowly ensued but with concern among some officials that property restitution led to new rounds of injustice (Stan 2006). The building where Ana-Maria lived was returned to its previous owner. Her rental contract was renewed for only five years, after which she became an illegitimate occupant. The property's sale was deferred until the market recovered from the global recession. Once sold, the house was razed to the ground, as the land itself, centrally located in Bucharest's historic and gentrifying district, held revenue potential. Ana-Maria and several other Roma families set up camp on the street beside where their home had stood. Without a defined residence, families received provisional identification documents popularly described as *fără spațiu* (without space), as they are specific to residents without an official address. *Fără spațiu* is an effective trope for the circumstances of many Roma and other low-income families within Bucharest's intractable framework of dispossession and slow violence. When we met Ana-Maria in the summer of 2016, she and her family had been living on the street for more than eighteen months in an improvised wooden and tarpaulin structure they had pieced together themselves.

To the degree that Bucharest's historic neighborhoods are gentrifying (Suditu 2014; Chelcea, Popescu, and Critea 2015), a combination of cronyism, capital accumulation, restituted ownership, and renewal of prime urban real estate (Chelcea 2006) pushes low-income tenants toward dispossession. Slow violence confounds the dispossession. Occupants are uncertain of eviction and allocation of state-subsidized peripheral social housing. The violence is slow because it takes place gradually and confers a form of placelessness on occupants—represented by *fără spațiu*, a residential category used by the city to administer belonging. These occupants in actuality contrive strong connections to places, other marginalized people, and various institutions (church groups, charities, and social organizations) trying to mitigate the precarity. An active and emotional citizenry emanates from these kinds of multifarious places and complex occupant lives, creating communities of care that sometimes push back against dispossession and exclusion (Aitken 2016; Askins 2016). Social justice is often constituted by processes of emotional citizenry, which are "embedded in the complexities of places, loves and feelings" and "beyond claims to and exclusions from nation-statehood" (Askins 2016, 515). Put simply, everyday encounters and place-based connections bolster security in ways that certain forms of violence and exclusivity cannot assail.

The stories of Roma families derived from observations and interviews conducted in the fall of 2015 and summer of 2016 illustrate dispossession that proceeds in slow but violent and forceful ways. We initially encountered families identified through a Roma community organization in one of Bucharest's historic districts. Families introduced us to their support groups (e.g., Roma churches and nongovernmental organizations) and more families were identified using snowball techniques. Although families represented in this article are Roma, we acknowledge in the first instance that some participants interviewed did not identify as Roma; second, that Roma is neither a monolithic nor singular ethnic category (Tcherenkov and Laederich 2004); and third, that other low-income groups live in similar contexts of precarity and slow violence.

In what follows, we start with a brief review of the process of property confiscation in communist Romania and the redistributive politics aimed to allay injustices, with the caveat that unpacking the complexity of these processes (Verdery 2003; Stan 2006, 2013) goes well beyond the remit of what we are trying to do here. We then suggest that an active and emotional citizenry residing within the complexities of inner-city places in Bucharest enables some dispossessed people to rally against the housing inequalities aggravated by emerging neoliberal forms of governance.[1] With help from what Wright (2015) called heterogeneous belongings that come about with "feeling-in-common" through a complex assemblage of actors (including nonprofit workers and charities), materials (wood, tarpaulins, and *fără spațiu* cards), and places, families, and individuals resist or oppose their dispossession by staying in place (Curti, Craine, and Aitken 2013; Gillespie 2016). Although dispossession and marginalization emanate from larger global processes (Sparke 2013), emotional citizenry coming from feeling-in-common can lead to progressive politics (Wright 2015) and mitigated dispossession.

Romanian Property Confiscation and Redistributive Politics

Property confiscation in communist Romania was a mechanism for curtailing profitable real estate ventures and punishing dissidents. Nationalization primarily, but not exclusively, targeted those who fled Romania, the former elite, and owners of multiple properties (Chelcea 2003; Verdery 2003). Our respondents are dispersed within the same historic area of Bucharest identified in Chelcea's (2006) study, where one third of families continued to inhabit nationalized property in the mid-2000s. Many confiscated properties, and especially those located centrally, were allocated to party members, military officers, and agents of the secret police (Stan 2006, 2013; Chelcea 2012). Redistribution of nationalized housing reduced spatial inequalities across social classes as educated professionals such as doctors and engineers shared apartments with lower status laborers (Chelcea 2012). Until the 1960s, when the communist state constructed apartments to house its labor supply, workers and their families were housed primarily in nationalized properties (including Roma, whose ethnicity was de-emphasized during the communist era). State-owned factories and enterprises continued to relocate their workers into confiscated homes, for instance, when families wanted more living space or lower rent.

Romania's violent revolution in December 1989 did not immediately lead to changes in political and economic structures. As in other Eastern European countries but at a slower pace, a neoliberal economy evolved in Romania (Birch and Mykhnenko 2009). Briefly explained here, national and local politics influenced the magnitude and rate of democratic and neoliberal development (Dawidson 2004; Stan 2006, 2013). Former communist leaders rebranded themselves as the Social Democrat Party and proposed a political "third way" (Stan 2010) and heterodox economics (Ban 2011). When this party returned to power in 2000, they were no longer inimical to the excesses of neoliberal and globalized economics (Ban 2011). These politicians pushed an ideology of social justice while crafting legislation to serve personal needs.[2] A privatization law passed in 1990 enabled 3 million tenants to purchase their apartments in state-constructed buildings at reduced prices, and in 1995, this option was extended to tenants of nationalized property (Stan 2006). While the parliament debated the fate of occupied properties not inhabited by their former owners, lower courts heard restitution claims and issued decisions, some of which were overturned due to the lack of official restitution legislation. Slow progress and confusing policies were attributed to, on the one hand, the protection of personal gains by political leaders (i.e., purchases or leases of nationalized properties at low rates) and, on the other, the lack of housing alternatives for tenants (Stan 2006).

In 2001, Parliament finally passed Law 10/2001 to guide property restitution. Tenants were protected to some degree. Law 10/2001 required owners of returned properties to extend rental contracts by five years and charge rent at a quarter of a household's annual income (Monitorul Oficial al Romaniei 2001).[3] By 2012, 170,000 files were registered reclaiming properties across the country, but only 9 percent had been resolved (Sultano 2012). According to the municipality's Web site, of the more than 43,000 property restitution claims, more than half (i.e., 24,142) are still unresolved (Legea 10-PMB 2017).

The process of reclaiming confiscated property is long, complicated, and uncertain for both owners and occupants. In what follows, we assume Verdery's (2003, 5) transitional process of "re-creating private property" and use Butler's lens of dispossession and

displacement to help understand the precarity of the Roma families in our study (Butler 2006; Butler and Athanasiou 2013). Property restitution is described as a "genealogical practice" for heirs (Chelcea 2003, 715) and "sentimental dramas" for former owners and current occupants (Zerilli 2006, 77). Zamfirescu (2015) argued that there are municipal authorities who take discriminatory action against current occupants, many of whom are poor and Roma. Following Butler—but also Glassman (2006) and Gillespie (2016)—we argue that the judicial and political landscape just described forces occupants into a state of liminal dispossession, where precarity is established and enhanced through destabilizing and dehumanizing processes.

Housing Insecurity and *Fără Spaţiu* Cards

In an assessment of Harvey's (2003) reworking of Marx's primitive accumulation into a more contemporary accumulation through dispossession, Glassman (2006) noted that new means of capitalist accumulation are often tied to dispossession of urban poor and ethnic minorities. He argued, further, that "extra-economic accumulation of (mainly) women's unpaid social reproductive labor makes the geography of struggle against global capitalism appear yet more complicated" (617). The complexities that Glassman alludes to represent the housing insecurity some of Bucharest's low-income Roma experience, and that insecurity is etched onto identification cards as *fără spaţiu*. Law 105/1996 for Population Record and Identity Cards, required applicants to provide documents verifying their location of residence; for instance, a title to the property, rental agreement, or signed document from the property owner indicating authorized occupancy (Monitorul Oficial al Romaniei 1996). Identification cards (both permanent and provisional) are available to all Romanian citizens starting at age fourteen. These documents are required to apply for social assistance services, including supplemental income and low-income subsidized housing. Given the requirement of annual renewal for provisional cards and confused by the application process, many informal occupants have neither permanent nor temporary identification documents.

For Corina, a young mother of three, who has lived informally for nearly ten years in restituted property, *fără spaţiu* cards represent and embody precarity:

> The temporary ID card is something normal for us. Because there are so many like this. Because there are

many who don't have homes. Temporary ID cards are [for] people who don't have homes. (Corina, 29 July 2016)

Corina clarified that her ID card lists "an address without space. So, the street is written but it is not written that I stay at this or that number. So [I am] without space." Owners of restituted properties rarely vouch for their known but informal occupants, who then are given provisional ID cards that require annual renewal. Official recognition of occupancy could legally complicate the property's future sale. Temporary ID cards communicate to others the vulnerability of the card holders, which might prompt discriminatory attitudes on the part of potential employers, landlords, or educators (Sastipen 2015). For occupants living in this state, uncertainty becomes ordinary until the prolonged slow violence of dispossession is interrupted by the trauma of displacement. These actions push legal liminality, spatial instability, and improper(tied) subjectivity.

Precarity and the Slow Violence of Gentrification: "Stay Here until the House Collapses over You"

Labor scholars and activists identify a new "precariat" class in advanced capitalism that reflects the nature of uncertain and insecure work, which is temporary and part time (Waitt 2011; Standing 2011). Although this literature focuses on neoliberal landscapes of production, many geographers are also attuned to how precarity shows up in reproduction and is lived in ways that are not bound by time or space (Ettlinger 2007; Clark 2015). Because reproduction and production are indelibly connected, the complex lives, events, environments, and places of people living precariously are embroiled in what Nixon (2013) described as slow violence. Nixon's focus is environmental degradation, but Kern (2016) pointed out that slow violence shows up in gentrifying urban settings where residents (and particularly renters) build complex attachments and displacement is a perennial threat (see also Gillespie 2016). Kern (2016) noted that the "rhythmic elements of everyday life are weighted with power precisely because they are so commonplace as to be unremarkable or beyond critique, but it is useful to examine how powerful groups imprint their own temporalities, especially in the context of gentrification" (445). Those who are unable to "access the temporalities of powerful groups" live on the precarious margins of changing urban landscapes, sometimes for years, as land is held in speculative

development; this lack of access enables the slow violence of gentrification (Kern 2016, 442).

Precarity is not just about loss of access to stable employment as suggested by Waitt (2011) and Standing (2011); then, it is about slow erosion and a larger swathe of dispossession that deterritorializes "the genealogy of the proper(tied) subject" (Butler and Athanasiou 2013, 14). By suggesting this, Butler and Athanasiou (2013) elaborated the notion of a self-contained, liberal subject with particular status ascribed to propertied citizenship. The precarious neoliberal subject loses that status through state-prescribed exceptions (Agamben 2001), mutations (Ong 2006), and erasures (Aitken 2016). Improper (temporary, provisional) citizenship in Romania, it seems, is related to provisional identification cards and is connected to being without space. Without space is not without complex place-based associations, including contacts in nongovernment organizations, church groups, and charities and what Askins (2016) referred to as friendship networks of emotional citizenry. Impropriety then, is a "refusal to stay in one's proper place ... [and] signals an act of radical reterritorialization, which might certainly include remaining in specific places" (Butler and Athanasiou 2013, 21).

Mariana's family history represents a deterritorialization, an open temporality, and the reproduction of precarity without (proper) space. In the late 1970s when Mariana was still in her twenties, she worked for a state-owned factory manufacturing prefabricated concrete slabs in Bucharest. The family's sole breadwinner, Mariana requested from her company, and was eventually granted, an apartment for herself and her four children, but her situation did not improve:

Since then our situation was seriously aggravated ... because my salary was not enough to pay for the apartment maintenance fee. My salary was not enough to pay for the electricity. My salary was not enough to support my children. I had four children, I was alone and young. (Mariana, 9 August 2016)

Despite steady and formal employment, Mariana's salary was simply not enough to support both her family and her home. She requested more affordable housing and was offered a shared living arrangement in a nationalized building with lower rent paid directly to the state. She relocated her family in 1986 and has since lived there. According to her son, Cornel:

We were raised this way, in poverty. There were days when we did not even have [anything] to put on the table. No, not days, weeks. And we would see to our work at school. Each of us did what we could. I was held back a year but I repeated. (Cornel, 9 August 2016)

Mariana's now-adult children are skilled laborers with children of their own but still live with her in an increasingly cramped apartment. In 2001, the court returned the nationalized building to the precommunist owner, who refused to draft a rental agreement with the occupants, pushing them into *fără spaţiu*. Since that time, the families have been living rent-free but without basic amenities such as running water, electricity, and natural gas for heat. The owner waits for a satisfactory real estate market boom and in the meantime allows these income-insecure families to occupy the property. After the property was returned, Mariana was reassured that she could stay put, temporarily: "They made a contract for two years. But those two years passed and then the owner said you could stay until the house collapses over you" (Mariana, 9 August 2016). Spurred not by an owner's responsibility or occupant's obligation but the basic need for sheltered security, Cornel does his best to "take care of the house so it won't fall." Mariana and her family stay put with the realization that an eviction notice could be delivered any day.

Dispossessed, Then Displaced: "Now We Have to Begin from Zero"

Moving beyond contexts of labor and employment, Butler (2006) deployed precarity to describe an ontological condition of vulnerability and interdependency. She elaborated how public representation determines whose lives are grievable and worthy of protection and whose are not. From this perspective, questions arise about the citizenship rights and the subject-hood of urban Roma in a Romania increasingly defined by political, social, and economic uncertainty. How are Roma lives made or unmade as "grievable" and "livable"? What kind of political possibility is created around the conditions and concepts of precarity (Worth 2016)?

Ana-Maria's story, with which we begin this article, is in the first instance a move from security during the communist era to dispossession and precarity in post-socialist Romania; second, it is a fight to stay put on the street of her demolished home (Figure 1); and third, with the help of friends in charities and nongovernmental organizations, it is a provocative call on the municipality to house them as vulnerable citizens. These three actions begin to situate a grievable and livable context for Ana-Maria and open the possibility of emotional citizenry. A few years after Ana-Maria's renewed contract, her family and twenty-six others received notices to evacuate. When asked how she felt about having to leave her home of twenty years, Ana-Maria responded:

As you can imagine, it was our life's work. It was the childhood of my children who were raised and born there. And it's hard. ... Your labor goes and you don't have anywhere else to go, after a life of living in that house, and you invested money. But it is not only about

Figure 1. Shacks on the street of the bulldozed building occupied by Ana-Maria and other disposed families. *Source*: Authors. (Color figure available online.)

investment, because you make investments in any house. It is a matter of where the children grew up, there are memories there, moments you lived over there. And it is hard, very hard. (Ana-Maria, 21 July 2016)

When asked about the emotional impact of eviction, Ana-Maria's teenage son was especially aware of the effects on his family's social network:

Well, that is where we stayed since we were little. We were used to staying there. We were accustomed to everyone around us. We got along well with everyone. That is it. After they kicked us out, some went to one place, others went somewhere else, and we were separated, us friends. (Andrei, 4 August 2016)

Bodies on the street are precarious and demand regard (Butler and Athanasiou 2013). The bodies of Andrei and Ana-Maria, and those of several neighbors, occupied their home's street for a year and half in constant worry of removal without reasonable alternatives. Their vulnerability enabled Ana-Maria and friends to create what Askins (2016) described as "emotional geographies of belonging" (520–21) that help build security and recognition in the face of confusing bureaucracies and unknown legalities. A sense of that which is possible (Butler 2006) empowers Ana-Maria, like Askins's (2016) refugees in northern England, to engage over time with friends in place-based encounters that promulgate interscale belonging through lines of connection and disconnection and fluid, hybrid identities, and political allegiances.

Bucharest's processes of displacement are not orchestrated to target Roma, nor is its inner-city gentrification designed to perpetuate a slow violence. That said, this minority's overwhelming precarity renders Roma particularly vulnerable (Berescu et al. 2006; Fleck and Rughinis 2008). In another corner of Bucharest's historic center, Corina and her neighbors were also stripped of their right to formally and legally reside where they had lived for decades. The property was returned to the former owner, who did not draft a new rental contract, but visited every few years reassuring more than twenty families they would not be evicted, and advising them to collect firewood for the winter (interview with Corina, 29 July 2016).

Many families occupied Corina's apartment building for decades, initially as formal tenants of the state. Others moved in as informal occupants from the start, after the property had already been restituted and as the slow violence of gentrification took hold in Bucharest:

[We were here] since 2007. How I came here? We came here exactly like this, the owner reclaimed my parent's home. We were evicted and we came here. And from here they are evicting us again. … Here, I found [this place] through acquaintances, friends. I came and they let me stay out of pity, basically. I do not have documents here. I am staying on my own count. Because I did not have anywhere to go with three children after we were evicted. (Corina, 29 July 2016)

More than twenty mostly Roma families currently live here, uncertain and uncomfortable but sheltered (Figure 2). About two dozen children play in the yard,

Figure 2. Precarity on a corner of Bucharest's historic city center. *Source:* Authors. (Color figure available online.)

only half of whom are enrolled in school. Their young parents hold precarious employment as unqualified laborers. Corina explained that she "had a work contract for four hours and [she] was working for eight," and her husband "works at a car wash. He doesn't have a salary, he works on tips and they schedule him nonstop; he stays three days, five days at work, day and night." Some young parents tested their luck working in Western Europe but ultimately returned home.

On a summer morning, the local police visited and demanded entry into the units, insisting that all families wait outside in the yard as rooms are searched for drugs. Without a court order, the police acted on the pretext of complaints from neighbors and reports of drug use and trafficking on the premises. Men and women were asked to stand separately while entry doors and storage shacks were knocked down. Families were given forty-eight hours to remove their belongings and themselves, but intervention from experienced local volunteer activists enabled them to continue living there. A month after the eviction attempt, families were still uselessly asking local police for updates on their situation. They continued to wait with fear of becoming homeless in the winter. Despite this numbing uncertainty, Corina was tearfully grateful for the volunteers' unassuming and crucial support:

> Thanks to that [volunteer], and I've said this many times, because if he didn't come then, we would have been abusively removed, the way they wanted. When that boy came and asked, "On what grounds are you evicting them? You came here to clean up, not to remove them." And no one said another word. (Corina, 29 July 2016)

Emotional citizenry enables Corina and the other occupants to stay put but with heightened feelings of precariousness.

Resistance Through Emotional Citizenry: "And Because We Stayed in Place, We Protested"

Corina and Mariana stayed put. Ana-Maria and her family were forcefully removed but, as we mentioned earlier, they did not move very far. Having no housing alternative and refusing to accept displacement, Ana-Maria and her family resisted by staying connected to a place directly adjacent to where the children were raised. Her decision was motivated by a perceived

right to housing, and she felt that staying close elaborated that right:

> We were not owners either, but we previously had documents [a rental agreement with the city's housing administration]. ... That's why we stayed in place. And because we stayed in place, we protested. ... Fine, it was also a matter of principle, but it was my right. To stay, for them to give me housing. ... I paid the state for years. (Ana-Maria, 21 July 2016)

Support from local and international volunteers created for Ana-Maria the kind of emotional citizenry that Askins argued is interscaled and affective. Even with this support, living on the street was not easy:

> I felt like a *boschetar* [street person]. And a lot of clothes caught on fire because the embers [from cooking and heating fires] would jump on us. Do you understand? It was suffering. I think this is a story that will remain engraved in our minds, like a trauma. Like a trauma, what we suffered. (Ana-Maria, 21 July 2016)

When their shacks were bulldozed in July 2016, Ana-Maria's family received affordable housing in a new city-owned apartment in the periphery of Bucharest. When asked how he feels about their new home, Andrei responded: "The new house is a new life. We started a new life in it. We arranged it with what is needed. With what God was able to give us. And we are grateful we have someplace to rest our head."

When we last met with them, it was hard not to conclude that the outcome for Ana-Maria and her family was positive. They were happy with the newness of the subsidized apartment, but they were nonetheless displaced from the social networks and place-based attachments of their previous neighborhood. It is only a matter of time before Corina and Mariana, and their families, are moved away from the gentrifying core of Bucharest. The slow violence of that gentrification is tied to the temporal imprinting of powerful groups (Kern 2016), who are not part of our study, and their administratively supported actions.

Butler and Athanasiou (2013) questioned how dispossession arises in conjunction with precarity and explored theoretically how those who lose property and land also lose broader contexts of citizenship and belonging. This article empirically elaborates these connections, suggesting ways in which citizenship and belonging are emotionally engaged against the exclusions of nation-statehood or what our Romanian example highlights as *fără spațiu*.

Conclusion

Evictions or attempted evictions, and the slow violence and augmented insecurity that follow, are traumatic. Our discussions with Roma families prompted feelings of fear, shock, and confusion, as well as anger over injustices and inequalities. Although we focus on their experiences, we briefly recall the dispossession and displacement of property owners in communist Romania. Mechanisms of transitional justice, specifically property restitution, modified the framework of property ownership. A remodeled political landscape eventually led to a new round of dispossession, of a slower and more prolonged kind. Tenants who rented legally from the state suddenly become informal occupants of restituted property. Everyone waits: Owners patiently wait for a satisfactory revival of the housing market, and occupants dreadfully wait for an inevitable eviction. We show that this case of dispossession in Bucharest, as it enhances the insecurity of low-income families, represents Nixon's (2013) notion of slow violence.

The study provides empirical evidence to support Butler and Athanasiou's (2013) assertion that those who lose property and land also lose broader contexts of citizenship and belonging. *Fără spațiu* cards highlight pointedly that being without space is also about a loss of status. We demonstrate that although this administrative taxonomy suggests placelessness, these informal occupants feel attached to their neighborhood. Family members elaborated an emotional citizenry through heightened feelings of strength in the face of more precarity and gratitude for the social connections and support that were part of their lives. We tentatively suggest with Askins (2016), then, that there are ways in which citizenship and belonging emotionally engage against the exclusions of a propertied nation-statehood. To the extent that property and possession are a hallmark of capitalism and neoliberalism, our case study provides some understanding of the possibility of a place-based, performative, and emotional citizenship from which the dispossessed can act through their precariousness to establish something different and perhaps better.

Funding

This research was supported by the June Burnett Endowment.

Notes

1. We assume that the evictions documented here are dispossessions. Although the occupants do not own their properties in the capitalist sense of ownership, they nonetheless gained rights to occupancy in the Lockean sense of becoming propertied through labor.
2. Although publicly acknowledging the need for lustration, they discouraged discussion of communist abuses or the suggestion of laws to remove former communists from public office and their homes (Stan 2013; Ciobanu 2015).
3. Our interviews suggest that this requirement is practiced inconsistently and never enforced.

References

Agamben, G. 2001. *Means without end: Notes on politics.* Minneapolis: University of Minnesota Press.

Aitken, S. C. 2016. Locked-in-place: Young people's immobilities and the Slovenian erasure. *Annals of the Association of American Geographers* 106 (2): 358–65.

Askins, K. 2016. Emotional citizenry: Everyday geographies of befriending, belonging and intercultural encounter. *Transactions of the Institute of British Geographers* 41 (4):515–27.

Ban, C. 2011. Neoliberalism in translation: Economic ideas and reforms in Spain and Romania. PhD diss., University of Maryland, College Park.

Berescu, C., M. Celac, O. Ciobanu, and C. Manolache. 2006. *Housing and extreme poverty: The case of Roma communities.* Bucharest: Ion Mincu University Press.

Birch, K., and V. Mykhnenko 2009. Varieties of neoliberalism? Restructuring in large industrially dependent regions across Western and Eastern Europe. *Journal of Economic Geography* 9 (3):355–80.

Butler, J. 2006. *Precarious life: The powers of mourning and violence.* London: Verso.

Butler, J., and A. Athanasiou 2013. *Dispossession: The performative in the political.* Hoboken, NJ: Wiley.

Chelcea, L. 2003. Ancestors, domestic groups, and the socialist state: Housing nationalization and restitution in Romania. *Comparative Studies in Society and History* 45 (4):714–40.

———. 2006. Marginal groups in central places: Gentrification, property rights and post-socialist primitive accumulation (Bucharest, Romania). In *Social changes and social sustainability in historical urban centres: The case of Central Europe,* ed. G. Enyedi and Z. Kovács, 127–46. Pecs, Hungary: Centre for Regional Studies of the Hungarian Academy of Sciences.

———. 2012. The "housing question" and the state-socialist answer: City, class and state remaking in 1950s Bucharest. *International Journal of Urban and Regional Research* 36 (2):281–96.

Chelcea, L., R. Popescu, and D. Critea 2015. Who are the gentrifiers and how do they change central city neighborhoods? Privatization, commodification, and gentrification in Bucharest. *Geografie* 120:113–33.

Ciobanu, M. 2015. The challenge of competing pasts. In *Post-communist transitional justice: Lessons from twenty-five years of experience,* ed. L. Stan and N. Nedelsky, 148–66. New York: Cambridge University Press.

Clark, J. 2015. "Just one drop": Geopolitics and the social reproduction of security in southeast Turkey. In *Precarious worlds: Contested geographies of social reproduction*, ed. K. Meehan and K. Strauss, 145–64. Athens: University of Georgia Press.

Curti, H. G., J. Craine, and S. C. Aitken 2013. *The fight to stay put: Social lessons through media imaginings of urban transformation and change*. Stuttgart, Germany: Franz Steiner Verlag Press.

Dawidson, K. E. 2004. Property fragmentation: Redistribution of land and housing during the Romanian democratisation process. PhD diss., Uppsala University, Uppsala, Sweden.

Ettlinger, N. 2007. Precarity unbound. *Alternatives* 32:319–40.

Fleck, G., and C. Rughinis 2008. *Come closer. Inclusion and exclusion of Roma in present-day Romanian society*. Bucharest, Hungary: Human Dynamics.

Gillespie, T. 2016. Accumulation by urban dispossession: Struggles over urban space in Accra, Ghana. *Transactions of the Institute of British Geographers* 41 (1): 66–77.

Glassman, J. 2006. Primitive accumulation, accumulation by dispossession, accumulation by 'extra-economic' means. *Progress in Human Geography* 30 (5):608–25.

Harvey, D. 2003. *The new imperialism*. Oxford, UK: Oxford University Press.

Kern, L. 2016. Rhythms of gentrification: Eventfulness and slow violence in a happening neighbourhood. *Cultural Geographies* 32 (3):441–57.

Legea 10-PMB [Law 10-Bucharest mayor's office]. 2017. Bucharest, Romania: Primaria Municipiului Bucuresti. Accessed March 19, 2017. http://www4.pmb.ro/wwwt/l112jur/iapag1jur.php.

Monitorul Oficial al Romaniei. 1996. Lege nr. 105 din 25 septembrie 1996 privind evidenta populatiei si cartea de identitate [Law Nr. 105 from 25 September 1996 regarding the population register and the identity card]. Accessed March 19, 2017. http://legislatie.just.ro/Public/DetaliiDocument/8556

———. 2001. Lege nr. 10 din 8 februarie 2001 privind regimul juridic al unor imobile preluate in mod abuziv in perioada 6 martie 1945–22 decembrie 1989 [Law Nr. 10 from 8 February 2001 regarding the legal status of real estate confiscated in an abusive manner in the period between March 6, 1945 and December 22, 1989]. Accessed March 19, 2017. http://www.monitoruljuridic.ro/act/lege-nr-10-din-8-februarie-2001-privind-regimul-juridic-al-unor-imobile-preluate-n-mod-abuziv-n-perioada-6-martie-1945-22-decembrie-1989-26680.html

Nixon, R. 2013. *Slow violence and the environmentalism of the poor*. Boston: Harvard University Press.

Ong, A. 2006. *Neoliberalism as exception: Mutations in citizenship and sovereignty*. Durham, NC: Duke University Press.

Ramet, S. P., ed. 2010. *Central and southeast European politics since 1989*. New York: Cambridge University Press.

Sastipen. 2015. Facilitarea accesul la documente de identitate pentru persoanele vulnerabile [Facilitating access to identification documents for vulnerable persons]. Accessed March 19, 2017. http://www.sastipen.ro/files/comunicate-presa/facilitarea%20accesului%20la%20documente%20de%20identitate.pdf

Sparke, M. 2013. From global dispossession to local repossession: Towards a worldly cultural geography of occupy activism. In *The companion to cultural geography*, ed. N. C. Johnson, R. Schein, and J. Winders, 387–408. New York: Wiley.

Stan, L. 2006. The roof over our heads: Property restitution in Romania. *Journal of Communist Studies and Transition Politics* 22 (2):180–205.

———. 2010. Romania: In the shadow of the past. In *Central and southeastern European politics since 1989*, ed. S. P. Ramet, 379–400. New York: Cambridge University Press.

———. 2013. Civil and post-communist transitional justice in Romania. In *Transitional justice and civil society in the Balkans*, ed. O. Simić, and Z. Volčič, 17–30. New York: Springer.

Standing, G. 2011. *The precariat: The new dangerous class*. London: Bloomsbury.

Sultano, M. 2012. Despagubirea fostilor proprietari. Scandal monstru intre Guvern, ANRP si Asociatiile de proprietary abuziv deposedati [Reimbursing former owners. Big scandal between the government, ANRP and the association for abusively dispossessed owners]. *Gandul*. April 24. Accessed October 31, 2016. http://www.gandul.info/stiri/despagubirea-fostilor-proprietari-scandal-monstru-intre-guvern-anrp-si-asociatiile-de-proprietari-abuziv-deposedati-9559437

Tcherenkov, L. and S., Laederich. 2004. *The Rroma: History, language, and groups*. Basel, Switzerland: Schwabe.

Verdery, K. 2003. *The vanishing hectare: Property and value in postsocialist Transylvania*. Ithaca, NY: Cornell University Press.

Waitt, L. 2011. A critical geography of precarity. *Geography Compass* 3 (1):412–33.

Worth, N. 2016. Feeling precarious: Millennial women and work. *Environment and Planning D: Society and Space* 34 (4):601–16.

Wright, S. 2015. More-than-human, emergent belongings: A weak theory approach. *Progress in Human Geography* 39:391–411.

Zamfirescu, I. M. 2015. Housing eviction, displacement and the missing social housing of Bucharest. *Calitatea Vieții* 2:140–54.

Zerilli, F. M. 2006. Sentiments and/as property rights: Restitution and conflict in postsocialist Romania. In *Postsocialism: Politics and Emotions in Central and Eastern Europe*, ed. M. Svasek, 74–94. New York: Berghahn Books.

JASMINE ARPAGIAN is a PhD student in the joint doctoral program at the Department of Geography, San Diego State University, San Diego, CA 92182, and University of California, Santa Barbara, CA 93106. E-mail: jarpagian@sdsu.edu. Her research interests include urban geography, disadvantaged families in Romania, and displacement.

STUART C. AITKEN is Professor of Geography and June Burnett Chair at San Diego State University, San Diego, CA 92182. E-mail: saitken@mail.sdsu.edu. His research interests include critical social theory, qualitative methods, children, youth and families, film, and masculinities.

14 Neoliberalizing Social Justice in Infrastructure Revitalization Planning

Analyzing Toronto's More Moss Park Project in Its Early Stages

David J. Roberts and John Paul Catungal

A public consultation process is currently underway to gather ideas on the revitalization of a park and community center in one of Toronto's most economically diverse neighborhoods. This project is a partnership between a lesbian, gay, bisexual, and transgender (LGBT)-focused community center, a private philanthropist, and the City of Toronto. In this article, we argue that More Moss Park is illustrative of the neoliberalization of social justice, in which social justice is touted as central to both the end goal of the project and the planning process that will shape it. We focus on three political moves that underwrite the neoliberalization of social justice in the project. The first is the technicalization of social justice as "know-how," a form of expertise that one of the main partners claims to have gained via its history of working for sexual minority communities and that it claims to be able to offer in other sociospatial contexts. The second is the normalization of an anonymous private donor as a necessary "silent" partner in urban development whose foremost concern is social justice in the form of neighborhood improvements for marginalized communities. The third is the use of crises of neighborhood insecurity and of budget shortfalls as planning problems whose solutions rest on the suspension of normal planning approaches, thus justifying the use of a public–private partnership. These moves illustrate the ways in which social justice has become neoliberalized not only through narrowing its scope but also through using it as ideological armature to mask marginalizations emerging from urban neoliberalism itself. *Key Words: homonormativity, LGBT, public–private partnership, social justice, urban planning.*

多伦多一处经济最为多样化的社区之一，正在进行公共咨询过程，以搜集有关公园与社区中心活化的想法。该计画是由聚焦女同性恋、男同性恋、双性恋和跨性别者 (LGBT) 的社区中心、私人慈善家，以及多伦多市政府所形成的伙伴关系。我们于本文中主张，莫尔莫斯公园能够说明社会正义的新自由主义化，其中社会正义被吹捧为该计画最终目标与达成该目标的规划过程之核心。我们聚焦支持该计画中社会正义的新自由主义化之三大政治行动。第一是将社会正义技术化为"专门知识"——主要伙伴中的一员宣称透过与性别少数社群一同工作的历史中获得、并且能够在其他社会空间脉络中提供的专业形式。第二是将匿名私人捐赠人视为城市发展中的必要"静默"伙伴，且视其最重要的考量为改善边缘化社群邻里的社会正义形式的常态化。第三则是利用邻里不安全的危机与预算赤字，作为需透过终止正常规划方法、从而正当化公—司伙伴运用进行解决的规划问题。这些行动，描绘了社会正义不但透过窄化其范畴、同时透过运用其作为意识形态工具来掩盖从城市新自由主义本身浮现的边缘化，从而迈向新自由主义化的方式。 关键词: *同性恋常规, LGBT, 公—私伙伴关系, 社会正义, 城市规划。*

Actualmente se lleva a cabo un proceso de consulta pública para explorar ideas sobre la revitalización de un parque y un centro comunitario en uno de los vecindarios de Toronto más diversificados desde el punto de vita económico. Este proyecto es una compañía entre un centro comunitario de orientación lesbiana, gay, bisexual y transgénero (LGBT), un filántropo privado y la Ciudad de Toronto. En este artículo sostenemos que el Parque More Moss es ilustrativo de la neoliberalización de la justicia social, en la cual la justicia social es promocionada como central tanto para la meta final del proyecto como para el proceso de planificación que lo configurará. Nos concentramos sobre tres movidas políticas que subrayan la neoliberalización de la justicia social en el proyecto. La primera es la "tecnicalización" de la justicia social como *know-how* (saber cómo), una forma de expertia que uno de los principales socios reclama haber ganado a través de su historia de trabajo para comunidades de minorías sexuales, la cual sostiene ser capaz de ofrecer en otros contextos socioespaciales. La segunda es la normalización de un donante privado anónimo como un socio "silencioso" necesario, en desarrollo urbano, cuya principal preocupación es la justicia social en la forma de mejoramientos vecinales para comunidades marginales. La tercera es el uso de crisis de inseguridad de barrio y de déficits presupuestales como problemas de planificación cuyas soluciones descansan en la suspensión de los enfoques de planificación normales, para justificar así el uso de una compañía de carácter público–privada. Estas movidas ilustran las maneras como la justicia social ha llegado a neoliberalizarse no solamente reduciendo su alcance, sino también dándole uso como armadura ideológica para enmascarar las marginalizaciones que emergen del propio neoliberalismo urbano. *Palabras clave: homonormatividad, LGBT, compañía público–privada, justicia social, planificación urbana.*

On 31 March 2016, more than 200 individuals gathered for a meeting at the John Innes Community Centre, in Toronto's Moss Park, one of the city's most economically diverse neighborhoods and home to many social services for the city's most marginalized residents. The purpose of the meeting was to announce a partnership between the City of Toronto, the 519 Community Centre, and an anonymous private donor and to launch a community consultation process designed to "shape a vision for new facilities and park space in Moss Park" (More Moss Park 2016). Those of us in attendance at this meeting were presented with an overview of the project and the planned public consultation process. Additionally, we were invited to provide input on the current usages of the park and community center and to articulate our hopes and wants for a future facility.

Although the 31 March meeting was billed as a launch, the project actually has a much longer genealogy. The project, now known as More Moss Park (MMP), is only the latest iteration of an older project, which was framed much differently in the beginning. The original project involved two of the now three partners (the 519 Community Centre and the anonymous private donor) envisioning a sports complex that would provide a central space for the various lesbian, gay, bisexual, and transgender (LGBT) sports and recreation leagues that are currently scattered across the city. After evaluating and deeming several sites inadequate for such a complex (Watson 2015), the 519 and the donor found themselves in conversations with the city on the possibility of a partnership to redevelop Moss Park and its community center. As the official MMP origin story goes, once sights were set on Moss Park as a location for this partnership, the mission, vision, and ultimate plan for the project had to be adapted to the realities of Moss Park and its established community (interview, executive with the 519, 13 December 2016). As a result, the LGBT focus of the project was modified into a more highly localized approach that looks directly to area-specific needs and a broader conception of social justice as central to the project.

Despite shifting goals, what has been consistent with the project has been its centering of social justice goals, first in terms of providing space for LGBT sport and recreation and then, as MMP, in terms of dealing with the infrastructure needs of a historically marginalized neighborhood. MMP thus provides an opportunity to examine the important and understudied enrollment of social justice in public–private partnerships in infrastructure development. We argue that this represents the neoliberalization of social justice. We identify three political moves that are central to this process: (1) the technicalization of social justice as a form of transferable planning expertise, (2) the normalization of private philanthropy to fund more socially just planning processes, and (3) the suspension of normal planning processes as the most efficient way to achieve social justice for marginalized communities. These three moves produce a neoliberalized form of social justice that then provides ideological and moral credence to the project, which, on paper, seeks to emplace marginalized people's desires and knowledges in the shaping of urban agendas. However, when considered alongside the neighborhood's longer histories of municipal disinvestment compared to other areas of the city, such moves discursively secure the creeping privatization of urban planning in the name of (concerns for) marginalized communities.

We draw in this article on semistructured key informant interviews, media coverage of the MMP project, and ethnographic observations in public meetings. One of us (David Roberts, who lives and works in Toronto) has attended numerous meetings and conducted (with the help of a research assistant) interviews with two of the three formal partners (minus the anonymous donor), with community organizers who were hired to carry out the public consultation process, and with representatives from groups opposed to the project. Additionally, we draw on media reportage on the project in mainstream news outlets (e.g., Canadian Broadcasting Corporation, *Toronto Star*, *National Post*), community-geared media (Now, Xtra), and other outlets (blogTO, insideToronto), along with official City of Toronto documents that relate directly to this project.

Given that the project is ongoing and will continue to evolve, we recognize that there are some distinct limitations to our article. By necessity, our analysis is focused on the project's early stages. We believe, however, that even these early stages already offer important insights into the politics of centering social justice in urban infrastructure development. Moreover, given that the forces at play in this case—development pressures, claims of austerity, public–private partnerships (P3s), social inclusion, gentrification and consultation processes—are far from unique to Moss Park or Toronto, insights from this case offer lessons that we hope will be useful in other contexts where, more generally, neoliberal planning is being pursued in and through the name of social justice.

The rest of the article unfolds in three sections, each of which examines specific modes through which the neoliberalization of social justice is enacted in MMP. These sections discuss, in turn, (1) how social justice is rendered technical as a form of planning expertise that transfers from one context of marginalization to another, (2) how privatization and philanthropy become idealized as ways to achieve social justice, and (3) how notions of crisis are mobilized to justify exceptional planning interventions in pursuit of social justice. The article concludes with a discussion of broader implications of the confluence of social justice discourse and P3s in the neoliberal city.

Social Justice as Know-How: Participation in Neoliberal Urban Processes through Expertise

The central role of the 519 as one of three project partners provides insight into the changing role that municipal LGBT services such as community centers are playing not only in sustaining supportive spaces and services for LGBT communities but also in the pursuit of infrastructure development and city building itself. According to Velasco (2013), "In 1975, The 519 Community Centre opened and marked the first time the City purchased a building for a community centre, after dedicated community members lobbied for the establishment of a meeting place." Since its initial opening, the 519 has been an important gathering space for community groups and social service providers.

The geography of the 519's work has, until more recently, focused on the relatively localized catchment area that centers around the Church and Wellesley neighborhood, also known as Toronto's Gay Village. Its forays beyond this area, first into the West Don-lands (in the early iteration of the project) and more recently into Moss Park, speak to a desire to exceed the confines of the traditional Gay Village by expressly supporting and actively pursuing the creation of permanent queer spaces and infrastructures elsewhere (as in the first iteration of this project) and, more recently, as we show later, by expanding beyond its queer focus toward expressly identifying a broader social justice expertise (cf. Doan 2015). Its ability to do this with support from, and indeed in partnership with, the City is conditioned in part by the still incomplete but nevertheless powerful formal recognition by the local state of (some) LGBT people as urban citizen subjects.

Coupled with this local "LGBT recognition" (cf. Duggan 2002), the 519 uses its history of creating spaces for marginalized communities to position itself as a proper and credible partner in the MMP project. It does so by arguing that this history endows it with a certain kind of "social justice expertise" that is transferable to other contexts. The MMP Web site notes:

> Marginalized communities have the clearest understanding of how to create inclusive spaces. The 519 has been working to provide service, space and leadership to vulnerable people within the LGBTQ2S community, including those experiencing discrimination, homelessness and poverty, for more than 40 years. Through this work it has gained a deep understanding of the needs of marginalized communities and is committed to ensuring their voices are heard and respected. (Frequently Asked Questions, n.d.)

As noted earlier, although near the Gay Village, Moss Park has a much different history, demographic, and urban identity compared to the sociospatial context from which the 519 honed its expertise. Nevertheless, the 519 harnesses its history and expertise as a form of credibility, its long-standing commitment to serving marginalized (gender and sexual) communities framed as a kind of technical know-how that can be applied to the case of Moss Park, despite its sociospatial specificity. This technicalization of social justice, premised as it is on equivalences of marginalization, is a form of neoliberalization insofar as it enables the 519 to trade in its LGBT-focused expertise for formal participation in state and market processes. It is thus an extension of Duggan's (2002, 179) notion of the "new homonormativity," which critiques neoliberal forms of gay and lesbian politics that align with and reproduce normalcy rather than disidentify from them to acquire state and market recognition. The 519's partnership with the City and the anonymous donor in the MMP project illustrates an extension of this process, not only in the form of its formal alliance with the local state and private philanthropy. Part of the 519's social justice expertise includes its history of making explicit use of market-based strategies in its approach to serving LGBT communities. It is noteworthy that the MMP project is not the 519's first foray into P3s. In 2004, it received a substantial ($750,000) donation from Salah Bachir, its Capital Campaign Chair at the time, to seed the expansion of the community center itself (City of Toronto 2004) at a cost of about $7 million, half of which was "raised by individual donors through The 519's Capital Campaign, while the remainder was donated by the City of Toronto and corporations" (Beneteau 2009). These kinds of experience and expertise in fundraising are surely useful as the 519 expands its purview in and through projects like MMP.

For critics of MMP, the ability of the 519 to claim a central role through invoking its social justice expertise is alarming. They question how the 519's commitment to social justice squares with the very real dangers of gentrification posed by the project (Lenskyl 2015; "No Pride in Gentrification" 2015; Errett 2016; Vendeville 2016). These are concerns that partners themselves recognize as a possibility that could result from the investment in community infrastructure (interviews with city planner and with 519 staff, 13 December 2016). Indeed, the prime location of Moss Park just east of Toronto's Downtown and its adjacency to the Distillery District and Regent Park—whose recent redevelopments, various scholars argue, facilitated gentrification and displacement (August 2008; James 2010; Kern 2010; Dunn 2012; Mathews 2014)—lends credence to worries that the 519 is aiding a future of gentrification for Moss Park through its claims of social justice expertise.

In response, the 519 and city planning staff see tailoring the consultation process to the specific context of Moss Park as a way to infuse the planning process with principles and practices of inclusion and equity, which they note can work against the exclusions of gentrification. City Councilor Kristyn Wong-Tam, who represents the ward that includes both Moss Park and the Gay Village, noted in response to queries about her support for the Moss Park project, "Gentrification does not include broad community consultation with 800 people" (Kenyon 2016). An interviewee expanded on the shifted approach to consultation as a social justice effort:

> It was really really important that we undertook a community consultation process that was intentional in terms of trying to engage with people who do not typically participate in public consultation processes. So we knew that we wanted to do something a little bit different and we knew that it was essential for us, certainly at the 519, that we designed a process to ensure that the voices of those folks were included and so we took a little different model in terms of just the typical public meeting conversation or consultation and adopted a sort of a community organizing strategy. (Interview, executive with the 519, 13 December 2016)

Rendering social justice as an issue of planning process exemplifies one way in which it is rendered technical, which allows consultation to be framed as a possible antithesis to gentrification. It belies the possibility of exclusion by inclusion, where well-meaning inclusionary measures like community consultations do little to halt the march of gentrification and displacement of vulnerable populations in areas under development. This was certainly the case in adjacent Regent Park, which, as James (2010) noted, included robust forms of consultation (including, similarly, the use of community animators) in the redevelopment process.

Given the central role of the 519 in MMP and its mobilization of LGBT-honed social justice expertise as broader technical social justice know-how, MMP represents, for us, a possible evolution of the process of gay gentrification. What differentiates this case from current scholarly accounts of this process is that a municipally funded LGBT community center is a possible driver of gentrification, not because the presence of LGBT people in poorer neighborhoods increases property values and commerce through cultural cache, sweat equity, and consumption, as is commonly theorized in the literature (for more complex accounts of this process, see Lauria and Knopp 1985; Nash 2013; Brown 2014; Doan, 2015; Gorman-Murray and Nash 2016). Instead, its role in this process pivots on its claim to technical know-how that can be transferred elsewhere. Central, then, to this process of possible gentrification is the neoliberalization of social justice through its technicalization, which enables the 519 to sustain its relationships to the state and to the market as a central actor in P3s for infrastructure development.

Not-So-Silent Partners and Social Justice for Purchase

At present, MMP is in the process of writing a project feasibility report that it planned to present to the City for evaluation in 2017. Yet, even at this relatively nascent stage, the project has shifted the dynamics of urban planning in the City of Toronto. The use of public–private partnerships and their potential to provide private funding for both infrastructure developments and attendant consultation processes is already being touted as a potential model for future infrastructure and community development in the city.

Over the last three decades, P3s have become an increasingly popular way to design, build, finance, operate, and maintain public infrastructure projects (Roberts and Siemiatycki 2015). Some of the most common arguments for these partnerships are the claims that the private sector brings to these partnerships knowledge, a level of efficiency, and certain skills that do not exist in the public sector (Miraftab 2004; Brinkerhoff and Brinkerhoff 2011). The argument goes that the private sector can and should push the public sector to act more like the private sector. This neoliberal argument pursues furthering market influence over government provisions of infrastructure and shifting the local state's role toward the entrepreneurial pursuit of private actors as funders and partners for development (Harvey 1989; Peck and Tickell 2002).

The case of MMP differs considerably from typical positionings of different partners within P3s. Here, the private partner has chosen to be anonymous and is described as "only really interested in good community outcomes related to this project ... which is obviously aligned with the 519 and our organizational values" (interview, executive with the 519, 13 December 2016). Here, rather than the private partner as the expert (a position that the 519 maintains), they are positioned as a silent partner whose motivation is not efficiency and profit but a firm commitment to social justice that is rooted in personal experience. Consequently, the case of MMP as a public–private partnership forces a reckoning with more-than-economic logics for private investments in community infrastructures.

It is difficult to know, however, the full extent of the donor's involvement or influence at this time; the label of silent partner works to obscure how the donor has been very central in shaping the contours of the project. Yet, it is clear that the involvement of the private donor has affected the direction of MMP. They have done so in several ways: first, by explicitly identifying LGBT and social justice causes as key targets for their investment; second, by shaping public knowledges about the project through anonymity as a condition of funding, which has implications for transparency and accountability in the planning process; and, finally, by tethering their donation to the 519 and thus ensuring the 519's centrality in any potential development that arises from the

investment. In addition to committing to fund 33 percent of the project (interview, executive with the 519, 13 December 2016), the donor has also underwritten the public consultation process led by the 519. As the interviewee noted:

> We have been incredibly blessed to have those resources to do this kind of community engagement work—to be able to develop a communications plan, to be able to think about how you build More Moss Park ... how you invest in community infrastructure and understanding the benefit and value that it brings in creating a socially inclusive spaces and how you build community. (Interview, executive with the 519, 13 December 2016)

Critiques of the standard official approach to public consultation, especially how they work to exclude marginalized voices, are well rehearsed (Forester 1988; Mitchell 2003; Innes and Booher 2004). For its part, the City recognizes the validity of these critiques, especially in a context with "so many conflicting goals and visions and issues," as one city official (13 December 2016) described Moss Park. In recognition of these deficiencies, MMP shifts away from the standard form of consultation in which the City usually engages. A City of Toronto planner described this shift this way: "Our standard process [is one in] which you have two public meetings and you tick them off the box; you have an open house and there is not really a discussion. Those things are fine when there is a common view of things and ... [where] the goals and visions are fairly well aligned" (interview, urban planner, City of Toronto, 13 December 2016). To facilitate MMP's alternative process, partners have hired MassLBP, a private consulting firm that focuses on facilitating more inclusive and effective public engagement processes (Mesiano-Crookston 2014) to host the public meetings and to design and conduct portions of the consultative process. MassLBP was brought in to supplement the planning practice of the City, specifically with the goal of producing a more socially just consultation process. Yet, the recognition that a socially just planning process must be premised on a need for a process more attuned to local needs and to the meaningful involvement of local actors introduces a financial challenge: How can cities afford such models given the municipal budget constraints (discussed more specifically in the next section)? In the case of MMP, the private partner provides the financial answer to this question.

Hence, although the public consultation process that has been undertaken as part of MMP is seemingly informed by a social justice commitment, this process, underwritten by the private philanthropic donor, essentially limits such forms of social justice to neighborhoods or communities that can afford to purchase it. We caution against touting the Moss Park case, as some city officials do, as a desirable model for future infrastructure initiatives, as it sets a dangerous precedent in exacerbating already existing conditions of uneven investment by signaling that those communities and projects that will be supported are ones that can find a private donor to (at least partially) fund their initiatives. The MMP case illustrates for us the dangers that emerge from relying on the benevolence of private funders to enact more socially just urban processes like meaningful community consultations, what might be called social justice for purchase. This, for us, is a textbook example of the neoliberalization of social justice in that private donors and partnerships are made necessary for urban residents to be able to enact urban citizenship in the form of meaningful participation in planning processes.

Masking Neoliberalism as Problem, Proclaiming It as Path to Social Justice

To justify Moss Park as the chosen site of intervention and a public–private partnership approach as the means of this intervention, project partners make use of two distinct but interrelated crisis discourses. The first represents Moss Park, both the park and the neighborhood that surrounds it, as in a type of neighborhood-level crisis: an unsafe space of infrastructural deficit, home to various threatening urban subjects, and thus in desperate need of urban planning interventions. The second concerns the crisis of planning to respond to the first. In the media and in interviews, Moss Park is understood as posing challenges to standard planning processes and the limited municipal capital budget, forcing project partners to turn to other planning and financing models. Following Martin (2003a, 2003b), we argue that crisis place frames, as social constructions and meaning-making practices, provide ideological armature for stakeholders to push forward with their urban visions in part by constituting a kind of singularizing understanding of Moss Park (see also Martin 2013). In the case of Moss Park, crisis place frames work to authorize both the public–private planning process that is being experimented with in Moss Park as well as the suspension of standard

planning processes, both of which are portrayed as necessary to fixing the neighborhood and thus to achieving social justice.

One oft-repeated trope among project's community animators is the universality of feelings about unsafety in Moss Park. As an executive with the 519 described it,

> Whether you talk to folks who are sex workers or . . . to folks in the residents' association, safety is an issue in that neighborhood. And there is a commonality that you can find through those processes—whether it is because of over-policing or it is because of people's concerns around drug dealers coming into the neighborhood and impacting their lives, there is still a commonality of safety. (Interview, executive with the 519, 13 December 2016)

Such sentiments have been echoed in media coverage of the neighborhood as well. As one recent newspaper article put it, "The corner of Queen Street East and Sherbourne [the intersection at the southeast boundary of Moss Park] is notorious for its drug use, sex workers and the nearby shelters keep the sidewalks crowded and the social services overloaded" (Csanady 2016). These tropes of danger, a common way of representing poor and especially racialized inner-city areas in North America, serve to authorize neighborhood "improvement" schemes that aid in gentrification and heightened securitization that tend to further disadvantage marginalized residents (Blomley and Sommers 1999; Catungal et al. 2009).

It is noteworthy that organizers do acknowledge that why the park feels unsafe differs wildly depending on the social location of the individual making the claim. Nevertheless, the common repetition of Moss Park as unsafe works to discursively produce the park as in need of intervention. As Matthew Cutler, then director of strategic partnership initiatives at the 519, noted, "The community centre is in great need for renovation. . . . It is a bit of a *blank canvas*. . . . We want to build a facility that is most useful and most effective for that neighbourhood and that community" (Watson 2015). This sentiment illustrates that crisis discourse facilitates interventions for the good of the neighborhood but does so through discourses (e.g., "blank canvas") that erase existing meanings and uses of the space. Crisis narratives do little to uncover the systemic neoliberal processes that work (and have worked) to produce Moss Park as unsafe, especially for poor and street-involved communities; these include chronic underfunding of social services in the city, underinvestment in the maintenance of the numerous public housing complexes in and near the neighborhood, and so on.

Despite consistent insistence that Moss Park is in crisis, the City claims an inability to make use of current resources and frameworks to intervene in the neighborhood. This incapacity to respond is framed as a second order of crisis. With the most recent municipal budget, the City highlights the $33 billion gap in the capital budget, representing projects that have been approved but remain unfunded (Rieti 2016). Through a discourse of budgetary crisis, the prospects of funding the project through the standard capital budget prioritization are quite dim, especially as the city has not identified "Moss Park—the John Innes Community Centre—as a place to reinvest in the next twenty years" (interview, urban planner, City of Toronto, 13 December 2016). Thus, the P3 process is framed as necessary: "We are just dealing with budget cuts this week. And it is not going to be pretty. We are going to have to rely on other ways of doing things and this [P3s] is a fabulous way" (interview, urban planner, City of Toronto, 13 December 2016). Crisis is thus understood as an opportunity to innovate new approaches to city planning. Indeed, the touting of the case of Moss Park, even in its early stages, as "a model for other kinds of projects where other agencies and donors come to the table and work with the City to create something that really needs to be there even though we do not have all the money for it" (interview, urban planner, City of Toronto, 13 December 2016) illustrates the utility of crisis framings in contexts of austerity urbanism and particularly for the neoliberalization of social justice (cf. Catungal 2015). Claims of urgency flowing from these crises allow municipal actors and their project partners to justify divergence from standard planning processes in the name of delivering social justice and fixing what ills neighborhoods. Ironically, it does so by masking neoliberal urbanism's very own hand in producing these crises in the first place.

Conclusion: Enrolling Social Justice in Neoliberal Urbanism

This article's examination of the early stages of the MMP redevelopment proposal has illustrated the enrollment of social justice discourses in an urban infrastructure project and planning process that is funded through a public–private partnership. MMP illustrates the ways in which institutions that serve marginalized (LGBT) communities participate actively in neoliberal forms of urban development by harnessing their claims to social justice expertise. Such expertise lends a moral façade to the continued and increasingly normalized pursuit of neoliberal approaches to urban development, in this case through P3s. These processes characterize the neoliberalization of social justice.

Three major lessons for urban planning and geography and for geographies of social justice can be gleaned from this case study. First, although seemingly a step forward in its recognition of the limits of standard practices of public consultation and their capacity to include the perspectives, needs, and visions of the city's most marginalized residents, MMP's reliance on a private, philanthropic donor to underwrite the public consultation leads to several steps backward on other aspects of the planning process. At a basic level, we caution that the use of a public–private model significantly reduces the transparency in decision making, when compared to a more public process. Equally important, this continues processes of neoliberal roll-back under the guise of doing what is necessary, under austere circumstances, for the greater good of marginalized neighborhoods. Given the heralding of MMP as a model to be replicated elsewhere, this will likely have a ripple effect across the city and affect other neighborhoods seeking social justice–informed public consultation processes. We argue that underpinning these processes is a narrow conception of social justice in which the ends justify the means, thus contributing to the normalization of neoliberal urbanism itself. In such an understanding, social justice is a neoliberal good that is available only to communities that can find willing donors, thus exacerbating already uneven urban landscapes of injustice and investment. Thus, any benefits that come with the use of a more inclusive, social justice–informed public consultation are purchased at the expense of these shifts.

Second, the central role of the 519 in the MMP case illustrates the dangers of institutionalizing marginalized communities into forms that are legible to and fit well within the local state's neoliberal agendas. In the form of state partner, the 519 and its history of working with LGBT people brings to the P3 a semblance of social justice in the form of LGBT legibility and inclusion in the planning process. Our research thus extends Doan's (2011) insistence on the need to query queer planning's relationship to heteronormativity, insisting that homonormative inclusion marks LGBT complicity in the violences of neoliberal urbanism. It also thus speaks to and extends more recent work that insists on queer inclusion within planning by highlighting that certain forms of (neoliberal) LGBT inclusion are not ideal or acceptable, not least because such inclusion is premised on the continued marginalization of poor communities through active participation and indeed partnership in municipally managed gentrification (Doan 2015; cf. Slater 2004).

Finally, the case of MMP demonstrates the ways in which social justice discourse can be used to limit the scope of what counts as urban citizenship and participation in the neoliberal city. We are especially concerned that wrapping urban planning processes in the mantle of social justice can be used to defang agonistic politics by seeking to enroll them in a process that has already been predetermined, through its framing, as socially just. Such practices aid not only in depoliticizing the urban but also in normalizing the place of private interests in shaping urban development agendas. As a consequence, any incremental benefit resulting from the embrace of a more socially just process overshadows the foreclosure of a myriad of alternative processes and more radically just urban futures that such agonistic politics might offer.

Acknowledgments

This article has benefited from the engaged reviews of the anonymous reviewers, whose astute comments strengthened our arguments. We are grateful to Daniel Mouret for assisting with data collection for this project and to our interviewees for generously taking their time to share their experiences, opinions, and work with us. We thank Nik Heynen for deftly and supportively shepherding the review process. All responsibility for claims in this article is solely ours.

References

August, M. 2008. Social mix and Canadian public housing redevelopment: Experiences in Toronto. *Canadian Journal of Urban Research* 17 (1):82–100.

Baptista, I. 2013. Practices of exception in urban governance: Reconfiguring power inside the state. *Urban Studies* 50 (1):39–54.

Beneteau, G. 2009. A sneak peek at the 519's new digs. *Daily Extra* June 17. Accessed January 22, 2017. http://www.dailyxtra.com/toronto/news-and-ideas/news/sneak-peek-at-the-519s-new-digs-7469.

Blomley, N., and J. Sommers. 1999. Mapping urban space: Governmentality and cartographic struggles in inner city Vancouver. In *Governable places: Readings on governmentality and crime control*, ed. R. C. Smandych, 261–87. Burlington, VT: Ashgate.

Brinkerhoff, D. W., and J. M. Brinkerhoff. 2011. Public–private partnerships: Perspectives on purposes, publicness, and good governance. *Public Administration and Development* 31 (1):2–14.

Brown, M. 2014. Gender and sexuality II: There goes the gayborhood? *Progress in Human Geography* 38 (3):457–65.

Cameron, A. 2006. Geographies of welfare and exclusion: Social inclusion and exception. *Progress in Human Geography* 30 (3):396–404.

Catungal, J. P. 2015. The racial politics of precarity: Understanding ethno-specific AIDS service organizations in neoliberal times. In *Planning and LGBTQ communities: The need for inclusive queer spaces*, ed. P. Doan, 235–48. London and New York: Routledge.

Catungal, J. P., D. Leslie, and Y. Hii. 2009. Geographies of displacement in the creative city: The case of Liberty Village, Toronto. *Urban Studies* 46 (5–6):1095–1114.

City of Toronto. 2004. Renovations of the 519 Community Centre unveiled. Accessed January 22, 2017. http://wx.toronto.ca/inter/it/newsrel.nsf/eb59b01aaf77114985257aa70063db4f/8d29b1ac45a419c785256eb600579dfb?OpenDocument.

———. 2014. 2011 Neighbourhood census/NHS Profile: 73. Moss Park. Accessed January 22, 2017. http://www1.toronto.ca/City%20Of%20Toronto/Social%20Development,%20Finance%20&%20Administration/Neighbourhood%20Profiles/pdf/2011/pdf4/cpa73.pdf.

Csanady, A. 2016. Toronto's rough Moss Park neighbourhood becoming the city's next gentrification battleground. *National Post* July 20. Accessed January 22, 2017. http://news.nationalpost.com/toronto/weve-been-waiting-a-long-time-for-some-change-moss-parks-fraught-lines-between-facelift-and-gentrification.

Doan, P., ed. 2011. *Queerying planning: Challenging heteronormative assumptions and reframing planning practice*. Burlington, VT: Ashgate.

———. 2015. *Planning and LGBTQ communities: The need for inclusive queer spaces*. London and New York: Routledge.

Duggan, L. 2002. The new homonormativity: The sexual politics of neoliberalism. In *Materializing democracy: Toward a revitalized cultural politics*, ed. D. Nelson and R. Castronovo, 175–94. Durham, NC: Duke University Press.

Dunn, J. R. 2012. "Socially mixed" public housing redevelopment as a destigmatization strategy in Toronto's Regent Park. *Du Bois Review: Social Science Research on Race* 9 (1):87–105.

Errett, J. 2016. The next battleground in fight over Moss Park gentrification is a community centre. *CBC News* June 29. Accessed January 22, 2017. http://www.cbc.ca/news/canada/toronto/moss-park-519-community-centre-1.3656134.

Forester, J. 1988. *Planning in the face of power*. Oakland: University of California Press.

Frequently asked questions. n.d. Accessed January 22, 2017. http://www.moremosspark.ca/information/faqs/.

Goldberg, D. T. 1993. *Racist culture: Philosophy and the politics of meaning*. Hoboken, NJ: Wiley-Blackwell.

Gorman-Murray, A., and C. Nash. 2016. Transformations in LGBT consumer landscapes and leisure spaces in the neoliberal city. *Urban Studies* 54 (3):786–805.

Harvey, D. 1989. From managerialism to entrepreneurialism: The transformation in urban governance in late capitalism. *Geografiska Annaler: Series B, Human Geography* 71 (1):3–17.

Innes, J. E., and D. E. Booher. 2004. Reframing public participation: Strategies for the 21st century. *Planning Theory & Practice* 5 (4):419–36.

James, R. K. 2010. From "slum clearance" to "revitalisation": Planning, expertise and moral regulation in Toronto's Regent Park. *Planning Perspectives* 25 (1):69–86.

Kenyon, M. 2016. Moss Park redevelopment provokes strong reaction at public meeting. *Now Toronto* June 8. Accessed January 22, 2017. https://nowtoronto.com/news/moss-park-redevelopment-provokes-strong-reaction-at-public-meeting/.

Kern, L. 2010. Gendering reurbanisation: Women and new-build gentrification in Toronto. *Population, Space and Place* 16 (5):363–79.

Lauria, M., and L. Knopp. 1985. Toward an analysis of the role of gay communities in the urban renaissance. *Urban Geography* 6 (2):152–69.

Lenskyl, H. J. 2015. Waving a rainbow flag over poor communities. *Now Magazine* October 26. Accessed January 22, 2017. https://nowtoronto.com/news/waving-a-rainbow-flag-over-poor-communities/.

Martin, D. 2003a. Enacting neighborhood. *Urban Geography* 24 (5):361–85.

———. 2003b. "Place-framing" as place-making: Constituting a neighborhood for organizing and activism. *Annals of the Association of American Geographers* 93 (3):730–50.

———. 2013. Place frames: Analyzing practice and production of place in contentious politics. In *Spaces of contention: Spatialities and social movements*, ed. W. Nicholls and B. Miller, 85–101. London and New York: Routledge.

Mathews, V. 2014. Incoherence and tension in culture-led redevelopment. *International Journal of Urban and Regional Research* 38 (3):1019–36.

McCann, E. J. 2002. The cultural politics of local economic development: Meaning-making, place-making, and the urban policy process. *Geoforum* 33 (3):385–98.

Mesiano-Crookston, J. 2014. About public consultation: An interview with Mass LBP. *The Public Policy & Governance Review* February 28. Accessed January 22, 2017. https://ppgreview.ca/2014/02/28/abouut-public-consultation-an-interview-with-mass-lbp/.

Miraftab, F. 2004. P3s: The Trojan horse of neoliberal development? *Journal of Planning Education and Research* 24 (1):89–101.

Mitchell, D. 2003. *The right to the city: Social justice and the fight for public space*. New York: Guilford.

More Moss Park. 2016. Consultation overview. Accessed January 22, 2017. http://www.moremosspark.ca/information/consultation-overview/.

Nash, C. J. 2013. The age of the "post-mo"? Toronto's gay village and a new generation. *Geoforum* 49:243–52.

No Pride in Gentrification. 2015. QueerTransCommunityDefense. Accessed January 22 2017. http://queertranscommunitydefence.blogspot.com/2015/12/no-pride-in-gentrification-community_13.html.

Omi, M., and H. Winant. 2014. *Racial formation in the United States*. London and New York: Routledge.

Peck, J., and A. Tickell. 2002. Neoliberalizing space. *Antipode* 34 (3):380–404.

Purdy, S. 2005. Framing Regent Park: The National Film Board of Canada and the construction of "outcast spaces" in the inner city, 1953 and 1994. *Media, Culture & Society* 27 (4):523–49.

Rieti, J. 2016. City of Toronto's 2017 preliminary budget unveiled. CBC December 6. Accessed January 22, 2017. http://www.cbc.ca/news/canada/toronto/toronto-budget-2017-1.3883264.

Roberts, D. J., and M. Siemiatycki. 2015. Fostering meaningful partnerships in public–private partnerships: Innovations in partnership design and process management to create value. *Environment and Planning C: Government and Policy* 33 (4):780–93.

Slater, T. 2004. Municipally managed gentrification in South Parkdale, Toronto. *The Canadian Geographer* 48 (3):303–25.

Velasco, A. 2013. A brief history of the Church Wellesley Village. *BlogTO* June 27. Accessed January 22 2017. http://www.blogto.com/city/2013/06/a_brief_history_of_the_church_wellesley_village/.

Vendeville, G. 2016. Moss Park development sparks gentrification concerns. *The Toronto Star* August 31. January 22, 2017. https://www.thestar.com/news/gta/2016/08/31/moss-park-development-sparks-gentrification-concerns.html.

Watson, H. G. 2015. Moss Park eyed as home for Toronto's LGBT sports centre. *Daily Extra* April 17. Accessed January 22, 2017. http://www.dailyxtra.com/toronto/news-and-ideas/news/moss-park-eyed-home-toronto's-lgbt-sports-centre-102162.

DAVID J. ROBERTS is Assistant Professor, Teaching Stream, in Urban Studies, Innis College, at the University of Toronto, Toronto, ON M5S 1J5, Canada. E-mail: d.roberts@utoronto.ca. His research interests include the geographies of race and racialization, urban infrastructure planning, and the politics of public participation in urban knowledge production and policymaking.

JOHN PAUL CATUNGAL is an Assistant Professor in the Institute for Gender, Race, Sexuality and Social Justice at the University of British Columbia, Musqueam Territories, Vancouver, BC V6T 1Z2, Canada. E-mail: catungal@mail.ubc.ca. His research interests include queer of color geographies, migration and diaspora studies, the politics of community organizing, campus and classroom climates, and critical approaches to the scholarship on teaching and learning.

15 Safe Cities and Queer Spaces

The Urban Politics of Radical LGBT Activism

Kian Goh

Lesbian, gay, bisexual, and transgender (LGBT) visibility is at a high. Gay marriage is a reality. Gay urban enclaves are threatened by their own success, historic icons of the movement subsumed by urban development. Yet violence and homelessness continue, and socioeconomic disparities are reinforced in LGBT communities, particularly among women, people of color, young and old, and gender-nonconforming. Overlapping identities and systems of oppression exacerbate the marginalization of LGBT-identified people, creating "unjust geographies" that intertwine race, class, gender, and sexuality. These queer struggles play out in gay centers and in urban areas far from those. How might researchers understand the complex and intersectional nature of queer marginalization in urban space today, situated within multiple modes of social and spatial oppression? How might those involved in the envisioning and making of cities contribute to the social movements still fighting for change and justice? Building on theories of critical geography and queer theory, this article explores the organizing work of queer activist organizations in two New York City neighborhoods, including the author's participatory role as a designer and activist: FIERCE's campaign for a queer youth center in the West Village and the Audre Lorde Project's safe neighborhood campaign in Bedford-Stuyvesant. Through an analysis of the strategies, politics, and spatial implications of such work, the article delineates the ways in which queer community organizers on the ground are fighting for social and spatial change, outside and despite dominant economic and sociopolitical structures. *Key Words: LGBT, New York City, queer space, radical activism, social justice.*

女同性恋、男同性恋、双性恋与跨性别 (LGBT) 的可视性正高涨。同性恋婚姻已是现实。同性恋的城市飞地，正因自身的成功而遭到威胁——同性恋运动的历史象征，被城市发展所吞噬。但暴力与无家可归仍持续存在，且 LGBT 社群中的社会经济不均仍继续强化，特别对女性、有色人种、年轻与老年，以及性别不符常规者而言。重叠的身份认同及压迫系统，使得被指认为 LGBT 人口的边缘化更加恶化，创造出与种族、阶级、性别与性向相互交缠的 "不正义地理"。这些酷儿斗争，在同性恋中心以及远离这些地带的城市区域中上演。研究者如何能够理解位于多重社会与空间压迫模式的当代城市空间中，酷儿边缘化的复杂与多重交织之本质？参与想像并打造城市者，如何对仍然为改变与正义奋斗的社会运动做出贡献？本文以批判地理学理论和酷儿理论为基础，探讨纽约市两座邻里中，酷儿倡议组织的组织工作，其中包含作者身为设计师和倡议者的参与角色: FIERCE 在西村的酷儿青年中心倡议，以及罗德计画 (Audre Lorde Project) 在贝德福 − 史蒂文生的安全邻里倡议。本文藉由分析上述工作的策略、政治与空间意涵，描绘在地酷儿社群组织者尽管面临主宰的经济与社会政治结构，在该结构之外争夺社会与空间变迁的方式。关键词: LGBT, 纽约市, 酷儿空间, 激进行动主义, 社会正义。

La visibilidad lesbiana, gay, bisexual y transgénero (LGBT) está en la cúspide. El matrimonio gay es una realidad. Los enclaves urbanos gay están amenazados por su propio éxito, como íconos históricos del movimiento incorporados por el desarrollo urbano. Con todo, la violencia y la privación de techo continúan, y las disparidades socioeconómicas se ven fortalecidas en las comunidades LGBT, en particular entre las mujeres, gente de color, joven y vieja, y entre los inconformistas en materia de género. Identidades superpuestas y sistemas de opresión exacerban la marginalización de la gente que se identifica como integrantes de la comunidad LGBT, creando "geografías injustas" en las que se entremezclan raza, clase, género y sexualidad. Estas luchas de raros y homosexuales hallan expresión abierta en los centros gay y en áreas urbanas alejadas de estos. ¿Cómo podrían los investigadores entender en el espacio urbano actual la naturaleza compleja e interseccional de la marginalización homosexual situada dentro de múltiples modos de opresión social y espacial? ¿Cómo podrían quienes están involucrados en el diseño y construcción de las ciudades contribuir a los movimientos sociales que todavía pelen por cambio y justicia? Edificando a partir de teorías de geografía crítica y teoría homosexual, este artículo explora el trabajo de organización de entidades activistas homosexuales en dos barriadas de la ciudad de Nueva York, incluyendo el papel participativo del autor como diseñador y activista: la campaña de FIERCE en favor de un centro para jóvenes raros en West Village y la campaña por un vecindario seguro del Proyecto Audre Lorde en Bedford-Stuyvesant. Por medio de un análisis de las estrategias, políticas e implicaciones espaciales de tal trabajo, el artículo delinea las maneras como los organizadores de la comunidad homosexual están batallando en el terreno por el cambio social y espacial, desde afuera y a pesar de las estructuras económicas y sociopolíticas dominantes. *Palabras clave: LGBT, Ciudad de Nueva York, espacio homosexual, activismo radical, justicia social.*

Until recently, one might have argued that the lesbian, gay, bisexual, and transgender (LGBT) rights movement has been monumentally successful in the United States. Gay marriage is now the law after the Supreme Court case *Obergefell v. Hodges*, decided in June 2015. Gays and lesbians may serve openly in the military after the repeal of the Don't Ask Don't Tell policy in 2011. Historically gay neighborhoods in cities, such as Chelsea in New York City, West Hollywood, the South End in Boston, and Dupont Circle in Washington, DC, are so successful in economic development terms that the main challenge often seems to be the protection of historical landmarks of gay liberation. These triumphs pose a sharp contrast to the histories of shame, violence, and hate that have marked the lives of people with marginalized sexualities and gender identities.

Yet the fight against equality and freedom for LGBT people appears to be regaining strength. Prominent acts of violence, such as the killing of Mark Carson on a West Village street in New York City in 2013, might be particularly shocking in that they appear to happen in neighborhoods and cities that have been considered "safe" for people of marginalized sexuality and gender identities. Physical violence and social dispossession against LGBT people have continued unabated, however, during the recent period of legislative wins. The two years of 2015 and 2016 were the most deadly on record for transgender people in the United States, overwhelmingly affecting transgender women of color (GLAAD 2016; National Coalition of Anti-Violence Programs [NCAVP] 2016). More broadly, social and economic disparities continue to adversely affect LGBT people of color and working-class populations (Sears and Badgett 2012; Badgett, Durso, and Schneebaum 2013). LGBT-identified youth are strikingly overrepresented among the broader homeless youth population (Ray 2006; Quintana, Rosenthal, and Krehely 2010; Durso and Gates 2012).

A surge of institutional and legislative campaigns to dial back or dismantle recently won rights and protections—and to impose new constraints—have emerged across the country. The so-called bathroom bill or House Bill 2 in North Carolina, aimed at persecuting transgender and gender-nonconforming people, is only the first to garner legislative success and popular media coverage. Since 2015, twenty-eight other states have considered such bills (National Conference of State Legislatures 2016). There are, as well, numerous "religious exemption" bills to legalize discrimination against LGBT people on religious grounds making their way through state and local legislatures. Republican electoral wins in 2016 have prompted conservative lawmakers to be optimistic about national-level legislation against LGBT people (American Civil Liberties Union 2016; Greenberg 2016; Revesz 2016). Legal scholars warn of the challenges of actually achieving a "lived equality" even with increased formal rights (Carpenter 2017). With the ongoing attacks on LGBT rights, as well as the broader threat of dismantling public and social services indicated by the Trump administration's election rhetoric and early actions, it is critical to understand the real-world struggles for alternatives, in particular by groups of people who have not generally been able to count on institutional and state protections.

To many, the city—more accurately urban life—presents a normative ideal, the potential of difference without exclusion, or living together (Young 1990; Sandercock 2006). Cities have been considered places of relative safety for those confronting institutionalized violence. In the aftermath of the 2016 election, many city governments proclaimed their continued commitment to inclusion and diversity. Cities are not always safe spaces, though. They are sites of disinvestment, marginalization, and inequality and spaces of capital accumulation and social contestation (cf. Brenner 2000; Leitner et al. 2007; Harvey 2008; Goldsmith and Blakely 2010). To the extent that the space of the city is sometimes safe for some people, it is due to specific actors, strategies, and processes. When are cities safe? How? It is in this moment, during which a more optimistic view of the city takes place alongside post–11 September 2001 fears of terrorism, increasing racial tensions, and xenophobia, as well as increasing alarm over the environmental threats caused by climate change, that the concept of safety in cities becomes particularly acute and contested.[1]

This article probes the urban spatial politics of radical queer activism and confronts the still "unjust geographies" (Soja 2010) of queer marginalization. It builds on research on queer space and safe space, particularly in the context of changing relationships between the LGBT movement and urban development. How has a new generation of activists constituted their claims on urban space? What kinds of spatial claims are made, and to what extent have they succeeded?[2] This article focuses on two examples of organizing in New York City: the work of Fabulous Independent Educated Radicals for Community Empowerment (FIERCE) to maintain access and safety for queer youth of color in the West Village of Manhattan and the work of the Audre Lorde Project (ALP) to create safe spaces without police intervention in Bedford-Stuyvesant, Brooklyn (Figure 1).

This research is based on a mixed-methods and participatory study. I have worked directly with both of these organizations. I served on the board of directors of ALP from 2007 to 2010, managing a capital campaign and contributing architectural design services and taking part in nonprofit organizational oversight. I collaborated with FIERCE as a pro bono architect and urban designer for its Our SPOT campaign, primarily from 2007 to 2009. For this research, I relied on participatory observation of the actions of organizers in both groups. As a board member with specific governance duties with ALP, I had formal knowledge of the operations, including reports from the working group responsible for the safe spaces campaign (although board members did not manage or direct the campaigns). I took part in meetings with FIERCE leadership during the development of the Our SPOT campaign and observed interactions between organizers and youth constituents during protests and actions at neighborhood meetings. While developing the research focus, I conducted in-depth, semistructured interviews with two key informants, lead organizers with each of the campaigns. I also conducted field visits and informal interviews at the sites in question (see also Goh 2015).

I consider this research an example of a *reflexive* model of knowledge production, based on "our own participation in the world we study" (Burawoy 1998, 5). As a queer woman of color and someone who is involved in academic research, a design profession, and activism, I bring both an insider status—and marginalized voice—and a distance informed and enforced by professional training and code of conduct and class and educational privilege. Here, the analysis builds from situated empirical observations and standard qualitative methods, tested against extant social science theories from critical geography, queer theory, and urban planning and design.

Queer Space

A now expansive body of research has asserted that space is heterosexually produced—and that there exist nonnormative practices by gender and sexual identity

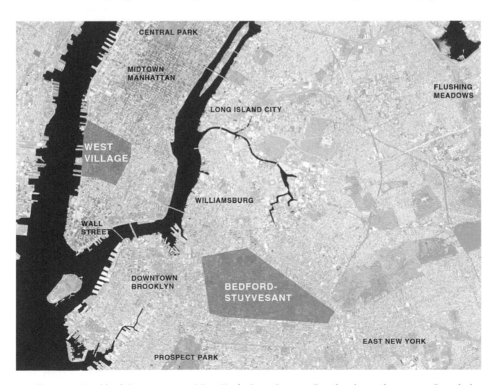

Figure 1. Map of West Village and Bedford-Stuyvesant in New York City. *Source:* Graphic by author using Google base map.

minorities that challenge this and actively "queer" space. Formative work by Binnie, Bell, Valentine, and Knopp, among others (see Knopp 1992, 1994; Bell and Valentine 1995; Binnie 1997), investigated the ways in which gay men and lesbians appropriated, marked, and claimed space, usually in cities. These investigations built on influential studies such as Castells's (1983) on gay men and collective political power in San Francisco, as well as cross-disciplinary accounts of the lives of gay men and women in the early years of gay liberation struggles.

Much of the earlier research on queer space focused on actions by individuals and groups that led to the formation of specific places in the city, so-called gay villages. In response to shifting LGBT politics and urban economic trajectories since the 1980s, a significant emerging literature looks through the lens of what might be called the political economy of sexuality to both reinforce and question dominant understandings of the gay village as the place of sexual minorities in space. Some researchers have attempted to explain the economic processes affecting such enclaves in strategies of urban regeneration and development (Collins 2004; Ruting 2008).

Among this varied research, an increasingly prominent strand is framed by a critical analysis of the ways in which gay and lesbian rights movements are subsumed into dominant cultural and political–economic systems. Conceptualizing *homonormativity*, Duggan (2002, 2003) explained how the post-1990s U.S. gay rights movement, in embracing consumer culture and institutions such as marriage, was assimilated into privatized, neoliberal economic agendas. Duggan (2002) decried the lack of a "vision of a collective, democratic public culture or of an ongoing engagement with contentious, cantankerous queer politics" (189). Developing the concept of homonormativity with a spatial lens, researchers have exposed the often exclusionary tendencies of gay urban spaces, particularly in terms of class, when harnessed into models of neighborhood redevelopment as a kind of urban "homo-entrepreneurialism" (Kanai 2014; Kanai and Kenttamaa-Squires 2015) and global city branding and appeals to cosmopolitanism (Bell and Binnie 2004; Binnie and Skeggs 2004).

Scholars have as well explored spaces in cities that can be ambiguous or multifaceted and variously inclusive. Gay cosmopolitanism and commodification, although exclusionary in one instance and often driven by and catering to white, gay men, might offer spaces of fluidity and inclusion for queers of marginalized ethic groups and lesbians (G. Brown 2006;

Podmore 2013b). Other researchers trace the movements and making of LGBT places beyond the gay village proper, observing queer space as both dispersing and reclustering (Whittemore and Smart 2016; Smart and Whittemore 2017), relational and mobile, constituted through power relations (Nash and Gorman-Murray 2014), and appearing beyond the "city" in its peripheries (Tongson 2011). Overall, these studies assert that the gay village, prone to commodification and branding, is multivalent and even in its general form is just one of many ways in which LGBT people make claims on space in and outside cities.

Alongside homonormativity, scholars have challenged uninterrogated binary readings both of space—as either heterosexist or queered—and gender—as either (lesbian) women or (gay) men. Halberstam (2005), referring to the intertwined temporal and spatial logics of "queer counterpublics"—the momentariness of queer time necessarily producing queer space—asserted the problematizing nature of transgender bodies in studies of time and space. This is not simply a matter of conceptual clarity or theoretical gaps. Such conceptual laxity neglects the uneven manifestations of acceptance and safety among LGBT people, including the vulnerability faced by transgender people even in lesbian and gay-friendly spaces (Doan 2007, 2010). Scholars have also critiqued research that excludes or neglects intersectional issues of gender identity, race, and class. These critiques build on Crenshaw's (1991) explanation of how binary either–or propositions in identity politics disallow an analysis of the compounded oppressions faced by women of color.

Attempting to move beyond homonormativity (see Podmore 2013a), Browne (2006) emphasized the need to question the "queer" of queer space, as a political space of slippages, constantly stabilizing and in need of continual investigation. Queer, like space, is always in contestation. Oswin (2008) attempted to recast queer geographies beyond investigations into the space of sexual otherness, or of simple inclusion. She proposed, instead, a queer approach to geography focused on understanding the making of norms across multiple identities: sexual, gendered, classed, and racialized. Oswin noted how much queer space research, although conscious of intersectional relationships between sexuality and race, remains relatively muted in its engagement with the latter.[3] For her, there exist broader issues at stake, including race and class as well as poverty, migrations and diasporas, and geopolitics and globalization.[4]

I position my own research in relation to and in alignment with these theories of queer space that look critically at the dominant political economy of cities in defining uneven outcomes within the broader LGBT movement. I focus on the multifaceted struggles among LGBT communities, particularly the spatial consequences of intersectional issues of race, class, and sexuality. Alongside the substantial research on the process of queer space development, I am interested in the actors and strategies involved in making concrete, physical claims on space (cf. G. Brown 2006; Nash and Bain 2007; Podmore 2013b). In discussing the cases in this article, I explore how a particular political positioning animates contestations and claims in distinct but related spatial contexts, within established gay villages and across other sites in cities.

This article seeks to accomplish two related objectives. First, it continues the explorations of "queer margins" (Browne 2006, 889) and extends the arguments that emphasize and insist on the interconnections between systems of oppression—particularly of race, class, and gender identity—when researching sexuality and space. Second, in a time when gay enclaves in cities might be considered normalized, it reengages issues of physical space at the scale of the urban, not simply explorations of sexuality in space but how territories are actively fought over, marked and signified, and controlled and built. I wish to enforce the intersectionalities, as it were, between conceptual and physical space. This builds on my own work on urban design, spatial planning, and queer space (Goh 2011, 2015), as well as broader research in planning and architecture (Colomina 1992; Betsky 1997; Frisch 2002; Doan 2011, 2015; Ehrenfeucht 2013).

Radical Queer Activism

The contradictions and contestations around queer politics and urban space can be observed in New York City. The West Village in Manhattan has figured prominently in the making of an explicitly queer urban history. It was the site of the Stonewall Rebellion in June 1969, when queers fought back against police raids (Duberman 1993), and the Christopher Street Liberation Day march, taking place one year after Stonewall, laying the foundation for gay parades in the city. Neighborhoods like the West Village served as prototypical gay neighborhood enclaves and now wear their status with pride, rainbow flags emblazoning storefronts and townhouses. Since the mid-1970s, when the city teetered on bankruptcy, enthusiastic urban development and rapidly rising rent have produced some of the most high-value real estate in the city (Figure 2), a trajectory reflecting much scholarship on gay villages (see, e.g., Bell and Binnie 2004; Sibalis 2004; Doan and Higgins 2011; M. Brown 2014).

In the mid-1990s, tensions erupted between older residents in the West Village and LGBT-identified youth, largely youth of color. Residents and business owners protested the presence of "unruly" queer youth, drawn to the area around Christopher Street and the adjacent Hudson River waterfront and demanded more police presence (see Davidson 2008; Mananzala 2011). These calls for control and policing took place amidst the ramping up of Mayor Giuliani–era citywide quality-of-life crime crackdowns on minor street offenses (Onishi 1994). "Gay Youth Gone Wild" proclaimed the headline for an op-ed written by leaders of

Figure 2. Christopher Street Pier and Hudson River Park, 2011. *Source:* Photograph by author.

the West Village resident street patrol and merchant association (Poster and Goldman 2005). Hanhardt (2013), in her research on gay safe spaces (including this case), revealed the ways in which the LGBT movement's strategies for safety during times of more overt antigay popular opinion and legislations, such as street patrols and neighborhood watch groups, can also be being turned against queer youth who are not perceived as belonging to those spaces. Hanhardt's research points to the complicity between LGBT movement building and real estate development. The status of gay safety in gay spaces in the city—in politics, economy, and popular culture—is historically constituted and contested, even between members of the LGBT community.

FIERCE, an organizing group for LGBT youth of color, was founded in 2000 in response to these emerging conditions. That year, the organization produced *Fenced Out*, a documentary film about the new challenges facing queer youth in the West Village. The film detailed the struggles of queer youth, including the constant harassment from security personnel along the Hudson River waterfront and Christopher Street Pier. It contrasted the personal narratives of queer youth finding community on the piers with the priorities of the city to spend $330 million for park redevelopment while neglecting services and support for homeless youth. Since then, FIERCE has fought to keep the pier, park, and adjacent neighborhood safe and accessible for its constituents, primarily queer youth of color. In 2007, FIERCE launched the Our SPOT campaign (see http://www.fiercenyc.org/campaigns/our-spot-cam paign). Its objective was to create a queer youth center in the West Village that would be accessible twenty-four hours a day on a drop-in basis (FIERCE 2008).

If the West Village reflects the prominent accounts of gay villages as sites of both urban regeneration and contestation, in other areas of the city far from established gay centers a somewhat different struggle plays out. Neighborhoods such as Bedford-Stuyvesant in Brooklyn, historically black working and middle class, confront rapid gentrification, financial distress among many longer term residents, and the displacement of people of color (Roberts 2011; NYU Furman Center 2016; see also Botein 2013). Adjacent Bushwick and Crown Heights face similar challenges. These neighborhoods have also witnessed assaults against immigrants, LGBT and perceived LGBT residents, and police violence (Figure 3).[5] In 2008, during the vicious attack against the Ecuadorian Sucuzhañay brothers in Bushwick that led to Jose Sucuzhañay's death, the attackers yelled anti-Latino and homophobic slurs (McFadden 2008; Lavers 2010). These intersectional struggles are taking place amidst increasing

Figure 3. Memorial wall mural in Bedford-Stuyvesant, Brooklyn, 2013. *Source:* Photograph by author.

inequality in the city (NYC Comptroller's Office 2012; Roberts 2014).

ALP (http://alp.org/), named after the black, feminist, lesbian poet, has been organizing for social and economic justice for LGBT people of color since its founding in 1994. ALP focuses on ensuring social and economic justice for the most marginalized members of its community, in particular transgender and gender-nonconforming people, the young, and the elderly. The Safe OUTside the System (SOS) Collective, an ALP working group, launched the Safe Neighborhood Campaign in Bedford-Stuyvesant in 2007, attempting to create community accountability around violence. Confronting an unjust criminal justice system in a neighborhood both violence-stricken and rapidly gentrifying, the SOS Collective activists challenged neighborhood institutions—businesses, schools, churches, and community centers—to become visibly identifiable "safe spaces." These spaces would serve as places of refuge from anti-LGBT physical violence and hate speech, without resorting to police intervention.

FIERCE and ALP are part of a network of progressive, radical queer activists that has developed over the last two decades in New York City. The two organizations, along with the now-defunct Queers for Economic Justice (primarily community organizers, as opposed to community service providers), and the Sylvia Rivera Law Project (combining legal services with community advocacy and organizing), have differing but overlapping constituents and objectives, with the overarching goals of transformative political change and social and economic justice. These organizations' theory of change positions LGBT identities as part of broader systems of oppression. They prioritize the especially marginalized members of their communities: queer people of color, transgender and gender-nonconfirming people, youth, immigrants, and working-class (DeFilippis and Anderson-Nathe 2017). They work from the point of view that the state has never been a supportive presence when one is queer, brown, black, and poor, and the city has never been that safe without the work of making safe spaces.

These organizations' political analysis is often in opposition to those of more mainstream gay and lesbian advocacy groups. They have objected to the prioritizing of marriage equality, arguing that the focus on such a singular issue does not offer change for many in the LGBT community suffering from economic injustices and racial discrimination and misallocates resources toward an inherently conservative institution.[6] They have also questioned the efforts around gays in the military, asserting that such a position cannot be separated from the broader issue of U.S. military imperialism.[7] Notably, many of these organizations have also taken a stand against hate crimes legislation, arguing that they cause more harm than good for communities of color who face an unjust criminal justice system.[8]

The positions and actions of these organizers are an extension of an earlier round of queer contestations, when activist groups such as ACT UP and Queer Nation harnessed direct action strategies to force awareness about the AIDS epidemic and discrimination against LGBT people while challenging prevailing political–economic structures (Berlant and Freeman 1992; Shepard and Hayduk 2002; Gould 2009). In contrast to notions of homonormativity, the work of FIERCE and ALP is "radical" in the sense that it is explicitly positioned against dominant power structures. The groups often work as allies with other antioppression groups, including antiracist, anticapitalist, counterglobalization, immigrant rights, and labor rights organizations. They are "queer" in that they hold anti-assimilationist stances, resonant with prior more militant LGBT struggles.[9]

Organizations like FIERCE and ALP bring both a renewed direct engagement to queer movement building and a broadly intersectional approach that includes race, class, and gender, as well as systemic oppressions linked to policing, immigration, and militarization. ALP's statement after the June 2016 Pulse nightclub shooting in Orlando, Florida, titled "Do Not Militarize Our Mourning," exemplifies its approach to political change. Invoking Audre Lorde's words from "A Litany for Survival" that "we were never meant to survive," ALP contested the media and political framing of the shooting as one of Islamic terrorism, decrying the use of violence against LGBT community members as an excuse to ramp up anti-Muslim legislation in the United States (ALP 2016). On a theoretical level, this shared political analysis—intersectional, rooted in transformative change, and deliberately against systemic power structures—positions the work of these organizations as the sort of contesting urban imaginaries and practices (cf. Leitner et al. 2007) that might fuel Duggan's "cantankerous politics" and Halberstam's "queer counterpublic."

Queer Urban Politics

FIERCE and ALP share political analyses and broad objectives, but they present distinct examples of organizing strategy and claims on urban space in their campaigns. FIERCE's organizing model centers on building leadership, political consciousness, and organizing skills among its members to fight police harassment and increase access to safe spaces (see http://fiercenyc.org/what-we-do). A former lead organizer with FIERCE recalls how he was brought into the organization as a youth member in 2007 and was taught knowledge and tools of organizing, before assuming his role as lead organizer in 2011. He learned, early on, "what the mainstream LGBT community doesn't tell you about our community"—about the struggles of transgender women of color, homeless black and brown people, and the fights against the New York Police Department (J. Blasco interview with author 29 January 2014). It is this model of ground-up organizing that first caught my eye, as a community member and ally to the youth work as well as a researcher of urban issues. I observed how the then-executive director of FIERCE mobilized a large group of young people, rallying them with concerted shouts of self-worth and affirmations when preparing for a community board meeting and then, during the meeting itself, amassing a group of silent bodies armed with protest signs.

Due to this work, FIERCE has garnered successes and a certain amount of respect. It now maintains a tenuous working relationship with the Hudson River Park Trust (HRPT), the city- and state-authorized entity that oversees the Hudson River Park, and with Community Board 2, representing the West Village. The lead organizer described West Village residents, in the early 2000s, "not even wanting to look at LGBT youth if they walked into those [community board] meetings." A decade later, "The chair people of Community Board 2 reach out to FIERCE because they want to talk about issues affecting LGBTQ youth. ... They want residents around the table to understand the impact of issues LGBTQ youth are facing" (J. Blasco interview with author 29 January 2014).

He cited the victories that FIERCE has won over the years, including later curfew hours and the provision of portable toilets along the Hudson River Park and permission to hold various events on the adjacent piers, including an annual Mini-Ball, begun in 2009, at which community members hold dance-offs. "It was important for us to be there," he stressed, noting the impact of the organizing strategies. "Because of shifts in the way FIERCE has done the work, this has become a reality now. ... It's not enough, but it is a lot."[10]

After finding out about proposed redevelopment plans for Pier 40, formerly a terminal for the Holland America cruise line, just south of the Christopher Street Pier, FIERCE decided to target the redevelopment efforts for the Our SPOT campaign for a youth drop-in center. Organizers protested proposals to prioritize high-value private development on the pier (including, in a proposal by The Related Companies, a theater for Cirque du Soleil) and advocated for public uses and community input in the process. Their protests pitted LGBT youth against the mode of urban development increasingly in favor in New York City, one in which private funds are counted on to provide public goods alongside private development. Whereas some members of HRPT were sympathetic to the organization's campaign demands, others were not. Henry Stern, a board member—and former New York City Parks commissioner—said it was "socialist" to suggest that the state or city help fund pier repairs (Rogers 2007).

FIERCE led the Our SPOT campaign for five years. During the campaign, the organization decided to back a competing development proposal by CampGroup LLC/Urban Dove, one believed to be more amenable to public uses. It organized a multiaction response to support the campaign. Activists held a public protest rally at Pier 40, with placards criticizing privatized development along the park. They conducted a community needs assessment survey, asking almost 300 queer youth about their needs and desires. FIERCE also produced its own proposal for a youth center to be part of the Pier 40 redevelopment (Figure 4). This proposal, jointly prepared by FIERCE, the Urban Justice Center, Super-Interesting! (an architecture and design firm), and students from Columbia University's Mailman School of Public Health, prioritized spaces for physical activity, emotional outlets, personal and professional development, and health care (FIERCE 2008). It made the case that directly addressing the needs of queer youth was fully consistent with the Trust's own mission for an inclusive park.

Whereas FIERCE's campaign focused on creating a physical space in the West Village, ALP's SOS Collective in Bedford-Stuyvesant was attempting a broader and less concrete approach to safety. They were envisioning "transformative justice," a model of community accountability (see Mogul, Ritchie, and Whitlock 2011; Whitlock 2012), at the scale of the neighborhood.

A former program coordinator of the SOS Collective explained that they had to come up with new approaches toward building community accountability at neighborhood scale:

The idea was to find a way to address violence without relying on the system. ... The thing that was

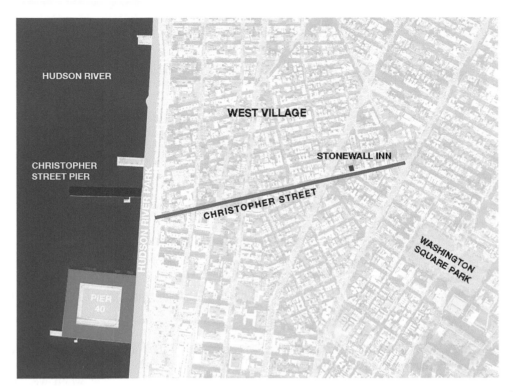

Figure 4. Map showing relationship between Christopher Street Pier, Pier 40, and the Hudson River Park on the west edge of Manhattan. *Source:* Graphic by author using Google base map.

challenging was that I made an assumption at the beginning that we would find something relatively similar, that we would adapt. And there really wasn't a lot that was that similar that was really adaptable for us. A lot of the work that was addressing violence without relying on the criminal legal system was happening in domestic violence work, in childhood sexual abuse, in violence ... among people who knew each other. ... Well, that's different from us.

Sista-2-Sista had a project that had cordoned off a space. ... YMPJ [Youth Ministries for Peace and Justice] had also done something around a safe space area ... and one of our members ... she was talking about how sometimes at colleges they have all these internal phone systems and you can, like, call them, and you can go someplace and be safer. And what if we applied an idea like that to the neighborhood. (E. Dixon, former program coordinator of the SOS Collective of ALP, interview with the author 17 January 2014)

The collective's plan involved three phases: first, garnering support and agreements from "friendly" neighborhood institutions, including providing training to staff and employees of those institutions; second, expanding the initiative to include many more establishments, ideally saturating the neighborhood; and, third, "restoring the harm" caused by violence, a concept derived from restorative justice theories, going

beyond prevention and actually making things better. The SOS organizer admitted that this third phase to create transformative change was difficult to conceptualize. She explained the target and metrics of the campaign, distinct from a more policy-oriented one:

We needed something concrete and measurable. We needed something to say, okay, we have made progress. So then we created the idea of, well, the target is 'violence as it runs through the neighborhood and how it operates within these institutions.' So, every new safe space could be considered a new victory, a new win, and some measure of progress. (E. Dixon interview with the author 17 January 2014)

In 2011, the collective had achieved ten such safe spaces in Bedford-Stuyvesant, ranging from community nonprofits, nightclubs, cafés, and bakeries, to a tattoo parlor (Figure 5). The spatial questions around making such safe spaces are immediately evident. At base, safe spaces should be present when needed, both dense enough and distributed enough. The SOS program coordinator noted that the main locational strategy was a loose correlation with areas with a history of violence. Organizers prioritized storefronts because they were visible, with street-level access. Other aspects, such as the scale of the neighborhood, the width of the street, the

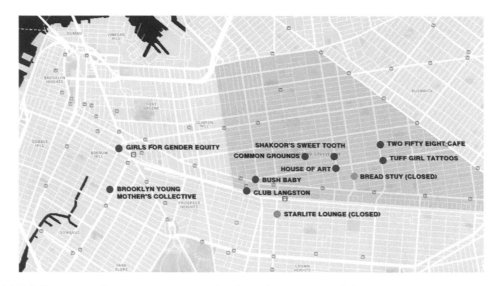

Figure 5. Map of SOS Collective's "safe spaces." *Source:* Graphic by author using Google base map and list of spaces cited on Audre Lorde Project brochure in 2011.

heights of the buildings, and elements such as street trees and lighting, however, remained unexplored (E. Dixon interview with the author 17 January 2014).

Systemic Challenges and Spatial Claims

In FIERCE's Our SPOT and ALP's Safe Neighborhoods campaigns, political and economic pressures continually challenged attempts to create alternative spaces that were outside of market and state institutional structures.

A testament to its organizing efforts, FIERCE was granted a seat on the HRPT Advisory Council in 2009, bolstering hopes for a successful Our SPOT campaign. In 2013, however, FIERCE ended the campaign. In an e-mail to supporters in November that year, it cited continued violence and police harassment as the reason for ending the campaign, and expanding its vision: "Establishing a 24-hour LGBTQ drop-in center cannot be fully realized without maintaining the West Village neighborhood as a safe space for our community—a space free of violence and harassment from the police, each other, and other communities" (E-mail message to supporters, titled "Join FIERCE to Celebrate the Our SPOT Campaign" 25 November 2013). FIERCE shifted its organizing to focus on citywide unjust policing and discrimination against queer youth of color. FIERCE's organizer explained this shift in focus:

By FIERCE closing [the campaign] out, we were never saying, 'Hey, there's no longer a need for a 24-hour

center.' We're saying we needed to open up capacity; we needed to open up opportunities to put people power behind other areas that when we focus on them, who knows, along the road, the vision is always to have those safe spaces for LGBTQ youth. But we are able to do some different things now. And use our creativity in different ways. (J. Blasco interview with the author 29 January 2014)

FIERCE's lead organizer asserted the importance of visionary long-term thinking: "In the past couple of years we've been bringing to the table in the West Village the concept of community accountability, and safety, and trust building. Which is a conversation that folks weren't really engaging in five plus years ago. But they are engaging in it now" (J. Blasco interview with the author, 29 January 2014).

In its effort to create safe spaces, FIERCE worked in a multilevel way. First, in the early stages of its campaigns, it emphasized the presence and power of youth constituents, displaying organization and continuing leadership building. It is through the power of this organization that FIERCE developed working relationships with formal institutions. Second, through those working relationships, it laid partial claims on space, asserting a presence along the park and piers. It pushed back curfew hours at the park and garnered agreements for temporary events and the right to continual presence. Third, it made efforts for a more concrete spatial claim. Through the Our SPOT campaign, it made the case for a long-term, sustained, physical presence.

The FIERCE campaign in the West Village shines a light on dominant modes of urban development in

New York City. The redevelopment of the Hudson River Park preceded more recent high-profile parks projects such the Brooklyn Bridge Park and the High Line,[11] where novel forms of private development are used to fund or enable green recreational spaces.[12] In a time when talk of urban sustainability has become mainstream, green, open space sits comfortably alongside exclusionary development. More broadly, this campaign renews questions about the relevance of the "gayborhood" as a space of safety for queer people. As FIERCE's experience shows, although the memory of a historically safe place for queer youth continues to draw them to the West Village and the Christopher Street Pier, the challenges they face there have their roots in broader injustices of racist policing that are perpetuated throughout the city. Understood at the intersection of race and sexuality, if queer youth of color are not safe in the city more broadly, then they are not safe in historically gay parts of the city.

These broader processes of urban development and policies of policing are, if anything, more pointed in Bedford-Stuyvesant. For ALP's SOS campaign, the continual challenge of maintaining the safe spaces in the network also underscores the difficulty faced by small community organizations working for marginalized groups, largely outside state-sponsored or institutionalized systems, in places of changing economics and demographics. Gentrification, according to the SOS activist, took its toll on many spaces (E. Dixon interview with the author 17 January 2014). The Starlite Lounge, one of the SOS safe spaces and widely known as the first black-owned gay bar in the borough, was forced to close in 2010 despite much community

Figure 6. Photograph of a Bedford-Stuyvesant café with SOS "safe space" sticker, 2013. *Source:* Photograph by author.

rallying to save it. I visited several of the safe spaces in late 2013. In one, a café, I noticed an SOS sticker on its window and brochures and flyers inside (Figure 6). The person working there that day, however, was not aware of the initiative. Other safe spaces had closed or appeared to have been renovated. The very nature of the campaign also posed significant challenges. Shopkeepers were hesitant about the idea of dealing with violence on their premises. SOS activists often had to shift their organizing focus when individual victims of violence asked for help, slowing down the broader organizing effort (E. Dixon interview with the author 17 January 2014).

Invocations of Jacobs's (1961) "eyes on the street" frequently accompany discussions of safety in cities. Jacobs, though, was not taking into consideration the complex interactions of race, class, and sexuality and gender presentation at play on the streets of present-day New York City, when eyes on the street might mean overly watchful homeowners calling the police on youth of color for loitering, or the continued aggression and ridicule shown to gender-nonconforming individuals. This is why SOS's three-phase organizing framework, with its focus on community accountability and training and education, is notable. Organizers were prototyping an approach to neighborhood safety beyond traditional concepts of surveillance and policing. The Safe Neighborhood Campaign attempts to mark physical space in a way that is distributed and communal. It is neither a targeted, concrete claim (like, e.g., a community center or FIERCE's efforts for Pier 40) nor one made via complicity with state institutions (e.g., community policing).

Urban Spatial Movement Building

The work of FIERCE and the SOS Collective of ALP involve basic notions of urban space and life, including issues of access, safety, and shelter. Although geared toward shared objectives, the two organizations' campaign strategies differ, as do their immediate contexts. FIERCE's campaign involves steps toward an increasingly concrete claim on space, following a longer term organizing strategy to build institutional power among constituents. The claim on a specifically targeted physical space is a response to the nature of the neighborhood, a historic gay enclave that now presents an insurmountably disparate power structure ensconced in real estate values. ALP's campaign involves a more distributed network of spaces, developed through agreements with a variety of partners, built on and sustained by education and training. This negotiated claim on space reflects the new and evolving contexts in that neighborhood, making contingent spaces with diffuse targets.

Both organizations' work offers roadmaps for taking on the unjust urban geographies of overlapping identities and multiple modes of oppression. Critically, they show us that specific claims on physical space matter but that these claims might be variously constituted through organizing strategies. Making queer safe spaces through spatial–political organizing is not simply about an appeal to queer identity. In taking on the modes and institutions of urban development and spatial marginalization, however, it poses a challenge to material relationships. It offers alternative social–spatial relations and the possibility of continued difference in the city. These are not success stories. The unfinished campaigns and ongoing struggles of organizing groups such as FIERCE and the ALP expose the unjust systemic spatial conditions of cities. They provide paths forward, approaches outside or in opposition to formal structures and institutional frameworks of urban governance, toward more just cities.

Acknowledgments

I thank the activists and organizers of FIERCE and the Audre Lorde Project for their inspiration, comradeship, and generosity and the anonymous reviewers for their incisive comments.

Notes

1. See, for instance, discussions about existential security and public safety in Marcuse (2006), climate change and urban economic security in Hodson and Marvin (2010), security and urban resilience in Coaffee, Wood, and Rogers (2009), and the intertwined politics of sexuality, race, and terrorism in Puar (2006, 2007).
2. In this article, I use LGBT when discussing lesbian, gay, bisexual, and transgender issues and communities more generally. I use queer as a distinctly political term, to point to sexualities and politics positioned outside of and in opposition to mainstream discourses of sexuality.
3. As Oswin (2008) wrote, "Queers are sexualized while non-white are raced and the need for an analysis of race and racism is considered necessary only when queers are non-white" (94). There is abundant work on issues of race and sexuality outside of the discipline of geography proper (see, e.g., DeFilippis et al. 2011; Bailey and Shabazz 2014a, 2014b).
4. See also Oswin (2015) for ongoing debates around and formulation of queer geographies.

5. The Anti-Violence Project documents specific incidences of anti-LGBT violence. The organization documented an alleged incidence of anti-LGBT police violence in Bedford-Stuyvesant in June 2013 (see GLAAD 2013; NCAVP 2014).
6. See http://www.beyondmarriage.org and Duggan (2011) for further discussion on these points.
7. See the Sylvia Rivera Law Project's statement on this at http://srlp.org/our-strategy/policy-advocacy/hate-crimes/.
8. These organizations promote a notion of "transformative justice," an alternative framework of community accountability (see Mogul, Ritchie, and Whitlock 2011; Whitlock 2012).
9. I note that queer here is my usage. FIERCE uses the term widely. ALP chooses a more deliberately explicit set of identities including lesbian, gay, bisexual, two-spirit, trans, and gender nonconforming, shortened to LGBTSTGNC in its materials.
10. See Hanhardt (2013, chapter 5) for more elaboration on FIERCE's formation as an organizing presence in the West Village and Hudson River Park.
11. At the High Line, the elevated park that connects the West Village to Chelsea and Hell's Kitchen further north, park managers tout the park's gay past but, as researchers point out, often at the expense of queer youth of color (Cataldi et al. 2011; Patrick 2014).
12. The 2013 amendment to the Hudson River Park Act, the state law that established the park and trust, permits the sale of air rights along the park for development not simply in adjacent parcels but up to one block east of the park boundaries (NYS Legislature 1998, 2013).

References

American Civil Liberties Union (ACLU). 2016. Anti-LGBT religious exemption legislation across the country. Accessed December 29, 2016. https://www.aclu.org/other/anti-lgbt-religious-exemption-legislation-across-country.

Audre Lorde Project (ALP). 2016. Do not militarize our mourning: Orlando and the ongoing tragedy against LGBTSTGNC POC. Accessed June 15, 2016. http://alp.org/do-not-militarize-our-mourning-orlando-and-ongoing-tragedy-against-lgbtstgnc-poc.

Badgett, M. V. L., L. E. Durso, and A. Schneebaum. 2013. *New patterns of poverty in the lesbian, gay, and bisexual community*. Los Angeles: The Williams Institute.

Bailey, M. M., and R. Shabazz. 2014a. Editorial: Gender and sexual geographies of blackness: Anti-black heterotopias (Part 1). *Gender, Place & Culture* 21 (3):316–21.

———. 2014b. Gender and sexual geographies of blackness: New black cartographies of resistance and survival (Part 2). *Gender, Place & Culture* 21 (4):449–52.

Bell, D., and J. Binnie. 2004. Authenticating queer space: Citizenship, urbanism and governance. *Urban Studies* 41 (9):1807–20.

Bell, D., and G. Valentine. 1995. *Mapping desire: Geographies of sexualities*. London and New York: Routledge.

Berlant, L., and E. Freeman. 1992. Queer nationality. *Boundary 2* 19 (1):149–80.

Betsky, A. 1997. *Queer space: Architecture and same-sex desire*. New York: William Morrow.

Binnie, J. 1997. Coming out of geography: Towards a queer epistemology? *Environment and Planning D* 15:223–38.

Binnie, J., and B. Skeggs. 2004. Cosmopolitan knowledge and the production and consumption of sexualized space: Manchester's gay village. *The Sociological Review* 52 (1):39–61.

Botein, H. 2013. From redlining to subprime lending: How neighborhood narratives mask financial distress in Bedford-Stuyvesant, Brooklyn. *Housing Policy Debate* 23 (4):714–37.

Brenner, N. 2000. The urban question as a scale question: Reflections on Henri Lefebvre, urban theory and the politics of scale. *International Journal of Urban and Regional Research* 24 (2):361–78.

Brown, G. 2006. Cosmopolitan camouflage: (Post-)gay space in Spitalfields, East London. In *Cosmopolitan urbanism*, ed. J. Binnie, 130–45. London and New York: Routledge.

———. 2007. Mutinous eruptions: Autonomous spaces of radical queer activism. *Environment and Planning A* 39 (11):2685–98.

Brown, M. 2014. Gender and sexuality II: There goes the gayborhood? *Progress in Human Geography* 38 (3):457–65.

Browne, K. 2006. Challenging queer geographies. *Antipode* 38 (5):885–93.

Burawoy, M. 1998. The extended case method. *Sociological Theory* 16 (1):4–33.

Carpenter, L. F. 2017. The next phase: Positioning the post-Obergefell LGBT rights movement to bridge the gap between formal and lived equality. *Stanford Journal of Civil Rights and Civil Liberties* 13:255.

Castells, M. 1983. *The city and the grassroots: A cross-cultural theory of urban social movements.* London: Arnold.

Cataldi, M., D. Kelley, H. Kuzmich, J. Maier-Rothe, and J. Tang. 2011. Residues of a dream world: The High Line, 2011. *Theory, Culture & Society* 28 (7–8):358–89.

Coaffee, J., D. M. Wood, and P. Rogers. 2009. *The everyday resilience of the city: How cities respond to terrorism and disaster.* Basingstoke, UK: Palgrave Macmillan.

Collins, A. 2004. Sexual dissidence, enterprise and assimilation: Bedfellows in urban regeneration. *Urban Studies* 41 (9):1789–1806.

Colomina, B. 1992. *Sexuality & space.* New York: Princeton Architectural Press.

Crenshaw, K. 1991. Mapping the margins: Intersectionality, identity politics, and violence against women of color. *Stanford Law Review* 43 (6):1241–99.

Davidson, M. 2008. Rethinking the movement: Trans youth activism in New York City and beyond. In *Queer youth cultures*, ed. S. Driver, 243–60. Albany, NY: SUNY Press.

DeFilippis, J. N., and B. Anderson-Nathe. 2017. Embodying margin to center: Intersectional activism among queer liberation organizations. In *LGBTQ politics: A critical reader*, ed. M. Brettschneider, S. Burgess, and C. Keating, 110–33. New York: NYU Press.

DeFilippis, J. N., L. Duggan, K. Farrow, and R. Kim. 2011. A new queer agenda. *Scholar & Feminist Online* 10 (1–2). Accessed March 24, 2014. http://sfonline.barnard.edu/a-new-queer-agenda/

Doan, P. L. 2007. Queers in the American city: Transgendered perceptions of urban space. *Gender, Place and Culture* 14 (1):57–74.

———. 2010. The tyranny of gendered spaces—Reflections from beyond the gender dichotomy. *Gender, Place and Culture* 17 (5):635–54.

———. 2011. Why question planning assumptions and practices about queer spaces. In *Queerying planning: Challenging heteronormative assumptions and reframing planning practice*, ed. P. L. Doan, 1–18. Farnham, UK: Ashgate.

———. 2015. *Planning and LGBTQ communities: The need for inclusive queer spaces.* London and New York: Routledge.

Doan, P. L., and H. Higgins. 2011. The demise of queer space? Resurgent gentrification and the assimilation of LGBT neighborhoods. *Journal of Planning Education and Research* 31 (1):6–2.

Duberman, M. B. 1993. *Stonewall.* New York: Dutton.

Duggan, L. 2002. The new homonormativity: The sexual politics of neoliberalism. In *Materializing democracy: Toward a revitalized cultural politics*, ed. R. Castronovo and D. D. Nelson, 175–94. Durham, NC: Duke University Press.

———. 2003. *The twilight of equality?: Neoliberalism, cultural politics, and the attack on democracy.* Boston: Beacon Press.

———. 2011. Beyond marriage: Democracy, equality, and kinship for a new century. *Scholar & Feminist Online* 10 (1–2). Accessed March 24, 2014. http://sfonline.barnard.edu/a-new-queer-agenda/beyond-marriage-democracy-equality-and-kinship-for-a-new-century/

Durso, L. E., and G. J. Gates. 2012. *Serving our youth: Findings from a national survey of services providers working with lesbian, gay, bisexual, and transgender youth who are homeless or at risk of becoming homeless.* Los Angeles: The Williams Institute with True Colors Fund and The Palette Fund.

Ehrenfeucht, R. 2013. Nonconformity and street design in West Hollywood, California. *Journal of Urban Design* 18 (1):59–77.

Fabulous Independent Educated Radicals for Community Empowerment (FIERCE). 2008. *LGBT youth center: Pier 40 recommendation.* New York: FIERCE. Accessed March 24, 2014. http://www.fiercenyc.org/resources/lgbt-youth-center-pier-40-recommendation.

Frisch, M. 2002. Planning as a heterosexist project. *Journal of Planning Education and Research* 21 (3):254–66.

GLAAD. 2013. *Anti-violence project speaks out against anti-gay violence from the NYPD.* GLAAD Blog. Accessed June 10, 2013. http://glaadblog.org/blog/anti-violence-project-speaks-out-against-anti-gay-violence-nypd.

———. 2016. *GLAAD calls for increased and accurate media coverage of transgender murders.* Accessed July 25, 2016. http://www.glaad.org/blog/glaad-calls-increased-and-accurate-media-coverage-transgender-murders.

Goh, K. 2011. Queer beacon: LGBT spaces in New York City. *Places: Design Observer* June 23. Accessed March 24, 2014. https://placesjournal.org/article/queer-beacon/

———. 2015. Place/out: Planning for radical queer activism. In *Planning and LGBTQ communities: The need for inclusive queer spaces*, ed. P. L. Doan, 217–34. London and New York: Routledge.

Goldsmith, W. W., and E. J. Blakely. 2010. *Separate societies: Poverty and inequality in U.S. cities*. Philadelphia: Temple University Press.

Gould, D. B. 2009. *Moving politics: Emotion and ACT UP's fight against AIDS*. Chicago: University of Chicago Press.

Greenberg, W. 2016. Brace yourselves for an onslaught of anti-LGBT proposals in 2017. *Mother Jones* December 22. Accessed December 28, 2016. http://www.motherjones.com/politics/2016/12/lgbt-advocates-gearing-another-year-fights

Halberstam, J. 2005. Queer temporality and postmodern geographies. In *In a queer time and place: Transgender bodies, subcultural lives*, 1–21. New York: New York University Press.

Hanhardt, C. B. 2013. *Safe space: Gay neighborhood history and the politics of violence*. Durham, NC: Duke University Press.

Harvey, D. 2008. The right to the city. *New Left Review* 53:23–40.

Hodson, M., and S. Marvin. 2010. *World cities and climate change: Producing urban ecological security*. Maidenhead, UK: Open University Press.

Jacobs, J. 1961. *The death and life of great American cities*. New York: Random House.

Kanai, J. M. 2014. Whither queer world cities? Homo-entrepreneurialism and beyond. *Geoforum* 56 (Suppl. C):1–5.

Kanai, J. M., and K. Kenttamaa-Squires. 2015. Remaking South Beach: Metropolitan gayborhood trajectories under homonormative entrepreneurialism. *Urban Geography* 36 (3):385–402.

Knopp, L. 1992. Sexuality and the spatial dynamics of capitalism. *Environment and Planning D: Society and Space* 10 (6):651–69.

———. 1994. Social justice, sexuality, and the city. *Urban Geography* 15 (7): 644–60.

Lavers, M. K. 2010. *Sucuzhañay case underscores LGBTs, immigrants remain particularly vulnerable to hate crimes*. EDGE Media Network. Accessed May 14, 2010. http://www.edgemedianetwork.com/news/national//105698

Leitner, H., E. S. Sheppard, K. Sziarto, and A. Maringanti. 2007. Contesting urban futures: Decentering neoliberalism. In *Contesting neoliberalism: Urban frontiers*, ed. H. Leitner, J. Peck, and E. S. Sheppard, 1–25. New York: Guilford.

Mananzala, R. 2011. The FIERCE fight for power and the preservation of public space in the West Village. *Scholar & Feminist Online* 10 (1–2). Accessed March 24, 2014. http://sfonline.barnard.edu/a-new-queer-agenda/

Marcuse, P. 2006. Security or safety in cities? The threat of terrorism after 9/11. *International Journal of Urban and Regional Research* 30 (4):919–29.

McFadden, R. D. 2008. Attack on Ecuadorean brothers investigated as hate crime. *The New York Times* December 8:A29.

Mogul, J. L., A. J. Ritchie, and K. Whitlock. 2011. *Queer (in)justice: The criminalization of LGBT people in the United States*. Boston: Beacon Press.

Nash, C. J., and A. Bain. 2007. "Reclaiming raunch"? Spatializing queer identities at Toronto women's bathhouse events. *Social & Cultural Geography* 8 (1):47–62.

Nash, C. J., and A. Gorman-Murray. 2014. LGBT neighbourhoods and "new mobilities": Towards understanding transformations in sexual and gendered urban landscapes. *International Journal of Urban and Regional Research* 38 (3):756–72.

National Coalition of Anti-Violence Programs (NCAVP). 2014. Lesbian, gay, bisexual, transgender, queer, and HIV-affected hate violence in 2013 (2014 release edition). Accessed January 5, 2017. http://www.avp.org/storage/documents/2013_ncavp_hvreport_final.pdf.

———. 2016. Lesbian, gay, bisexual, transgender, queer, and HIV-Affected hate violence in 2015 (2016 release edition). Accessed December 14, 2016. http://avp.org/resources/avp-resources/520-2015-report-on-lesbian-gay-bisexual-transgender-queer-and-hiv-affected-hate-violence.

National Conference of State Legislatures. 2016. "Bathroom bill" legislative tracking. Accessed August 30, 2016. http://www.ncsl.org/research/education/-bathroom-bill-legislative-tracking635951130.aspx.

New York City Comptroller's Office. 2012. *Income inequality in New York City*. New York: New York City Comptroller's Office. Accessed July 2, 2015. http://comptroller.nyc.gov/wp-content/uploads/documents/NYC_IncomeInequality_v17.pdf.

New York State Legislature 592. 1998. The Hudson River Park Act. Ch. 592, S. 7845.

———. 2013. Amendment to the Hudson River Park Act.

NYU Furman Center. 2016. *State of New York City's housing and neighborhoods in 2015*. New York: NYU Furman Center. Accessed December 24, 2016. http://furmancenter.org/files/sotc/NYUFurmanCenter_SOCin2015_9JUNE2016.pdf.

Onishi, N. 1994. Police announce crackdown on quality-of-life offenses. *The New York Times* March 13:33.

Oswin, N. 2008. Critical geographies and the uses of sexuality: Deconstructing queer space. *Progress in Human Geography* 32 (1):89–103.

———. 2015. World, city, queer. *Antipode* 47 (3):557–65.

Patrick, D. J. 2014. The matter of displacement: A queer urban ecology of New York City's High Line. *Social & Cultural Geography* 15 (8):920–41.

Podmore, J. A. 2013a. Critical commentary: Sexualities landscapes beyond homonormativity. *Geoforum* 49:263–67.

———. 2013b. Lesbians as village "queers": The transformation of Montreal's lesbian nightlife in the 1990s. *ACME: An International Journal for Critical Geographies* 12 (2):220–49.

Poster, D., and E. Goldman. 2005. Gay youth gone wild: Something has got to change. *The Villager* 76 (18). Accessed March 24, 2014. http://thevillager.com/villager_125/gayyouthgonewild.html

Puar, J. K. 2006. Mapping US homonormativities. *Gender, Place and Culture* 13 (1):67–88.

———. 2007. *Terrorist assemblages: Homonationalism in queer times*. Durham, NC: Duke University Press.

Quintana, N. S., J. Rosenthal, and J. Krehely. 2010. *On the streets: The federal response to gay and transgender homeless youth*. Washington, DC: Center for American Progress.

Ray, N. 2006. *Lesbian, gay, bisexual and transgender youth: An epidemic of homelessness*. New York: National Gay and Lesbian Task Force Policy Institute and the National Coalition for the Homeless.

Revesz, R. 2016. Republicans confident they will pass Anti-LGBT bills after Donald Trump becomes president. *The Independent* December 12. Accessed December 20, 2016. http://www.independent.co.uk/news/world/americas/republicans-utah-mike-lee-fada-religious-freedom-donald-trump-ted-cruz-anti-lgbt-a7470876.html

Roberts, S. 2011. Striking change in Bedford-Stuyvesant as the white population soars. *New York Times* August 5:A19.

———. 2014. Gap between Manhattan's rich and poor is greatest in U.S., census finds. *New York Times* September 18:A28.

Rogers, J. 2007. Hudson Park HUAC: Pier plan foes all "socialists." *The Villager* 77 (10). Accessed March 24, 2014. http://thevillager.com/villager_223/hudsonpar-khuac.html

Ruting, B. 2008. Economic transformations of gay urban spaces: Revisiting Collins' evolutionary gay district model. *Australian Geographer* 39 (3):259–69.

Sandercock, L. 2006. Cosmopolitan urbanism: A love song to our mongrel cities. In *Cosmopolitan urbanism*, ed. J. Binnie, 37–52. London and New York: Routledge.

Sears, B., and L. Badgett. 2012. *Beyond stereotypes: Poverty in the LGBT community.* Accessed March 14, 2014. http://williamsinstitute.law.ucla.edu/headlines/beyond-stereotypes-poverty-in-the-lgbt-community/#sthash.Cj80G7e1.dpbs.

Shepard, B. H., and R. Hayduk. 2002. *From ACT UP to the WTO: Urban protest and community building in the era of globalization.* London: Verso.

Sibalis, M. 2004. Urban space and homosexuality: The example of the Marais, Paris' gay ghetto. *Urban Studies* 41 (9):1739–58.

Smart, M. J., and A. H. Whittemore. 2017. There goes the gaybourhood? Dispersion and clustering in a gay and lesbian real estate market in Dallas TX, 1986–2012. *Urban Studies* 54 (3):600–615.

Soja, E. W. 2010. *Seeking spatial justice.* Minneapolis: University of Minnesota Press.

Tongson, K. 2011. *Relocations: Queer suburban imaginaries.* New York: New York University Press.

Whitlock, K. 2012. *Reconsidering hate: Policy and politics at the intersection.* Somerville, MA: Political Research Associates.

Whittemore, A. H., and M. J. Smart. 2016. Mapping gay and lesbian neighborhoods using home advertisements: Change and continuity in the Dallas–Fort Worth metropolitan statistical area over three decades. *Environment and Planning A* 48 (1):192–210.

Young, I. M. 1990. City life and difference. In *Justice and the politics of difference*, 226–56. Princeton, NJ: Princeton University Press.

KIAN GOH is an Assistant Professor in the Department of Urban Planning at the Luskin School of Public Affairs, University of California, Los Angeles, Los Angeles, CA 90095. E-mail: kiangoh@ucla.edu. Her research interests include urban design, spatial politics, and social mobilization in the context of environmental change and global urbanization.

16 Disciplining Deserving Subjects through Social Assistance

Migration and the Diversification of Precarity in Singapore

Junjia Ye and Brenda S. A. Yeoh

Cities are not only associated with incommensurable human diversity but also play a pivotal role in generating, assembling, and mobilizing differences. Alongside neoliberal processes that drive migrant-led diversification in global cities, we are witnessing growing inequality and precarity in populations of long-term residents that are themselves heterogeneous. Indeed, the diversification of peoples in the global city is also paralleled by the diversification of precarity. Yet, the ways in which new configurations of difference are producing more nuanced if still shadowy subjects of citizenship deserve more conceptual and contextualized attention. Although much has been written on the management of migration, far less attention has been focused on the management of multiplying forms of precarity resulting from insecure sociolegal status, disadvantaged labor market position, and deeply inscribed social prejudice. Even less has been documented on how these forms of management set up specific vernaculars about and subjects of citizenship, migrancy, and precariousness. This article addresses social inequality and the relationality of subject making in the context of diversification in Singapore, a city-state that has a particular historical understanding of diversity through a fixed formulaic "multiracialism." Drawing on state narratives and interview data, we analyze organized social support for both migrants and citizens both by state organizations and nongovernmental organizations to demonstrate the limits and possibilities of change and continuity in the production of precarity in the diversifying city. In doing so, we aim to extend scholarship of the global city beyond the well-debated issue of social polarization in the global city and to highlight the diversity and relationality of precariousness in a contemporary non-Western global city. *Keywords: citizenship, diversity, migration, precarity, subjectivities.*

城市并非仅关乎无法相互比较的人类多样性, 同时也扮演着生产凑组并动员差异的关键要角。除了驱动全球城市由移民带来的多样化之新自由主义过程之外, 我们正在见证本身便相当异质的长期居住人口中的不均与不安定的增加。全球城市的人口多样化, 的确同时伴随着不安定性的多样化。但崭新的差异组合生产更为细缓且可能仍难以捉摸的公民主体之方式, 却值得更为概念化且脉络化的关照。尽管已有诸多研究书写移民的管理, 但却鲜少关注聚焦不确定的社会法律身份、弱势的劳动市场地位, 以及深刻铭刻的社会偏见所造成的多重不安定形式之管理。这些管理形式如何建立有关公民权主体、移民与不安定性的特定行话, 更鲜少受到记载。本文处理新加坡——一个透过固定的 "多元文化主义" 公式对多样性具有特定历史理解的城市国家——的多样化脉络中, 社会不均与主体打造的关系性。我们运用国家叙事和访谈数据, 分析国家组织和非政府组织对于移民与公民的组织性社会支持, 证实在多样化的城市生产不安定中, 改变与持续的机会和限制。我们藉由这麼做, 旨在将全球城市的文献延伸至全球城市中的社会极化此一受到大幅辩论的议题之外, 并强调当代非西方的全球城市中, 不安定性的多样化与关系性。关键词: 公民权, 多样性, 移民, 不安定, 主体性。

Las ciudades no solo están asociadas con una inconmensurable diversidad humana, sino que también juegan un papel crucial en generar, ensamblar y movilizar diferencias. Junto con los procesos neoliberales que controlan la diversificación inducida por el migrante en las ciudades globales, estamos siendo testigos de un avance de la desigualdad y la precariedad en las poblaciones de residentes antiguos, en sí mismos muy heterogéneos. En verdad, la diversificación de pueblos en la ciudad global va también de la mano con la diversificación de la precariedad. Con todo, los modos como las nuevas configuraciones de la diferencia están produciendo sujetos de ciudadanía, más matizados, aunque todavía imprecisos, merecen atención más conceptual y contextualizada. Aunque mucho es lo que se ha escrito sobre el manejo de la migración, menor lo es la atención puesta en el manejo de las formas multiplicadas de precariedad que resultan de un estatus sociolegal inseguro, la posición desventajosa del mercado laboral y el prejuicio social grabado profundamente. Aun menos documentado es el modo como estas formas de manejo establecen lenguajes coloquiales específicos acerca de los sujetos de ciudadanía, migración y precariedad. Este artículo aboca la desigualdad social y la relacionalidad de construir sujeto en el contexto de la diversificación en Singapur, un estado–ciudad que tiene un particular entendimiento histórico de la diversidad a través de un predecible "multiracialismo" inalterable. Apoyándonos en narrativas del estado y en datos de entrevistas, analizamos la ayuda social organizada para migrantes y ciudadanos, para ambos casos por organizaciones del estado y organizaciones no gubernamentales, para demostrar los límites y posibilidades de cambio y continuidad en la producción de precariedad en la ciudad diversificante. Al hacer esto, nos proponemos extender la erudición de la ciudad global más allá de la cuestión, por demás bien discutida, de la polarización social en la ciudad global, y para destacar la condición de lo diverso y la relacionalidad de la precariedad en una ciudad global contemporánea no occidental. *Palabras clave: ciudadanía, diversidad, migración, precariedad, subjetividades.*

The challenge of poverty in Southeast Asia is often assumed to be located in the region's poorest countries. Recent work, however, has shown that by 2008 over 80 percent of Asia's poor were living in rapidly growing economies (Wan and Sebastian 2011). This regional incarnation of the "new geography of global poverty" points to the urgent need to give attention to the "poor" in "successful" countries, who tend to remain obscured from view by developmentalist narratives (Rigg 2016, 4). The geographical dimensions of poverty in the region also need to take into account the increasing intraregional flows of low-skilled migrant workers, fueled by growing socioeconomic disparities among places as a result of uneven development and also facilitated by the availability of low-cost travel and a burgeoning migration industry.

In this context, the rapid influx of not just internal but also cross-border migrants into the region's growing cities is raising new questions for the study of sociospatial difference in urban life. Neoliberal processes that drive migrant-led diversification in these cities are also contouring the growing inequality and precarity in populations comprised of long-term residents, altering how difference is envisaged and experienced. The ways in which these more complex configurations of difference are, in turn, producing new and more nuanced forms of sociospatial inequalities in the globalizing city, however, have largely remained understudied in the growing literature on diversity in Asia.

In this article, we are interested in the ways in which the diversification of people in the city is also paralleled by the diversification of precarity. Difference in the global city is hence framed in terms of poverty expressed as precarity that is relational, rather than absolute. We illustrate the sociopolitical relations that generate precarity as a key axis of differentiation. Referring to those who experience precariousness, the concept includes both subjective and structural dimensions: It conjures life worlds that are inflected with a sense of uncertainty and instability (Waite 2009) and is also predicated on the interdependent care of others as well as institutionalized forms of recognition (Shaw and Byler 2016). Precarity engendered by contingent work conditions is spatially extensive and long-standing (Ettlinger 2007). Low-wage labor migrants increasingly make up a large share of the world's precariat (Waite 2009; Standing 2011; Lewis et al. 2015). Recent work has also shown that focusing on the "precarity" as a category of economic exclusion is oversimplistic and that precarity encompasses a much broader swath of life resulting from social and political exclusions. Recognizing this breadth of what constitutes precarity, we focus on the production and sustenance of precarity through practices and discourses of differential inclusion. We demonstrate that diverse peoples are incorporated through uneven modes of governance, ordering, and management, generating the precarious subject. More specifically, we show that these forms of difference making are multiply sustained through processes of racialization and citizen making through public assistance in the global city. The focus here is on the sociopolitical relations that generate and govern precarity and precarious subjects in conditions of migrant-led diversification.

Specifically, we interrogate how citizenship and race are deployed in the context of rapidly diversifying Singapore through the analysis of organized social assistance for both migrants and locals provided by state and nonstate organizations. We draw on empirical data from Singapore to identify how public assistance is administered and, consequently, the implications for how beneficiaries are identified and managed. Our argument is developed from an analysis of policy documents and ministerial speeches and interviews conducted with staff and volunteers of state-linked organizations and nongovernmental organizations (NGOs) in Singapore in 2014 and 2015. Examining both state-provided and community-run spaces of assistance, this analysis provides insight into how race, class, and citizenship interact in aggregate ways to contour access to assistance.

As a Southeast Asian city-state that has succeeded in alleviating absolute poverty,[1] Singapore's progress toward becoming a livable city with a high-quality environment to live, work, and play has attracted the attention of city planners around the world. At the same time, as a country that has eschewed both a minimum wage and an official poverty line, Singapore has one of the world's highest Gini coefficients, logged at 0.478 in 2014 (Chan 2014). For all of its successes, Singapore contains staggering contrasts of wealth, poverty, and power. Such diversity is further compounded by the large influx of temporary migrants from across the globe of different skill levels, from professional and managerial elites to low-skilled workers. In particular, the city relies heavily on increasing numbers of low-wage foreign workers—about a million constituting 27 percent of the workforce—to do the jobs that locals cannot be persuaded to do (Koh et al. 2016). In this sense, the Singapore case demonstrates the need to reconceptualize socioeconomic inequality and precarity in the global city to take into account the place of transnational migrant workers who service and support the needs of the emerging middle classes.

In this light, the next part of the article provides a contextual sketch of the interactions of race, class, and citizenship in contouring the landscape of public social and economic assistance in Singapore. This is followed by a third section that illustrates the way in which the migrant–local divide is reinforced through state-provided assistance and inflected by constructions of citizenship and race. In the penultimate part of the article, we turn to encounters between middle-class staff and volunteers at community-based social support organizations and their clientele living on the socioeconomic margins in the city to flesh out the politics of help. The Conclusion brings together these interrelated sets of processes through which we trace change and continuity in the construction of precariousness in the city.

Situating Economic and Social Support in Singapore

The political nexus of citizenship, race, and poverty is most clearly demonstrated through the realm of public assistance. As many feminist scholars have shown, welfare policies situate and shape needs and degrees of deservedness (Fraser 1994; Schram 2000; Scheneider and Ingram 2005; Lawson and Elwood 2014; Teo 2015). These in turn shape the norms and values of society and are important sites of governance processes (Lawson and Elwood 2014; Teo 2014). Crucially, the definition and implementation of welfare regimes define and limit citizens' sense of their relationship to the state and to others in society (Somers 2008). Furthermore, it is not just how much is spent, absolutely and relatively, but also how spending on support is oriented and the rationalities around which reforms are designed (Teo 2014). Even as public spending increases to fund a growing range of welfare services, particularly in response to growing inequality or in times of a recession, specific programs, campaigns, measures, and policies operate in ways that mask certain constraints on the vulnerable. As Sennett (2003) pointed out, the "act of giving needn't in itself carry the positive charge of a cooperative act. Giving to others can be a way of manipulating them, or it can serve the more personal need to affirm something in ourselves" (136). Problematizing apparent generosity, or largesse, Sennett went on to argue that at the core of any form of welfare is the double-edged gift. At "one extreme is a gift freely given, at the other is the manipulative gift. The first embodies that aspect of character focused on the sheer fact that others lack something, that they are in need; the other act of giving uses it only as a means to gain power over them" (137–38). To be included within welfare often requires that one be subjected to particular conditions or fit into preexisting normativities (including, as in this case, racialized categories). Welfare, even with diversification, can continue to filter out different groups of people. Even as there is continuity in various forms of state-led social assistance, the way in which these are mobilized also accounts for partition, filtering, and hierarchization. In the case of Singapore, the following three dimensions are significant in shaping the public assistance landscape.

First, to understand how the processes of citizenship and race work together to manage boundary making in deservedness, we consider the city-state's organization of difference in its historical context. Following independence in 1965, the governing elites faced the challenge of imagining a common space that could act as a nucleus of nationhood for the city-state. To build a nation-state out of an ethnically diverse population with a complex background of economic, political, social, and cultural differences, the state promulgated an overarching national identity based on the ideology of "multiracialism" (Lai 1995, 17). This measure officially gives separate but equal status to the Chinese, Malays, Indians, and "Others"[2] (CMIO) and informs official policies on various issues related to the economy, language, culture, religion, housing, and community life (Lai 1995; Perry, Kong, and

Yeoh 1997) and, indeed, welfare. Thus, the framework of race became part of the national imagination from the birth of the new nation-state, encouraging Singaporeans of various backgrounds to imagine themselves as one "race" within a multiracial people. The management of citizenship through multiracialism in the Singapore context relies on the simplification and essentialism of race. Race and ethnic identity are also clearly denoted on every Singaporean's identity card. This particular vernacular of multiracialism is conveyed, experienced, and spatialized as commonplace in the everyday lives of Singaporeans. As we discuss later, the formation of race-based self-help groups is an extrapolation of multiracialism as a founding ideology in the city-state.

A second dynamic is entwined with the calculated integration of foreigners, reproducing a relational and graduated continuum of laboring bodies in Singapore. The strategic reliance on "foreign workers" is part and parcel of the dominant neoliberal discourse of globalization as an "inevitable and virtuous growth dynamic" (Coe and Kelly 2002, 348). Foreigners' access to rights and privileges is mainly differentiated by skills status and by the perceived desirability of these skills to the achievement of national goals. Varied access is institutionalized by the issuance of a range of work passes and permits that fall broadly into the employment pass and the work permit categories. The former category, usually referred to as foreign talent in the Singapore context, includes highly skilled professional workers, entrepreneurs, and investors who are part of the face of cosmopolitanism in Singapore (Ye 2016). This group of migrants enjoys a number of rights and privileges, including eligibility for dependents' passes and access to greater job mobility. Far greater in number, however, are work permit holders, most of whom are concentrated in the manufacturing, counstruction, shipbuilding, and domestic industries. With a steady annual increase, there are presently 992,700 work permit holders in Singapore, constituting about 27 percent of the total labor force. This pool is also broken down further by nationalities, with rules and regulations set by the Ministry of Manpower (MOM 2017), permitting only certain nationalities to access work in particular industries (Ye 2013). These low-wage temporary migrant workers face much more constrained conditions and are neither eligible to bring dependents nor access pathways to permanent residency or citizenship (Yeoh 2004). Of this group that hold work permits, the largest groups are foreign construction workers, many of whom are men from Bangladesh, China, India, and Myanmar, followed by female migrant domestic workers from Indonesia, the Philippines, and Myanmar. These socioeconomic divisions manifest tangibly in the segregated landscapes inhabited by temporary migrant workers. Shipyard and construction workers, for example, work on sites away from interactions with the public and are accommodated in purpose-built dormitories that are commercially run, industrial, or warehouse premises that have been partly converted to house workers, temporary quarters on work sites, harbor crafts (e.g., ships and marine vessels), and, to a lesser extent, public housing. The majority of such accommodations are segregated from residential areas where locals live. Their circumscribed positions in Singapore are further reinforced by their highly limited access to state-organized social support, as discussed later.

Third, economic and social assistance in Singapore is shaped not only by state-led management of difference and diversity through race and citizenship but also by the city-state's emphasis on values of meritocracy and self-reliance and aspirations of social mobility. As Prime Minister Lee Hsien Loong illustrated during the National Day Rally in 2014:

> [Singapore's first president] Encik Yusof showed that in Singapore, you can rise to the top if you work hard. He stood for enduring values that underpin Singapore's success—meritocracy, multiracialism, modernization. (Prime Minister's Office 2014)

Within this context of meritocracy and a multiracial citizenry imagined as "separate but equal," there are a variety of state-led social and economic support services for Singaporean citizens. Social support is inflected by both citizenship as well as race and class through the national narrative of multiracialism, migration, and labor policies.[3] The principles embedded within the Singapore state's approach toward issues of state support include an emphasis on the importance of self-reliance through formal employment; reliance on family members before nonfamily members; the significant role of NGOs known as voluntary welfare organizations; and the state as supporter of last resort (Teo 2014). It bears pointing out that state support programs in Singapore do not only target the poor. Middle-class Singaporeans are also recipients of economic help from state agencies in the form of monetary incentives and grants in buying flats, in school fees, and in child care. Many of the state's assistance measures also come in the form of labor market interventions. Most of these schemes are run by the MOM, the Workforce Development Agency, and the Inland Revenue Authority and include programs such as the Skills

Program for Upgrading and Resilience,[4] which subsidizes *employers* for their employees who enroll in training courses to increase productivity, and the Wage Credit Scheme[5] where the government cofunds 40 percent of salary increases for Singaporean employees over a period of time. The Progressive Wage Model was introduced to the landscaping and cleaning sectors of the labor market in mid-2015 to prevent employers from outsourcing. As National Trade Union Center Assistant Secretary-General Zainal Sapari said:

> With the Progressive Wage Model, we are hoping that we level the playing field in terms of the wages paid to the workers and if they have to compete, they really have to compete based on their productivity, based on their track record. We want to give the assurance to the service buyers that the Progressive Wage Model is not about increasing salaries, it's about ensuring that the workers are properly skilled, about ensuring that there will be higher level of productivity and hopefully they will get a better service out of it. (Saad 2015)

Productivity, in this sense, forms the normative criterion that naturalizes a hierarchy of value. Through these programs of upgrading people's skills, service provision, and staggering wages in tandem with economic output, the increase of state support has, in effect, rationalized economic productivity as part of a gradation of worthiness and deservingness.[6] Assistance is hence shaped to support desired labor market transitions and access to assistance is both predicated on and reinforces what it means to be a citizen. Furthermore, several forms of support are also, in practice, operating cost-cutting measures more directly beneficial to companies, rather than to workers. This is coherent with the state's principles of self-reliance through formal, paid employment where those eligible to receive aid are valued because they fit within the normative terms of citizenship. Assistance also becomes a zone that reestablishes the exclusionary boundaries of citizenship for the large number of low-wage migrant workers who have no access to state assistance and remain outside the frame of deservedness.

Apart from state schemes, a form of assistance that is consistent with the state's social management is rendered by community self-help groups organized around the national multiracialism schema. The sociopolitical salience of meritocracy and the "separate but equal" mode of multiracialism in Singapore, in essence, precludes the questioning of race and vulnerability in the provision of assistance. We turn now to discuss the work of race-based self-help groups in hardening the lines that divide citizens from noncitizens.

Class, Race, and Citizenship along Assistance Trajectories

Formed in the early 1990s, the importance of race-based self-help groups is evident in the steady increase of funding they have received from the government.[7] In 2014, for example, the Singapore Indian Development Association (SINDA) received up to $3.4 million, up from $1.7 million previously, and the Eurasian Association received up to $400,000, a doubling from the previous figure of $$200,000. The Malay self-help groups MENDAKI and Association of Muslim Professionals (AMP) received grants of $5 million, up from the previous $4 million, and the Chinese Development Assistance Council (CDAC) received a one-time sum of $10,000 for a four-year period from 2014 (Chuan 2014). Although each group caters to their corresponding ethnic community, common among them is the approach to welfare through emphases on education, strengthening family ties, and employability. The beneficiaries of these self-help groups are mainly lower income citizens. For example, MENDAKI's programs largely cater to the lower 30 percent of the Malay and Muslim population (see http://www.mendaki.org.sg/about-mendaki/general).

Although each of these self-help groups adheres closely to the ideology of multiracialism, it must be remembered that race is deployed in decidedly strategic terms here. Help is only accessible to Singaporeans and, to an extent, permanent residents who fall within the state's CMIO ethnic and racial organization of society (despite the fact that there are now large numbers of especially low-wage migrants from China and India within the country's borders). As the following quotes demonstrate, the politics of welfare inclusion is strongly contingent on citizenship and race.

> Because we want to help our local Singaporeans first. . . . Because some of the funding that we get are from the government, so we have to be sure that we have the right target. (Informant A from a race-based self-help group)

> When we say Singaporeans and PRs, and we are talking about Singaporean Indian and Indian PRs, for us, it's the government's interpretation of "Indian" which is anyone from the subcontinent. (Informant B from a race-based self-help group)

Precarity management is strongly informed by race and citizenship management. These raced and classed politics of boundary-making within social assistance produce graduated modes of citizenship that situate precarity politics in Singapore. This powerful imagination of who constitutes the precarious is materially consequential as only those recognized as citizens qualify for assistance. The quotes just presented illustrate that the increased diversity of people leading precarious lives is not reflected in the provision of organized help toward legal labor migrants to Singapore. Paradoxically, then, the ethos of self-reliance through formal employment in the realm of assistance excludes a large number of low-wage migrant workers. The fixation on productivity as a proxy for assistance is situated within a particular framework of race and citizenship.

As Informant C, who is currently working for a community development center (CDC) for Singaporeans but who used to work for a migrant NGO, said:

> [The migrant workers] don't have any help services. Actually for Singaporeans they can go to CDC, FSCs [Family Service Centers], religious organizations, self-help group like CDAC, MENDAKI, whereas the migrants will just go to NGOs. I don't think [any of the race-based self-help groups] will help them [migrants].

Economic growth and absolute increase in funds for public assistance hence constitute the very forms of differential inclusion that render certain people much more precarious than others. Strikingly, low-wage migrants are not recognized as poor within this dominant frame of assistance. These ongoing reinforcements of race, class, and citizenship entrench commonly held beliefs and norms about who constitute poor others and how they need to be managed. In other words, how assistance is organized and distributed shows that race is intimately tied to and plays a crucial organizational role within the imagining and making of citizenship and precarity in the global city.

Disciplining Norms through Assistance

Social support is inflected by both citizenship and race, through the national narratives of multiracialism, migration and labor policies, and cross-class encounters. In this section, we expand on the politics of help and precarity through the dynamic cross-class encounters between middle-class staff and volunteers of social support organizations and new citizens of the nation-state. Where newcomers do receive assistance from state-linked organizations, this appears to be more focused on the issue of social integration measures that shape them into deserving citizens through norms of deservedness, rather than economic redistribution.

The state's integration efforts and citizenship norms of multiracialism and diversity appear to have been internalized within some welfare organizations.

> Then we bring them out for this learning journey, where we get them actively involved through participating and bonding with the locals through local events: lantern festival, Chinese New Year celebration, Hari Raya[8] and all that. And of course, our ultimate aim is to get them involved in volunteerism. No point getting a pink IC[9] and be part of Singaporean. But most important, you must get yourself emotionally involved, because this is your country and no better way than volunteerism. (Informant D, volunteer with a local CDC)

> For me [and] my friends, [we feel the] issue is integration. Not only back home, in office they integrate with locals. Most time is spent with colleagues. (Informant E, new migrant to Singapore who is volunteering with a local community development group)

The explicit tutelage of integration by the state is adopted by assistance organizations and their volunteers. This reflects a centralized response to ongoing diversification that addresses national interests rather than improved labor standards and greater equality for all newcomers. As such, new arrivals become subject to governmental reform. Assistance becomes a set of relations of power, practices, technologies, and rationalities that regulate subjectivities involved in both the government of others and self-government and to render this regulation normal. As Ettlinger (2007) argued, "At issue is legitimizing the social relations that permit and nurture similarities and differences within and among groups, (rather than) the boundaries that homogenize and separate" (340). The notion of deservedness hence not only shapes people's access to economic assistance but also contours citizenship and belonging. This latter point is especially cogent with regard to new citizens and is reinforced through encounters such as the interactive "learning journeys." Measures and practices to integrate attempt to shape the deserving new arrival, reaffirming the boundaries of citizenship.

We argue that a purely economic reading of precarity is reductive. The production of precarity and the precarious subject is far more expansive and relational. It emerges through the governance of the matrix of race, citizenship, and labor relations and by calibrating access to social assistance and community safety nets permissible to noncitizens (low-waged migrant) and new citizens (newcomers). The precarious subject is positioned just so, precisely because of his or her relationship to layers of differential inclusion.

Conclusion: Normative Structuring of Precarity in Singapore

At a time when cities around the world are experiencing austerity cuts in welfare and various forms of public assistance, the Singaporean state continues to invest in public support. As Derrida wrote, though, "There is residual violence of the hospitable gesture, which always takes place in a scene of power" (as cited in Leung and Stone 2009, 193). We have illustrated how the aggregate politics of race, class, citizenship, and migration together shape the contours of precarity in Singapore. The politics of these contours are located at the intersections of citizenship and race at the policy level as well as reproduced through the interactions between staff and volunteers of assistance organizations. Taken together, these practices and discourses produce the precarious subject in the zone of assistance.

How citizens and citizenship are defined and experienced in relation to other people and the state are particularly urgent to address because this shapes the definition of precarity and experience of living in the global city. The structuring of assistance is also divided by race, as seen by the overarching ideology of multiracialism that situates self-help groups. Increasingly, populations in global cities are composed of significant numbers of noncitizens. For migrants at the lower echelons of the hierarchy, their adverse incorporation into the labor market and affective encounters with middle-class volunteers through assistance often contribute to their precarity. These sociopolitical relationships situate precarity as a site of struggle.

As seen, migrants' lack of access to social support further compounds migrants' marginality in the city. Such is the consequence of parameters of social support that are inflected not only by citizenship and migrancy but by race and class as well. Belonging and deservedness in the global city are multiply determined and are constituted through the work of transforming, ordering, and governing identities, norms, and boundaries. The global city features modes of inclusion and integration that produce new sociospatial patterns of differentiation and "outside-ness." This article has demonstrated the imperative of rethinking how we understand the implications of exclusion in the precarious present.

Funding

This work was supported by the Humanities and Social Sciences Seed Funding (R-109-000-177-646), National University of Singapore.

Notes

1. This refers to what Rigg (2016, 9) called "Poverty 1.0: the residual poor," who are challenged by the lack of food, health, facilities, education, clean water, and other "basic needs." The measurement of who constitutes the residual poor draws on absolute and monetary terms based on poverty lines.
2. This is a group composed of other ethnic minorities in Singapore such as Eurasians.
3. For a detailed explication of the principles that underlie these measures, see Teo (2014).
4. See http://www.wda.gov.sg/content/wdawebsite/L209-001About-Us/L219-PressReleases/06_Jan_2009.html (accessed 2 October 2015).
5. See https://www.iras.gov.sg/IRASHome/Schemes/Businesses/Wage-Credit-Scheme–WCS-/ (accessed 2 October 2015).
6. Having said this, there has also been a greater attention paid to aiding poor citizens. Comcare Long Term Assistance, run by the Ministry of Social and Family Development, is a scheme that provides cash assistance to low-income households who have no access to stable sources of income (see http://app.msf.gov.sg/ComCare/Find-The-Assistance-You-Need/Permanently-Unable-to-Work, accessed 2 October 2015). Child care subsidies are also dependent on income, with the lowest 20 percent of household income receiving about 90 percent of subsidies (see http://app.msf.gov.sg/Assistance/Child-Care-Infant-Care-Subsidy, accessed 2 October 2015). These forms of state assistance are only available to Singaporean citizens, with limited eligibility for permanent residents.
7. For further historical context of race-based self-help groups in Singapore, see, for example Chua (2007) and Hill and Lian (1995).
8. This is the day that marks the end of the month-long Ramadan fasting.
9. This refers to the National Registration Identity Card, which citizens and permanent residents hold in Singapore.

References

Cacho, L. M. 2012. *Social death: Racialized rightlessness and the criminalization of the unprotected.* New York: New York University Press.

Chan, R. 2014. Eye on the economy: Income + wealth inequality = more trouble for society. *The Straits Times* February 11. Accessed August 3, 2015. http://www.straitstimes.com/singapore/income-wealth-inequality-more-trouble-for-society.

Chua, B. H. 2007. Political culturalism, representation and the People's Action Party of Singapore. *Democratization* 14 (5):911–27.

Chuan, T. Y. 2014. Self help groups get more government funding help. *The Straits Times* August 31. Accessed October 2, 2015. http://www.straitstimes.com/singapore/self-help-groups-get-more-government-funding-help.

Coe, N., and P. F. Kelly. 2002. Languages of labour: Representational strategies in Singapore's labour control regime. *Political Geography* 21:341–71.

Espiritu, Y. L. 2003. *Homebound: Filipino American lives across cultures, communities and countries.* Berkeley: University of California Press.

Ettlinger, N. 2007. Precarity unbound. *Alternatives* 32:319–40.

Fraser, N. 1994. After the family wage: Gender equity and the welfare state. *Political Theory* 22 (4):591–618.

Hill, M., and K. W. Lian. 1995. *The politics of nation-building and citizenship in Singapore.* London and New York: Routledge.

Koh, C. Y., C. Goh, K. Wee, and B. S. A. Yeoh. 2016. Drivers of migration policy reform: The day off policy for migrant domestic workers in Singapore. *Global Social Policy* 17 (2):188–205.

Lai, A. E. 1995. *Meanings of multiethnicity: A case-study of ethnicity and ethnic relations in Singapore.* Oxford, UK: Oxford University Press.

Lawson, V., and S. Elwood. 2014. Encountering poverty: Space, class and poverty politics. *Antipode* 46 (1):209–28.

Leung, G., and M. Stone. 2009. Otherwise than hospitality: A disputation on the relation of ethics to law and politics. *Law and Critique* 20 (2):193–206.

Lewis, H., P. Dwyer, S. Hodkinson, and L. Waite. 2015. Hyper-precarious lives: Migrants, work and forced labour in the Global North. *Progress in Human Geography* 39 (5):580–600.

Mezzadra, S. 2011. The gaze of autonomy: Capitalism, migration, and social struggles. In *The contested politics of mobility: Borderzones and irregularity*, ed. V. Squire, 121–43. London and New York: Routledge.

Mezzadra, S., and B. Neilson. 2012. Between inclusion and exclusion: On the topology of global space and borders. *Theory, Culture and Society* 29 (4–5):58–75.

Ministry of Manpower. 2017. Foreign workforce numbers. Accessed May 16, 2017. http://www.mom.gov.sg/documents-and-publications/foreign-workforce-numbers.

Olds, K., and H. W. C. Yeung. 2004. Pathways to global city formation: A view from the developmental city-state of Singapore. *Review of International Political-Economy* 11 (3):489–521.

Ong, A. 2000. Graduated sovereignty in South-East Asia. *Theory, Culture, and Society* 17 (4):55–75.

———. 2006. *Neoliberalism as exception: Mutations in citizenship and sovereignty.* Durham, NC: Duke University Press.

Perry, M., L. Kong, and B. Yeoh. 1997. *Singapore: A developmental city state.* Chichester, UK: Wiley.

Prime Minister's Office. 2014. Prime Minister Lee Hsien Loong's National Day Rally 2014 speech (English). Accessed October 1, 2015. http://www.pmo.gov.sg/mediacentre/prime-minister-lee-hsien-loongs-national-day-rally-2014-speech-english.

Rigg, J. 2016. *Challenging Southeast Asian development: The shadows of success.* London and New York: Routledge.

Saad, I. 2015. Progressive Wage Model for landscape industry launched. Channel NewsAsia. Accessed October 3, 2015. http://www.channelnewsasia.com/news/business/singapore/progressive-wage-model/1804132.html.

Scheneider, A. L., and H. M. Ingram, eds. 2005. *Deserving and entitled: Social constructions and public policy.* New York: SUNY Press.

Schram, S. 2000. *After welfare.* New York: New York University Press.

Sennett, R. 2003. *Respect in a world of inequality.* New York: Norton.

Shaw, J. E., and D. Byler. 2016. Curated collection: Precarity. *Cultural Anthropology.* Accessed July 18, 2017. http://culanth.org/curated_collections/21-precarity.

Somers, M. R. 2008. *Genealogies of citizenship: Markets, statelessness, and the right to have rights.* Cambridge, UK: Cambridge University Press.

Standing, G. 2011. *The precariat: The new dangerous class.* London: Bloomsbury.

Teo, Y. Y. 2014. Interrogating the limits of welfare reforms in Singapore. *Development and Change* 46 (1):95–120.

———. 2015. Differentiated deservedness: Governance through familialist social policies in Singapore. *TRaNS: Trans-Regional and National Studies of Southeast Asia* 3 (1):73–93.

Waite, L. 2009. A place and space for a critical geography of precarity? *Geography Compass* 3 (1):412–33.

Wan, G., and I. Sebastian. 2011. *Poverty in Asia and the Pacific: An update.* Manila, Philippines: Asia Development Bank.

Ye, J. 2013. Migrant masculinities: Bangladeshi men in Singapore's labour force. *Gender, Place and Culture* 21 (8):1012–28.

———. 2014. Labour recruitment and its class and gender intersections: A comparative analysis of workers in Singapore's segmented labour force. *Geoforum* 51:183–90.

———. 2016. *Class inequality in the global city: Migrants, workers and cosmopolitanism in Singapore.* Basingstoke, UK: Palgrave Macmillan.

Yeoh, B. S. A. 2006. Bifurcated labour: The unequal incorporation of transmigrants in Singapore. *Tijdschrift Voor Economische en Sociale Geografie* 97 (1):26–37.

JUNJIA YE is Assistant Professor in Human Geography at Nanyang Technological University, Singapore 637332. E-mail: jjye@ntu.edu.sg. Her research interests lie at the intersections of cultural diversity, critical cosmopolitanism, class, gender studies, and the political–economic development of urban Southeast Asia.

BRENDA S. A. YEOH is Professor (Provost's Chair) in the Department of Geography as well as Research Leader of the Asian Migration Cluster at the Asia Research Institute, National University of Singapore, Kent Ridge, Singapore, 117570. E-mail: geoysa@nus.edu.sg. Her research interests include the politics of space in colonial and postcolonial cities, and she has considerable experience working on a wide range of migration research in Asia, including key themes such as cosmopolitanism and highly skilled talent migration; gender, social reproduction, and care migration; migration, national identity, and citizenship issues; globalizing universities and international student mobilities; and cultural politics, family dynamics, and international marriage migrants.

17 Occupy Hong Kong? *Gweilo* Citizenship and Social Justice

Michael Joseph Richardson

The 2017 election of Hong Kong's chief executive has been the catalyst for recent campaigns for social justice. The date marks twenty years since the handover of British colonial rule to China (through the 1984 Sino–British Joint Declaration) and democracy itself is again being questioned. Ultimately, Hong Kongers are concerned with universal suffrage and specifically that the chief executive is elected from just 1,200 members of an electoral committee in a city of more than 7 million people (Census 2011). Occupy Hong Kong took hold of several areas of the city in 2014, with campaigners employing the use of nonviolence and civil disobedience to challenge social and political injustice; their mantra was "Occupy Central with Love and Peace" (OCLP). Through a postcolonial lens, this article analyzes the political engagement of fifteen white Hong Kong city workers. The biographies of the research participants differ: Some are permanent residents[1] who are the children of pre-1997 expatriates,[2] and others more contemporary economic migrants. Underpinning this research is the Cantonese term *gweilo*, which is particularly useful in explaining "whiteness" in Hong Kong, and I use it to investigate claims about their apparent apathy. Its nuanced definitions and meanings are especially significant in the postcolonial era and contribute to broader discussions of citizenship and social justice in the city. *Key Words: citizenship, Hong Kong, occupy, postcolonial, social justice.*

香港 2017 年的特首选举, 是晚近社会正义运动的催化剂。该日标示着香港 (自 1984 年中英联合声明之后) 从英国殖民回归中国后的二十年, 而民主本身再次受到质问。香港最终关心的是普选议题, 特别是在此般超过七百万人口 (2011 年普查) 的城市中, 特首仅由一千两百位选举委员会的成员选出。2014 年的香港佔中行动佔领了该城市的若干区域, 而其倡议者运用非暴力和公民不服从的手段, 挑战社会与政治不公; 他们的口号是 "和平佔中" (OCLP)。本文透过后殖民的视角, 检视十五位香港白人工作者的政治参与。本研究受访者的背景各异: 有的是 1997 年前来到香港的外籍人士的孩子、并成为永久居民, 其他则是当代的经济移民。支撑此一研究的是广东话中的 "鬼佬", 该词语对于解释香港中的 "白人性" 特别有用, 而我运用此概念来探讨对于他们显着的冷漠之宣称。其细缘的定义与意义在后殖民的年代中特别重要, 并对广泛的公民权与城市中的社会正义之讨论做出贡献。 *关键词: 公民权, 香港, 佔领, 后殖民, 社会正义。*

La elección del jefe del ejecutivo de Hong Kong en 2017 ha sido el catalizador de recientes campañas de justicia social. Esta fecha marca veinte años desde cuando el control británico fue cedido a China (por medio de la Declaración Conjunta Chino–Británica de 1984), y de nuevo la propia democracia está en entredicho. Al fin de cuentas, a los habitantes de Hong Kong les preocupa el sufragio universal y específicamente que el jefe del ejecutivo se elija dentro de un comité electoral de apenas 1.200 miembros en una ciudad de más de 7 millones de habitantes (censo de 2011). Ocupar Hong Kong tomó control de varias áreas de la ciudad en 2014, en un episodio durante el cual quienes adelantaron la campaña emplearon el uso de la no violencia y la desobediencia civil para retar la injusticia social y política; su mantra fue "Ocupar Central con Amor y Paz" (OCLP). A través de una lente poscolonial, este artículo analiza el combate político de quince trabajadores blancos de la ciudad de Hong Kong. Las biografías de los participantes en la investigación difieren entre sí: algunos son residentes permanentes,[1] hijos de expatriados pre-1997,[2] en tanto otros son migrantes más contemporáneos, de motivaciones económicas. Esta investigación se apuntala en el término cantonés *gweilo*, el cual es particularmente útil para explicar "blancura" en Hong Kong, y lo uso para investigar reclamos sobre la aparente apatía de los blancos. Sus variadas definiciones y significados son especialmente significativos en la era poscolonial y contribuyen a dar más amplitud a las discusiones sobre ciudadanía y justicia social en la ciudad. *Palabras clave: ciudadanía, Hong Kong, ocupar, poscolonial, justicia social.*

Occupy Hong Kong focused on the rights to universal suffrage, a fair and transparent electoral system, and ultimately an open democratic process. The campaign was marked by the use of yellow umbrellas, protective face masks, and protective eye goggles. These symbols of pacifism were employed against the violent use of pepper spray and tear gas by police. #Occupy Central became the epithet for the movement via social media with the umbrella as its emblem. Prodemocracy movements have been active in Hong Kong for some time, with the 2017 election of Hong Kong's chief executive being the latest catalyst in campaigns for social justice (Tse 2015, 2016). Newly appointed Chief Executive Carrie Lam has been greeted as another puppet of Beijing, after serving as the second in command to the previous chief executive, Leung Chun-ying.

In the twenty years since the handover of British colonial rule to China in 1997 through the 1984 Sino–British Joint Declaration, democracy has been repeatedly questioned. The people of Hong Kong have been striving for universal suffrage, challenging a process that sees the chief executive elected from just 1,200 members of an electoral committee in a city of more than 7 million people (Census 2011). Occupy Hong Kong took hold of several areas of the city in 2014 through the use of nonviolence and civil disobedience in a campaign against social and political injustice. Ortmann (2015) asserted that the "very slow progress of democratic reforms in Hong Kong is, however, due to the ruling elite" (32), made up of the authoritarian government in China and Hong Kong's senior business figures. Both share concerns over greater autonomy in the Special Administrative Region (SAR), albeit for different reasons. Chinese authorities fear that greater autonomy in Hong Kong will add pressure to its own one-party leadership, whereas business leaders are concerned that more representative politics will weaken their hold in the city through a greater distribution of wealth and more stringent regulation. Even in early commentaries of Occupy Hong Kong, questions were raised about the emphasis placed on inequality within the movement:

> Popular analyses of burgeoning political agitation around universal suffrage in Hong Kong often side-step an inconvenient reality—that the underlying story is not simply about relations with the mainland or concerns over its authoritarian ways, but rather about massive social inequality and the diminishing opportunities available to many Hong Kongers. (Carroll 2014)

Of course, investigating the relationship among inequality, citizenship, and social justice is not new within geography. Staeheli and Nagel's (2012) work on the Arab Awakening and Horschelmann and El Rafaie's (2014) study of transnational citizenship draw such connections; in doing so, they highlight how the promotion of civic engagement is often critiqued for its underlying neoliberal agendas (Basok and Ilcan 2006; Fraser 2007). By focusing discussion in this article on empirical contributions of white city workers, the work aims to shed light from an elite (neoliberal) perspective.

In investigating the absence of "whiteness" in Occupy Hong Kong, a threefold analysis emerged to explain a lack of political engagement. This postcolonial and racial lens on urban social justice in Hong Kong builds, in particular, on the work of Knowles and Harper (2009) and Leonard (2010). Intersectional analyses have been conducted elsewhere, albeit with a religious rather than racial focus (for this, see Tse 2015, 2016; Cloke, Sutherland, and Williams 2016). More specifically, though, this article responds to calls from Fechter and Walsh (2010, in a special issue of the *Journal of Ethnic and Migration Studies*) for further research into (1) who else inhabits transnational space beside expatriates and locals, (2) the perspectives of locals on expatriate experience, and (3) the changing status of white expatriates in postcolonial contexts. Cranston (2017) began to answer these key questions, in particular, on the axiomatic nature of expatriate identification. She went on to state, "We should not see the term 'expatriate' as axiomatic in describing this type of mobility, as we need to pay attention to the political context in which the term is enmeshed" (Cranston 2017, 1). In bringing together the work on elite transnational migration to investigate citizenship and social justice, this article explicates the political context.

This article explains the role of *gweilo* in relation to social justice and develops understandings of the contested nature of expatriate experience. This contributes to what Rogaly and Taylor (2010, 1336) highlighted as the "racialization evident in discussions of migration." The crux of the article's contribution builds on what Cranston raised herself in her recent article. She pointed out that work in this field (transnational mobility and migration) tends to focus first on "exploring how an expatriate identity is produced in relation to encounter, and second, modelling the expatriate as a category" (Cranston 2017, 3) and that crucial to productive analysis is to discuss this identification and categorization relative to other migrant groups. It is for these reasons that my own research is mindful of the self-declaring (axiomatic) nature; through the longitudinal nature of the work, it corroborates stated opinions, experiences, and encounters with Hong Kong Chinese and asks questions of what forms citizenship takes among these white city workers.

What follows in this article is structured around four sections and a conclusion. The first considers occupying here, there, and everywhere as the particularities of Occupy Hong Kong are set against a wider discourse of Occupy movements. The second explains the explicit role that *gweilos* play in understanding racial identity politics in Hong Kong. The third presents the threefold analysis that emerged from the empirical data (postcolonial burden, arrogant apathy, and pragmatic resignation). The fourth puts forward *gweilo* citizenship as a means to successfully integrate white city workers into Hong Kong society. This key contribution tackles the call for future research as set out by Fechter and Walsh (2010) and the development of these ideas by Cranston (2014, 2016, 2017). It is through the promotion of *gweilo citizenship*—a more politically engaged and socially responsible residency—where active participation and affective citizenry are seen to address inequality and tackle injustice. The work expands notions of citizenship and social justice by accounting for the geographies of social change in the postcolonial city.

Occupying Here, There, and Everywhere

Occupy Hong Kong gained the external label of Umbrella Revolution (due to campaigners' symbolic use of umbrellas in the face of police challenge) and officially lasted from 28 September to 15 December 2014, although physical remnants of Occupy Hong Kong were still very much part of the urban fabric when I returned to the city to lead an undergraduate field trip in March 2015. The ephemeral aspects were starting to fade (Figure 1), although the embodied occupation was still present (Figure 2). As Halvorsen (2015) noted, materiality has already been acknowledged as significant within work on the geographies of occupation, namely, through the work of Arenas (2014) and Vasudevan (2015). Indeed, before this, in a special issue of *Social Movement Studies*, Pickerill and Krinsky (2012) outlined eight reasons Occupy matters, including the making of space, the crafting and repeating of slogans, and the ritualizing and institutionalizing of protest. This making, crafting, and ritualizing of protest was almost exclusively ethnic Chinese.

In homage to Juris's (2012) work on #Occupy Everywhere, this section positions the particularities of Occupy Hong Kong against the backdrop of the global Occupy movement. As Juris (2012) advised, to better understand the effectiveness of #Occupy movements they "will have to be empirically assessed

Figure 1. Occupy poster in Mong Kok, Hong Kong. The traditional Chinese characters read in Cantonese as "I want true universal suffrage." Photo by author. (Color figure available online.)

Figure 2. Occupy encampment in Central Hong Kong. Written in both English and Cantonese, "Umbrella Movement." Photo by author. (Color figure available online.)

through ongoing comparative ethnographic research" (271). In reviewing relevant literature, I return to the first Occupy movement, in Wall Street, New York, in 2011. Roberts (2012) explained:

OWS's [Occupy Wall Street's] modus operandi was not new. Many of its organizers had been involved in anti-globalization demonstrations at trade and economic summits over the preceding 15 years. It was in the antiglobalization movement that activists refined their skill in using new communications technologies to build loosely structured protest networks. (755)

OWS was born, then, out of a volatile political climate that followed the Arab Spring as well as the Spanish *indignados* (antiausterity) movement (Abellan et al. 2012). Whether via Twitter revolutionists or a traditional street protestors, Occupy held a significant presence and continues to carry traction within the current political climate (Deluca, Lawson, and Sun 2012). The term has become ubiquitous with social movements everywhere and carries with it an interpellation, as Althusser (2001) would attest, that which brings into being. The movement creates a sense of global citizenship, mobilizing, for example, much more localized issues such as protests against student tuition fees (Hopkins and Todd 2015). #Occupy encourages active citizenship (see later section on *gweilo* citizenship) either in support of or in opposition to, discrediting a passive response by its very provocative stance. Yet it is this passivity that shaped much of my research findings. Those not occupying Hong Kong were not doing so as conscientious objectors; rather, their elite status (enabled through white privilege) created an environment where disengagement was simply normalized and deemed apolitical.

Butler (2015) eloquently stated, "Revolution sometimes happens because everyone refuses to go home" (as cited in Shaw 2017, 117) and the production of these spaces are, as Pickerill and Chatterton (2006) claimed, "spatio-temporal strategies" (735). Within occupations of urban public space is, however, the issue of rights to the city. The ability to claim public space is always power laden, and citizens of Hong Kong have a unique relationship with its public space. Hong Kong is well versed in the occupation of public space (namely, Filipino domestic workers on their day off every Sunday), although it is less used to the occupation of its public spaces by Hong Kong Chinese. Mention here of the more mundane and everyday occupations of public space in Hong Kong speaks to Fechter and Walsh's (2010) question of who else

inhabits transnational space. These "weapons of the weak" (Scott 1985, 1) are employed every Sunday, yet in Occupy Hong Kong it is the middle-class and professional composition of the social movement that ensured that it gained traction. Equally white city workers, labeled in other studies as "mobile professionals," "privileged migrants" (Fechter and Walsh 2010, 1199), or "skilled international migrants" (Cranston 2017, 1), tend to congregate in private, elite spaces within the city. What is important to note in Occupy Hong Kong is the collective mobilization of well-educated, politically oriented Hong Kong Chinese. This alters power structures, making these occupations the weapons of the well-educated. Social justice was being called for by good, hard-working Hong Kong citizens, which makes them harder to ignore and more difficult to silence.

A student strike in late September 2014 (objecting to Beijing's decision to impose restrictions on the election of the chief executive of Hong Kong) proved to be the stimulus. A demonstration had been planned for over a year beforehand, however, when Occupy Central with Love and Peace (OCLP) first announced (January 2013) an occupation of a main road in the city's commercial district (Central) if the government refused to implement universal suffrage. This had been promised by Chinese government authorities as early as 2007. OCLP was organized by academics Benny Tai, an Associate Professor of Law at the University of Hong Kong, and Chan Kin-man, an Associate Professor of Sociology at the Chinese University of Hong Kong, along with Reverend Chu Yiu-ming, Minister for the Chai Wan Baptist Church in Hong Kong and Chairman of the Hong Kong Democracy Network. Nevertheless, Ortmann (2015) noted:

The Umbrella Movement is deeply rooted within Hong Kong's political history and its protracted democratization process. While the British colonial administrators envisioned democratization immediately after World War II, opposition from the Chinese government brought these early attempts to an end. (33)

What followed OCLP's conception was a series of public debates, a well-received—yet unofficial— referendum, scenes of civil disobedience, and what became the largest mass protest movement Hong Kong has ever witnessed. I was gripped, not least because of the undergraduate field trip I run every year to the city but more widely because the movement gained international recognition and was widely praised for its promotion of positive and peaceful protest. Hong Kongers I spoke with proudly boasted of the significance of their peaceful protest, which was only heightened in the wake of recent more violent protests elsewhere around the world (e.g., UK riots in 2011; Black Lives Matter protests across the United States in 2016). Indeed, the priority of OCLP sets it apart from earlier Occupy movements that were marked by such violence (Gitlin 2012; Roberts 2012). Chris Hedges (2012), a prominent figure within OWS, wrote about the disunity and disruption caused by Black Bloc anarchists and described them as "the cancer in Occupy"—believing that the criminal behavior holds back the progressive ambitions of Occupy more broadly.

As has been explained, although the official occupation of Central Hong Kong lasted for seventy-nine days, its presence was felt for significantly longer in Central as well as across the city in the areas of Admiralty, Wan Chai, Causeway Bay, and Mong Kok. Indeed, legacies of this movement continue to shape Hong Kong and dominate the political landscape between Hong Kong and the outside world (Hoyng and Es 2014; Ortmann 2015). Interestingly, a more militant, localist (Hong Kong Chinese) movement has taken on a greater significance in the aftermath of Occupy Hong Kong, where points of tension often revolve, however, around border politics and the lax governance of mainland Chinese tourists.

The Role of *Gweilo*

The Hong Kong movement was primarily driven by demands for universal suffrage and challenging the knock-on effects of the business-oriented—unelected—Hong Kong leadership. These were blamed for the exacerbation of social inequalities and, more specifically, pushing living costs ever higher and blocking routes to social mobility. Consequently, a vicious cycle is formed by the economic, and often white, elite; their presence in the city is a major factor in increasing social polarization. Ironically, it is the very pressures of long working hours and corporate neoliberal culture of "Asia's global city" that led many white city workers to grow tired of the disruption that Occupy Hong Kong brought to their lives.

The empirical evidence in this article is threefold, with the political engagement of white city workers in Hong Kong mirrored by their nuanced appropriations of the term *gweilo*. A later section analyzes the research findings in relation to social justice through an exploration of active and affective citizenship. By way of introduction to the analysis, however, the first theme that emerged is that of *postcolonial burden*, a phenomenon

marked by white privilege as an inhibiting factor to political engagement. This position becomes manifest as a contested identity, one where some efforts are made to integrate (e.g., the speaking of Cantonese or cultural consumption).

Second, and more critically, there is what I perceived as *arrogant apathy*, an identity that is epitomized by the manufacture of a craft beer (see Figure 3). In addition to the literal definition provided in Figure 3, participants spoke of more colloquial understandings of the term *gweilo* such as "white devil," "white monkey," "crazy white person," "Western foreigner," "ghost person," or "ghost foreigner." One participant (white British who was born in Hong Kong), who felt the postcolonial burden particularly heavily, stated:

> For me it means a spoilt, overprivileged, wealthy, egotistical white man.

Moreover, and irrespective of an exact definition, the appropriation demonstrates an embrace of a derogatory term, which in turn, is rendered impotent; such is the power of white privilege.

The third strand of analysis is conceptualized as *pragmatic resignation*. This stance is marked mainly through

gwei·lo

(鬼佬)(gweɪləʊ),*n.*,*pl.*-**gweilos**
1. *lit.* Cantonese term meaning 'ghost chap'. **2.** *hist.* Cantonese slang term used to describe barbarian hedonistic invaders of Canton in the 16th Century. **3.** *mod.* Slang term used by many in the first and third person to describe foreigners. **5.** *trademark.* Used to denote exceptional craft beer brewed in Hong Kong with passion, flare and above all, modesty. *phrase* A chilled and full-bodied gweilo can be surprisingly sophisticated.

Figure 3. Label of Gweilo Beer. Reproduced with permission. *Source:* http://gweilobeer.com/.

an economic view of life in Hong Kong (inherently linked to the working identities of many white residents, especially those without permanent resident status). The pragmatic resignation—or acceptance—that the Occupy movement was futile stems from an attitude that the sun has already set on the Hong Kong they have known and valued (Phillips 2014). When speaking with an academic colleague (who is Hong Kong Chinese), however, I was told that in this tense political climate, there was no hope for implementing change and that he was already looking at work opportunities in other countries. From an economic perspective, white city workers pointed to significant statements such as Hong Kong and Shanghai Banking Corporation (HSBC) maintaining its headquarters in London and rejecting a move back to Hong Kong (BBC 2016). Furthermore, the aforementioned colleague pointed to the control at the top levels of Hong Kong universities already based in Beijing. Those who adopt this pragmatic position perceive Occupy Hong Kong as profligate. This position already sees Beijing's influence as being too strong, with no way to reverse the process.

Sensitivities to these concerns are inevitably heightened within the postcolonial context; researching as a white British man in Hong Kong elevates an awareness of my embodied presence. More appositely, the recent work of Griffiths (2017, 2) raises the notion of postcolonial positionality, the turn inward to further investigate postcolonial theory, or, put simply, "postcoloniality" (see also Madge 1993; Potter 1993; Siddaway 2000; Besio 2003). This approach, itself informed by postcolonial feminism (Sharp 2009), helps to explain the white privilege of myself as researcher and also crucially of my participants. Positionality has long been debated; indeed, I have written elsewhere about the importance of the shifting nature of researcher positionalities (Richardson 2015a, 2015b) and note that some self-reflexive writing has been critiqued as self-indulgent (Kobayashi 2003; Mansson McGinty, Sziarto, and Seymour-Jorn 2013). I clarify its purpose in this research as helping to articulate the perceived privilege that white academic status permits and how it affects a study of the city's gweilos. As Longhurst, Ho, and Johnston (2008) stated:

> In discussing our own bodies as researchers and our participants' bodies, we can begin to establish relationships. We situate ourselves not as autonomous, rational academics, but as people who sometimes experience irrational emotions including during the course of the research. Emotions matter. This enables geographers to begin to talk from an embodied place, rather than from a place on high. (213)

In this research, where I conducted participant observation with white city workers in their homes, workplaces, and social spaces before, during, and after Occupy Hong Kong, relationships were of course established. Based primarily on participant assumptions, this "place on high" was significantly lowered. I, like my participants, was subject to the label of gweilo. I visited their offices, homes, and social spaces around the city (both city bars and coffee shops) as well as private members' sports clubs. I was afforded the same privileges as my participants as I experienced the city. Similar reflections are drawn in Cranston's (2017, 5) work in Singapore, where she discussed the "white man's wave," an intercultural encounter where whiteness permits access to private spaces.

Participants read a lot into my identity, in particular by drawing on notions of what Bourdieu (1984) called social and cultural capital: educational attainment, sporting interests, and embodied characteristics. Through an analysis of these assumptions, an embodied politics emerges as important to the interrogation of social and cultural capital. My white, British, male academic status saw me often read as an equal. The labeling as gweilo therefore takes place internally, as well as externally across racial divides. Although not adopting the oppositional subject positions (and racialized diaspora perspectives) of Jazeel (2007) or Noxolo (2009), what marks this research is the assumed homogeneity of researcher and participant. Although acknowledging that my very presence in Hong Kong positions me somewhere on what Besio (2003, 28) called the "coloniser–colonised" continuum, I claim some critical distance. Like Griffiths (2017, 4), I would state that the horrors of the British Empire are "not my skeletons," although I heed the advice of Griffiths—and Reay (1997) before him—in warning of the overassertion of classed difference. I instead seek to articulate difference on a more practical level. At times, discussions I observed—and was expected to participate in, especially when revolving around working in Hong Kong—were conducted as if in a foreign language. The numbers, percentages, and figures woven into the everyday speak of working life were alien to me. I could converse freely with the two teachers of the study, but with the thirteen others who worked for, and as, the city's business elite, this was lessened dramatically (a difference also noted in the work of Knowles and Harper [2009] and Leonard [2010]).

Even as I sat in a coffee shop in Hong Kong making revisions to this article, I was overhearing conversations spoken in English with content almost as impenetrable to me as the baristas' Cantonese. My presence in the company of these white city workers therefore "makes the familiar, strange" (Mannay 2010, 91). I look like a *gweilo*, but I cannot contribute to the city in the way I am expected to. This led, I argue, to open and frank discussions of citizenship and social justice that would otherwise have not emerged in a less critical analysis of racial homogeneity.

Themes of Analysis

As earlier stated, the critical discussion in this article is structured on three key standpoints that emerged from the data collected: postcolonial burden, arrogant apathy, and pragmatic resignation. Each section is covered with empirical observations as well as informed by relevant literature.

Postcolonial Burden

With regard to Occupy Hong Kong, the position of postcolonial burden is performed by empathetic onlookers. An attitude emerges that the Occupy movement is not their fight, although equally, it is opined that their presence would be damaging to the movement. This idea insists that a more prominent white presence in the movement could enable a mainland Chinese critique that Occupy Hong Kong is a hangover from an imperial past. There is also a feeling among some participants that ethnic Chinese Hong Kongers would not welcome their involvement due to what is claimed by some participants to be an inferiority complex.

An oft-cited example is the use of spoken English in the workplace. When a participant leaves the room, his or her colleagues revert back to talking in Cantonese (although this in itself has its own level of complexity with the increasing pressure that Mandarin presents). Participants who feel a postcolonial burden recognize offense in the term *gweilo* and, although resenting it, understand it. They can explain their distance from the movement and lack of public engagement in the following terms:

> Since arriving in HK it is very clear that although I may work and live in HK, any further interaction with the locals beyond work would be quite limited. Culturally the locals are very different. They like different food, have different interests, have been educated differently, have been conquered and ruled by different countries (British, Japanese, Chinese) and

all very recently. Most live with their entire families due to the cost of living. They spend their weekends watching movies and "resting," whereas the expats will try to get out of their apartment, drinking or playing sport.

> I am very mindful that I am in their country. I think that there are a lot of locals that despise the expat community, we generally get paid more, take better jobs, and we have the ability to leave and go home if things go wrong. We drive up the cost of living (housing, food, transport). That being said, there are a lot of HK locals that are very appreciative of the expats. The British colonialization has led to vast fortunes and good business opportunities for most. (White Australian, lives and works in Hong Kong)

This extract typifies the position of postcolonial burden. Responses are engulfed in a sense of guilt, an awareness of their privilege. This position of postcolonial burden is more likely to be adopted by permanent Hong Kong residents, although as the preceding example serves to explain, the pressure can still be felt by newer migrants. This participant has been in Hong Kong less than five years and he feels comfortable in Hong Kong, and this is in no small part due to his white privilege. The same levels of workplace isolation and racially segregated social life would make—and indeed do make—life very difficult for nonwhite migrants.

What also becomes clear through my research is a hierarchy that is linked to length of residency in Hong Kong. Leonard (2010) identified this also and pointed to prehandover workplace perks (more favorable work contracts, free education for children, spousal allowances, etc.) as leading to superior attitudes over newer workers. Within my research this differentiation stems from individuals who are from families who have cross-generational experience of the city. These people have grown up, attended school, or worked in the city for most of their lives:

> It's tough, as unfortunately a lot of white people think they are better than locals as they have moved to Hong Kong to live the good life, and some people believe that locals are a lower grade/class. It raises the argument why do people who have moved to Hong Kong from the Western world call themselves "expats" and anyone who moves from a poorer country are considered "immigrants"? There unfortunately is a presence of a superiority complex amongst foreigners who have moved to Hong Kong with very high salaries who get drunk in Lan Kwai Fong and Wan Chai and think they own Hong Kong. On the contrast there are locals who believe that *gweilos* have everything presented at their feet and that

they've never had to work for anything in their life. So it is a very delicate situation and one that is hard to find a balance, especially for someone like my brother and myself. We see Hong Kong as home but we aren't seen as locals by Chinese locals and we are seen as expats by all; but we don't feel like expats. (White British, born in Hong Kong)

These participants are aware of differences between themselves and newer expats (Walsh 2006, 2011; Leonard 2007; Cranston 2014, 2016, 2017) and are keen to stress the distinction, but they also claim that this goes unnoticed by most Hong Kong Chinese. Indeed, what I argue is an exclusive racialized (and to a lesser extent classed) identity is best captured by a recent article in *The Guardian*:

It's strange to hear some people in Hong Kong described as expats, but not others. Anyone with roots in a western country is considered an expat ... Filipino domestic helpers are just guests, even if they've been here for decades. Mandarin-speaking mainland Chinese are rarely regarded as expats. ... It's a double standard woven into official policy. (Koutonin 2015)

This is where political engagement can prove particularly problematic for those participants who feel a postcolonial burden. Some are keen to distance themselves from the label of expat, which carries with it an elitist notion. In a way different to Leonard's (2010) findings, length of residency created an investment in the city beyond monetary terms. The prehandover perks, without doubt, can be generously quantified, but their effect as "tied employment" begins to develop a greater emotional citizenry. As the preceding example indicates, this is more the preserve of longer term residents over their newer compatriots. Currently, it is much more likely to become ingrained for the longer term migrants through schooling, housing, and permanent resident status (Ku and Pun 2004). Yet despite an empathetic outlook and the beginnings of emotional citizenry, this fails to translate into political action. The inaction of these individuals led to a compliance with the oppressive system and continues to mark Hong Kong society as unjust.

Arrogant Apathy

The position of arrogant apathy that emerged from the research fits with the stereotypical Western playground attitude that is rife in colonial discourse (Phillips 2014). Participants here would point to the "natives" (to mean ethnic Chinese Hong Kongers) as not wanting their involvement and would intimate that their Western integration would not be welcomed. The attitude is, in my view, ignorant to structural barriers that prevent integration. It is also more damaging to the social movement than the guilt or ambivalence of a postcolonial burden. Equally, the position enables individuals to portray that they are excluded from engagement and social integration rather than acknowledging their elitist, apathetic, and self-exclusionary practices:

I've been here for over eighteen months and I only tried dim sum for the first time last week. Thing is, the natives don't really want us to integrate. They don't really drink either and so we don't have much common ground. I socialize mainly through the expat community but I can't say that I stray far from that community at all having not spoken any other languages apart from Australian and American! (White British, lives and works in Hong Kong)

These individuals see no offense in the term *gweilo*, as no harm is felt by it. Such is their elite standpoint; they see themselves as untouchable. They are, in many ways, "bubbled" and carry this social vacuum throughout their lives in Hong Kong, as the preceding flippant remarks indicate. They have no emotional connection to the city. These types of responses were limited to a few individuals but are nevertheless significant, especially when remembering that they were elicited through a questioning of solidarity with a host society. Ortmann (2015) wrote of the "symbiosis" of the political and business elites and that because "the colonial government did not enjoy the legitimacy of an elected government, they sought to co-opt this business elite instead" (37). Certainly the attitudes of some individuals carry this legacy, and this in essence is what the arrogant apathetic standpoint captures.

White city workers have enjoyed great opportunities in Hong Kong; fears that this would disappear in a postcolonial setting have been allayed and they continue to command large salaries and enjoy an elite existence (notwithstanding some of the restrictions to this as pointed out by Leonard 2010). This research offers similar stories of social mobility and class progression for white expatriate workers who managed to gain employment and confirms that race supersedes class in dividing society in Hong Kong. Whereas apathy runs throughout the sample, arrogance does not.

Pragmatic Resignation

For some, Hong Kong was and will only ever be seen as a Western playground. The city was there to be used and it was fun while it lasted; so the narrative goes. Hong Kong is not (yet) dead for these individuals, but their contingency plans, escape routes, and next ventures are already lined up. The upwardly mobile nature of these white city workers promotes a conscious disengagement from Occupy Hong Kong, although this is more about nurturing a skilled migrant trajectory than any critical comment on the social movement.

A key point to note is that although public support for the seventy-nine-day Occupy Central remained high, among my participants, their empathy waned primarily due to everyday practical issues (most commonly a prolonged commute). Similar experiences have been felt in and among Occupy movements elsewhere. For example, Gessen et al. (2011) described neighboring residents of the Zuccotti Park site in New York driven "apeshit crazy" (57) by what Roberts (2012) explained as "an unrelenting circle of drummers [who] alienated sympathetic neighbors" (757). Even the collective, Writers for the 99% (2012, 64) acknowledged "a real sense of mutual antipathy" that emerged between local residents and OWS protestors. Additionally, in my research, participants talked of a "tiresome" campaign that affected their access to public transport, "which is not what you want after a long day at work."

On a visit to one participant's workplace (shortly after the main Occupy encampments had been officially dispersed) the issue of commuting was raised. As we stood at the base of a prominent skyscraper in Admiralty, we were already towering above Queensway, the main road through Hong Kong's central business district. Both as a symbolic—and literal—vantage point, the research participant, through his bemoaning of the banality of commuting, highlighted the extent to which elite white city workers' priorities do not align with this campaign for social justice. Compromise and sacrifice are the means to progressive politics and were required during Occupy Hong Kong, yet they seem alien (and the antithesis) to most white city workers. Here it is the uncompromising selfishness of big business that is king.

In this sense then, it is the embodied practice of occupation and the disruption caused by the emplacement of these bodies that challenges the metabolism of the urban environment. With mobility as the "pulse of the community" (Burgess, cited in Sheller and Urry 2000, 740), immobility challenges not only the metaphorical heartbeat of the city but the physical movements of its people and therefore the support of nonprotesting residents.

Toward a *Gweilo* Citizenship?

Whether citizenship is conceptualized as a constitutional-legal status tied to certain rights and responsibilities or practiced by people as they negotiate their identity formation and spaces or communities of belonging, or even as embodied, emotional, and affective, it can serve a role in tackling social justice (Ku and Pun 2004). According to Johnson (2010), "The broad concept of 'affective citizenship' used here emphasises that the recognition and encouragement of emotions has long been part of the very way in which citizenship itself is constructed" (496). Johnson continued by explaining that the affective qualities of citizenship are explored through the ways in which intimate emotional relationships are governed by the state; as well as how citizens are encouraged to feel about each other in public domains. Drawing from Askins's (2016) work on emotional citizenry and intentional spaces of intercultural encounter, it is the everyday geographies of befriending and belonging where "such citizenry is engendered through local spaces though not place-bound, and enacted via desires to belong and quests for recognition" (525).

Conventional notions of citizenship posit that political awareness along with emotional intelligence is valuable for social cohesion and that when societies lack "active" citizens, they face problems. Turner (2004, xvi), for example, commented, "Modern Americans watch more television, read fewer newspapers, and undertake less voluntary service, producing an erosion of trust and active citizenship" (xvi). The context of postcolonial Hong Kong, though, presents a different scenario, one that is shaped by a fusion of neoliberal urban governance along with Chinese authoritarianism and that leads to a different conceptualization of an active citizen. Here a narrower definition exists: one that helps explain the dominant emphasis on economic contribution and that one white city workers embody most accurately. "Homo economicus" (Turner 2004, xvi)—the rational subject who contributes to society by working hard—was inscribed into Hong Kong life through an elitist British nationality bill that (posthandover) only permitted 50,000 Hong Kong residents' right of abode in the

United Kingdom. Based on strict criteria—of education, occupational status, and age—active and affective citizenship became politicized notions (cf. Fraser 2007). I argue that it is this postcolonial and politicized context that continues to maintain *gweilo* sensibilities in Hong Kong today.

Citizenship in global cities is multilayered and Hong Kong's postcolonial perspective only renders it more nuanced. Concurrently, questions of race, class, and social justice are being asked and continue to shape the inequalities within the city. This article has asked what kinds of citizenship exists among the city's *gweilos*. It has asked what kinds of postcolonial subjects are being produced in these moments of rapid social change. Ultimately what the research has shown is that as yet, white city workers (even those who are permanent residents and Hong Kong citizens in their own right) do not make a significant contribution to achieving social justice in Hong Kong. Meanwhile, their very presence in the city continues to play a role in maintaining a sense of injustice.

Gweilos and ethnic Chinese Hong Kongers actually share some common ground and present potential for greater collaboration. Both are marked by their identity construction in the global city and are united by a *sinophobia*.[3] There is a burgeoning discourse surrounding civic and political freedoms that play out in the banality of city life in Hong Kong and both ethnic Chinese Hong Kongers and white city workers are keen to stress distance between themselves and mainland Chinese. Mundane urbanisms of etiquette, education, and civility are emphasized as being uniquely Hong Kongese qualities in comparison to the unruly, unkempt, and uncivilized behaviors of mainlanders in the city. China's booming economic development (which admittedly has slowed in recent times) has brought more mainland "tourists" into the city, although as Rowen (2016) recently pointed out:

> In the complicated sovereign and territorial topology of this "contingent state" (Callahan 2004), tourism is political instrument, provocation to protest, and stage of high-stakes struggle over ethnic identity, national borders, and state territory. (386)

Mainland tourists have been blamed for further increasing living costs and cross-border trade that has caused food and other product shortages. In the example that Garrett and Ho (2014) cited—the advertisement of mainland tourists depicted as locusts in Hong Kong—we see the epitome of anti-Chinese resentment. Rowen (2016) stated that these demonstrations

"reflected an incipient nativism" (389). It is through the embodied and intersectional analysis of issues of social justice that we see such opportunity for unity around this "local" Hong Kong identity. As Rowen (2016) concluded:

> Although many were careful to articulate their demands in the terms of demands for electoral reforms, cultural and embodied difference was still a persistent theme. "It's nice to be here with each other with just Hong Kong people. I don't think I've heard so much pure Cantonese in weeks," said a twenty-six-year-old journalist. "This is like the Hong Kong of my youth," said a forty-five-year-old salon worker. (389)

Although there is clear negativity through the expression of anti-Chinese sentiment, the promotion of localized identities presents Hong Kong society with an opportunity:

> A focus on emotions and space, as mutually co-constituting social relations, productively opens up a range of tensions: across ethnic segregation and romanticised notions of "cohesion," socio-cultural difference and ethnicity, institutional and personal relations. (Askins 2016, 525)

The localist movement is marked by controversy, with issues of more militant protest from Hong Kong Chinese; as a reaction, the pro-Beijing government has called for (and succeeded with) the removal of significant localist figures from the democratically elected Legislative Council.[4] Local Hong Kongers, both white and ethnic Chinese, could use these shared values to work more productively in developing a progressive citizenship in their global city. As just one example, the threat to the Cantonese language is significant and if white city workers made greater efforts to speak it, a stronger resistance could be presented to the seemingly relentless pressures of Mandarin. Perhaps this is too naive a hope or, as Askins pointed out earlier, "romanticised," given the influence of Mandarin in international finance. Galvanizing a localized identity can still take place, however, if the affective qualities of civility and order—such key features of Hong Kong—are proactively protected. On a more pragmatic level, should Cantonese be more vehemently defended through wider usage by *gweilos*, then unpopular secondments to Beijing offices would be less necessary; with a strong rationale behind these, the development of institutional knowledge of Mandarin for Hong Kong's multinational corporations.

Conclusion

Social justice in Hong Kong centers on struggles for democracy, representation, and rights to the city, and it is fueled by inequality and injustice. Through revanchist notions, ethnic Chinese Hong Kongers have been successful in inspiring a social movement to tackle such issues. Although this movement has "failed" to achieve its goals (which the appointment of the new chief executive signifies), its reigniting of citizenship struggles represents clear progress. Should white city workers, especially those who are permanent residents, develop their affective citizenship—through greater political engagement—then greater unity can be championed. I argue that should a coherent *gweilo* citizenship (as active and affective) unify ethnic Chinese Hong Kongers and white city workers more prominently, then an even more representative movement, focused on tackling injustice, could form and take root.

Acknowledgments

Thanks go to Professor Nik Heynen and Jennifer Cassidento for their patience and guidance as an editorial team. Thanks also to the anonymous reviewers who have without doubt improved the writing and argument of this article. Special mention is due to Dr. Raksha Pande and Dr. Graeme Mearns for comments on earlier drafts and to my dad for his invaluable proofreading skills. Above all, thanks to the people of Hong Kong who have participated in this research and who have inspired me in my own pursuits of social justice.

Notes

1. Granted permanent residency until the age of twenty-one, the children of expatriates must then establish their own resident status. They can do so by proving that they have lived lawfully for seven years in Hong Kong and have Hong Kong as their permanent place of residence.
2. Expatriates are defined by Fechter and Walsh (2010) as "citizens of 'Western' nation-states who are involved in temporary migration processes to destinations outside 'the West'" (1197). As Knowles and Harper (2009) explained, "Expat is a widely contested term" (12). Like Knowles and Harper, however, its usage is appropriate to this work, as some participants in the study use this term to refer to themselves. Others use the term to define themselves against it. Furthermore, in an intersectional approach, Bonnett's (2004) work helps explain how *white* and *Western* are often used

synonymously and these are both characteristics of the expatriate.
3. Sinophobia is a paradoxical imaginary that labels Chinese people as both weak and effeminate yet simultaneously a threatening other. Useful discussions of the term, albeit from a Mongolian perspective, are evident within the recent work of Bille (2016).
4. As Cheng (2017) reported on the Hong Kong Free Press Web site, prodemocracy campaigners are again being charged with criminal offenses for their role in the Umbrella Revolution. Despite already serving their sentences, they face fresh sentencing after the Court of Appeals decision. Clearly, due to the ongoing nature of these cases, this article cannot accurately account for them at the time of writing. Their mention here, though, goes some way to explaining the deep and pervasive sense of injustice in Hong Kong.

References

Abellán, J., J. Sequera, and M. Janoschka. 2012. Occupying the #Hotelmadrid: A laboratory for urban resistance. *Social Movement Studies* 11:320–26.

Althusser, L. 2001. *Lenin and philosophy and other essays.* New York: Monthly Review Press.

Arenas, I. 2014. Assembling the multitude: Material geographies of social movements from Oaxaca to Occupy. *Environment and Planning D: Society and Space* 32 (3):433–49.

Askins, K. 2016. Emotional citizenry: Everyday geographies of befriending, belonging and intercultural encounter. *Transactions of the Institute of British Geographers* 41 (4):515–27.

Basok, T., and S. Ilcan. 2006. In the name of human rights: Global organisations and participating citizens. *Citizenship Studies* 10:309–28.

BBC News. 2016. HSBC to keep headquarters in London. Accessed January 4, 2017. http://www.bbc.co.uk/news/business-35575793.

Besio, K. 2003. Steppin' in it: Postcoloniality in northern Pakistan. *Area* 35:24–33.

Bille, F. 2016. *Sinophobia: Anxiety, violence, and the making of Mongolian identity.* Honolulu: University of Hawaii Press.

Bonnett, A. 2004. *The idea of the west: Culture, politics and history.* Basingstoke, UK: Palgrave

Bourdieu, P. 1984. *Distinction.* London and New York: Routledge.

Burgess, E. [1925] 1970. The growth of the city: An introduction to a research project. In *The city*, ed. R. Park, E. Burgess, and R. McKenzie, 35–41. Chicago: University of Chicago Press.

Butler, J. 2015. *Notes toward a performative theory of assembly.* Cambridge, MA: Harvard University Press.

Callahan, W. A. 2004. *Contingent states: Greater China and transnational relations.* Minneapolis: University of Minnesota Press.

Carroll, T. 2014. Hong Kong's pro-democracy movement is about inequality: The elite knows it. *The Guardian Online.* Accessed April 12, 2016. https://www.theguardian.com/commentisfree/2014/jul/28/hong-kongs-pro-democracy-movement-is-about-inequality-the-elite-knows-it.

Census. 2011. 2011 population census: Graphic guide. Accessed October 30, 2017. http://www.census2011.gov.hk/pdf/graphic-guide.pdf.

Cheng, K. 2017. Hong Kong justice chief insisted on pursuing harsher sentences for pro-democracy activists. *Hong Kong Free Press*. Accessed August 18, 2017. https://www.hongkongfp.com/2017/08/18/hong-kong-%ef%bb%bfjustice-chief-insisted-pursuing-harsher-sentences-pro-democracy-activists-report%ef%bb%bf/.

Cloke, P., C. Sutherland, and A. Williams. 2016. Postsecularity, political resistance, and protest in the Occupy movement. *Antipode* 48 (3):497–523.

Cranston, S. 2014. Reflections of doing the expat show: Performing the global mobility industry. *Environment and Planning A* 46 (5):268–78.

———. 2016. Producing migrant encounter: Learning to be a British expatriate in Singapore through the global mobility industry. *Environment and Planning D: Society and Space* 34 (4):655–71.

———. 2017. Expatriate as a "good" migrant: Thinking through skilled international migrant categories. *Population, Space and Place*. 23 (6):1–12.

DeLuca, K. M., S. Lawson, and Y. Sun. 2012. Occupy Wall Street on the public screens of social media: The many framings of the birth of a protest movement. *Communication, Culture and Critique* 5 (4):483–509.

Fechter, A.-M., and K. Walsh. 2010. Examining "expatriate" continuities: Postcolonial approaches to mobile professionals. *Journal of Ethnic and Migrations Studies* 36 (8):1197–1210.

Fraser, E. 2007. Depoliticising citizenship. *British Journal of Educational Studies* 55:249–63.

Garrett, D., and W. C. Ho. 2014. Hong Kong at the brink: Emerging forms of political participation in the new social movement. In *New trends of political participation in Hong Kong*, ed. J. Y. S. Cheng, 347–84. Hong Kong: City University of Hong Kong Press.

Gessen, K., S. Leonard, C. Blumenkranz, M. Greif, A. Taylor, S. Resnick, N. Saval, and E. Schmitt. 2011. *Occupy! Scenes from occupied America*. New York: Verso.

Gitlin, T. 2012. *Occupy nation: The roots, the spirit, and the promise of Occupy Wall Street*. New York: It Books.

Griffiths, M. 2017. From heterogeneous worlds: Western privilege, class and positionality in the South. *Area*. 49 (1):2–8.

Halvorsen, S. 2015. Encountering Occupy London: Boundary making and the territoriality of urban activism. *Environment and Planning D: Society and Space* 33:314–30.

Hedges, C. 2012. The cancer in Occupy. Accessed May 24, 2017. http://www.truthdig.com/report/item/the_cancer_of_occupy_20120206.

Hopkins, P., and L. Todd. 2015. Creating an intentionally dialogic space: Student activism and the Newcastle Occupation 2010. *Political Geography* 46:31–40.

Horschelmann, K., and E. El Rafaie. 2014. Transnational citizenship, dissent and the political geographies of youth. *Transactions of the Institute of British Geographers* 39 (3):444–56.

Hoyng, R., and M. Es. 2014. Umbrella revolution: The academy reflects on Hong Kong's struggle. openDemocracy. Accessed August 18, 2017. https://www.opendemocracy.net/rolien-hoyng-murat-es/umbrella-revolution-academy-reflects-on-hong-kong%e2%80%99s-struggle.

Jazeel, T. 2007. Awkward geographies: Spatializing academic responsibility, encountering Sri Lanka. *Singapore Journal of Tropical Geography* 28:287–99.

Johnson, C. 2010. The politics of affective citizenship: From Blair to Obama. *Citizenship Studies* 14 (5):495–509.

Juris, J. S. 2012. Reflections on #Occupy everywhere: Social media, public, space, and emerging logics of aggregation. *American Ethnologist* 39 (2):259–79.

Knowles, C., and D. Harper. 2009. *Hong Kong: Migrant lives, landscapes and journeys*. Chicago: University of Chicago Press.

Kobayashi, A. 2003. GPC ten years on: Is self-reflexivity enough? *Gender, Place and Culture: A Journal of Feminist Geography* 10:345–49.

Koutonin, M. R. 2015. Why are white people expats and the rest of us are immigrants? *The Guardian Online*. Accessed March 13, 2015. https://www.theguardian.com/global-development-professionals-network/2015/mar/13/white-people-expats-immigrants-migration?CMP=share_btn_link.

Ku, A. S., and N. Pun. 2004. *Remaking citizenship in Hong Kong: Community, nation and the global city*. London and New York: RoutledgeCurzon.

Leonard, P. 2007. Migrating identities: Gender, whiteness and Britishness in postcolonial Hong Kong. *Gender, Place and Culture* 15 (1):45–60.

———. 2010. Work, identity and change? Post/conlonial encounters in Hong Kong. *Journal of Ethnic and Migration Studies* 36 (8):1247–63.

Longhurst, R., E. Ho, and L. Johnston. 2008. Using "the body" as an "instrument of research": Kimch'i and Pavlova. *Area* 40 (2):208–17.

Madge, C. 1993. Boundary disputes: Comments on Siddaway(1992). *Area* 25:294–99.

Mannay, D. 2010. Making the familiar strange: Can visual research methods render the familiar setting more perceptible? *Qualitative Research* 10 (1):91–111.

Mansson McGinty, A., K. Sziarto, and C. Seymour-Jorn. 2013. Researching within and against Islamophobia: A collaboration project with Muslim communities. *Social & Cultural Geography* 14 (1):1–22.

Noxolo, P. 2009. "My paper, my paper": Reflections on the embodied production of postcolonial geographical responsibility in academic writing. *Geoforum* 40:55–65.

Ortmann, S. 2015. The umbrella movement and Hong Kong's protracted democraticization process. *Asian Affairs* 46 (1):32–50.

Phillips, R. 2014. Colonialism and post-colonialism. In *Introducing human geographies*. 3rd ed., ed. P. Cloke, P. Crang, and M. Goodwin, 493–508. London and New York: Routledge.

Pickerill, J., and P. Chatterton. 2006. Notes towards autonomous geographies: Creation, resistance and self-management as survival tactics. *Progress in Human Geography* 30:730–46.

Pickerill, J., and J. Krinsky. 2012. Why does Occupy matter? *Social Movement Studies* 11 (3–4):279–87.

Potter, R. 1993. Little England and little geography: Reflections on third world teaching and research. *Area* 25:291–94.

Reay, D. 1997. The double-bind of the 'working class' feminist academic: The success of failure or the failure of success. In *Class matters: 'working class' women's perspectives on social class*, ed. P. Mahony and C. Zmroczek, 19–30. London: Taylor & Francis.

Richardson, M. J. 2015a. Embodied intergenerationality: Family position, place and masculinity. *Gender, Place and Culture: A Journal of Feminist Geography* 22 (2):157–71.

———. 2015b. Theatre as safe space? Performing intergenerational narratives with men of Irish descent. *Social and Cultural Geography* 16 (6):615–33.

Roberts, A. 2012. Why the Occupy movement failed. *Public Administration Review* 72 (5):754–62.

Rogaly, B., and B. Taylor. 2010. They called them communists then: What d'you call 'em now? ... Insurgents? Narratives of British military expatriates in the context of new imperialism. *Journal of Ethnic and Migration Studies* 36 (8):1335–52.

Rowen, I. 2016. The geopolitics of tourism: Mobilities, territory, and protest in China, Taiwan, and Hong Kong. *Annals of the American Association of Geographers* 106 (2):385–93.

Scott, J. C. 1985. *Weapons of the weak: Everyday forms of peasant resistance.* New Haven, CT: Yale University Press.

Sharp, J. 2009. *Geographies of postcolonialism.* London: Sage.

Shaw, R. 2017. Pushed to the margins of the city: The urban night as a timespace of protest at Nuit Debout, Paris. *Political Geography* 59:117–25.

Sheller, M., and J. Urry. 2000. The city and the car. *International Journal of Urban and Regional Research* 24 (4):737–57.

Siddaway, J. 2000. Postcolonial geographies: An exploratory essay. *Progress in Human Geography* 24:591–612.

Staeheli, L., and C. R. Nagel. 2012. Whose awakening is it? Youth and the geopolitics of civic engagement in the "Arab Awakening." *European Urban and Regional Studies* 20 (1):115–19.

Tse, J. K. H. 2015. Under the umbrella: Grounded Christian theologies and democratic working alliances in Hong Kong. *Review of Religion and Chinese Society* 2 (1):109–42.

Tse, J. K. H., and Y. T. Jonathan. 2016. *Theological reflections on the Hong Kong umbrella revolution movement.* New York: Palgrave Macmillan.

Turner, B. S. 2004. Foreword. In *Remaking citizenship in Hong Kong: Community, nation and the global city*, ed. A. S. Ku and N. Pun, xiv–xxi. London and New York: RoutledgeCurzon.

Vasudevan, A. 2015. The autonomous city: Towards a critical geography of occupation. *Progress in Human Geography* 39 (3):316–37.

Walsh, K. 2006. "Dad says I am tied to a shooting star!" Grounding research on British expatriate belonging. *Area* 38 (3):268–78.

———. 2011. Migrant masculinities and domestic space: British home making practices in Dubai. *Transactions of the Institute of British Geographers* 36 (4):516–29.

Writers for the 99%. 2012. *Occupying Wall Street: The inside story of an action that changed America.* New York: O/R Books.

MICHAEL JOSEPH RICHARDSON is a Lecturer of Human Geography in the School of Geography, Politics and Sociology, Newcastle University, Daysh Building, Newcastle Upon Tyne NE1 7RU, UK. E-mail: michael.richardson@ncl.ac.uk. His research interests include citizenship in the postcolonial city of Hong Kong as well as the geographies of gender and intergenerational justice.

18 Land Justice as a Historical Diagnostic

Thinking with Detroit

Sara Safransky

Debates around urban land—who owns it, who can access it, who decides, and on what basis—are intensifying in the United States. Fifty years after the end of legally sanctioned segregation, rising rents in cities across the country are displacing poor people, particularly people of color. In this article, I consider debates around land in Detroit. Building on work in critical race studies, indigenous studies, and decolonial theory, as well insights from community activists, I introduce and develop what I call a "historical diagnostic." This justice-oriented analytical approach illuminates the racialized dispossession that haunts land struggles and foregrounds the historical antecedents to and aspirations of contemporary land justice movements. Drawing on research conducted in Detroit between 2010 and 2012, I analyze instances when the moral economy of land becomes visible, including a truth and reconciliation process, the period when the state of Michigan placed the city under emergency management, and a tax foreclosure auction. An examination of these events reveals alternative ways of knowing and being in relation to land that we might build upon to confront displacement in cities today. *Key Words: the land question, moral economy of land, racialized dispossession, truth and reconciliation, urban commons.*

在美国，有关城市土地的辩论 —— 谁拥有土地、谁能使用、谁决定、以及根据什麼基础 —— 正逐渐加剧。终止合法进行隔离五十年后，全国各大城市不断上涨的租金，持续造成穷人流离失所，特别是有色人种。我于本文中考量底特律的土地辩论。我植基于批判种族研究、原住民研究、去殖民理论，以及社区行动者的洞见，引介并发展我称之为 "历史诊断" 的概念。此一以正义为导向的分析方法，描绘出纠缠着土地争议的种族化迫迁，并强调当代土地正义运动的历史前身与灵感。我运用 2010 年至 2012 年间在底特律进行的研究，分析土地的道德经济成为可见的境况，包含真相与和解过程、密西根州将该城市至于危机管理之期间，以及因未缴税而取消赎回权的财产拍卖。对这些事件的检视，揭露出我们认知并与土地产生关系、并以此为基础来应对当下城市中的迫迁的另类方式。 关键词： 土地问题，土地的道德经济，种族化迫迁，真相与和解，城市公有地。

Los debates alrededor de la tierra urbana—sobre quién la posee, quién tiene acceso a la misma, quién decide, y con qué bases—se están intensificando en los Estados Unidos. Cincuenta años después de que se castigara la segregación, el aumento de la renta de la tierra a través del país está desplazando a los pobres, en particular a la gente de color. En este artículo, tomo en cuenta los debates sobre la tierra que se están presentando en Detroit. Edificando desde el trabajo desarrollado sobre estudios críticos de raza, estudios indígenas y teoría descolonizadora, lo mismo que a partir de perspectivas esgrimidas por los activistas de la comunidad, presento y desarrollo lo que yo llamo un "diagnóstico histórico". Este enfoque analítico orientado por la justicia ilustra las desposesiones racializadas que acompañan la lucha por la tierra y pregonan los antecedentes históricos de los movimientos contemporáneos de justicia por la tierra, y sus aspiraciones. Basándome en investigación efectuada en Detroit entre 2010 y 2012, analizo los casos en los que la economía moral de la tierra se hace visible, incluyendo un proceso de verdad y reconciliación, en un período durante el cual la ciudad fue puesta bajo administración de emergencia por el estado de Michigan y sometida a una subasta de ejecución hipotecaria por impuestos. El examen de estos eventos pone de manifiesto maneras alternativas de saber y ser en relación con la tierra sobre los cuales podríamos construir para confrontar el desplazamiento en las ciudades de nuestros días. *Palabras clave: la cuestión de la tierra, economía moral de la tierra, desposesión racializada, verdad y reconciliación, bienes comunes urbanos.*

"It's the last day to pay," the man said to a passerby marveling at the line spilling out of the Wayne County Treasurer building in downtown Detroit. The gray winter day seemed to match the mood of the residents clutching envelopes and folders stuffed with documents as they waited for hours in hopes of keeping their homes from being auctioned. Once inside, a police officer directed homeowners, most of whom were African American, where to go to "make arrangements." "When you get to the eighth floor, you will get a number," the officer yelled. "Keep that number! Then go to the fifth floor." There residents filed into lines where they waited to settle their debts or get on payment plans to spare their homes. Each October, Wayne County, which encompasses Detroit, holds the country's largest tax foreclosure auction. Houses can be sold for as little as $500. In 2015, as many as 100,000 residents risked eviction.[1]

In the twentieth century, Detroit was not only famous for putting the world on wheels but also shaping an American Dream that celebrated homeownership. In the twenty-first century, the city stands as an exemplar of housing precarity and urban land crisis. Today, fewer than 700,000 residents occupy Detroit, which was built for almost 2 million. Following decades of deindustrialization and white flight, the city, which has a population that is 83 percent African American (U.S. Census 2010), was one of the hardest hit during the 2008 and 2009 subprime mortgage crisis. Bank foreclosures were compounded by Wayne County's tax foreclosure policy, fueling a speculative real estate market with investors snatching up property at rock-bottom prices. At the time of writing, city officials characterize an astounding 150,000 parcels—one third of the city's landed area—as vacant or abandoned. Much of this land has become de facto public through the state's tax reversion policy. The municipal government has been engaged in a concerted effort to develop land acquisition, disposition, and regularization policies as part of a broader fiscal austerity and revitalization agenda. Yet the ownership, value, and political meaning of so-called vacant land is disputed by people who live on the land, care for it, and imagine different futures on it.

Between 2010 and 2012, I spent fifteen months in Detroit studying how various actors staked claims to the city's "abandoned" land. It was a period of intense uncertainty and friction. A contentious city-wide planning process called the Detroit Works Project (DWP) had just been launched by a public–private consortium. The DWP sought to "rightsize" and "green" Detroit by dividing it into market-based zones (Safransky 2014; Akers 2016; Montgomery 2016). The resulting plan or strategic framework slated "distressed" zones for disinvestment and reimagined occupied neighborhoods as ponds and farms (Detroit Future City 2012). Simultaneously, a state-imposed emergency manager began instituting severe austerity and privatization measures, precipitating the city's bankruptcy in 2013. In response to these political-economic and territorial reconfigurations, some activists began fighting not just against urban displacement but for land justice. They joined a diverse and growing chorus of organizations, domestic and international, that have begun resisting urban and rural land consolidation, gentrification, black land loss, and the loss of Native land rights under the banner of land justice.

What do twenty-first-century social movements mean when they advocate for urban land justice? Since its classical formulation in agrarian political economy, the land question has been a catch-all phrase for concerns about land and resource distribution: Who owns what, who decides who gets what, and on what basis (Bernstein 2010; Peluso and Lund 2011)? These questions have been the subject of scholarship in the postcolonial Global South where rural social movements from the Landless Workers' Movement in Brazil (e.g., Wolford 2010) to the Landless People's Movement in South Africa (e.g., Ntsebeza and Hall 2007) have demanded agrarian and land reforms for decades, but they have received less attention in Global North cities. The racial and cultural politics of land and property in the United States are, like South Africa or Brazil, haunted by colonial conquest, historical racialized dispossession, and a state that has perpetuated white property privilege (Harris 1993). In urban geography, the land question tends to be approached from a political-economic perspective that illuminates how capital circulates through cityscapes (Smith 2002; Harvey 2006). Such analyses are critical for understanding the relationship between the production of urban space and the expansion of capitalism, as well as contemporary urban displacements. They often underemphasize, however, the politics of race and difference (Young 1998; Kholsa 2005). Political-economic frameworks alone do not give us the tools to capture the powerful feelings of historical loss and injustice associated with urban land struggles, nor do they capture the consciousness, aspirations, and claims of resistance movements on their own terms. In other words, when twenty-first-century movements talk about urban land, they are often not just talking about capital and class but also about race and colonialism.

In the 2015 Association of American Geographers plenary "What Is Urban about Critical Urban Theory?" Ananya Roy (2016) argued that "today's urban question is a land question" and emphasized the importance of attending to historical difference in the study of urban problems. Roy argued that agrarian pasts and rural land regimes are implicated in urban development today and that the not-urban, rural, or agrarian is a necessary supplement to the urbanization-of-everything theories, like "planetary urbanization" (Brenner and Schmid 2014). Following Derickson's (2009) call for "non-totalizing" theory, Roy argued that such conceptualizations of the urban should be accompanied by methodological attention to uncertainty and "undecidability" (cf. Mouffe 2000) and invites us to read the urban "from the standpoint of an absence" (inspired by critical theorist and feminist Nancy Fraser; Roy 2016; see also Roy n.d.). Building on Roy's work, I consider the land question in Detroit from the standpoint of absence.

Thinking with community activists in Detroit, I argue for a "historical diagnostic" of the urban land question. Here, *diagnostic* refers to the practices or techniques that residents use to diagnose contemporary problems and *historical* signifies their concern with how "history," to quote James Baldwin (1968), "is not even the past, it's the present." A historical diagnostic seeks to link contemporary land struggles to the historical continuum of racial capitalism (Robinson 1983) and racial liberalism (Mills 2008) and capture the ways in which social movements try to reckon with state violence by staking alternative claims to land. This approach emerged during my fieldwork out of a sense of urgency to account for the old racial tensions and historical traumas around land loss as well as feelings of uncertainty, resentment, and indignation that were surfacing as dramatic territorial reconfigurations in Detroit pushed the land question front and center in political debates. In 2011, I helped organize a participatory research project called Uniting Detroiters that brought together residents, activists, scholars, students, progressive social justice organizations, and neighborhood groups to study and discuss the emerging development agenda in Detroit, how it fit into broader national and global trends, and identify local challenges to and opportunities for transformative social change (Newman and Safransky 2014; Campbell et al. forthcoming).[2]

During a series of community meetings, we engaged in conversation and debate about the land question: To whom did Detroit's abandoned lands belong? How were they being redistributed? By what processes? Who decided? These conversations, as well as interviews that we conducted as part of the project, revealed that official categorizations of land as vacant and abandoned often contrasted with how residents materially used and cared for the land, imbued it with affective meaning, and staked claims to it. They also revealed ways of knowing and being in relationship to land at odds with dominant conceptualizations of land as surplus, exchange value, and something to be owned. Finally, they suggested that to make sense of the racial antagonisms and fierce resistance to new land governance policies, it was important to attend to both political economy and the moral economic dimensions of the urban land question as well as develop ways of thinking about "land beyond property" (Goeman 2015, 87).

In this article, I engage in a historical diagnostic of the land question in Detroit by analyzing instances when expectations of what is right and just are violated and the moral economy of land flares up. E. P. Thompson (1971) developed the concept of moral economy to explain the customs, traditions, and ethical norms that led to widespread food riots in the English countryside in the late eighteenth century when enclosures dispossessed peasants. Like Thompson's moral economy, a historical diagnostic is interested in the deep historical meanings, emotions, norms, and moral and ethical beliefs that clash in contestations around land, whether in the form of mass mobilization, everyday resistance, or public outrage, but it seeks to understand how they arise from the standpoint of historical racialized property relations and freedom dreams that are often absent from political-economic theorizations of land problems whether within the academy or policy and planning circles.[3] This entails accounting for how calls for land justice in Detroit are often as much about staking claims to alternative forms of sovereignty, political subjectivity, and personhood as they are about affordable housing or rights to landed property.

The article proceeds in three parts. The first part develops the case for a historical diagnostic. Next, I theorize the concept of property and land beyond property. The third part of the article turns to three cases where the moral economy of land flares up, including a truth and reconciliation process, the state takeover of the city, and the Wayne County land auction. In the conclusion, I turn to increasingly urgent land questions facing U.S. cities and consider the stakes of a historical diagnostic as a justice-seeking mode of inquiry and tactical response.

Racial Justice and the City

Roy's call to attend to historical difference in the study of urban transformation was a challenge to urban geographers. In the Global North, Marxist political economy has dominated geographical studies of urban space since the early 1970s when David Harvey ([1973] 2009) published *Social Justice and the City*. The landmark text drew attention to how capitalism structures contemporary urbanism and social inequality. In so doing, it expanded the scope of urban geography beyond mapping and modeling of urban spatial patterns to explain the political-economic processes behind them. One consequence of this text's lasting impact on urban geography has been a tendency to emphasize class contradictions as the driver of history and urban change in a manner that deemphasizes racial antagonisms. This legacy begs a larger question concerning how we conceptualize historical injustice in our analyses of urban problems and, for the purposes of this article, how such conceptualizations shape our approach to the land question in the city today.

When Harvey completed *Social Justice and the City* in 1973, he was living in Hampden, Baltimore. Like other industrial urban areas of the time, Hampden was experiencing factory and mill closures, economic decline and rising unemployment, and outmigration in the wake of intense urban uprisings across the country. From 1964 to 1967, every major central city in the United States with a sizable black population experienced civil disorder. There were 329 major rebellions in 257 different cities (Woodward 2003). The largest was in Detroit, where on 23 July 1967, a police raid on an after-hours bar quickly escalated to widespread protest and destruction. Within a week, 17,000 armed officials patrolled the city, more than 7,000 people were arrested, and forty-three people were dead. The uprisings radically reshaped the material landscape of the city. Property damage was extensive, with more than 2,500 buildings looted, burned, or destroyed. Although white flight was underway prior to the uprisings, it increased rapidly in the years following. Between 1967 and 1969, 173,000 residents left the city (Fine 1989). After Martin Luther King Jr.'s assassination on 4 April 1968, there were another 200 uprisings in 172 cities, including one in Baltimore in which six people died, more than 700 were injured, and 5,800 were arrested (Woodward 2003; Elfenbein, Hollowak, and Nix 2011). Inner cities across the country were literally and figuratively on fire.

Surprisingly, however, in *Social Justice and the City*, there is little mention or analysis of the "problem of the color line" (Du Bois [1903] 2007) in the United States or the uprisings as a response. It is particularly striking given how the rebellions dramatically reshaped urban politics and patterns of urbanization, serving as a reminder that the standpoint from which we theorize is critically important.

Within urban studies, there is an increasingly vocal movement informed by critical race theory,

postcolonial and decolonial theory, queer theory, subaltern studies, and feminist theory that seeks "to 'provincialize' urban theory born out of observations of European and North American cities" (Derickson 2016, 2). Instead, it aims to attend to lived experience of difference in place and the knowledge that emerges as people negotiate complex histories and struggle against oppression and for alternative possibilities. It is here that I locate my call for a historical diagnostic.

The way I am thinking about a historical diagnostic builds on sociologist Avery Gordon's call for "alternative diagnostics" to postmodern forms of analysis that account for the political-economic, institutional, and affective dimensions of modern forms of dispossession as well as imaginaries of "what has been done and what is to be done otherwise" (Gordon 2008, 18). Gordon used haunting as an analytical framework to study how abusive systems of power, particularly those that seem to be over (e.g., slavery), become a "seething presence," interrupting any neat separations of past, present, and future. When it comes to land, a historical diagnostic seeks to become attuned to the moral claims that historical dispossessions make on the present (Bird Rose 2004). In its pursuit of justice, a historical diagnostic is concerned with the task of illuminating hidden histories that point toward alternatives, decolonization, and the challenge of recuperation.[4] To this end, it is particularly concerned with uplifting the ways in which people of color have struggled for land and their contributions to the production of space, especially the ways in which such struggles bequeath ways of knowing and being that are not merely responsive but propositional.

Engaging in a historical diagnostic of the land question in the United States, then, is about attending to those utterances, viewpoints, feelings, and deferred dreams that exploded during the rebellions of the 1960s, exposing and challenging a pillar of social and political organization established during the colonial and antebellum eras that divided territory and society by race and put land ownership and, therefore, the means of production in the hands of settler colonial whites.

Seeing Land beyond Property: The Moral Economy of Land

Property relations established under colonialism and slavery and perpetuated under legal and extralegal segregation practices—from historical rural black land theft to Jim Crow segregation, racialized federal housing programs, and the reverse redlining of the 2008–2009 subprime mortgage crisis—remain a huge obstacle to democratization. Given that the majority of Americans hold and pass on most of their wealth in the form of land and home equity, land dispossession has severely reduced the ability of blacks and other people of color to accumulate wealth via property ownership (Oliver and Shapiro 2006).[5] Moreover, programs that have sought to correct such injustices by redistributing land or granting reparations—such as the Freedman's Bureau that promised to redistribute abandoned Southern land to emancipated African Americans or the Pigford class action lawsuit of 1977 that sought to reconcile the U.S. Department of Agriculture's racist lending practices from the 1930s—have failed to be realized, or compensation has been so limited that only a small number of people who can prove that they were dispossessed for a finite period of time have been compensated (Daniel 2013; Goldstein 2014).

Such compensatory promises of justice have worked, as Alyosha Goldstein (2014) argued, to *close off* and *contain* the colonial past and its history of racist discriminatory laws in a way that reproduces uneven geographies and social orders. In terms of the contemporary land question, there remains a challenge of how to hold open this history in a way that foregrounds, first, how property relations in the United States are thoroughly saturated in racism and, second, how resistance movements have fought for alternative ways of thinking and being in relationship to land. Taking up this challenge requires deep thinking about our taken-for-granted conceptualizations of property and land.

Property is typically invoked as a material object that one owns. Yet, as critical property theorists have shown, it is better understood as a bundle of negotiated social, political, legal, and economic relationships that confer value through exclusion (Hann 1998; Merrill 1998). I think of this as the relationality of landed property. A focus on relationality helps illuminate how property regimes do much more than mediate the distribution of land (Blomley 2003, 2008, 2010). They structure our relationship to the state. They order bodies in space. They bring into being political identities. They also shape how we think about belonging in relationship to one another. In this way, a focus on relationality brings attention to how property is subject forming (Strathern 1999; Pottage and Mundy 2004), suggesting, as Grace Kyungwon Hong (2014) wrote, that "propertied subjectivity is not universal." This is

to say, that propertied subjectivity is not inherent in human nature but that private property models have lived effects, particularly promoting an individual legal, autonomous subject and intersubjective severalty.

Historically, in the United States, property ownership was the path to citizenship but, of course, only those considered white and male could vote. In other words, possessive individualism structured racialized citizenship. Not only does the state enable the existence of private property but it exists, in part, to protect it (Locke 1704). This history has led, as Cheryl Harris (1993) demonstrated, to a "property interest in whiteness" and continued inequities. It has also "occluded, rendered deviant, or erased" (Hong 2014) alternative claims to land and property that challenge liberal notions of personhood, citizenship, and governance such as cooperative forms of stewardship and notions of reciprocity, for example, the idea that we should care for the land because it, in turn, cares for us (e.g., Nembhard 2014; Bandele and Myers 2017). At the same time, histories of racialized dispossession and the particular conditions of black property ownership in the United States have given rise to distinct spatial imaginaries and relationships to land (Armstrong 1994). For example, George Lipsitz (1994) termed this the "black spatial imaginary," which, among other attributes, favors use value over exchange value and supports public spaces and services. Likewise, Katherine McKittrick (2011) wrote about "a black sense of place," which illuminates how "bondage did not foreclose black geographies but incited alternative mapping practices" (949). And bell hooks (2009) reflected on how the relationship among blackness, culture, and the Kentucky landscape where she spent her early childhood forged within her a distinct sense of belonging.[6]

Attending to the narrative dimension of land claims, particularly those put forth by protest movements (Roy n.d.), illuminates the meaning that land holds for them (Tuck, Guess, and Sultan 2014). Bearing witness to these claims necessitates that we develop, following Mishuana Goeman (2008), ways of seeing land beyond property. Recognizing land beyond property and territory involves understanding land as a "meaning making process rather than a claimed object" (Goeman 2015, 72–73). Here land becomes a repository for people's experiences, aspirations, identities, memories, and visions for alternative futures. It is a site of ritual and ancestral communication. Methodologically, Goeman (2015) suggested focusing on land as a "storied" site of struggle and resistance (cf. also La Paperson 2014).

A focus on meaning making also indicates a particular way of reading history in relationship to the land question. As Goeman (2008) wrote, "Deconstructing the discourse of property and reformulating the political vitality of a storied land means reaching back across generations, critically examining our use of the word land in the present, and reaching forward to create a healthier relationship for future generations" (24). Goeman's concept of storied land is similar to the distinction Rob Nixon made between official and vernacular landscapes. Official landscapes are those of planners and bureaucrats (think the property grid), whereas vernacular landscapes are those shaped by the affective multigenerational maps of communities with, as he wrote, "all the hindsight and foresight that entails" (Nixon 2011, 18). My interest in Nixon's and Goeman's concepts of storied land and vernacular landscapes is not simply that I think they offer a more accurate description of our relationship to land but that they open up the land question beyond political-economic analysis and, in so doing, help us see ways to repair what scholar and activist Coulthard (2007) referred to as "the structural and psycho-affective facets of colonial domination" (456).

Truth, Reconciliation, and Rebellion

"A shared history is needed to claim and accept truths," said Naomi Tutu, the daughter of Desmond Tutu. She was addressing a crowd of approximately 300 people gathered in a large ballroom at the Cobo Hall convention center in November 2011 for the inaugural event of the Metropolitan Detroit Truth and Reconciliation Commission on Racial Inequality (MTRC). The MTRC, which was modeled on the South African Truth and Reconciliation Commission established after apartheid, was charged with investigating the historical roots of race-based opportunity in the Detroit metropolitan region, specifically the legacies of segregation and housing discrimination (MTRC 2011; Inwood, Alderman, and Barron 2016). The MTRC raises a critical set of questions about historical injustice and reconciliation in regard to the urban land question, not just in Detroit but in the United States.

One of the goals of the commission was to revisit the 1967 Detroit uprising and consider its impacts today.[7] The postwar struggle for Detroit is a critical historical conjuncture for understanding the city's contemporary land crisis and the structure of feeling that shapes ongoing black struggles for land, property, citizenship rights, and liberation. Cultural theorist Raymond Williams (1976) developed the concept of "structure of feeling" to capture how meanings and values are lived and felt in ways that are simultaneously structured and fleeting, inevitably moving forward while always being historically and politically informed. Whereas earlier riots were often sparked by whites defending property and jobs through violence against blacks, the uprisings in the 1960s were a response by the black community to a multitude of factors making life untenable: the racism and brutality of the white police force, the murders of key leaders of the freedom struggle and civilians alike, the promise and failure of the Great Society era and urban renewal, increasing unemployment as car factories moved toward automation and outsourcing and left the city, and segregation and redlining.

Oppressive policing tactics set off the 1967 rebellion, but in the ensuing days it turned into an assault on those who controlled housing and commerce in the community. In contrast to the 1940s, most "rioters" did not direct violence at civilians. Rather, they targeted, as Ahmad Rahman (2008) argued, "the most visible symbols of capitalism and racism" (184): property and the firefighters and policeman who were its protectors. Property—whether landed buildings or commercial goods—embodied unequal power relations and segregation in the city and the spatial isolation of African Americans. The 1967 rebellion was, Rahman wrote, "an extremely destructive attempt by the black community to violate those boundaries of 'place,' raising the question of who would rule, and under what condition" (189).

Prior to the rebellion, black radicals in Detroit had started seeing themselves as part of a global struggle against imperialism and for decolonization. New alliances linked Detroit, Cuba, and China, signaling the freedom movement's commitment to international solidarity and a growing understanding among African American intellectuals and activists that the urban ghetto was an internal colony and that land was central to the struggle for self-determination (Cruse 1968; Clark 1965; Carmichael and Hamilton [1967] 1992; Blauner 1969). Efforts to establish a territorial base for the black community were, as Russell Rickford (2017) argued, "one of the period's defining political developments" (956). For example, when Malcom X delivered his famous "Message to the Grassroots" in Detroit in 1963, he argued that "land is the basis of freedom, justice, and equality." In an essay titled "The Land Question," Eldridge Cleaver wrote, "Black people are a stolen people held in a colonial status on stolen land, and any analysis which does not acknowledge the colonial status of black people cannot hope to deal with the real problem" (Cleaver 1970, 186, cited in Fung 2014, 164).

Calls by Malcolm X, Cleaver, and others to establish a land base for the black freedom struggle were taken up in different ways. Although black agrarianism and efforts to establish rural land bases were prominent, others saw the city as the key site for land reclamation and domain of black politics. Detroit was an important center of activity. In 1966, James and Grace Lee Boggs, black radical activists, public intellectuals, and Detroit residents, published a revision of the rural black belt thesis and called for people to reclaim the city as the "black man's land" (Boggs and Boggs 1966). Two years later, 500 radicals convened at the Black Government Conference held at the Shrine of the Black Madonna church in Detroit and signed a Declaration of Independence with the aim of creating the Republic of New Afrika (RNA), an independent black nation that would occupy five Southern states within the United States (Alabama, Georgia, Louisiana, Mississippi, and South Carolina) where Afrikan citizenship could be realized (Berger and Dunbar-Ortiz 2010). The RNA rejected U.S. political structures and citizenship. They sought to operationalize Amiri Baraka's assertion that "black is a country" by "liberating" land in the rural South and claiming land in Northern cities.

Meanwhile, the Black Panther Party (BPP), which established a chapter in Detroit in 1968, deployed a territorial strategy that sought to develop solidarity networks (global in reach) and locally dispersed power centers. Like the RNA, the BPP made demands for land in their Ten Point program. They called for the overdue debt of "forty acres and two mules," for "land, bread, housing, education, clothing, justice, and peace," and for a United Nations–supervised referendum for the "black colonial subjects" to determine their "national destiny." Rather than seeking to establish a national land base, however, they focused on reclaiming institutional spaces (e.g., housing projects, schools, community centers, and prisons) and developing what they thought of as city-center communes with the goal of making liberated territories (Reyes 2009). The BPP's extensive survival programs were envisioned as a way to escape the oppression of U.S. empire through everyday social reproduction, mutual

aid, the establishment of a political base of resistance, and the production of alternative forms of community (Hilliard 2002).

The increasing emphasis on the land question among these black groups and others—like the American Indian Movement, the Chicano group Crusade for Justice, and the Chicano Mexicano and Puerto Rican group Movimiento de Liberacion Nacional—and their growing coalitional politics points to a shared understanding that the United States functioned as a capitalist imperialist system that exploited people of color at home and abroad and that the nation's settler colonial roots needed to be overturned (Berger and Dunbar-Ortiz 2010; Fung 2014). As Native American activist and scholar Vine Deloria Jr. (1969) wrote in an essay titled "The Red and the Black," "No movement can sustain itself, no people can continue, no government can function, and no religion can become a reality except to be bound to a land area of its own" (179). These groups saw land control as foundational. More than an end unto itself, land was envisioned as a means of creating society anew. While activists battled with the state through armed conflict and legal challenges, suffering death, political imprisonment, and the disintegration of organizational capacity, the land question waned, but it never died, nor did the impulse that land was the material grounds for self-determination and survival (Berger 2009; Rickford 2017).

These historical struggles condition the structure of feeling that surrounds debates over the land question today. They show up in white nostalgia for the "old" Detroit and unrelenting antiblack racism that emanates from the suburbs. They present themselves in the symbolic and cultural value of Detroit as a majority African American city. They pulse in residents' expressed love for the city and the way they claim its radical history as a source of power and resilience. They condition the ways racialized groups negotiate rising insecurity in an era of finance capital and austerity politics. They inflect calls for land justice and claims that the city has a right to remain majority black in the face of urban shrinkage and gentrification. They also shape debates over how to deal with the burden of historical violence in the present.

Historical Debt and Reconstruction

Although much hope surrounded the MTRC, it also elicited skepticism and even resistance and resentment from some community activists, particularly given its launch amidst the controversial state takeover of the city, which many saw as an attempt to loot the collective black legacy of Detroit. I draw attention to the skepticism triggered by the MTRC because it raises a question of how we might more adequately grapple not just with the legacies of segregation, as was the stated task of the truth commission, but with the "abusive colonial structure itself" (Coulthard 2007) and the immense violence exacted on indigenous, black, Latino, and other communities of color in the United States through the historical construction of private property and the racialization of space.

The MTRC, which was the only second-ever truth commission in the United States (the first was in Greensboro, North Carolina), emerged as part of a global industry advocating official and nonofficial apologies as a way to heal harms resulting from historical violence based on the assumption that truth and forgiveness help build communities anew (Coulthard 2014; Stauffer 2015). Although many grassroots activists participated in the inaugural event, others critiqued the commission for their outreach efforts and for lacking the teeth to actually do anything. Some also expressed concern that the commission would move too quickly past historical racialized violence to unity, particularly when new modes of governance and urban investment were leading to widespread dispossession and hardship for residents, epitomized by the looming threat of emergency management (Inwood, Alderman, and Barron 2016).

At the time the MTRC launched, five cities in Michigan had been appointed emergency managers. They all had majority black populations in a state where only 14.3 percent of the populations identified as black or African American (U.S. Census 2010). In Michigan, emergency managers are given sweeping powers over city finances and operations. Locally elected representatives by and large lose their power to make decisions. It is worth underscoring that Detroit's fall to emergency management in 2013 meant that over half of African Americans in Michigan essentially had their voting rights nullified, calling up a long and ugly history in the United States of white efforts to suppress the black vote that includes slaughtering African Americans for simply discussing voting and instituting poll taxes that required citizens to pay a fee to vote (Anderson 2016).

It is in this context that the memory of such unfinished struggles to claim space, citizenship, and reconstruct society overdetermined the structure of feeling in public meetings about the state takeover. Meeting rooms were often filled to capacity with discontented and concerned citizens associated with a range of groups, from African American church leaders to union representatives, seasoned organizers who were active during the civil rights and black power era, and a younger generation of community activists working on a range of issues from foreclosures to water shutoffs. During the public comment period, older activists frequently invoked the names of local black freedom fighters and emphasized the threat the takeover posed to black self-determination. One black woman who prayed before the city council exemplified these concerns and feelings of collective heartbreak: "Do not let them take away our home rule," she cried out, "our dignity. Let us stand on our own ground."

Others compared emergency management to slavery. They shamed city council members, arguing that the black political class had forgotten where they came from. Housing rights activists condemned fraudulent bank practices and racist, predatory mortgages that resulted in foreclosures. Others critiqued tax breaks given to developers. "Land and water mean wealth and power for the people," one person said. On occasion, demonstrators threatened to burn the city down again, referencing 1967, and argued that the state's violence should be met "by any means necessary," echoing Malcolm X's call to take up arms if necessary. Instead of arguing for residents to arm themselves to defend their communities, however, they urged them to "hit the streets" to fight for their jobs, pensions, homes, and schools. They also joined in protest songs, drowning out meeting proceedings with verses of "We Shall Overcome."

Their protests always seemed like they were about waging a war against forgetting the past as much as demonstrating against policies in the present. The indignation expressed at the loss of black home rule served to show how the neoliberal agenda and new austerity regimes mapped onto earlier forms of racial and colonial subjugation and how this past was not a bygone era that could be reconciled because it was a prologue to the present. In truth and reconciliation processes, philosopher Jill Stauffer (2015) has argued that resentment and resistance might actually be more restorative than forgiveness and that forgiveness might not be a goal worth reaching. "[W]e need to understand how to make judgments," she wrote, "about what can be repaired, what should be repaired, what cannot be repaired, and, perhaps, what should be left broken" (Stauffer 2015, 35).

In challenging Detroit's fiscal crisis, residents and community activists inverted the meaning of the city's indebtedness by arguing that the suburbs had a debt to the city and that the city's debt was not to the banks or the state but a historical obligation to their ancestors to carry on the liberation struggle. This kind of inversion of the debt relationship exposed, first, how property interests in whiteness that have been reified by law and privilege are perpetuated by debt; second, the inadequacy of how care and moral obligation are built into the legal and economic structures of racial liberalism; and, finally, that the people who came before them had dreams that went beyond "limited emancipation," to use Hartman's (2007) words. Honoring this debt suggested refusing the state takeover as a force that was antagonistic to black power and self-rule. It also meant channeling resentment and indignation into the cultivation of relationships, infrastructures, and community-organizing strategies that would support an antiracist urban commons, an essential part of which was securing the right of blacks and other people of color to "stay put" (Rickford 2017) in the city.

A Land Justice Paradigm

In October 2015, I received an e-mail with the subject heading "Help Us Keep Our Homes." For years, community activists have called, unsuccessfully, for a moratorium on the Wayne County tax foreclosure auction. When the auction happens, some residents who are unable to settle their debts use a variety of tactics to defend their homes and land, including staging eviction defenses and sabotaging their own property, sometimes out of rage and other times to ward off potential buyers. Another tactic is to buy back your home through the auction, which the Keep Our Homes campaign was using. The goal of the campaign, which was led by two organizations committed to antiracist and black-centered leadership, the Detroit People's Platform and the Storehouse for Hope, was to save fourteen occupied homes, keep families in them, and secure them permanently through a community land trust. (They raised enough money to purchase fifteen homes.) It emerged as part of a larger grassroots activist effort in Detroit to articulate a land justice paradigm, which involved maintaining Detroit as a majority African American city and changing how we think about home, private property, governance, and

citizenship in the twenty-first century as a necessary step in the cultivation of an antiracist urban commons.

Detroit's land justice movement was galvanized by debates over the planned ruralization of the city, in particular a contentious proposal by a white billionaire investor named John Hantz to build the world's largest urban farm in the center of the city as a way to create scarcity in land markets and drive up value. Hantz's project violated the moral economy of Detroit's vibrant food justice community, many of whose members saw farming in the city as being as much about establishing alternative forms of development as growing vegetables (Safransky 2017).

Hantz's project ignited vociferous resistance because it stood for something much larger than the 1,800 parcels that the city sold him for a mere $500,000. The project (which Hantz referred to as a "legacy project" that he hoped to pass on to his daughter) was the largest land sale in the city's history. As such, it served to perpetuate white supremacy in property relations in a city where the gap between African American and white homeownership was growing.[8] Moreover, it signaled a broader assault on democracy, the rise of a neoliberal development paradigm that disregarded black life, and the city's implicit support of land speculation. In short, the project stood for an approach to the land question that threatened how many Detroiters, in the face of state and corporate abandonment, had been working to establish visionary ways of addressing needs in their communities from gardens that fostered food security to the restorations of homes, parks, and schools that made neighborhoods more livable and grassroots campaigns to support local economies that would create a kind of city where people could live in "dignity, mutual respect, and love" (Detroiters Resisting Emergency Management 2014).

The People's Platform, among others, saw the mobilization against Hantz as an opening to extend land conversations beyond agrarian land tenure to how land-use decisions were being made, who benefited, and imagining new relationships to land. They saw the community land trust model, which creates a structure for shared ownership over space according to a defined mission and values, as one way to gain greater land control and begin to articulate a land justice paradigm. The Keep Our Homes campaign emerged from this pursuit. But why focus on homes?

In the United States, homeownership has been a symbol of democracy and a mechanism of political order. In Detroit, Henry Ford strived to turn his employees into good workers and good citizens. Private property and homeownership were central to the latter. From the 1930s onward, the federal government (in conjunction with real estate brokers, building contractors, and manufacturers of house-related equipment) also propped up this vision by supporting thirty-year mortgages, extending homeowners' insurance, offering tax incentives, and encouraging land-use and zoning policies for single-family detached houses. Although local, state, and federal policies created a severely racialized homeownership landscape, high rates of African American homeownership in twentieth-century Detroit contributed to the rise of the city as a symbolic place of black home rule and sovereignty that was now under threat. "We were a community of block clubs," as Linda Campbell, the director of Building Movement Detroit and partner on the Uniting Detroiters project, put it.

> I think at one time we were at 300 plus block clubs across the city. A lot of that was about getting my sidewalk fixed and beautification, but it was also a way that people felt connected to their neighborhoods and their communities, and felt power around that.

It is the cultivation of such power that comes from being able to claim space and "stay put," less than the preservation of individual homeownership, that motivated the "Keep Our Homes" campaign. It reflected an understanding to quote one activist involved in the Uniting Detroiters conversations,

> A home is more than a relationship to governance and taxes. It's your physical basis for your relationship to the rest of the community.

Thus, the Keep Our Homes campaign recognized that the loss of housing in Detroit was not simply about the houses themselves but was about losing relationships and community infrastructure (the closing of schools, recreation centers, and churches that often follow the loss of homes). It also recognized that the loss of these relationships translated into a loss of humanity and that the only way out was to curb the displacement of black life and at the same time articulate notions of self, community, and space that did not buy into the universalizing tendency of propertied citizenship and possessive individualism. Land justice, in this sense, then, was not just about community land control but also about imagining new relationships to land as the necessary foundation on which to reconstruct a new citizenship and new humanity.

Conclusion

In the 1960s and 1970s, a number of social movements in the United States saw land as the material basis for the struggle for collective self-determination. Today, as displacement in U.S. cities intensifies, the land question is once again gaining urgency. The 2008 foreclosure crisis continues to reverberate in home losses, widening wealth gaps, and the revival of "contract for deed" lending aimed at people who do not qualify for mortgages, particularly black and Latino homebuyers. At the same time, the reversal of white flight and return of upwardly mobile residents has caused a revaluation of land in urban centers. In Nashville, for example, where I live, there is a severe affordable housing crisis. From 2013–2017, the *average* home price in Nashville increased by 37 percent. Meanwhile, Nashville lost 5,300 affordable rental units in two years due to developers buying apartment complexes and increasing rents (Coleman and Ries 2017).

Although the challenges facing urban poor and working-class residents, particularly people of color, are formidable, there is also a resurgence of activism nationwide around urban land. Resistance takes the form of antieviction defenses, land reclamations, campaigns to organize tenant unions and increase renter power, and transnational alliances. For example, in 2006, the organization Take Back the Land established Umoja Village, a shanty town on public land in Miami-Dade County, Florida, where affordable housing was destroyed for a new condo development (Rameau 2008). In 2007, the Right to the City Alliance began work on gentrification and the displacement of low-income people, people of color, and marginalized lesbian, gay, bisexual, transgender, and queer communities from neighborhoods. They now have forty-nine member organizations across the country (Right to the City Alliance 2017). In 2009, the Western Cape Anti-Eviction Campaign in South Africa inspired the founding of the Anti-Eviction Campaign affiliates in Chicago and Los Angeles, which defend families facing eviction and take over vacant, bank-owned homes for homeless families (Chicago Anti-Eviction Campaign 2009). Most recently, on Juneteenth 2017, African American Independence Day, the BlackOut Collective and Movement Generation launched the Land and Liberation Initiative, calling for reclamation and arguing that "[l]and is essential in the fight for self-determination and liberation for Black folks" (Black Land and Labor Initiative 2017).

For those of us concerned with social justice in the city, the crises of affordability and accompanying land rights activism present a political and ethical imperative to develop a more robust research agenda on the urban land question in the United States. Such an agenda would involve collective analysis of how new urban orders under construction by the state, the market, and philanthropic foundations are actively reinscribing inequality and racial segregation into the materiality of the U.S. city. It would equally attend to the visions and aspirations of those organizing to take back the land and consider how strategic research alliances within and beyond the academy might be developed to uplift and support these efforts.

In this article, I have introduced a historical diagnostic as a justice-oriented analytical approach that aims to expose the history of racialized property relations as well as alternative land epistemologies, ontologies, and structures of organization of resistance movements. In this way, a historical diagnostic of the land question suggests the importance of bending down and "listening to the land," as Guyanese poet Martin Carter (1977) suggested. It seeks to listen to those who have sought radically new ways of belonging and being in relationship to one another and the earth. It also seeks to amplify alternative structures of value that we might build on to confront the land question in cities today.

Acknowledgments

Many thanks to all of those in Detroit whose interviews, stories, and actions shaped this piece and who have taught me so much. I am grateful for the questions and comments I received on this article when I presented a version of it as part of the Detroit School Series at the University of Michigan. Finally, thanks to Ashley Carse and two anonymous reviewers for your generous feedback and Nik Heynen and Jennifer Cassidento for your editorial support and guidance.

Funding

The National Science Foundation (Award #1203239), the Wenner-Gren Foundation, and the American Council of Learned Societies provided financial support for research and writing.

Notes

1. The opening scene comes from a description of the auction in an article by Laura Gottesdiener (2015). For more on the history of the auction, see Akers (2015).
2. As part of the Uniting Detroiters project, we produced a documentary video called *A People's Story of Detroit* (available on YouTube) and a book called *A People's Atlas of Detroit* (forthcoming from Wayne State University Press).
3. Heynen (2016a, 2016b) made a similar point in his call for "abolition ecology."
4. A historical diagnostic is inspired by calls for "recuperative histories" from Bird Rose (2004), "secretive histories" from McKittrick (2013), and "legacies of ethical witnessing" from Ioanide (2014).
5. The largest gap in wealth transferences is between blacks and whites. It is also important to point out variation among black households, however. As Martin (2009) argued, the limited ability of blacks to transfer wealth from one generation to the next through the accumulation of property and other assets is a particular experience of being African American (vs. other black ethnicities) in the United States. She found that African Americans had the lowest likelihood of interest, dividends, and rental income of all black ethnic groups.
6. See also the Black/Land Project (http://www.blackland project.org), which gathers and analyzes stories about the relationship between black people, land, and place.
7. When talking about the uprisings of the 1960s, the distinction in terminology between riots and rebellions is important. *Riots* signal irrationality, whereas *rebellion* suggests a political response from blacks in the North facing de facto segregation and institutional racism. Moreover, calling the 1960s uprisings riots masks their difference with race riots of earlier decades that erupted as whites exacted raw violence on blacks fleeing the Jim Crow South in the name of defending white property. Consider, for example, the 1943 riot in Detroit that was sparked in part by a dispute the previous year over the siting of a black housing project called Sojourner Truth Homes in a white neighborhood. The Federal Housing Administration fueled white rage when it announced that it would not back mortgages in nearby neighborhoods, suggesting the role that the federal government played in de jure housing segregation, white flight, and the creation of a discriminatory marketplace (Freund 2007; Rothstein 2017). When black families tried to move in, white mobs numbering in the thousands assaulted them. Eventually, more than 1,000 city and state police and 1,600 members of the Michigan National Guard came to keep the peace as six black families moved in.
8. The nation's homeownership rates since 2007 have stabilized. According to a report by Harvard University's Joint Center for Housing, however, African American homeownership rates have not rebounded equally. The gap is particularly pronounced in Metro Detroit, where in 2015 African Americans had a 42 percent homeownership rate compared to 77 percent for whites. Between 2010 and 2015, homeownership rates for African Americans in the region declined by 11.6 percent compared to 3.0 percent for whites (Har 2017).

References

Akers, J. M. 2015. Making markets: Think tank legislation and private property in Detroit. *Urban Geography* 34 (8):1070–95.

———. 2016. Emerging market city. *Environment and Planning A* 47:1842–58.

Anderson, C. 2016. *White rage*. New York: Bloomsbury.

Armstrong, M. 1994. African Americans and property ownership: Creating our own meanings, redefining our relationships. *Berkeley Journal of African-American Law & Policy* 1 (1):79–88.

Baldwin, J. 1968. Speech at National Press Club. Accessed January 22, 2016. https://www.youtube.com/watch?v=CTjY4rZFY5c.

Bandele, O., and G. Myers. 2017. The roots of black agrarianism. In *Land justice: Re-imagining land, food, and the commons*, ed. J. Williams and E. H. Giminez, 19–39. Oakland, CA: Food First Books.

Berger, D. 2009. The Malcolm X doctrine: The Republic of New Afrika and national liberation on U.S. soil. In *New world coming*, ed. K. Dubinsky, S. Lord, S. Mills, S. Rutherford, and C. Krull, 46–55. Toronto: Between the Lines.

Berger, D., and R. Dunbar-Ortiz. 2010. "The struggle is for land!": Race, territory, and national liberation. In *The hidden 1970s: Histories of radicalism*, ed. D. Berger, 57–76. Piscataway, NJ: Rutgers University Press.

Bernstein, H. 2010. *Class dynamics of agrarian change*. Halifax, NS, Canada: Fernwood.

Bird Rose, D. 2004. *Reports from a wild country: Ethic for decolonization*. Sydney, Australia: University of New South Wales Press.

Black Land & Labor Initiative. 2017. Action toolkit. Accessed September 13, 2017. http://blacklandandliberation.org.

Bird Rose, D. 2004. *Reports from a wild country: Ethic for decolonization*. Sydney: University of New South Wales Press.

Blauner, R. 1969. Internal colonialism and ghetto revolt. *Society for the Study of Social Problems* 16 (4):393–408.

Blomley, N. 2003. Law, property and the geography of violence: The frontier, the survey, and the grid. *Annals of the Association of American Geographers* 93 (1):121–41.

———. 2008. Enclosure, common right, and the property of the poor. *Social & Legal Studies* 17 (3):311–31.

———. 2010. Cuts, flows, and the geographies of property. *Law, Culture and the Humanities* 7 (2):203–16.

Brenner, N., and C. Schmid. 2014. Planetary urbanization. In *Implosions/explosions: Towards a study of planetary urbanization*, ed. N. Brenner, 160–65. Berlin: Jovis.

Campbell, L., A. Newman, S. Safransky, and T. Stallman. Forthcoming. *A people's atlas of Detroit*. Detroit, MI: Wayne State University Press.

Carmichael, S., and C. Hamilton. [1967] 1992. *Black power: The politics of liberation in America*. New York: Vintage.

Chicago Anti-Eviction Campaign. 2009. Taking our human right to housing into our own hands: Chicago forms an anti-eviction campaign. Accessed September 22, 2017. http://chicagoantieviction.org/2009/11/taking-our-human-right-to-housing-into.html.

Clark, K. 1965. *Dark ghetto*. New York: Harper & Row.

Cleaver, E. 1970. The land question. In *What country have I? Political writings by black Americans*, ed. H. J. Storing. New York: St. Martin's.

Coleman, P., and S. Ries. 2017. Lack of affordable housing tarnishes Nashville's image. *Tennessean*. October 26. Accessed November 9, 2017. http://www.tennessean.com/story/opinion/2017/10/26/lack-affordable-housing-tarnishes-nashvilles-image/106967398/.

Coulthard, G. 2007. Subjects of empire: Indigenous peoples and the "politics of recognition" in Canada. *Contemporary Political Theory* 6:437–60.

———. 2014. *Red skin, white masks: Rejecting the colonial politics of recognition*. New York: Taylor & Francis.

Cruse, H. 1968. *Rebellion or revolution?* New York: Morrow.

Daniel, P. 2013. *Dispossession: Discrimination against African American farmers in the age of civil rights*. Chapel Hill: University of North Carolina Press.

Deloria, Jr., V. 1969. *Custer died for your sins: An Indian manifesto*. New York: Macmillan.

Derickson, K. 2009. Toward a non-totalizing critique of capitalism. *Geographical Bulletin* 50:3–15.

———. 2016. Urban geography II : Urban geography in the age of Ferguson. *Progress in Human Geography* 41 (2):230–44.

Detroit Future City. 2012. Strategic framework plan. Accessed September 15, 2017. https://detroitfuturecity.com/wp-content/uploads/2014/12/DFC_Full_2nd.pdf.

Detroiters Resisting Emergency Management. 2014. Whose city? Our city. Accessed September 15, 2017. http://www.d-rem.org/wp-content/uploads/2014/10/Whose-City-Our-City.pdf.

Du Bois, W. E. B. [1903] 2007. *The souls of black folk*. New York: Oxford University Press.

Elfenbein, J., T. Hollowak, and E. Nix. 2011. *Baltimore '68*. Philadelphia: Temple University Press.

Fine, S. 1989. *Violence in the model city: The Cavanagh administration, race relations, and the Detroit riot of 1967*. East Lansing: Michigan State University Press.

Freund, D. 2007. *Colored property: State policy & white racial politics in suburban America*. Chicago: University of Chicago Press.

Fung, C. 2014. "This isn't your battle or your land": The Native American occupation of Alcatraz in the Asian-American political imagination. *College Literature* 41 (4):149–73.

Goeman, M. 2008. From place to territories and back again: Centering storied land in the discussion of indigenous nation-building. *International Journal of Critical Indigenous Studies* 1 (1):23–34.

———. 2015. Land as life: Unsettling the logics of containment. In *Native studies keywords*, ed. S. N. Teves, A. Smith, and M. Raheja, 71–89. Tucson: University of Arizona Press.

Goldstein, A. 2014. Finance and foreclosure in the colonial present. *Radical History Review* 118:42–63.

Gordon, A. 2008. *Ghostly matters*. Minneapolis: University of Minnesota Press.

Gottesdiener, L. 2015. A foreclosure conveyor belt: The continuing depopulation of Detroit. TomDispatch. April 19. Accessed September 20, 2017. http://www.tomdispatch.com/blog/175983/.

Hann, C. M. 1998. The embeddedness of property. In *Property relations: Renewing the anthropological tradition*, ed. C. M. Hann, 1–47. New York: Cambridge University Press.

Har, J. 2017. Black homeowners lag in housing market recovery. *Detroit News*, July 8. Accessed August 11, 2017. http://www.detroitnews.com/story/news/local/michigan/2017/07/08/black-homeowners-struggle-housing-market-recovers/103532856/.

Harris, C. I. 1993. Whiteness as property. *Harvard Law Review* 106 (8):1707–91.

Hartman, S. 2007. *Lose your mother*. New York: Farrar, Straus and Giroux.

Harvey, D. 2006. *Spaces of global capitalism*. New York: Verso.

———. [1973] 2009. *Social justice and the city*. Athens: University of Georgia Press.

Heynen, N. 2016a. Toward an abolition ecology. *Abolition Journal*, December 29. Accessed September 15, 2017. https://abolitionjournal.org/toward-an-abolition-ecology/.

———. 2016b. Urban political ecology II: The abolitionist century. *Progress in Human Geography* 40 (6):839–45.

Hilliard, D. 2002. Introduction. In *The Huey P. Newton reader*, ed. D. Hilliard and D. Weise, 9–19. New York: Seven Stories Press.

Hong, G. K. 2014. Property. In *Keywords for American cultural studies*, ed. B. Burgett and G. Hendler. New York: NYU Press. Accessed November 10, 2017. 1. http://keywords.nyupress.org/american-cultural-studies/essay/property/.

hooks, b. 2009. *Belonging: A culture of place*. London and New York: Routledge.

Inwood, J., D. Alderman, and M. Barron. 2016. Addressing structural violence through U.S. reconciliation commissions: The case study of Greensboro, NC and Detroit, MI. *Political Geography* 52:57–64.

Ioanide, P. 2014. The alchemy of race and affect: "White innocence" and public secrets in the post-civil rights era. *Kalfou* 1 (1):151–68.

Khosla, P. 2005. Privatization, segregation and dispossession in Western urban space: An antiracist, Marxist–feminist reading of David Harvey. PhD thesis, Faculty of Environmental Studies, York University, Toronto.

La Paperson. 2014. A ghetto land pedagogy: An antidote for settler environmentalism. *Environmental Education Research* 20 (1):115–30.

Lipsitz, G. 1994. The racialization of space and the spatialization of race. *Landscape Journal* 26 (1):10–23.

Locke, J. 1704. *The second treatise of government*. Indianapolis, IN: Bobbs-Merrill.

Martin, L. L. 2009. Black asset ownership: Does ethnicity matter? *Social Science Research* 38 (2):312–23.

McKittrick, K. 2011. On plantations, prisons, and a black sense of place. *Social & Cultural Geography* 12 (8):947–63.

———. 2013. Plantation futures. *Small Axe: A Caribbean Journal of Criticism* 42:1–15.

Merrill, T. W. 1998. Property and the right to exclude. *Nebraska Law Review* 77 (4):730–55.

Metropolitan Truth and Reconciliation Commission. 2011. Charter for the Metropolitan Detroit Truth and Reconciliation Commission on Racial Inequality. Accessed January 22, 2017. http://www.miroundtable.org/Roundtabledownloads/FinalMandate2011.pdf.

Mills, C. W. 2008. Racial liberalism. *PMLA* 123 (5):1380–97.

Montgomery, A. 2016. Reappearance of the public: Placemaking, minoritization and resistance in Detroit.

International Journal of Urban and Regional Research 40 (4):776–99.

Mouffe, C. 2000. *The democratic paradox*. New York: Verso.

Nembhard, J. G. 2014. *Collective courage: A history of African American cooperative economic thought and practice*. University Park: The Pennsylvania State University Press.

Newman, A., and S. Safransky. 2014. Remapping the Motor City and the politics of austerity. *Anthropology Now* 6 (3):17–28.

Nixon, R. 2011. *Slow violence and the environmentalism of the poor*. Cambridge, MA: The President and Fellows of Harvard College.

Ntsebeza, L., and R. Hall. 2007. *The land question in South Africa: The challenge of transformation and redistribution*. Cape Town, South Africa: Human Sciences Research Council.

Oliver, M., and T. M. Shapiro. 2006. *Black wealth, white wealth: A new perspective on racial inequality*. London and New York: Routledge.

Peluso, N. L., and C. Lund. 2011. New frontiers of land control: Introduction. *Journal of Peasant Studies* 38 (4):667–81.

Pottage, A., and M. Mundy. 2004. *Law, anthropology, and the constitution of the social: Making persons and things*. New York: Cambridge University Press.

Rahman, A. 2008. Marching blind: The rise and fall of the Black Panther Party in Detroit. In *Liberated territory*, ed. Y. Williams and J. Lazerow, 181–231. Durham, NC: Duke University Press.

Rameau, M. 2008. *Take back the land: Land, gentrification and the Umoja village shantytown*. Miami, FL: Nia Press.

Reyes, A. 2009. Can't go home again: Sovereign entanglements and the black radical tradition in the twentieth century. PhD dissertation, Duke University, Durham, NC.

Rickford, R. 2017. "We can't grow food on all this concrete": The land question, agrarianism, and black nationalist thought in the late 1960s and 1970s. *Journal of American History* 103 (4):956–80.

Right to the City Alliance. 2017. Accessed September 22, 2017. http://righttothecity.org/about/member-organizations/.

Robinson, C. 1983. *Black Marxism*. Chapel Hill: University of North Carolina Press.

Rothstein, R. 2017. *The color of law: A forgotten history of how our government segregated America*. New York: Liveright.

Roy, A. 2016. What is urban about critical urban theory? *Urban Geography* 37 (6):810–23.

———. n.d. The land question. Accessed January 22, 2017. https://lsecities.net/media/objects/articles/the-land-question/en-gb/.

Safransky, S. 2014. Greening the urban frontier: Race, property, and resettlement in Detroit. *Geoforum* 56:237–48.

———. 2017. Rethinking land struggle in the post-industrial city. *Antipode* 49 (4):1079–1100.

Smith, N. 2002. New globalism, new urbanism: Gentrification as global urban strategy. *Antipode* 34 (3):427–50.

Stauffer, J. 2015. *Ethical loneliness: The injustice of not being heard*. New York: Columbia University Press.

Strathern, M. 1999. Potential property: Intellectual rights and property in persons. In *Property, substance, and effects*, ed. M. Strathern, 161–78. London: Althone.

Thompson, E. P. 1971. The moral economy of the English crowd in the eighteenth century. *Past and Present* 50:76–136.

Tuck, E., A. Guess, and H. Sultan. 2014. Not nowhere: Collaborating on selfsame land. *Decolonization: Indigeneity, Education & Society* 26 (June):1–11.

Tuck, E., M. Smith, A. M. Guess, T. Benjamin, and B. K. Jones. 2014. Geotheorizing black/land: Contestations and contingent collaborations. *Departures in Critical Qualitative Research* 3 (1):52–74.

U.S. Census Bureau. 2010. State and country quickfacts. Data derived from Population Estimates and American Community Survey.

Williams, R. 1976. *Keywords: A vocabulary of culture and society.* New York: Oxford University Press.

Wolford, W. 2010. *This land is ours: Social mobilization and the meanings of land in Brazil.* Durham, NC: Duke University Press.

Woodard, K. 2003. It's nation time in NewArk: Amiri Baraka and the black power experiment in Newark, New Jersey. In *Freedom north: Black freedom struggles outside the south, 1940–1980,* ed. J. F. Theoharis and K. Woodard, 287–312. New York: Palgrave Macmillan.

Young, I. M. 1998. Harvey's complaint with race and gender struggles: A critical response. *Antipode* 30 (1):36–42.

SARA SAFRANSKY is an Assistant Professor in the Department of Human & Organizational Development at Vanderbilt University, Nashville, TN 37203. E-mail: sara.e.safransky@vanderbilt.edu. Her research interests include land justice, governmental technologies of planning, the politics of memory, social justice movements, and participatory research.

19 Wrangling Settler Colonialism in the Urban U.S. West

Indigenous and Mexican American Struggles for Social Justice

Laura Barraclough

This article contributes to geographical scholarship on settler colonialism by exploring its urban valences in the U.S. West. Urban development in the U.S. West has long been guided by ideologies of the cowboy and frontier, which seek to explain and justify settler occupation of indigenous and Mexican land. Produced overwhelmingly in cities, frontier narratives have shaped urban landscapes and organized urban industries in powerful ways. As a result, indigenous and migrant populations in the urban U.S. West must wrestle with frontier myths—and the institutions that produce them—in their struggles for social justice. In this article, I examine two recent struggles over urban land use, economic development, and public space in the U.S. West that engaged settler myths of the frontier. These include the purchase, relocation, and operation of Rawhide Wild West Town in Phoenix, Arizona, by the Gila River Indian Community and failed efforts by *charros* (Mexican cowboys) to use a major sports facility in Las Vegas, Nevada. Together, these cases show that U.S. settler colonialism is far from settled but that settler nostalgia for the frontier—and the urban institutions that produce it—imposes real limits on the abilities of indigenous and marginalized peoples to pursue and achieve their visions of social justice. *Key Words: cowboys, frontier, rodeo, settler colonialism, U.S. West.*

本文藉由探讨美国西部的城市效价, 对于迁佔者殖民主义的地理学研究作出贡献。长期以来, 美国西部的城市发展, 由牛仔与边境的意识形态所主导, 并企图解释与合理化对原住民及墨西哥土地的迁佔。边境的叙事压倒性地来自城市, 并以强而有力的方式, 形塑了城市地景和组织城市产业。因此美国西部城市的原住民和移民人口, 在争取社会正义时, 便必须与边境的迷思 —— 以及生产这些迷思的制度 —— 进行争夺。我于本文中, 检视晚近在美国西部涉及边境迁佔者迷思的城市土地使用、经济发展和公共空间的争夺。这些争夺, 包含希拉河原住民社群在亚历桑那凤凰城的生皮革西部荒野城镇中所进行的购置、再安置与经营, 以及charros (墨西哥牛仔) 无法在内华达州拉斯维加斯中运用主要运动设施之案例。这些案例共同显示, 美国的迁佔者殖民主义并非固定不变, 但边境的迁佔怀旧 —— 以及生产此般怀旧的城市制度 —— 却对原住民和边缘化的人群追求并达成自身的社会正义愿景之能力, 加诸了实际的限制。 *关键词: 牛仔, 边境, 牛仔竞技大会, 迁佔者殖民主义, 美国西部。*

Este artículo contribuye a la erudición geográfica sobre colonialismo con pobladores explorando sus valencias urbanas en el Oeste de los EE.UU. El desarrollo urbano de aquella región fue guiado durante mucho tiempo por las ideologías del vaquero y de la frontera, con las que se pretende explicar y justificar la ocupación de tierras indígenas y mejicanas. Construidas a manos llenas en las ciudades, las narrativas de frontera han configurado de manera vigorosa los paisajes urbanos y las industrias urbanas organizadas. En consecuencia, las poblaciones indígenas y de migrantes en el Oeste urbano de los Estados Unidos tienen que lidiar con los mitos de frontera—y con las instituciones que los producen—en sus luchas por la justicia social. En este artículo, examino dos campañas recientes sobre uso del suelo urbano, desarrollo económico y espacio público en el Oeste norteamericano en las que se abocaron mitos de los colonos de frontera. Entre estos se incluye la compra, relocalización y operación del Pueblo Rústico del Oeste Salvaje [Rawhide Wild West Town] de Phoenix, Arizona, por la Comunidad India del Río Gila, y los fallidos esfuerzos de charros (vaqueros mejicanos) por usar unas instalaciones deportivas de valor en Las Vegas, Nevada. Conjuntamente, estos dos casos muestran que el colonialismo norteamericano con colonos está lejos de haberse asentado, si bien la nostalgia del colono por la frontera—y por las instituciones urbanas que la generan—impone límites reales a las capacidades de los pueblos indígenas y marginados para perseguir y lograr sus visiones de justicia social. *Palabras clave: vaqueros, frontera, rodeo, colonialismo con colonos, Oeste norteamericano.*

Urban development in the U.S. West, a region structured by settler colonialism, has long been dominated by ideologies of the cowboy and the frontier (Barraclough 2011). As in other settler colonial contexts, these ideologies function as imperial transition narratives—stories that justify the settler occupation of indigenous land and the unequal distribution of resources in the settlers' favor (Razack 2002). They have been produced overwhelmingly in cities, which have both the necessary concentration of political, economic, and cultural institutions and the incentive to pursue urban development in regionally specific ways (Barraclough 2011). As a result, settler narratives of the frontier have come to organize urban political economies in the region. They are the cultural foundation of urban industries such as film and television, country-western music, professional rodeo, and tourism (La Chapelle 2007; Philpott 2013), and they have profoundly shaped the urban landscape through real estate and historic preservation practices; the creation of public art; and the design, regulation, and use of public space, among other spatial processes (Morley 2006; Barraclough 2011; Brosnan and Scott 2011; Baker 2014).

As a result, indigenous and migrant populations in the U.S. West, who have also become disproportionately urban since World War II, must wrestle with settler ideologies of the frontier—and the urban institutions that produce them—in their ongoing efforts toward decolonization, self-determination, and cultural citizenship. In this article, I examine two recent cases in which American Indians and Mexican Americans in the urban U.S. West have engaged settler imaginaries of the frontier in their struggles for social justice. First, I consider the Gila River Indian Community (GRIC) 2005 purchase of Rawhide Wild West Town, a western-style theme park originally located in the upscale, racially exclusive suburb of Scottsdale, north of Phoenix. The GRIC relocated the park to its nearby reservation, where it has become both an important revenue source and a site for the subtle rearticulation of regional history through an indigenous lens. Through its purchase of Rawhide, the GRIC sought to put settler imaginaries of the frontier to work for indigenous economic development, although its impacts on the alleviation of tribal poverty have so far been limited. Second, I trace efforts by urban *charros* (Mexican cowboys) to host Mexican-style rodeos in Nevada since about 2010. The charros were primarily interested in finding appropriate locales to hold their events but found themselves caught in a legal battle between the state's professional (read: Anglo American) rodeo industry and its animal welfare activists. The result was a 2013 state law that banned several events in the Mexican rodeo while leaving comparable events in the American-style rodeo untouched. Not only was rodeo racialized

(again) as white, but Mexican Americans were excluded from several of Nevada's premiere urban recreational spaces.

This article contributes to emergent geographical scholarship on settler colonialism (e.g., Bonds and Inwood 2015; Rose-Redwood 2016; Pulido forthcoming) by exploring its urban valences in the U.S. West. Similar to how Blomley (2003) and Shaw (2007) identified the centrality of urban property regimes and logics to settler colonialism in Canada and Australia, I highlight the institutional power vested in settler imaginaries of the frontier as an important structuring framework within which struggles for social justice in the urban U.S. West unfold. As I have argued elsewhere (Barraclough 2011), urban growth machine interests in the U.S. West have long used processes of "rural urbanism"—the intentional urban production of rural landscapes—to simultaneously secure a white settler society on conquered indigenous and Mexican land and to guide regional development. Yet, as critical indigenous studies scholars have argued (Thrush 2007; Peters and Andersen 2013; Simpson 2014; Kēhaulani Kauanui 2016), this process can never be fully secured. Urban indigenous and Mexican American communities, among others, have engaged in creative efforts to transform the content, form, and meaning of frontier mythology in ways that advance their goals of sovereignty, self-determination, economic mobility, and cultural citizenship. Nonetheless, as we shall see, settler institutions that remain vested in propagating nostalgia for the frontier impose real limits on the abilities of conquered and marginalized groups to pursue and achieve their visions of social justice.

The Old and New Rawhide Wild West Towns in Metropolitan Phoenix

After World War II, Phoenix and its suburbs became the fastest growing of the United States' twenty-five most populous metropolitan areas (Frantz 2012; Needham 2014). As in other Western cities, Phoenix's growth—and the deep social and environmental inequalities on which it rested (Ross 2011)—was guided and explained by ideologies of the frontier. These ideologies were made manifest in the urban landscape through the placement of wagon wheels, stagecoaches, wooden sidewalks, hitching posts, and related aesthetic markers. Especially notable were efforts in suburban Scottsdale, north of Phoenix, where in 1951 the Chamber of Commerce adopted the "West's Most Western Town" as the city's slogan and established a Western design theme for downtown. Many of the area's proliferating residential subdivisions likewise embraced an "Old West" aesthetic (City of Scottsdale 2016). These decisions inscribed a white settler imaginary of the Western frontier into Scottsdale's built environment and nurtured a civic identity based on conquest of the U.S. West for a suburban population that was almost exclusively white and upper middle class (Romig 2005).

One of metropolitan Phoenix's most prominent actors in this regard was Jim Paul, a real estate developer and amateur history buff who was responsible for some of the city's first subdivisions. Paul was enamored with Anglo fantasies of the Old West. He had moved to Arizona in 1928, when he "fell in love with the place. I loved the cowboys and Indians and everything about it." In addition to an influential career in real estate, Paul raised cattle and horses recreationally. He was also cofounder of two civic organizations, the Scottsdale Boys Club and the Verde Vaqueros, which he described as "an excuse for grown men to play cowboys" in their urban contexts (Paul 1982).

In addition to tract homes, Paul was the developer of Rawhide Wild West Town, an amusement park in North Scottsdale. Rawhide was one of many Wild West–style amusement parks that opened across the U.S. West (and beyond) after World War II; other examples—still operating today—include Frontier City in Oklahoma City and Knott's Berry Farm in Orange County, California. These commercial sites of leisure appealed to white settler audiences by invoking a supposedly shared historical experience of conquest. When Rawhide opened to the public in November 1971, the *Scottsdale Daily Progress* described it as a recreation of the historic lifestyle endured by "hardy pioneers who conquered an inhospitable land, trekking uncharted mountains and plains and fighting savage Indians; cowboys driving vast herds of cattle to market and singing songs around a campfire at night" (Erickson 1971, 12).

Rawhide was not only an exercise in white settler nostalgia; it was also a vehicle for profit. Admission was free, but the park had many ways to capture revenue, including souvenir shops, "chuck wagon" restaurants, stagecoach rides, and opportunities to take photos with a costumed "Indian" (Erickson 1971). Later owner-operators added a live-action theater and petting zoo, upgraded the merchandise, and offered higher quality entertainment, including rodeos and country-western concerts. Rawhide employed 300 people and attracted between 850,000 and 900,000 visitors each year, making it the state's largest privately owned tourist attraction and a major engine of metropolitan growth ("Rawhide Gets a Dusting" 1998).

By the 1990s, however, rampant development had made the operation of a Wild West–style theme park in North Scottsdale untenable. As the *Yuma Daily Sun* ("Rawhide Gets a Dusting" 1998) explained, Rawhide "was once a remote location, but development has been pushing houses and shopping centers practically right up to its unpaved streets" (27). By June 2004, co-owner Jerome Hirsch had sold the land Rawhide occupied, which was to be developed as a mixed-use project, and solicited proposals to relocate its structures (Ropp 2004).

Later that year, Rawhide's new owner was announced: the GRIC (or the Community), a sovereign indigenous nation and one of twenty-one federally recognized tribes in Arizona (Ropp and Beard 2004). The GRIC includes two tribal groups, the Pimas (Akimel O'Odham) and the Maricopas (Pee-Posh), who share a 372,000-acre reservation, created in 1859, that borders the cities of Phoenix, Chandler, and Tempe (Intertribal Council of Arizona 2016; see Figure 1). At the time of the purchase, the GRIC's reservation was the largest undeveloped landmass in the United States contiguous to a major metropolitan area, poising the GRIC to become a major player in regional economic development (Ross 2011).

The GRIC's purchase of Rawhide was part of its comprehensive vision to develop the Wild Horse Pass, a 2,400-acre economic development zone founded by the Community in 1998. The Wild Horse Pass was enabled by the Arizona Water Settlement Act of 2004, a landmark law that reversed the decades-long drought imposed on the GRIC by settler theft of water (Ross 2011; DeJong 2014). The settlement enabled the GRIC to create capitalist enterprises that would allow them to pursue their long-term vision of social justice: restoring the sustainable and culturally appropriate agriculture for which their ancestors were well known (Ross 2011; DeJong 2014). As of 2005, the Wild Horse Pass included two luxury hotels, two casinos, an industrial park, golf course, spa, and equestrian center. Collectively, these enterprises employed nearly 3,000 people and represented $300 million in investment (Beard 2007). Tribal officials envisioned Rawhide as "bring[ing] critical mass" to the Wild Horse Pass (Schad 2005) by putting settler myths of the frontier to work for indigenous economic development. As the *Ahwatukee Republic* ("Illusion of the Old West"

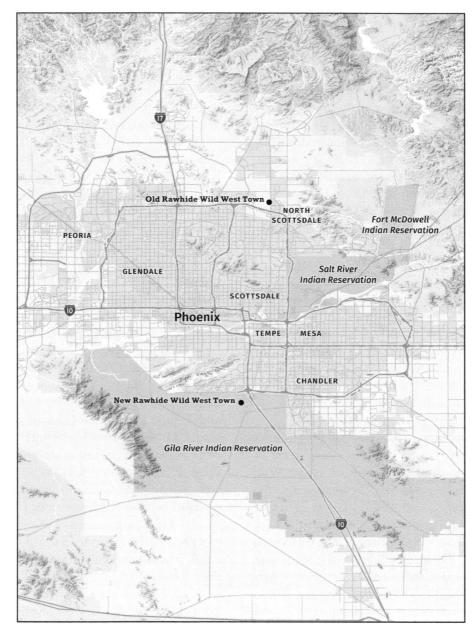

Figure 1. Map of metropolitan Phoenix, including adjacent Indian reservations, with the locations of the old and new Rawhide Wild West Towns. *Source:* Map by John Emerson.

2006), a local suburban newspaper, editorialized, "there's nothing old-tymey or sepia-toned about the big money that our Wild West heritage draws in tourism. It's pure 21st-century capitalism" (18).

In many ways, the GRIC left unchanged the white settler fantasies that had infused the original Rawhide and guided metropolitan Phoenix's development for decades. After all, those myths had proven themselves to be not only culturally durable but also highly profitable. The Rawhide purchase agreement of 2005 stipulated that most of the building façades, props, and signs from the old Rawhide be relocated to the GRIC

reservation, where they would be installed on top of new structures with updated plumbing, electrical, and fire sprinkler systems. The Wild Horse Pass Development Authority hired a "theming expert" to make these new structures look appropriately aged; for example, by painting with whisk brooms instead of brushes and using wood from infected trees (Beard 2005a). Some alterations were made to the park's design, such as widening the main street to better accommodate traffic, but most of the structures, stores, restaurants, services, and live-action shows remained identical to the original location (see Figures 2 and 3).

Figure 2. Entrance to Rawhide Wild West Town in its new location on the Gila River Indian Community reservation. *Source:* Photo by author, 2016. (Color figure available online.)

In other important ways, though, the GRIC adapted Rawhide to celebrate the enduring presence and cultural vitality of native peoples amidst U.S. settler colonialism. Rawhide's grand opening ceremony in March 2006 featured a ballet folklórico performance, basket dancers, and a blessing conducted by Jennifer Allison-Ray, lieutenant governor of the GRIC (Beard 2006). The Indian Artists of America show, held annually at Rawhide beginning in January 2006, featured not only traditional artistry, like baskets and pottery, but also a fashion show highlighting work by contemporary native designers (Arnold 2006). By 2009, managers had added a farmers' market, an organic garden that supplies Rawhide's restaurants and hosts workshops for native

children, and a re-created Pima-Maricopa village that evokes the GRIC's preconquest life (see Figure 4; Creno 2009; "Planting Seeds of Hard-Earned Success" 2009). For at least some indigenous employees, these elements allow for the strategic incorporation of native music, art, dance, and history into the prevailing settler imaginary of the U.S. West. As tribal member Mona White, an employee in the photo emporium where visitors have their pictures taken in "Old West" costumes, explained, "We welcome everybody to the reservation. That way they can learn about our history" (Wanke 2005, B-2).

Equally important, the new Rawhide represents a major economic development strategy with the

Figure 3. Main Street at Rawhide Wild West Town on the Gila River Indian Community reservation. *Source:* Photo by author, 2016. (Color figure available online.)

Figure 4. Recreation of a historic Pima Maricopa village at the new Rawhide Wild West Town on the Gila River Indian Community reservation. *Source:* Photo by author, 2016. (Color figure available online.)

potential to alleviate tribal poverty and diversify the economic opportunities available to Community members. Although the GRIC maintained 75 percent of the employees from the original Rawhide location, it also significantly expanded employment of tribal members (Beard 2005b). Daily attendance has remained steady in the new location, with nearly 1 million visitors to the park annually. Among Rawhide's most profitable elements are the 250 conventions, weddings, and other special events hosted there each year, the economic benefits of which now accrue to the GRIC rather than the City of Scottsdale (Beard 2005c).

Despite Rawhide's success in its new location and the GRIC's growing political and economic power, the Community continues to struggle. In 2010, the GRIC's median household income was $28,779, far below that of the county ($51,310) and the state ($50,448). Poverty rates are extremely high: Nearly half of GRIC residents are poor, compared to 13 percent of county residents and 15 percent of Arizona residents; poverty is even higher among children (61 percent). Not surprisingly, households in the Community are three times more likely (33 percent) to receive food stamps than residents of the county or state (10 percent each), and they receive considerably more public assistance income overall (12 percent, compared to 3 percent in the county and 2 percent statewide; Arizona Rural Policy Institute n.d.). These material conditions are the entrenched effects of conquest and settler colonialism. They have not yet been substantially altered by the GRIC's economic development initiatives, although the development of the Community's vision is just beginning and its long-term effects remain to be seen.

Thus, the new Rawhide reflects the complexities and contradictions of settler colonialism itself. On the one hand, it demonstrates the resilience, adaptability, and creativity of urban indigenous peoples, especially as they have assumed greater power in shaping the urban political economies of the U.S. West. On the other hand, it does so at cost: The settler fantasies that continue to structure Rawhide's tourist experience risk celebrating the very processes of conquest, violence, and dispossession that have produced tribal poverty and disenfranchisement. These contradictions represent the "dilemma of the resource curse" with which indigenous peoples have long been familiar (Ross 2011, 237) but, in this case, the resource in question is the economic power vested in white settler mythologies of the frontier.

Cowboys and *Charros* in Urban Nevada

The GRIC is not the only set of actors seeking to recast settler myths of the frontier for their own purposes; so, too, are the Mexican American men who honor the Spanish and Mexican origins of rodeo by competing and performing as *charros* (Mexican cowboys). *Charrería*, or Mexican rodeo, emerged in the seventeenth and eighteenth centuries under Spanish colonialism, as the indigenous and *mestizo* (mixed-race) workers who labored on Mexico's colonial *haciendas* innovated techniques for roping, riding, and ranching. These techniques were ritualized and shared during the *rodeos* (round-ups) in which cattle were gathered and branded (Sands 1993; Nájera-Ramírez 1994). By the late 1700s, rodeos and the cattle industry had spread across the Mexican North (now the U.S. Southwest), where they evolved to include westward-moving white settlers, among others. Thereafter, indigenous, Mexican, white, and black ranch workers together constituted a multiethnic, multilingual economy and social order (LeCompte 1985).

After U.S. conquest in 1848 and the subsequent dispossession of most Mexican landowners, alongside continuing Indian Wars and the creation of reservations throughout the nineteenth century, ranching increasingly became the purview of white settlers. Settler-ranchers created political associations and lobbying groups charged with advancing their economic and spatial agendas. By the early twentieth century, rodeo had also come to be controlled by corporate organizations that whitewashed its indigenous and Mexican origins (LeCompte 1985). Notably, they did so from urban locations as diverse as New York, Boston, Denver, Dallas, Los Angeles, Oklahoma City, and Colorado Springs (Professional Rodeo Cowboys Association 2017).

Thus, for the region's Mexican American inhabitants, reclaiming the Spanish and Mexican origins of rodeo and ranching became an important exercise in cultural citizenship. This project gained steam after World War II, when Mexican Americans—like Americans of most ethnic and racial groups—urbanized in large numbers seeking work in defense and manufacturing. Many achieved middle-class status, purchased homes, and started small businesses (Telles and Ortiz 2009). Following similar developments in urban Mexico (Sands 1993; Nájera-Ramírez 1994), they also began to practice *charrería*. They established *charro* associations, built *lienzos* (arenas), and performed in parades and other civic events in Western cities (Barraclough 2012). Through these activities, they revised the region's whitewashed histories of ranching and rodeo to include Mexicans and Mexican Americans, even as

they overlooked indigenous and African American contributions to ranching and rodeo, reflecting the contradictory position of Chicanas and Chicanos in relationship to U.S. settler colonialism more broadly (Pulido forthcoming).

Along the way, Mexican American *charros* have necessarily wrangled with the institutional power vested in settler myths of the frontier, in this case through their conflicts with both the corporate rodeo industry and animal welfare activists. Animal welfare activists have long sought to restrict all forms of rodeo, as well as other animal practices they deem cruel. The corporate rodeo industry has been a difficult beast to tackle, however, because of its importance to the political economy of many Western cities (Beers 2006). Not so for Mexican American *charros*, who have far less political power: Their events offered a toehold for animal welfare activists seeking to ban all rodeos. Since the early 1990s, animal welfare activists have worked to ban the *manganas*, two events in the Mexican rodeo that involve bringing a running mare to the ground in a shoulder roll by roping her front legs—a practice that *charros* call wrangling and activists call horse tripping. After California banned this practice in 1994, *charros* in the United States voluntarily agreed to change the way they practice the *manganas*: Although they still rope the running horse's legs, they no longer bring her to the ground. This concession has not shielded them from legislative action, however. As of this writing, more than a dozen states and counties have banned the *manganas*; some have also banned a third event, the *coleadero* (which involves flipping a steer by its tail). In virtually all cases, comparable events in U.S.-style rodeo have been deliberately exempted because of the economic power of corporate rodeo lobbyists, as well as the centrality of whitewashed myths of rodeo to U.S. national identity (Barraclough 2014).

These issues came to the fore most recently in several Nevada cities where growing Latino populations have used *charrería* to claim urban space and exert civic belonging. In 2011, State Senator Allison Copening, who represents Las Vegas and Clark County, introduced a bill to ban "horse tripping" statewide (Nevada Senate Committee on Natural Resources 2011a). The impetus was an upcoming *charreada* to be held in Winnemucca, a small city in northern Nevada that had become one third Hispanic and Latino in the previous decade (Suburban Stats 2016). Punishment would range from a jail sentence of up to six months and a fine of $1,000 for a first offense to a felony conviction, punishable by imprisonment and a fine of up to $10,000 for a third offense (Nevada Senate Committee on Natural Resources 2011a).

Copening's bill was referred to the Committee on Natural Resources, which heard testimony in the state capital, Carson City, on 6 April 2011. Animal welfare activists decried the practice as animal abuse, and the state's *charros* protested the need for the bill, given their voluntary changes to the *manganas* twenty years prior, as well as its clear ethnic targeting. Alejandro Galindo, president of the Las Vegas Charros Association, also sought to reframe regional history to account for the prominent roles of Mexican ranchers. He said, "Someone has erroneously translated manganas to mean horse tripping. Manganas is not horse tripping, manganas is the act of horse wrangling. Wrangling is part of the West" (Nevada Senate Committee on Natural Resources 2011a).

The decisive opposition came not from the *charros* but rather from Nevada's enormously influential corporate (Anglo-American) rodeo and ranching industries. In the decade prior, those industries had formed advocacy groups with the specific purpose of lobbying urban audiences (Gordon 2006; Hawkes, Lillywhite, and Libbin 2006). At the hearing, those advocates argued that the "horse tripping" bill could be interpreted to eliminate their events and rangeland practices; there would be major repercussions, they warned, for the economies of Nevada cities. Senators John Lee and Tom Collins both worried aloud that the legislation might affect the National Finals Rodeo held annually in Las Vegas. Phillip Hacker, an intern to Senator Lee, then testified that the National Finals Rodeo and the Reno Rodeo had a combined annual economic impact of more than $85 million, before warning:

> Passing this will send a message to the rodeo community that Nevada does not stand with them. Mind you this is a community which brings hundreds of thousands of visitors and contributes millions of dollars to the economies of communities throughout the state ... we can ill afford to burn bridges with people who reliably contribute to the state's economy. (Nevada Senate Committee on Natural Resources 2011a)

Although Copening clarified that corporate rodeo and ranching would not be affected, the Natural Resources Committee was not convinced, and a motion to pass the bill failed for lack of a second. Nonetheless, several senators pledged that if evidence of horse tripping were

to emerge, they would pass a statewide ban (Nevada Senate Committee on Natural Resources 2011b).

Several weeks later, the Winnemucca *charreada* proceeded as planned, and participants described it as a peaceful event in which no people or animals were seriously injured. Shortly thereafter, however, a Reno television station aired a video created by an activist who had attended. Although filmed with a mobile phone and of poor quality, the video claimed to provide evidence that horses had indeed been "tripped"—a claim refuted by participating *charros* as well as other spectators. Nonetheless, members of the Committee on Natural Resources were furious, believing that they had been lied to. Committee Chairman Mark Manendo spearheaded a new bill to ban the *manganas* and the *coleadero*. The new bill passed both houses, and in June 2013, Governor Brian Sandoval signed Nevada's "horse tripping" bill into law (Animal Law Coalition 2013).

The new law allowed local governments to issue conditional permits for events with "horse roping"—an ambiguous loophole meant to ensure that events associated with the corporate rodeo and ranching industries could proceed at the discretion of municipal governments. That loophole created a quandary for the World Series of Charrería, however, which was scheduled to take place 26 through 29 September 2013 at Las Vegas's South Point Hotel Resort and Casino (see Figure 5). The event's planned location at South Point, which hosts some of the country's most prestigious equestrian events, was a milestone for U.S.-based *charros*. The commissioners of Clark County, which includes Las Vegas, had to decide whether to use the loophole to allow the World Series of Charrería to proceed with all nine traditional events. After several hours of "impassioned debate," commissioners voted six to one against such a proposal; essentially, they rejected the *charros'* bid to be considered equivalent to U.S.-style rodeo. The Nevada charros explored other options, including holding the event without the *manganas* and *colaeaderos* or relocating it to another state, but ultimately the entire event was scrapped (Botkin 2013).

The legal tussles over *charrería* in Nevada were negotiated in four of its cities—Winnemucca, Las Vegas, Reno, and Carson City—where the institutional power vested in white settlers' version of rodeo ensured its protection, whereas *charros* and their version of rodeo were marginalized. The outcome of these processes affected not only the *charros* but also the thousands of Latino inhabitants, businesspeople, and tourists who attend and enjoy *charreadas* in

Figure 5. Bus stop advertisement for the World Series of Charrería 2013 at South Point Casino in Las Vegas, which was canceled after the passage of a Nevada state law banning three events in the *charreada* (Mexican rodeo). *Source*: Graphic Illusion Design Studio of Puerto Vallarta, Mexico, reproduced with permission. (Color figure available online.)

Nevada's cities. After the event's cancellation, Galindo asserted that 20,000 visitors, primarily Latinos, were anticipated to attend the World Series of Charrería, yet their contributions to Las Vegas's economy had not once been acknowledged in the way that those associated with corporate rodeo were. He argued that, through Clark County's failure to grant an available exemption, Latinos had been told that they were not welcome in one of the city's premiere recreational spaces. Finally, Galindo remarked on the nativist and exclusionary history embedded in the law: "The only thing American about American rodeo is 'American' in front of it because *rodeo* is Spanish for 'roundup.' . . . One culture is able to rodeo but yet the other one can't" (Botkin 2013). Galindo's analysis points to the ways in which rodeo continues to operate as economic powerhouse, political constituency, and white settler fantasy in urban Nevada in ways that marginalize the very people who helped to create it: Mexican Americans as well as indigenous and black peoples of the Americas (LeCompte 1985).

Conclusion

The examples examined here illustrate some of the many ways in which indigenous people and Mexican Americans in the urban U.S. West have worked toward social justice through their efforts to transform white settler myths of the cowboy, rodeo, and the frontier.

Through these initiatives, they have won some moderate gains. The GRIC's purchase of Rawhide Wild West Town is part of a broad economic development strategy that shows promise in the long term for indigenous-led economic development, the restoration of sustainable agriculture, and tribal self-sufficiency (Ross 2011). For *charros* in Nevada and elsewhere, the animal welfare disputes have had the positive effect of encouraging them to form national associations, seek new political alliances, and generate public awareness of their history and craft (Federación Mexicana de Charrería—USA 2017).

Even so, these cases highlight the pervasive institutional power vested in settler imaginaries of conquest, westward expansion, and the frontier and the ways in which that power continues to shape the region's cities. After all, narratives of the frontier not only narrate regional history through a white settler lens but also organize urban economies, political alliances, and built environments in ways that remain profoundly influential. Given this fact, marginalized and conquered groups are reticent to challenge the legitimacy or desirability of settler narratives altogether. Instead, they mostly push for inclusion within settler narratives or try to direct the frontier's economic and political rewards to their favor. Consequently, the romance and nostalgia attached to white settler imaginaries of conquest remain driving forces behind the making of urban life and landscape in the U.S. West, and the material impacts of settler colonialism are largely reproduced—although not without contestation.

Acknowledgments

I wish to thank Laura Pulido and two anonymous reviewers for their very helpful comments and suggestions on earlier versions of this article. I also received useful feedback from participants in the Critical Encounters lecture series in American Studies at Yale University. Finally, thanks to Nik Heynen for his editorial support and to my mother, Bette Barraclough, for providing child care for my son during the Arizona research.

References

Animal Law Coalition. 2013. NV horse tripping bill signed into law. 4 June 2013. Accessed July 27, 2017. https://animallawcoalition.com/nv-horse-tripping-bill-closer-to-becoming-law/.

Arizona Rural Policy Institute. n.d.. *Demographic analysis of the Gila River Indian Community using 2010 Census and 2010 American Community Survey estimates.* Flagstaff: Northern Arizona University Center for Business Outreach.

Arnold, E. 2006. A stylish weekend. *Ahwatukee Republic* 27 January:6A.

Baker, A. 2014. From rural South to metropolitan sunbelt: Creating a cowboy identity in the shadow of Houston. *Southwestern Historical Quarterly* 118 (1):1–22.

Barraclough, L. 2011. *Making the San Fernando Valley: Rural landscapes, urban development, and white privilege.* Athens: University of Georgia Press.

———. 2012. Contested cowboys: Ethnic Mexican charros and the struggle for suburban public space in 1970s Los Angeles. *Aztlán: A Journal of Chicano Studies* 37:95–124.

———. 2014. "Horse tripping": Animal welfare legislation and the production of ethnic Mexican illegality. *Ethnic and Racial Studies* 37:2110–28.

Beard, B. 2005a. Old West rising again. *Arizona Republic* 26 November:2.

———. 2005b. Rawhide breaks ground. *Scottsdale Republic* 22 April:S8.

———. 2005c. West Fest trails Rawhide to SE Valley. *Arizona Republic* 17 September:2.

———. 2006. Rawhide's grand opening scheduled for March 25. *Ahwatukee Republic* 15 March:11.

———. 2007. Gila River community vital business partner. *Arizona Republic* 27 (January):CH–7.

Beers, D. 2006. *For the prevention of cruelty: The history and legacy of animal rights activism in the United States.* Athens: Swallow Press/Ohio University Press.

Blomley, N. 2003. *Unsettling the city: Urban land and the politics of property.* London and New York: Routledge.

Bonds, A., and J. Inwood. 2015. Beyond white privilege: Geographies of white supremacy and settler colonialism. *Progress in Human Geography* 40 (6):1–19.

Botkin, B. 2013. Mexican rodeo canceled at South Point Arena. *Las Vegas Review-Journal* August 23. Accessed July 27, 2017. http://www.reviewjournal.com/news/las-vegas/mexican-rodeo-canceled-south-point-arena.

Brace, L. 2004. *The politics of property: Labour, freedom, belonging.* Edinburgh, UK: Edinburgh University Press.

Brosnan, K., and A. L. Scott, eds. 2011. *City dreams, country schemes: Community and identity in the American West.* Reno: University of Nevada Press.

City of Scottsdale. 2016. About Scottsdale: West's most Western town. Accessed July 27. http://www.scottsdaleaz.gov/about/history.

Creno, C. 2009. Gila River Indian Community grows into tourist destination. *Arizona Republic* 18 December: 10.

DeJong, D. 2014. Navigating the maze: The Gila River Indian Community Water Settlement Act of 2004 and administrative challenges. *American Indian Quarterly* 38 (1):60–81.

Erickson, M. 1971. The Wild West as recreated up north. *Scottsdale Daily Progress* 31 December:12–13.

Federación Mexicana de Charrería A.C.—U.S.A. 2017. About us. Accessed July 27, 2017. http://charrosfederationusa.com/about/.

Frantz, K. 2012. The Salt River Indian Reservation: Land use conflicts and aspects of socioeconomic change on the outskirts of Metro-Phoenix, Arizona. *GeoJournal* 77 (6):777–90.

Gordon, K. 2006. Taking the reins: Nevada and Idaho have taken a proactive stance on selling ranch stewardship to the public. *Rangelands* 28 (4):33–34.

Hawkes, J., J. Lillywhite, and J. Libbin. 2006. A sporting alternative: Sport cattle may help cattle growers round up their profits. *Rangelands* 28 (6):15–17.

Illusion of the Old West packing fun at Rawhide. 2006. *Ahwatukee Republic* 17 March:18.

Intertribal Council of Arizona. 2016. Introductory information: Gila River Indian Community. Accessed July 27, 2017. http://itcaonline.com/?page_id = 1158.

Kēhaulani Kauanui, J. 2016. "A structure, not an event": Settler colonialism and enduring indigeneity. *Lateral: Journal of the Cultural Studies Association* 5 (1). Accessed October 2, 2017. https://doi.org/10.25158/L5.1.7.

La Chapelle, P. 2007. *Proud to be an Okie: Cultural politics, country music, and migration to Southern California.* Berkeley: University of California Press.

LeCompte, M. L. 1985. The Hispanic influence on the history of rodeo, 1823–1922. *Journal of Sport History* 12 (1):21–38.

Morley, J. M. 2006. *Historic preservation and the imagined West: Albuquerque, Denver, and Seattle.* Lawrence: University Press of Kansas.

Nájera-Ramírez, O. 1994. Engendering nationalism: Identity, discourse, and the Mexican charro. *Anthropological Quarterly* 67:1–14.

Needham, A. 2014. *Power lines: Phoenix and the making of the modern Southwest.* Princeton, NJ: Princeton University Press.

Nevada Senate Committee on Natural Resources. 2011a. Minutes of hearing held 6 April, Carson City, NV.

———. 2011b. Minutes of hearing held 15 April, Carson City, NV.

Paul, J. 1982. Interview by Jennings Morse. 20 January, Scottsdale Historical Society, Scottsdale, AZ.

Peters, E., and C. Andersen. 2013. *Indigenous in the city: Contemporary identities and cultural innovation.* Vancouver, Canada: University of British Columbia Press.

Philpott, W. 2013. *Vacationland: Tourism and environment in the Colorado high country.* Seattle: University of Washington Press.

Planting seeds of hard-earned success. 2009. *Arizona Republic* 5 August:6.

Professional Rodeo Cowboys Association. 2017. History of rodeo, history of the PRCA. Accessed May 4, 2017. http://prorodeo.com/prorodeo/rodeo/history-of-the-prca.

Pulido, L. Forthcoming. Geographies of race and ethnicity III: Settler colonialism and nonnative people of color. *Progress in Human Geography.* Advance online publication. doi:10.1177/0309132516686011

Rawhide gets a dusting. 1998. *Yuma Daily Sun* 28 June:27.

Razack, S., ed. 2002. *Race, space, and the law: Unmapping a white settler society.* Toronto, Canada: Between the Lines Press.

Romig, K. 2005. The upper Sonoran lifestyle: Gated communities in Scottsdale, Arizona. *City and Community* 4 (1):67–86.

Ropp, T. 2004. Rawhide, Gila River officials expected to meet Wednesday. *Scottsdale Republic* 14 December:S5.

Ropp, T., and B. Beard. 2004. Rawhide move expected to become official today. *Scottsdale Republic* 16 December:1.

Rose-Redwood, R. 2016. "Reclaim, rename, reoccupy": Decolonizing place and the reclaiming of PKOLS. *ACME: An International E-Journal for Critical Geographies* 15 (1):187–206.

Ross, A. 2011. *Bird on fire: Lessons from the world's least sustainable city.* New York: Oxford University Press.

Sands, K. 1993. *Charrería Mexicana: An equestrian folk tradition.* Tucson: University of Arizona Press.

Schad, C. 2005. Rawhide bringing "critical mass" to area. *Arizona Republic* 5 January:11.

Shaw, W. 2007. *Cities of whiteness.* New York: Wiley-Blackwell.

Simpson, A. 2014. *Mohawk interruptus: Political life across the borders of settler states.* Durham, NC: Duke University Press.

Suburban Stats. 2016. Population demographics for Winnemucca, Nevada in 2016 and 2015. Accessed July 27, 2017. https://suburbanstats.org/population/nevada/how-many-people-live-in-winnemucca.

Telles, E., and V. Ortiz. 2009. *Generations of exclusion: Mexican Americans, assimilation, and race.* New York: Russell Sage Foundation.

Thrush, C. 2007. *Native Seattle: Histories from the crossing-over place.* Seattle: University of Washington Press.

Wanke, J. 2005. Rawhide unveils its new Old West. *Arizona Republic* 17 December:B-1, B-2.

LAURA BARRACLOUGH is Associate Professor of American Studies at Yale University, New Haven, CT 06520. E-mail: laura.barraclough@yale.edu. Her research examines how race, class, immigration, and colonialism are produced and negotiated through spatial practices in cities of the U.S. West.

20 The Legacy Effect

Understanding How Segregation and Environmental Injustice Unfold over Time in Baltimore

Morgan Grove, Laura Ogden, Steward Pickett, Chris Boone, Geoff Buckley, Dexter H. Locke, Charlie Lord, and Billy Hall

Legacies of social and environmental injustices can leave an imprint on the present and constrain transitions for more sustainable futures. In this article, we ask this question: What is the relationship of environmental inequality and histories of segregation? The answer for Baltimore is complex, where past practices of de jure and de facto segregation have created social and environmental legacies that persist on the landscape today. To answer this question, we examine the interactions among past and current environmental injustices in Baltimore from the late 1880s to the present using nearly twenty years of social and environmental justice research from the Baltimore Ecosystem Study (BES), a long-term social–ecological research project. Our research demonstrates that patterns and procedures in the city's early history of formal and informal segregation, followed by "redlining" in the 1930s, have left indelible patterns of social and environmental inequalities. These patterns are manifest in the distribution of environmental disamenities such as polluting industries, urban heat islands, and vulnerability to flooding, and they are also evident in the distribution of environmental amenities such as parks and trees. Further, our work shows how these legacies are complicated by changing perceptions of what counts as an environmental disamenity and amenity. Ultimately, we argue that the interactions among historical patterns, processes, and procedures over the long term are crucial for understanding environmental injustices of the past and present and for constructing sustainable cities for the future. *Key Words: Baltimore, distributive justice, environmental justice, procedural justice, segregation.*

社会与环境不正义的遗产, 能够在当下留下深刻的印记, 并对转变成为更具可持续性的未来产生限制。我们于本文中质问此一问题: 环境不公与隔离历史之间的关系为何? 对巴尔的摩而言, 答案相当复杂, 因其过往法律上与实际的隔离, 已创造了今日在地景上续存的社会及环境遗产。为了回答此一问题, 我们运用巴尔的摩生态系统研究 (BES) 这个长期的社会生态研究计画近乎二十年的社会与环境不公研究, 检视自 1880 年代晚期至今, 巴尔的摩的过往与当下环境不公之间的互动。我们的研究显示, 该城市早期正式与非正式的隔离历史模式与过程, 伴随着 1930 年代"拒绝贷款区"的划设, 已遗留了难以磨灭的社会与环境不公模式。这些模式, 在诸如污染工业、城市热岛和面对洪灾的脆弱性等不友善环境的分佈上十分显着, 且同时在诸如公园与植栽等友善环境的分佈上相当明显。再者, 我们的研究显示, 这些遗产如何受到有关何谓环境不友善与环境友善的认知改变而复杂化。我们最终主张, 长期的历史模式、过程与程序, 对于理解过往与当下的环境不正义以及打造未来可持续发展的城市而言至关重要。 关键词: 巴尔的摩, 分配正义, 环境正义, 程序正义, 隔离。

Los legados de las injusticias sociales y ambientales pueden trasmitir una huella al presente y obstaculizar las transiciones a futuros más sustentables. En este artículo formulamos esta pregunta: ¿Cuál es la relación entre la desigualdad ambiental y las historias de la segregación? Para Baltimore, la respuesta es compleja, donde las prácticas pasadas de la segregación de jure y de facto han generado legados que persisten en el paisaje actual. Para responder esta pregunta, examinamos las interacciones entre las injusticias ambientales pasadas y presentes de Baltimore desde los años 1880 hasta la actualidad, utilizando cerca de veinte años de investigación sobre justicia ambiental y social del Estudio del Ecosistema de Baltimore (BES), un proyecto de investigación socio-ecológica a largo plazo. Nuestra investigación demuestra que los patrones y procedimientos en la historia temprana de segregación formal e informal de la ciudad, seguida por la "discriminación" de los años 1930, han dejado patrones indelebles de desigualdades sociales y ambientales. Estos patrones se manifiestan en la distribución de incomodidades ambientales como industrias contaminantes, islotes urbanos de calor y vulnerabilidad a las inundaciones, las cuales también son evidentes en la distribución de atractivos ambientales, tales como parques y arbolado. Adicionalmente, nuestro trabajo muestra cómo estos legados se complican por las percepciones cambiantes sobre lo que cuenta como una incomodidad ambiental o un atractivo. Por último, argüimos que las interacciones entre los patrones, procesos y procedimientos históricos a plazo largo son cruciales para entender las injusticias ambientales del pasado y del presente y para construir ciudades sostenibles para el futuro. *Palabras clave: Baltimore, justicia distributiva, justicia ambiental, justicia procedimental, segregación.*

In this article, we argue that Baltimore's history of racial and economic segregation has produced patterns of environmental disamenities and amenities that sometimes counter expectations about the sociospatial characteristics of environmental justice (EJ). EJ emerged in the United States in the 1980s as a social movement and field of scholarship in response to the disproportionate exposure of people of color to environmental hazards (Boone 2010). Classic EJ theory has repeatedly demonstrated the correlations among race, class, and the distribution of environmental hazards and benefits (United Church of Christ 1987; Bullard 1990; Colten and Skinner 1996). White privilege has sociospatial characteristics as well, as Pulido (2015) theorized, resulting in landscapes where communities of color are disproportionately exposed to environmental hazards and white neighborhoods are insulated from those risks.

Counter to classic expectations, Baltimore reveals exceptions: Whites live closer to polluting industries and African Americans have greater access to parks, for instance. Twenty years of EJ research in Baltimore suggests that the city's history of de jure and de facto segregation, followed by "redlining" in the 1930s, has left legacies that account for the sociospatial distribution of environmental amenities and disamenities in the city today. Attention to these legacies helps us understand the complex temporal dynamics of urban EJ, complexities sometimes missing from ahistorical analyses.

Baltimore's racial and economic segregation mirrors the history of other cities in the United States.[1] After the Civil War, Baltimore at the ward level exhibited a mixture of small, affordable houses and grander homes for managers and entrepreneurs, members of different classes, and whites and African Americans living among each other (Duneier 2016). After Reconstruction, with the hardening of racial discrimination and the arrival of new immigrant groups, Baltimore's social mosaic became less diverse at the ward level, with households of similar class and identical race most frequently together. At the same time, new legal and procedural mechanisms furthered ward-level patterns of segregation and social differentiation in Baltimore. As we describe here, these mechanisms included the first, and short-lived, municipal segregation ordinance in the United States; subsequent deed covenants; and the discriminatory activities of neighborhood improvement associations to compensate for the ordinance's invalidation. Sociospatial practices of exclusion, often at the ward level, produced a fundamentally heterogeneous city.

Our EJ research examines historic and current patterns and processes of social injustice by focusing on interactions among distributive and procedural justice and disamenities and amenities over time. Further, we situate our EJ research in the larger context of research on the city's environmental heterogeneity over the long term, a focus of the Baltimore Ecosystem Study (BES) Long-Term Ecological Research project established in 1997. The ecology of Baltimore is highly differentiated. Situated on the Fall Line, Baltimore straddles the hilly Piedmont with its deep stream valleys and the sandy coastal plain with its estuarine edge. The city is dissected by three major streams. The heterogeneity of geology, native vegetation, and drainage are fundamental features of the region. BES has revealed the spatial patterns of soil contamination by heavy metals; the pollution of different stream reaches by road deicers, pharmaceuticals, and personal care products; and nitrate from leaky city sewers and suburban septic systems. It has documented the effects of restoration on streams, the impacts of storm water retention basins, and the impacts of green infrastructure. These heterogeneities provide an unusually rich understanding of the ecological dimensions of environmental inequity, which subsequently affect EJ. At the same time, our research on EJ is important to understanding the dynamics of the Baltimore region as a social–ecological system over the long term.

Our research in Baltimore incorporates insights from EJ scholarship (e.g., Shrader-Frechette 2002),

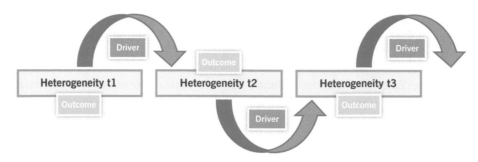

Figure 1. Dynamic heterogeneity is a framework that investigates how a given socioecological state is composed of both drivers and outcomes of heterogeneity. Although considering heterogeneity to be both drivers and outcomes might seem circular, this representation explains why that is not the case. Causation has a temporal dimension that separates the status of the socioecological state as outcomes and drivers and acknowledges the role of social, economic, and ecological phenomenon as both a cause and consequence. Thus, drivers responding to pattern at t_1 produce a new spatial pattern at t_2 and a subsequent driver responding to the spatial pattern at t_2 produces a new pattern for t_3. (Color figure available online.)

urban political ecology (Swyngedouw and Heynen 2003; Heynen et al. 2006; Heynen 2014), and ecosystem ecology (Cadenasso, Pickett, and Grove 2006; Pickett et al. 2011). It is operationalized through the analytic framework of dynamic heterogeneity (*sensu* Pickett et al. 2016). The dynamic heterogeneity framework posits that a pattern of spatial heterogeneity at a given point in time can be affected by social and environmental events, interventions, and actions, resulting in a new pattern of heterogeneity (Figure 1). Further, spatial heterogeneity and social–environmental events can cascade through time, generating social and environmental legacies and lags that could shape future outcomes. Because EJ is about spatial heterogeneity or distributive justice and outcomes on one hand and drivers of procedural justice on the other hand, dynamic heterogeneity can help frame the interactions of these two components of EJ in terms of spatial and temporal explanations of EJ over time (Figure 2).

A critical feature of the dynamic heterogeneity framework is that it can be used as an analytical tool to unpack long-term interactions among distributive justice, which involves patterns of disamenities and amenities, and procedural justice by de jure and de facto allocation of disamenities and amenities. Examples of disamenities include polluting industries, urban heat islands, and vulnerability to flooding, and examples of amenities include such features as parks and trees. Unpacking and disentangling these long-term interactions can reveal, for instance, the complex relationships among patterns of amenities and procedures of disamenities and, conversely, patterns of disamenities and procedures of amenities (Figure 2). Further, by investigating the dynamics among patterns and procedures, we can uncover social and environmental legacies and changing perceptions that might be associated with both distributive and procedural justice.

Our research in Baltimore emphasizes the need to examine EJ as more than a snapshot of correlations at one point in time. We understand the heterogeneity of urban landscapes as the legacy and result of uneven

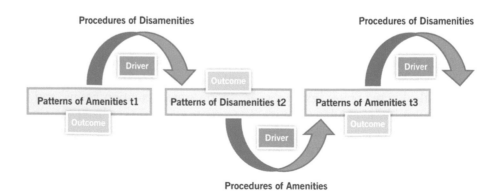

Figure 2. The dynamic heterogeneity framework can be used to explore interactions among distributive and procedural justice and amenities and disamenities. For example, patterns of amenities at t_1 could affect the allocation of disamenities that produce patterns of disamenities at t_2. Subsequently, patterns at t_2 could affect the allocation of amenities for patterns of amenities at t_3. (Color figure available online.)

development and social injustices. Geographers and other urban theorists have built on Marxist political economy to examine how capital accumulation in cities produces patterns of racialized segregation and the uneven distribution of social and environmental amenities. In general, urban political economic theory suggests that land use in cities is driven by the maximum accumulation of capital (Harvey 1978, 1987) or "highest and best use" (Blomley 2004, 4). For example, Smith (1987) argued that patterns of uneven development are the instantiation of a capitalist logic, where patterns of uneven investment enable gentrification to transform poor neighborhoods and displace residents (Smith 1987; Heynen et al. 2006). Baltimore's long history of state-sponsored neglect and decades of segregation have created neighborhoods seemingly abandoned by capitalist investment, requiring attention to the legacies of uneven development to understand contemporary processes and patterns of environmental inequality. The term *dynamic* in dynamic heterogeneity invokes changes, metastability, feedbacks, path dependencies, and legacies manifest in the past, present, and potential futures. Procedural justice, the de jure and de facto allocation of disamenities and amenities, can be significant instrumentally to the dynamics of urban ecological systems.

The Political Economy of Segregation in Baltimore

Baltimore is located on the Chesapeake Bay. It was once the second leading port of entry for immigrants to the United States and a major manufacturing center. In the late 1800s, Baltimore, like most industrial cities, faced many social–environmental challenges. Private sanitation services dumped household and industrial sewage into the harbor, compounding the problems caused by ship discharge. More than 20,000 cesspits drained illegally into the Jones' Falls, one of the three major streams in the city. As the city rapidly industrialized and expanded in the late nineteenth century, elected and civic leaders began to recognize that its modest system of parks and squares did not meet the needs of its residents and that those amenities were not equally distributed (Korth and Buckley 2006).

A sequence of three planning paradigms responded to Baltimore's environmental inequalities and hazards: Progressivist reconstruction after the fire of 1904, urban renewal in the 1950s to 1970s, and urban sustainability since the 2000s. Baltimore's devastating 1904 fire, which destroyed seventy city blocks of

the downtown area (140 acres), enabled a progressive approach to redevelopment (Euchner 1991). Over the next decade, the city would employ comprehensive approaches to water supply and sewers (Boone 2003), roads (Buckley, Boone, and Grove 2017), a park system (Crooks 1968; Korth and Buckley 2006), city arboriculture (Boone et al. 2009), and zoning (Lord and Norquist 2010). Urban renewal, beginning with the Federal Housing Act of 1949, funded slum removal and neighborhood revitalization, new home construction, and the development of open space and landscaping. More recently, the sustainability period was ushered in by the adoption of a forward-looking sustainability plan in 2009, targeting six areas of concern: cleanliness, pollution prevention, resource conservation, greening, environmental education, and green economy. During each planning period, concerns for development, environmental health, and equality have been central, although shifting political and economic realities have challenged implementation, including declines in manufacturing, industrialization, population, and rail transportation since 1950.

Baltimore persistently constrained black families to dense and relatively expensive housing through deed restrictions, covenants, and municipal ordinances (Power 1983; Olson 1997). After the Civil War, African Americans lived throughout the city. By the early twentieth century, however, block-by-block segregation began to give way to sizable hemmed-in ghettos in East Baltimore, West Baltimore, and South Baltimore (Power 1983). In May 1911, following hostilities after a black lawyer moved into a white neighborhood, Baltimore enacted the first municipal segregation ordinance in the United States, which was authored by progressives who agreed that "blacks should be quarantined in isolated slums in order to reduce the incidents of civil disturbance, to prevent the spread of communicable disease into the nearby white neighborhoods, and to protect property values among the white majority" (Power 1983, 301).

In 1917, the ordinance fell after a decision by the U.S. Supreme Court. In response to the Supreme Court's decision, the mayor set out to replace the de jure segregation with de facto segregation, "enforced by a conspiracy in restraint of rental or sale" of housing to blacks on blocks that had been set out as white neighborhoods (Power 1983, 318). The plan was to use white property associations, the real estate board, the health department, and the city building inspector to ensure that African Americans left the neighborhoods where they were in the minority and did not enter those neighborhoods that were already white. Over time, the conspiracy grew and formalized, with white neighborhood associations adopting racial segregation as a top priority and neighborhood protection associations passing restrictive covenants that prohibited the rental or sale of properties to blacks. Such deed restrictions imposed by property owners persisted well into the 1970s, even after the Fair Housing Act of 1964. By 1925, seventeen neighborhood improvement associations had held meetings to discuss various perceived threats to their neighborhoods and coordinated their efforts across neighborhoods to share information and enforce restrictive practices (Buckley and Boone 2011).

During this same period, banks and the federal government began to use race in a way that isolated black residents in certain neighborhoods. In 1937, the Federal Home Owners' Loan Corporation (HOLC) was charged with refinancing homes in danger of foreclosure. The HOLC assigned a security grade to each neighborhood based on the perceived risk of default, using criteria that included occupation of residents, annual income, predominant nationality, percentage of "negro families," percentage of families on relief, and "threat of

infiltration of foreign born, negro or lower grade population" (Grove et al. 2015, 138–39). The neighborhoods deemed at highest risk were labeled hazardous and mapped in red—hence the term *redlining*[2] (Figure 3). The effect of these various policies, programs, laws, and practices has been to segregate African Americans into areas of West and East Baltimore and some sections of South Baltimore, institutionalizing a racially based legacy of disinvestment characterized by overcrowding, poor housing quality, encroachment of industrial uses, and noise from nearby businesses.

Unexpected Outcomes, Legacies, and Changing Perceptions

Here we exemplify how Baltimore's contemporary social–environmental landscape is the result of changing patterns of racial and economic segregation. At different periods in Baltimore's history, practices of social exclusion coproduced patterns of environmental inequality and racism, such as environmentally related zoning variances. Yet there is sometimes a lag between the

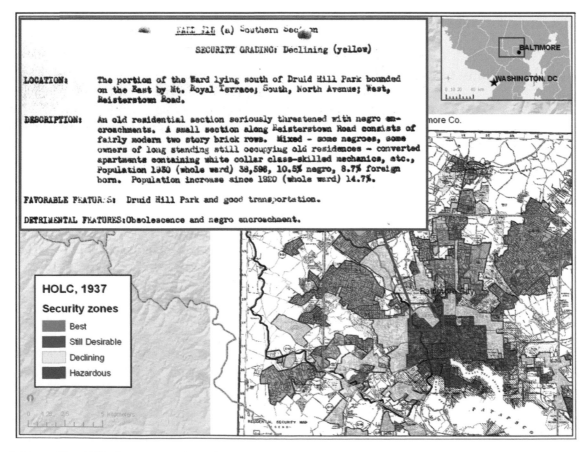

Figure 3. Baltimore City, 1937 Home Owners' Loan Corporation map (Grove et al. 2015). The first sentence of the description states, "An old residential section seriously threatened with negro encroachment." HOLC = Home Owners' Loan Corporation. (Color figure available online.)

temporal dynamics of social and environmental change, resulting in contemporary sociospatial distributions of environmental benefits and hazards in Baltimore that do not always conform to classic EJ explanations. Further complicating the contemporary landscape are changing perceptions of what counts as an environmental amenity. Thus, EJ research in Baltimore illustrates the importance of a historical approach to understanding the city's dynamic heterogeneity. History reveals how legacies of past practices of segregation account for contemporary sociospatial distributions of environmental hazards and benefits.

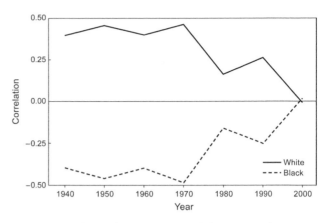

Figure 4. Correlation between race and distance to disamenities (adapted from Lord and Norquist 2010).

Distributive Justice and Disamenities

Our analyses of EJ in Baltimore have generated both expected and unexpected patterns. Several expected patterns of disamenities emerge: Nonconforming, environmentally related zoning variances were disproportionately approved for African American neighborhoods and disapproved for white neighborhoods between 1930

and 1970 (Lord and Norquist 2010; Figure 4). The concentration of vacant lots and abandoned buildings and the absence of trees in neighborhoods reflect the HOLC's redlining (Grove et al. 2015; Figure 5). The absence of trees produces collateral social and environmental disamenities, including higher levels of crime (Troy, Grove, and O'Neil-Dunne 2012; Troy, Nunery, and Grove

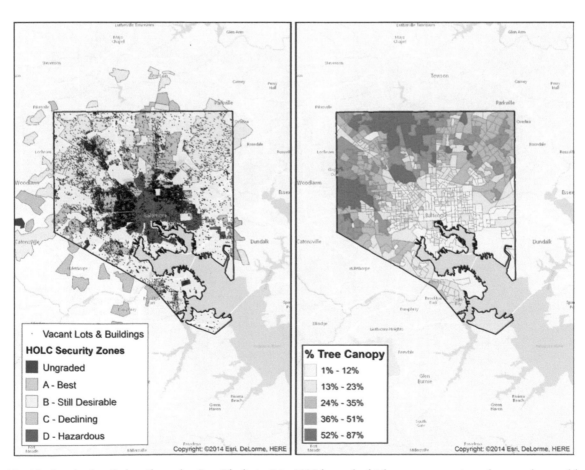

Figure 5. Neighborhoods classified as "hazardous" or "declining" in 1938 have the highest concentration of vacant lots and buildings in 2012 and the lowest percentages of canopy cover (Grove et al. 2015). (Color figure available online.)

2016) and higher temperatures and vulnerability to heat waves (Huang et al. 2011; Huang and Cadenasso 2016).

The unexpected patterns of EJ in Baltimore are understandable through the legacies of racial and economic segregation. The current pattern of Toxic Release Inventory (TRI) sites, representing industrial land uses and pollution, is predominantly associated with majority-white neighborhoods, contrary to the usual pattern of association with minority communities (Figures 6 and 7). The amenity of living close to the factories where they worked was reserved mainly for whites. Although some blacks could secure the lowest paid factory jobs, they had to travel long distances to work. Thus, housing near factories later came to be identified as neighborhoods with TRI sites (Boone 2002).

Distributive Justice and Amenities

It is not only disamenities that are of concern for EJ. Indeed, access to amenities, including access to parks and tree canopy cover, is also a concern. Once again, we find unexpected results that are understandable through the lens of Baltimore's history of racial and economic segregation. Today, African American neighborhoods enjoy greater access to parks within walking distance than white neighborhoods, although whites generally have access to more acres (Boone et al. 2009). The amenity of proximity to parks was not justly achieved, as African Americans were historically excluded from participating in recreational activities even when they lived close to a

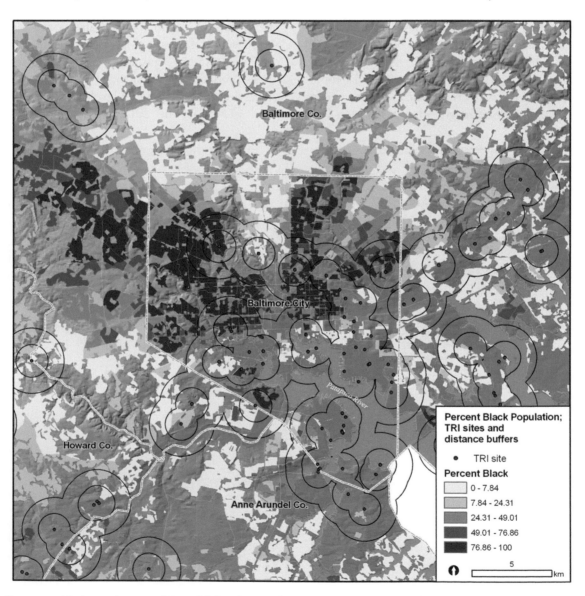

Figure 6. Percentage black population and 1- and 2-km distance buffers from TRI sites for Baltimore City and surrounding areas (Boone 2002). TRI = Toxic Release Inventory. (Color figure available online.)

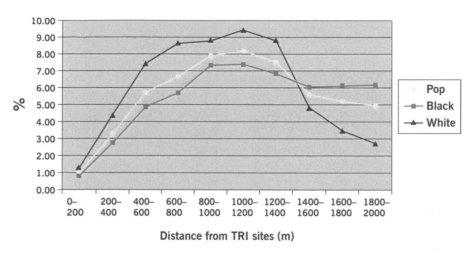

Baltimore City
Percent of Race Category Population

Figure 7. Percentage of race category populations by distance zones from TRI sites, Baltimore City (Boone 2002). TRI = Toxic Release Inventory. (Color figure available online.)

neighborhood park. Only after large numbers of whites migrated to the suburbs, often prompted by white real estate agents engaged in block busting, was the African American population able to disperse and gain access to all of the city's parks and recreational facilities, including golf courses and swimming pools (Wells, Buckley, and Boone 2008). Thus, the current distribution of parks is more associated with the legacies of white privilege and an "inherited landscape" (Boone et al. 2009).

The legacies of white privilege and inherited landscapes are also present in the current distribution of urban tree canopy. Urban tree canopy is positively associated with the percentage of African Americans currently living in a neighborhood (Grove et al. 2006; Troy et al. 2007; Boone et al. 2010). This is attributable in part to the legacy of the white neighborhood improvement associations that worked to establish the city's forestry division and to plant trees in their neighborhoods (Buckley and Boone 2011). Just as African Americans inherited greater access to parks, they also inherited greater tree canopy cover as they dispersed into new neighborhoods.

Procedural Justice and Disamenities

Patterns of disamenities are the stock in trade of EJ research, but it is critical to examine the mechanisms by which the distribution of environmental inequities come to exist. Lord and Norquist (2010) studied how

racial biases in the decisions about nonconforming zoning variances associated with environmental disamenities varied between 1940 and 2000. For each decade from 1940 to 1990, race and the distance to disamenities were correlated: The higher the percentage of African American residents, the closer to disamenities, and the higher the percentage of white residents, the farther away the disamenities. Beginning in 1970, however, the correlation between race and proximity to disamenities began to weaken, and in 2000 there was no correlation, coinciding with the period during which the city became predominantly African American and the zoning variance approval process was reformed (Figure 4).

Several procedural factors produced this bias, causing a disproportionate number of zoning variances to be approved for redlined neighborhoods (Lord and Norquist 2010). First, the legacy of prior neighborhood conditions such as obsolescence of housing stock, encroachment of industrial uses, or noise from businesses, which had already been concentrated in redlined neighborhoods, provided a rationale for zoning variance approvals. Second, businesses purposefully sought to locate their nonconforming operations in African American neighborhoods because they thought those neighborhoods were poor, poorly organized, and unlikely to resist. When African American neighborhoods did resist, the neighborhoods had a high rate of success. Where African American neighborhoods could afford a lawyer, the zoning board disallowed the variance in a majority of cases (Lord and Norquist 2010). This suggests that lack of economic and legal resources in African American neighborhoods often put them at a disadvantage.

Procedural Justice and Amenities

Our research has examined a variety of processes affecting the allocation of amenities, including segregated parks and park uses, biases in park acquisition, and tree planting programs since the early 1900s. A major social movement in Baltimore, and many other cities in the United States, was African Americans' resistance to exclusionary practices that prohibited them from using the city's golf courses, pools, tennis courts, and beaches.

Baltimore began to construct municipal golf courses in the 1920s, all of which were "white only." In 1934, Baltimore's Board of Public Park Commissioners (BPPC) allowed African Americans to golf at Carroll Park—a poorly designed and maintained course with only nine holes—after the Monumental Golf Club of Baltimore, an African American organization, challenged the "white only" policy. In part, BPPC made this change because they did not believe that an increased African American presence in the predominantly industrial area would have negative impacts on nearby white neighborhoods. The surrounding neighborhoods strenuously objected, though, and the BPPC was forced to rescind its decision. Soon after, a compromise was reached by splitting course privileges and providing whites exclusive rights on Tuesdays, Thursdays, and Saturdays and second and fourth Sundays. Two years later, African Americans were granted unrestricted access to the course but remained barred from the other municipal courses (Wells et al. 2008).

The Monumental Golf Club eventually sued the city for access to the other courses, claiming that Carroll Park did not meet the "separate but equal" standard established by the U.S. Supreme Court because of the course's inferior qualities. A subtext to the lawsuit was to go after the segregationists' wallets, assuming that they would rather grant African Americans access to the city's other golf courses than spend money to upgrade Carroll Park (Olson 1997). With persistence and threats of lawsuits, the BPPC finally granted full access to all of the municipal golf courses, regardless of race or ethnicity. Again, white neighborhood associations petitioned the BPPC, arguing that white neighborhoods needed to be protected from African American "invasion" and the detriment to property values surrounding the newly desegregated

courses (Wells et al. 2008). Eventually, the BPPC relaxed its rules against mixed play among the races on its golf courses, baseball diamonds, tennis courts, and other athletic fields (Gibson and Yoes 2004).

During this period, the Baltimore Board of Public Recreation concluded that the city had inadequate acreage in parks, especially for children's playgrounds (Figure 8), and that the "colored community is lacking in areas and facilities quite out of proportion to the ratio of its numbers to the total population" (Pangburn and Allen 1941, ix). The Board had also received a large financial gift from a private donor for public park acquisition. Whereas many advocated for the purchase of small playgrounds in East Baltimore, others sought a large park in West Baltimore. Theodore Marburg, Chairman of the Municipal Arts Society, contacted his old friend Frederick Law Olmsted, Jr. for advice.

Although previous reports in 1904 and 1926 made it clear that Olmsted wished to provide the city with "a roughly equitable distribution" of parks and recreation facilities "for all its citizens" (Korth and Buckley 2006, 2), Olmsted replied to Marbury that although playgrounds in East Baltimore were needed, such an expenditure would be risky given the condition and instability of the neighborhoods. The city heeded Olmsted's advice and made the first of several purchases of large tracts of stream valley lands in West Baltimore in 1941 (Korth and Buckley 2006). Once again, the legacy of preexisting conditions of deteriorated neighborhoods justified the procedural decision not to invest in amenities.

Urban tree cover is typically seen as an environmental benefit. During the Progressive era, neighborhood associations lobbied for tree planting and maintenance (Buckley 2010; Buckley and Boone 2011). This process continued during the urban renewal programs in the 1950s and 1960s. For instance, the Bolton Hill neighborhood—which had declined due to block busting—advocated for tree planting investments based on its preexisting conditions and former prominence as one of the wealthiest neighborhoods in Baltimore (Merse, Buckley, and Boone 2009). At the same time that Bolton Hill was planting trees, other neighborhood associations were actively discouraging tree planting programs. When tree planting started in predominantly white neighborhoods in East Baltimore, the City's Forestry Division discovered that many residents did not perceive trees as an amenity and opposed tree planting in their neighborhoods (Buckley 2010).

GROUND BREAKING CEREMONIES
CLOVERDALE PLAYGROUND
JULY 13, 1949

Figure 8. Baltimore Mayor Thomas D'Alesandro at the groundbreaking ceremony for Cloverdale Playground, 13 July 1949. Separate facilities for African Americans were often enforced through de jure and de facto segregation. University of Baltimore Special Collections & Archives, Thomas D'Alesandro, Jr. Collection, Series IV-C, Box 4. (Reproduced courtesy of the University of Baltimore.)

Some predominantly African American neighborhoods in East Baltimore remain virtually treeless despite the best efforts of city foresters since the City's urban renewal programs. Starting in the 1960s, large numbers of African Americans migrated to this section of East Baltimore and inherited this landscape devoid of trees. This demographic shift did not signal a change in attitude toward urban trees, however. As the city has implemented its sustainability plan, Battaglia et al. (2014) found that although residents are not opposed to more trees, they question the ability of the city to manage its existing street trees and voice concern about significant negative costs of these "so-called amenities," including gentrification. They argue that the city has more important disamenities that it should solve before it embarks on a tree planting campaign. Similar results are found in the city's tree giveaway programs, where addresses of Baltimore participants in free tree giveaway programs were most likely to be from the most affluent neighborhoods with the highest rates of canopy cover. In contrast, less affluent neighborhoods with much lower rates of canopy cover were much less likely to participate in these free programs (Locke and Grove 2016). Although the issues affecting participation can be complex, these experiences indicate that perceptions of disamenities and amenities might vary over time and among different social groups.

Conclusion

We have recognized EJ as a normative stance and scholarly pursuit. As a normative stance, EJ is fundamental to sustainability and transitions for the future. What might we learn from the past to achieve current and future sustainability goals? Sustainability is in essence a set of desired outcomes and associated with future patterns of distributive justice. Sustainability policies and planning, however, might also look at the current distribution of outcomes associated with sustainability goals and how past social and environmental procedures, processes, and legacies have produced the current distribution. Without these types of understandings, the rationale and ability to achieve future sustainability outcomes might be severely limited.

For more than 100 years, policymakers and planners have called for the equitable distribution of environmental benefits in the City of Baltimore. We propose that more attention also needs to be paid to transitions:

how the equitable distribution of sustainability outcome goals is produced. Transitions are associated with procedural justice, which includes fairness in the application of environmental and other laws; development and maintenance of fair institutions; fairness in decision making; and recognition, enfranchisement, and removal of barriers to participation by marginalized groups as stakeholders in decisions (Boone 2008, 2010; Schlosberg 2009; Boone and Klinsky 2016). This might be summarized in terms of principles: (1) expand and enable partnerships (who participates), (2) empower people and foster participation (how people participate), and (3) enable good governance (the even application of laws). More basically, to paraphrase Inez Robb, a local activist and member of the Baltimore Sustainability Commission, "we need to change from the City doing things *to* the people, from the City doing things *for* the people, to the City doing things *with* the people" (Robb 2016). Understanding the barriers and missed opportunities in procedural justice in the past could help build practices and institutions for just procedures for the present and future.

The long-term view and attention to each of the components in our analytical framework of dynamic heterogeneity (Figure 2) makes manifest the potential for dynamic interactions among distributive and procedural justice and among disamenities and amenities, including legacies of racial biases and changing perceptions. Understanding these dynamics requires both social and environmental explanations. We do not witness a social system acting on a passive environmental system. An inclusive perspective of both social and biophysical sciences is required. Promising new avenues for EJ scholarship could build on theories, data, and methods from other fields and approaches, including land use law, industrial and

housing location theory, land economics, hazards and vulnerability, political ecology, public health, and environmental sciences. In particular, EJ scholarship could benefit from putting more "environment," meaning an understanding of ecological processes, into EJ (Pickett, Boone, and Cadenasso 2007; Boone 2008). Knowledge about how ecosystems are structured and function—the spread of diseases or flows of water contaminants—would allow EJ researchers to tap into rich bodies of knowledge that could lead to improved understanding of the processes of environmental inequality. EJ collaborations with biological and physical scientists might also create opportunities to work with new and underexplored data sets in EJ research, such as soil surveys for lead or atmospheric dynamics for air pollution models. Conversely, knowledge about EJ theories, data, and methods could be crucial to a general understanding of patterns and processes of urban social–ecological systems over the long term.

Although our analytical framework can be useful for capturing dynamic interactions and social and environmental chains of explanation of the past and present (Figure 9), we propose that it could also be used to advance EJ as a forward-looking, actionable science that is concerned with producing just, sustainable, and resilient futures (Childers et al. 2015). The need for such an endeavor is not unique to Baltimore. Although the history of racial segregation and redlining is ubiquitous in many postindustrial cities, so is the desire to create more sustainable urban futures for all.

Finally, we propose that there is a need to systematically consider the long-term role and legacies of segregation of people and place to understand EJ for the past, present, and future. Patterns, processes,

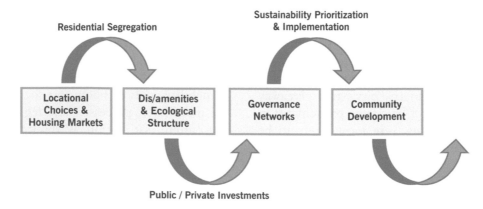

Figure 9. Patterns and processes of distributive and procedural justice contributing to historical and contemporary environmental injustices viewed through the dynamic heterogeneity framework. (Color figure available online.)

and systems of racial segregation have produced social and environmental places of environmental injustice that are different in type and not degree. Current urban ecological research often implicitly assumes "landscapes of choice," with households choosing where to live and how to manage their lands, unconstrained by biases of race, ethnicity, or religion. We propose, however, that there is also the need to understand and address the ecology of people and place where choice is not, or was not, present. We might call this an *ecology of segregation*. Understanding its people, place, legacies, and long-term dynamics might be crucial to achieving more sustainable and resilient cities.

Acknowledgments

We thank two anonymous reviewers and the editor for their constructive suggestions. They added clarity to the article and identified novel directions.

Funding

This material is based on work supported by the U.S. Department of Agriculture (USDA) Forest Service and the National Science Foundation (NSF) under Grant No. DEB-1027188 to the Baltimore Ecosystem Study, a project of the NSF Long-Term Ecological Research (LTER) Program, and to the National Social–Environmental Synthesis Center under Grant No. DBI-1052875. Any opinions, findings, and conclusions or recommendations expressed in this material are those of the author(s) and do not necessarily reflect the views of the USDA Forest Service or the National Science Foundation.

Notes

1. We note that other scholars have examined segregation in Baltimore and its relationships to public health (Roberts 2009; Markowitz and Rosner 2013), recreation (Wiltse 2007), policing (Alexander 2012), and government at federal and local levels as public actors in segregation (Rothstein 2017). These phenomena are critical to a systematic study of segregation and environmental justice.
2. For an assembly of HOLC maps for the United States, see Nelson et al. (n.d.).

References

Alexander, M. 2012. *The new Jim Crow: Mass incarceration in the age of colorblindness.* New York: The New Press.

Battaglia, M., G. L. Buckley, M. Galvin, and J. M. Grove. 2014. It's not easy going green: Obstacles to tree-planting programs in East Baltimore. *Cities and the Environment* (CATE) 7 (2):6.

Blomley, N. K. 2004. *Unsettling the city: Urban land and the politics of property.* London and New York: Routledge.

Boone, C. G. 2002. An assessment and explanation of environmental inequity in Baltimore. *Urban Geography* 23 (6):581–95.

———. 2003. Obstacles to infrastructure provision: The struggle to build comprehensive sewer works in Baltimore. *Historical Geography* 31:151–68.

———. 2008. Environmental justice as process and new avenues for research. *Environmental Justice* 1 (3):149–54.

———. 2010. Environmental justice, sustainability and vulnerability. *International Journal of Urban Sustainable Development* 2 (1–2):135–40.

Boone, C. G., G. L. Buckley, J. M. Grove, and C. Sister. 2009. Parks & people: An environmental justice inquiry in Baltimore, Maryland. *Annals of the Association of American Geographers* 99 (3):767–87.

Boone, C. G., M. L. Cadenasso, J. M. Grove, K. Schwartz, and G. L. Buckley. 2010. Landscape, vegetation characteristics, and group identity in an urban and suburban watershed: Why the 60s matter. *Urban Ecosystems* 13 (3):255–71.

Boone, C. G. and S. Klinsky. 2016. Environmental justice and transitions to a sustainable urban future. In *Handbook of urbanization and global environmental change*, ed. K. C. Seto, W. D. Solecki, and C. A. Griffith, 327–35. London and New York: Routledge.

Buckley, G. L. 2010. *America's conservation impulse: A century of saving trees in the Old Line state.* Chicago: Columbia College and the Center for American Places.

Buckley, G. L., and C. G. Boone. 2011. "To promote the material and moral welfare of the community": Neighborhood improvement associations in Baltimore, Maryland, 1900–1945. In *Environmental and social justice in the city: Historical perspectives*, ed. G. Massard-Guilbaud and R. Rodger, 43–65. Cambridge, UK: White Horse Press.

Buckley, G. L., C. G. Boone, and J. M. Grove. 2017. The greening of Baltimore's asphalt schoolyards. *The Geographical Review* 107 (3):516–35.

Bullard, R. D. 1990. *Dumping in Dixie: Race, class, and environmental quality.* Boulder, CO: Westview.

Cadenasso, M. L., S. T. A. Pickett, and J. M. Grove. 2006. Integrative approaches to investigating human-natural systems: The Baltimore ecosystem study. *Natures Sciences Sociétés* 14 (1):4–14.

Childers, D. L., M. L. Cadenasso, J. M. Grove, V. Marshall, B. McGrath, and S. T. A. Pickett. 2015. An ecology for cities: A transformational nexus of design and ecology to advance climate change resilience and urban sustainability. *Sustainability* 7 (4):3774–91.

Colten, C. E., and P. N. Skinner. 1996. *The road to Love Canal: Managing industrial waste before EPA.* Austin: The University of Texas Press.

Crooks, J. B. 1968. *Politics and progress: The rise of urban progressivism in Baltimore, 1895–1921.* Baton Rouge: Louisiana State University Press.

Duneier, M., 2016. *Ghetto: The invention of a place, the history of an idea.* New York: Macmillan.

Euchner, C. C. 1991. The politics of urban expansion: Baltimore and the sewerage question, 1859–1905. *Maryland Historical Magazine* 86 (3):270–91.

Gibson, L. S., and S. Yoes. 2004. The battle for the links: "Nine holes could never be equal to eighteen." *Baltimore Afro-American* February 14–20:3–4.

Grove, J. M., M. L. Cadenasso, S. T. A. Pickett, W. R. Burch, Jr., and G. E. Machlis. 2015. *The Baltimore School of Urban Ecology: Space, scale, and time for the study of cities.* New Haven, CT: Yale University Press.

Grove, J. M., A. R. Troy, J. P. M. O'Neil-Dunne, W. R. Burch, Jr., M. L. Cadenasso, and S. T. A. Pickett. 2006. Characterization of households and its implications for the vegetation of urban ecosystems. *Ecosystems* 9:578–97.

Harvey, D. 1978. The urban process under capitalism: A framework for analysis. *International Journal of Urban and Regional Research* 2:101–31.

———. 1987. The urbanization of capital: Studies in the history and theory of capitalist urbanization. *Science and Society* 51 (1):121–25.

Heynen, N. 2014. Urban political ecology I: The urban century. *Progress in Human Geography* 38 (4):598–604.

Heynen, N., M. Kaika, and E. Swyngedouw, eds. 2006. *In the nature of cities: Urban political ecology and the politics of urban metabolism.* London and New York: Routledge.

Housing Act of 1949 (Title V of Public Law 81–171 (7/15/49).

Huang, G., and M. L. Cadenasso. 2016. People, landscape, and urban heat island: Dynamics among neighborhood social conditions, land cover and surface temperatures. *Landscape Ecology* 31 (10):1–9.

Huang, G., W. Zhou, and M. L. Cadenasso. 2011. Is everyone hot in the city? Spatial pattern of land surface temperatures, land cover and neighborhood socioeconomic characteristics in Baltimore, MD. *Journal of Environmental Management* 92:1753–59.

Korth, C. A., and G. L. Buckley. 2006. Leakin Park: Frederick Law Olmsted, Jr.'s critical advice. *The Olmstedian* 16 (1, Fall). Accessed September 29, 2017. http://www.olmstedmaryland.org/research/

Locke, D. H., and J. M. Grove. 2016. Doing the hard work where it's easiest? Examining the relationships between urban greening programs and social and ecological characteristics. *Applied Spatial Analysis and Policy* 9 (1):77–96.

Lord, C., and K. Norquist. 2010. Cities as emergent systems: Race as a rule in organized complexity. *Environmental Law* 40:551–97.

Markowitz, G., and D. Rosner. 2013. *The politics of science: Lead wars and the fate of America's children.* Berkeley: University of California Press.

Merse, C. L., G. L. Buckley, and C. G. Boone. 2009. Street trees and urban renewal: A Baltimore case study. *The Geographical Bulletin* 50 (2):65–81.

Nelson, R. K., L. Winling, R. Marciano, N. Connolly, et al. n.d. Mapping inequality. In *American Panorama*, ed. R. K. Nelson and E. L. Ayers. Accessed May 14, 2017. https://dsl.richmond.edu/panorama/redlining/#loc=4/36.77/-96.86&opacity=0.8.

Olson, S. H. 1997. *Baltimore: The building of an American city.* Baltimore, MD: Johns Hopkins University Press.

Pangburn, W., and W. Allen. 1941. *Long range recreation plan, City of Baltimore, Maryland.* Baltimore, MD: Department of Public Recreation.

Pickett, S. T. A., C. G. Boone, and M. L. Cadenasso. 2007. Relationships of environmental justice to ecological theory. *Bulletin of the Ecological Society of America* 88 (2):166–70.

Pickett, S. T. A., M. L. Cadenasso, J. M. Grove, C. G. Boone, P. M. Groffman, E. Irwin, S. Kaushal, et al. 2011. Urban ecological systems: Scientific foundations and a decade of progress. *Journal of Environmental Management* 92:331–62.

Pickett, S. T. A., M. L. Cadenasso, E. J. Rosi-Marshall, K. T. Belt, P. M. Groffman, J. M. Grove, E. G. Irwin, et al. 2016. Dynamic heterogeneity: A framework to promote ecological integration and hypothesis generation in urban systems. *Urban Ecosystems* 20 (1):1–14.

Power, G. 1983. Apartheid Baltimore style: The residential segregation ordinances of 1910–1913. *Maryland Law Review* 42:289–328.

Pulido, L. 2015. Geographies of race and ethnicity 1: White supremacy vs. white privilege in environmental racism research. *Progress in Human Geography* 39 (6):809–17.

Robb, I. 2016. Public statement at Baltimore City Green Network kick-off meeting, Baltimore City Department of Planning, Baltimore, MD.

Roberts, S. 2009. *Infectious fear: Politics, disease, and the health effects of segregation.* Chapel Hill: University of North Carolina Press.

Rothstein, R. 2017. *The color of law: A forgotten history of how our government segregated America.* New York: Liveright.

Schlosberg, D. 2009. *Defining environmental justice: Theories, movements, and nature.* New York: Oxford University Press.

Shrader-Frechette, K. 2002. *Environmental justice: Creating equity, reclaiming democracy.* New York: Oxford University Press.

Smith, N. 1987. Gentrification and the rent gap. *Annals of the Association of American Geographers* 77:462–65.

Swyngedouw, E., and N. C. Heynen. 2003. Urban political ecology, justice and the politics of scale. *Antipode* 35:898–918.

Troy, A. R., J. M. Grove, and J. P. M. O'Neil-Dunne. 2012. The relationship between tree canopy and crime rates across an urban–rural gradient in the Greater Baltimore Region. *Landscape and Urban Planning* 106 (3):262–70.

Troy, A. R., J. M. Grove, J. P. M. O'Neil-Dunne, S. T. A. Pickett, and M. L. Cadenasso. 2007. Predicting opportunities for greening and patterns of vegetation on private urban lands. *Environmental Management* 40 (3):394–412.

Troy, A. R., A. Nunery, and J. M. Grove. 2016. The relationship between residential yard management and neighborhood crime: An analysis from Baltimore City and County. *Landscape and Urban Planning* 147:78–87.

United Church of Christ Commission for Racial Justice. 1987. *Toxic wastes and race in the United States: A national report on the racial and socio-economic characteristics of communities with hazardous waste sites.* Public Data Access. New York: United Church of Christ Commission for Racial Justice.

Wells, J., G. L. Buckley, and C. G. Boone. 2008. Separate but equal? Desegregating Baltimore's golf courses. *The Geographical Review* 98 (2):151–70.

Wiltse, J. 2007. *Contested waters: A social history of swimming pools in America.* Chapel Hill: University of North Carolina Press.

MORGAN GROVE is the team leader for the U.S. Department of Agriculture Forest Service's Baltimore Field Station, Baltimore, MD 21228. E-mail: morgangrove@fs.fed.us. His research interests include the social and ecological dynamics of residential landscapes and neighborhoods, particularly in the context of environmental justice.

LAURA OGDEN is an Associate Professor in the Department of Anthropology at Dartmouth College, Hanover, NH 03755-3529. E-mail: laura.a.ogden@dartmouth.edu. Her research interests include the political ecology and ethnography of rural and urban landscapes.

STEWARD PICKETT is a Senior Scientist at the Cary Institute of Ecosystem Studies, Millbrook, NY 12545-0129. E-mail: picketts@caryinstitute.org. His research interests include spatial heterogeneity and dynamics of social-ecological systems in urban areas.

CHRIS BOONE is Professor and Dean of the School of Sustainability at Arizona State University, Tempe, AZ 85287-2402. E-mail: christopher.g.boone@asu.edu. His research includes ongoing debates in sustainable urbanization, environmental justice, vulnerability, and global environmental change.

GEOFF BUCKLEY is Professor in the Department of Geography at Ohio University, Athens, OH 45701-2978. E-mail: buckleg1@ohiou.edu. His research focuses on environmental history and the legacy effects of past decisions, practices, and processes of urban systems.

DEXTER H. LOCKE is a Postdoctoral Fellow at the National Socio-Environmental Synthesis Center, Annapolis, MD 21401. E-mail: dexter.locke@gmail.com. His research focuses on the social-environmental dynamics of residential landscapes.

CHARLIE LORD is an environmental lawyer and Principal in Renew Energy Partners, Boston, MA 02111. E-mail: Lord@renewep.com. His research and activism has focused on environmental justice in Boston and Baltimore.

BILLY HALL is a Postdoctoral Fellow at the National Socio-Environmental Synthesis Center, Annapolis, MD 21401. E-mail: billyhallthe3rd@gmail.com. His research has focused on the relationship between historic segregation and food deserts in cities.

21 "This Port Is Killing People"

Sustainability without Justice in the Neo-Keynesian Green City

Juan De Lara

This article examines how regional policymakers in Southern California deployed a green growth strategy that cemented racial, environmental, and class precariousness into the region's ecological fabric. It uses participant observation and extant data to show how environmentalist statecraft provided ideological cover for a type of neo-Keynesian logistics growth regime that used infrastructure spending to stimulate the economy without addressing underlying issues of racial, economic, and environmental justice. Urban political ecology and racial capitalism are used as theoretical frameworks to stretch the boundaries of how sustainability is conceptualized and to challenge assumptions behind a green capitalism framework. Finally, the article examines how labor and environmental justice activists used what Sze et al. (2009, 836) called "cultural and ecological discourses" to challenge the green capitalist agenda by incorporating subaltern spatial imaginaries. *Key Words: environmental racism, logistics, ports, racial capitalism, social movements, sustainable development, urban ecology, urbanism.*

本文检视南加州的区域政策制定者, 如何部署绿色成长策略, 将种族、环境与阶级的不安定固定于该区域的生态纹理之中。本研究运用参与式观察与当下的数据, 展现环境保护主义的国家机器, 如何提供意识形态, 掩盖一种运用基础建设花费来刺激经济, 但却未能应对其所涉及的种族、经济和环境正义之根本问题的新凯因斯罗吉特成长体制。本文运用城市政治生态学与种族资本主义, 作为延伸可持续性如何概念化之疆界、并挑战绿色资本主义架构背后的预设之理论架构。最后, 本文检视劳动与环境正义行动者, 如何运用 Sze 等人 (2009, 863) 称为的 "文化与生态论述", 透过纳入从属的空间想像, 挑战绿色资本主义的议程。 *关键词: 环境种族歧视, 罗吉特, 港口, 种族资本主义, 社会运动, 可持续发展, 城市生态学, 城市主义。*

Este artículo examina el tema del modo como los legisladores de California del Sur desplegaron una estrategia de crecimiento verde con la que afianzaron la precariedad racial, ambiental y clasista en la fábrica ecológica de la región. Se usa la observación participativa y los datos existentes para mostrar cómo el arte de gobernar ambientalista suministró la cubierta ideológica para un tipo de régimen de crecimiento logístico neo-keynesiano que utilizó el gasto en infraestructura para estimular la economía, sin abocar los asuntos subyacentes de justicia racial, económica y ambiental. La ecología política urbana y el capitalismo racial sirvieron de una suerte de marco teórico para estirar los límites como se conceptualiza la sostenibilidad y para retar los supuestos que se esconden detrás de un marco de capitalismo verde. Por último, el artículo examina el modo como los activistas de justicia laboral y ambiental usaron lo que Sze et al. (2009, 836) denominaron "discursos culturales y ecológicos," con lo cual desafiar la agenda capitalista verde, incorporando imaginarios espaciales subalternos. *Palabras clave: racismo ambiental, logística, puertos, capitalismo racial, movimientos sociales, desarrollo sustentable, ecología urbana, urbanismo.*

"This port is killing people and we've got to cut it out as fast as we can."[1] The statement marked a dramatic beginning to S. David Freeman's tenure as president of the Port of Los Angeles Harbor Commission from 2006 to 2009. Freeman's comparison of the ports to a diesel death machine was an acknowledgment by regulators that they were responsible for managing an apparatus that caused 13,360 cancer cases per year in the four counties that make up the South Coast Air Basin (South Coast Air Quality Management District 2008).[2] The ports were especially toxic to Black and Latinx residents who lived near logistics hubs in South Los Angeles and the Inland Empire. What was surprising about Freeman's concern over the "diesel death zone" was his claim that the only viable way to reduce environmental damage was to expand port capacity.[3]

In this article, I examine how regional policymakers in Southern California deployed a green growth strategy that cemented racial, environmental, and class precariousness into the region's ecological fabric.[4] I advance a theoretical framework that centers race, capital, and political ecology in the analysis of urbanization by illustrating how the systems, institutions, and infrastructure that made green growth possible depended on state projects that included the management of racial and class difference. I do so by interrogating how logistics as a sustainable development model was produced and contested through the debate on the Clean Air Action Plan and the Clean Truck Program. When I say *logistics*, I mean the political, economic, social, and natural infrastructures required for the circulation of commodities.

I use participant observation and analyze extant data to show how environmentalist statecraft provided ideological cover for a type of neo-Keynesian logistics growth regime that adopted infrastructure spending to stimulate the economy without addressing underlying issues of racial, economic, and environmental justice. As I describe, port leaders claimed that their plans would spur job creation and economic growth without adding to the deadly toll exacted by diesel pollution. I assess some of the assumptions in these claims to show how the ideological tethering of environmentalism to capitalist accumulation affected the strategic choices that were available to public policymakers. The article also examines how labor and environmental justice activists used what Sze et al. (2009) called "cultural and ecological discourses" (836) to challenge the green capitalist agenda.

My analysis builds on the work of environmental justice scholars and activists who have broadened the scope of urban political ecology by insisting that political economy and the environment cannot be adequately explained by ahistorical definitions of power that exclude race (Braz 2006; Mirpuri, Feldman, and Roberts 2009). I show why it is necessary to move beyond distributive models of environmental justice that focus too much on process and procedure without examining the social, cultural, and institutional conditions that produce inequality (Schlosberg 2004; G. Walker 2009; Pulido 2015). Instead, I use a racial capitalism framework to outline how green policy solutions that strip labor from any definition of sustainability only reinforce the existing social structures of capitalism, which are dripping with a deadly mix of difference, power, and dispossession (Robinson 1983). It is a response to Gilmore's (2002) call for geographers to "develop a research agenda that centers on race as a condition of existence and as a category of analysis" (22). I argue that this deeper reading of green growth policies and urban ecology is critical to our understanding of how racial and spatial difference are produced and sustained (Swyngedouw 1996; Davis 1998; Gandy 2002,

2004; Keil 2003; Swyngedouw and Heynen 2003). In fact, Southern California's logistics regime will reveal how contemporary articulations of global capitalism have a critical "logic which works in and through [racial and class] specificity" (King 1997, 29).

A critical urban ecology framework that treats race seriously yields two key results. First, it recognizes that laboring bodies are vital to the production of urban natures. Specifically, scholars have used Marx's notion of metabolism to describe how urban environments are produced through complex and interdependent relationships between humans and nature (Foster 2000; Moore 2000; Prudham 2005; Heynen, Kaika, and Swyngedouw 2006). As Heynen, Kaika, and Swyngedouw (2006) noted, "In the most general sense, 'labouring' is seen exactly as the specifically human form through which the metabolic process is mobilized and organized" (7). I show how this metabolic process is mobilized and organized through the social relations of racial capitalism and urban ecology. For instance, the environment as killing machine cannot be separated from the logistics economy as an apparatus that subjects racialized others to financial precarity and premature death. This is especially true for the truck drivers who are discussed later. Second, this more intersectional and critical reading places locally embedded histories in conversation with multiscalar global processes. I put Southern California's multilayered histories of race and class into conversation with a brand of green capitalist redevelopment that casts global logistics as a solution to the region's economic and environmental crisis. The result is an analytical toolkit that combines urban political ecology, racial capitalism, and global commodity chain frameworks to analyze the intersections among race, space, and capital.

Performing Logistics

Logistics proves to be an important analytical lens because it represents a specific modality through which the hyperrationalization and scientific management of space has enabled capitalism to flourish. Likewise, Southern California's racial transformation and spatial reorganization around global capitalism provides a rich intersectional landscape to examine important questions about urbanization, race, and class. This local–global nexus manifested itself in three key ways. First, Southern California became the largest gateway for imported goods in the United States during the post-1970s neoliberal period. Second, it was one of the first major regions in the country to adopt a green growth agenda as a solution to the unfolding environmental and economic crisis caused by global capitalist restructuring. Third, the region provides a valuable perspective on the geographies of race and uneven development because it underwent a dramatic demographic shift from a majority white population during the 1970s to an ethnic plurality led by a growing immigrant and native-born Latinx population beginning in the 1980s (Allen and Turner 1997).

Given this socioeconomic context, political and business leaders mounted an elaborate campaign to frame logistics as a solution to the region's looming deindustrialization and environmental crisis during the 1990s. This included a series of studies commissioned by the Southern California Association of Governments (SCAG; The Tioga Group 2003; Husing 2004; Southern California Association of Governments 2005). The reports represented one element of the statecraft that was needed to transform the region into a logistics economy. Such reports help us understand the underlying ideologies and spatial politics that produce specific environments. They reveal how social and economic concepts and categories are produced and what "they serve to produce" (Harvey 1973, 298).

SCAG's planning documents and regional studies served as aspirational "representations of space" (Lefebvre 1991, 38). The elaborate logistics networks and infrastructure project plans laid out in these plans became the codes, signs, and narratives that mobilized diverse actors to dream up and enact "new meanings or possibilities for spatial practices" (Harvey 1987, 265). They provided a cognitive roadmap for future growth. Consequently, these regional mappings produced the "future economic orientations and practices" on which Southern California's logistics landscape was built (Bailey and Caprotti 2014, 1805). I have detailed elsewhere how this unfolded and do not repeat it here (De Lara 2018). In short, all of the planning and dreaming resulted in the passage of two significant green growth policies that I now highlight.

On 20 November 2006, harbor commissioners for the ports of Long Beach (POLB) and Los Angeles (POLA) passed the San Pedro Bay Ports Clean Air Action Plan (CAAP). The document called for a 45 percent reduction in port-related pollution within five years of implementation.[5] It also included the framework for a Clean Truck Program (CTP) that reduced diesel emissions by replacing or retrofitting the more than 16,000 drayage trucks that move shipping containers from the docks to inland railyards and warehouses. Port officials took this unprecedented action because they were legally and morally responsible for an apparatus that was killing people (Port of Los

Angeles and Port of Long Beach 2006). In fact, the adjoining Long Beach and Los Angeles ports (San Pedro Bay Ports) created health risks for more than 2 million residents. They also generated over $590 billion in annual medical expenses (The Port of Los Angeles n.d.). Both the CAAP and the CTP assumed that the ports could continue to expand without incurring further health and environmental damages. This decoupling of expansion from pollution meant that logistics boosters could push for unbridled growth without being saddled with the political complications caused by opposition to deadly diesel emissions.

Although the political discourse surrounding the CAAP and CTP focused on public health and safety, the plans played another key political function for port leaders because they cleared the way for policymakers to pursue a green ports development strategy that had been planned since the 1980s when signs of deindustrialization began to affect the local economy (Soja, Morales, and Wolff 2009). Deindustrialization produced a crisis discourse among policymakers who argued for strategic interventions to lessen the financial shocks caused by mounting job losses. Regional leaders argued that they could rebuild the dilapidated manufacturing sector by investing in logistics infrastructure (The Alameda Corridor Project: Its Successes and Challenges 2001; Erie 2004; Southern California Association of Governments 2005; De Lara 2012; Kyser n.d.). They pointed to the more than 500,000 manufacturing jobs that were lost between 1990 and 2010—a decline of 46 percent—to argue that global trade was one of the few options left for the region (Husing 2004). According to this logic, responsible governance meant growing the market while minimizing the environmental damage.

Los Angeles Mayor Antonio Villaraigosa (2005–2013) was one of leading forces behind port expansion. His political performance of progrowth policies as a form of responsible government was critical to the production of a logistics-based green capitalism. In fact, the production of a pro-ports development agenda required actors to build the ideological infrastructures that positioned green capitalism as a viable and responsible enterprise (Prudham 2009). Political actors often use morality and cultural value to infuse their agendas with claims that extend beyond accumulation for accumulation's sake. The mayor argued, for example, that the ports made significant social and economic contributions to Southern California and thus deserved public support. This idea that logistics provided added value to the general public enabled port boosters to claim that green growth represented a social and thus moral good.

Villaraigosa's decision to support logistics above other development paths marked a new period of urban entrepreneurialism, as political leaders invested in trade infrastructure to solidify their regional competitive advantage over other port regions (Brenner 2000). He pushed harbor commissioners to "accommodate the rising tide of trade" that he claimed would "transform our cities into the global cities of the 21st century."[6] What Villaraigosa called the "rising tide of trade" was in fact a performance of speculation; that is, his naturalization of an inevitable growth in global trade was based on speculative data projections commissioned by the state to show why support for port expansion was necessary (De Lara 2018). The projections, which later proved to be inaccurate, promised significant trade growth if regional leaders invested in the necessary infrastructure to accommodate future needs (Southern California Association of Governments 2005; Leachman 2007).

Mayor Villaraigosa's performance of logistics as a necessary element of responsible governance embodied the "development power and crisis tendencies" of speculative accumulation, as well as the relationship among performance, statecraft, and space (R. A. Walker 2004, 436). I raise this point to argue that discourse plays a key role in what scholars of contentious politics describe as a "cluster of performances that constitutes a particular campaign" (Tilly 2012, 8). The green growth discourse articulated by Villaraigosa was an especially powerful development campaign because it was paired with the material force of the entrepreneurial state. For instance, public spending on logistics infrastructure helped to expand container volume from 9.5 million TEUs in 1999 to 13.1 million by 2004, an increase of 38.2 percent.[7]

Mayor Villaraigosa performed another critical role in the spatial politics of development. He was the city's first Latinx mayor in more than a century. His candidacy galvanized a progressive coalition of Los Angeles's burgeoning immigrant, community, and labor organizations (Milkman 2006; Pastor, Benner, and Matsuoka 2009). Villaraigosa used this popular support to transform logistics into a moral geography of regional development. He argued that logistics could provide jobs and therefore help to alleviate the racial and class effects of uneven development (N. Smith 1984). This moral appeal was evident in the following exchange with port commissioners as he disdainfully dismissed the notion that 500,000 logistics jobs were insignificant: "There are some that believe that's not that big of a thing. Well I'm gonna tell you something, tell that to the people in Watts who are looking for economic development. Tell that to people who have made their lives and their futures here in Long Beach and in San Pedro they'll tell you that jobs are important." His reference to Watts was coded language meant to represent poor and working-class Black and Latinx neighborhoods in South Los Angeles. He continued this moral narrative by claiming that port-related jobs "give people pride. They give people a sense of the possible. They allow them to maintain their families." Villaraigosa's description was an emotional plea that reframed these abandoned areas as worthy of investment by appealing to a capitalist moral geography of green development (Amin and Thrift 2007). His progressive vision constituted a "political strategy of global resource management and ecological modernization" that leveraged state resources to provide for the region's devalued and abandoned poor communities (Brand 2016, 109).

Port supporters knew that capturing future trans-Pacific trade required Long Beach and Los Angeles to draw more of Southern California's hinterland into their geographic orbit. They needed, as Long Beach Mayor Bob Foster argued, "a regional solution to a regional problem." Implementing such a solution meant greater regional cooperation. Prior to the CAAP, Long Beach and Los Angeles regularly competed for port business. Mayor Villaraigosa acknowledged the fractious relationship when he told a joint meeting between the two ports, "These two commissions which for the longest time were competitors, were adversaries from time to time, are now realizing that together we could do so much more for this region." Although the joint effort to develop a CAAP did not mark an end to port competition, it did create political space for both ports to mobilize public assets to expand regional trade corridors. Greater cooperation laid the institutional framework for a new round of neo-Keynesian investment in regional trade infrastructure.

Crisis, Green Growth, and the Neo-Keynesian State

Even as members of the port leadership congratulated themselves on the passage of the CAAP, there was a tacit recognition that their actions were subsidizing a deadly industry. Some community members and social movement organizations questioned the use of public funds for private development. Long Beach Harbor Commissioner James Hankla tried to preempt such criticism by claiming that investment in green growth was "a legitimate cost of doing business" but that "the ports cannot and will not subsidize the cost of cleaner transportation indefinitely." Hankla's corporatist language rationalized public expenditures "for cleaner technologies and fuels" that he claimed rewarded "the true pioneers in the industry but only for short periods of time." Such language evoked a central ethos of green capitalism. At its core, green capitalism involves an ideological commitment to the idea that market strategies will fix environmental problems. One consequence of this commitment, according to Prudham (2009), is that "environmental politics become semiotically and ideologically tethered to the reproduction of the conditions of accumulation" (1596).

I now show how this happened in Southern California when state actors tried to capitalize on ecological crisis by incorporating the deadly environmental health problems of the ports into the logics of accumulation and growth. For example, Mayor Villaraigosa argued that the region was confronted with "complex environmental and economic challenges" but that these challenges provided "an opportunity for us to define the boundaries of what is possible." Regional leaders like Villaraigosa celebrated the adoption of port policy by declaring, "Today we're expanding those boundaries with the Clean Air Action Plan. We're calling for cleaner ships, cleaner trucks, cleaner cargo equipment, cleaner harbor craft, and locomotive engines."

It marked a key moment of adaptation during which the diesel-based systems of logistics were able to shed their deprecated and deadly constraints to adapt and reproduce themselves as legitimate systems of accumulation (Perez 2015; Brand 2016). Such adaptations have enabled capitalism to survive and reproduce itself as a system without being "confined to a single logic" of accumulation (Lefebvre 1991, 162). If logistics and the green ports campaign became the new strategy of accumulation for global goods, then regional infrastructure and public finance were the modalities through which this new strategy was territorialized.

Freeman articulated how this new logic of accumulation represented an ecological solution to the ongoing environmental crisis. He explained, "You can't stop growing and then clean things up," because "growth is the leverage the ports have" (Mongelluzzo 2007).[8] Freeman's green growth doctrine was consistent with his definition of what port authorities were supposed to do. According to him, harbor

commissioners were "trustees ... responsible for the goods movement" industry to run "smoothly." This corporatist ethos was front and center when Mayor Villaraigosa made growth a civic responsibility, as he told port commissioners that although "all stakeholders continue to hold our ports accountable for being green I call on the ports to also be accountable for making our ports grow." Villaraigosa's mandate was for public servants to act as market provocateurs while leaving the work of lowering the diesel death toll to other members of the community. Meanwhile, Long Beach Mayor Bob Foster (2006–2014) warned that increasing infrastructure and thus boosting port cargo without attending to environmental pollution would create a "major health crisis for generations." Limiting cargo was not a feasible answer. Instead, his solution was for port growth and environmental protections to "move forward in lock step at the same time ... [because] they should be viewed as one project."

Regional leaders understood that a green economy strategy provided new opportunities for local investment. More specifically, Foster acknowledged, "If you link them [infrastructure spending and environmental mitigation], the funds will come." He was referring to the "only two realistic funding sources" for infrastructure expansion, "the federal government or a charge that finds its way on the costs of goods sold."[9] Harbor commissioners agreed to such a fee in January 2008, when they passed a container tax to raise funds for clean truck replacement. The container fee served two functions. One was political theater that demonstrated the state's commitment to cleaning up the environment. It was a statement that, according to Freeman, was meant to "make sure ... that every economic interest out there understands that in order for their commercial activity to flourish that they have to, we have to spend the appropriate resources on the environment." Freeman was visibly aghast at "the idea that we would go forward with infrastructure without cleaning up." He called this suggestion "abhorrent to everything that we've done."

Although port leaders framed both the CAAP and the container fee as a symbol of good government that shifted some of the costs to private companies, there was a second motivation behind their actions. Revenues from the container fees also provided a down payment that enabled the regime to pursue state and federal funds. Dr. Robert Kanter, POLB Managing Director of Environmental Affairs & Planning, noted, "A large portion of this money would be used to leverage other sources such as 1B funding." POLA Executive Director Geraldine Knatz stated that "the reason why we are doing this fee today is because we really want to position ourselves to be the best applications for that bond money." Knatz claimed that the container fee would "strengthen our application." Richard Steinke, the Executive Director for the POLB—was more explicit. He said, "This is the action that allows us to determine what money we're gonna get from Prop 1B. It positions us well for money." Each of these examples illustrates how local state actors used green growth discourse to create new spaces for logistics.

Sustainability without Justice: Bringing Labor Back In

While state actors created new pro-port policies, environmental justice organizations and mainstream environmental groups used the courts and the ballot box to contest major logistics projects during the 1990s, including a successful injunction against construction of the China Shipping Terminal (Hastings 2002). Such legal tactics and political lobbying were part of a campaign by social movement organizations to destabilize the commonsense definitions of sustainability and economic justice that were touted by port boosters (Brand 2016). These efforts suggest that the labor of logistics cannot be separated from the urban political ecologies that are necessary for accumulation to do its work. In fact, logistics needs to be understood as a specific modality of capitalist spatial reorganization that grew in importance with the introduction of just-in-time global production networks (Hesse and Rodrigue 2006). Logistics development is therefore tethered to specific social relations of production and distribution; this includes laboring bodies. More specifically, it was the cyborg gaze of the logistician—which combined bodies and machines to produce efficient commodity chains—that enabled global capitalism to reconfigure its territoriality (Cowen 2014; De Lara 2018). In fact, the logistics revolution was critical to the reterritorialization of global capitalism at the turn of the twenty-first century, when corporate executives from companies like Walmart and Toyota recognized that logistics gave them the ability to master and control the space–time compression that defined globalization (Harvey 1990; Watts 1992; E. B. Smith, Mendoza, and Ciscel 2005; Cowen 2010). It is foolish, therefore, to believe that discussions of Southern California's environment can be stripped of the larger social and natural connections—including human labor—that mutually construct the region's urban ecology.

Even as logistics scientists created new ways to expand the territoriality of capitalism, locally embedded social movements mounted campaigns to challenge these global mappings. Debate over the CAAP provided a political space for development of counternarratives to emerge. I introduce the voices of immigrant and Latinx port truckers to explore how they used their laboring bodies and their trucks to question dominant green growth narratives that tried to erase them from conversations about sustainability. I borrow this epistemological technique from queer theorists who have used positionality and performance to produce antinormative possibilities (Muñoz 1999).

A number of different stakeholder groups challenged some of the basic assumptions that were included in the CAAP and the CTP. The notion that environmental mitigation represented a sustainable ecological model was one such point of contention. Vice Mayor of Long Beach Bonnie Lowenthal expressed concerns over what she considered to be the inadequate politics of mitigation. She told port commissioners during a public hearing that "mitigation actually is harm reduction. So to mitigate a problem is not enough . . . we need a net improvement in the air quality in this region." Lowenthal's comment rebuked the idea that port growth and negative environmental health disparities could be easily uncoupled because future harm reduction could not undo the damage that had already been done. Such arguments echoed critics of the green capitalism approach, who claim that a system that reproduces accumulation for the sake of accumulation is inherently unsustainable because the systemic need for growth is predicated on the exponential use and depletion of finite resources and human labor (Conca, Princen, and Maniates 2001; Prudham 2009; Blackwater 2012; Ferguson 2015; Brand 2016).

Another challenge to the idea that logistics development was sustainable came from those who argued that the replacement of dirty diesel trucks did not address the underlying power imbalances and patterns of uneven development that routinely exposed low-wage workers and marginalized publics to premature death under Southern California's racialized capitalist regime (Gilmore 2002). Even some political leaders acknowledged that the deadly consequences of environmental racism and classism were written into the spatial logics of logistics space. Long Beach Mayor Bob Foster noted before a port commission meeting that "the costs of these programs, the costs of these impacts are already in the system but the wrong people are paying them. Our kids in Long Beach are contracting asthma at record levels so the rest of America can buy cheaper TVs." When Foster referred to "our kids" he was referring to a city in which 70 percent of the population was Latinx, Black, or some other non-white ethnic group.[10]

One of the most powerful critiques of progrowth policies came from logistics industry workers who challenged the dominant green economy model by insisting that their labor be included in the definition of sustainability. Approximately 17,000 trucks serviced the San Pedro Bay Ports at the time that the plan was passed in 2008. Most were owned by individual operators or by very small companies. In fact, 20 percent of the largest operators accounted for less than 50 percent of drayage truck traffic. The result was what a port consultant called a "long tail of very small companies" in a market that shifted operating costs onto small and less capitalized drivers who operated on small margins because the "fiercely competitive" market pushed prices down.[11] Consultants and planning experts claimed that most truck operators were "not in a position to invest in the new greener trucks" that were mandated by the CTP.[12]

Several truck drivers showed up to testify during harbor commission meetings, especially on the day that commissioners were going to vote on the CAAP. One driver approached the microphone and said, "My name is Luis Ceja. … I'm the one that brings the things out to the warehouse, the one that brings the stuff to your stores." Mr. Ceja had been a truck driver for more than seven years. His introduction as "the one that brings the stuff to your stores" was an eloquent way to make himself visible within the circuit of actors that constitute the regional logistics ecology. This deep connection between Mr. Ceja, his labor, and his truck formed an affective bond that tied him to the environment. For example, he was visibly upset when he told commissioners, "I hate my truck because the pollution makes people sick … my truck is responsible for the pollution that my family and all the families around here" are exposed to. For immigrant truck drivers like Luis Ceja, the ecological connections among his work, deadly diesel, and global logistics were an everyday reality.

Another driver named Narcisso Roman told commissioners, "I'm concerned that my truck contributes to pollution. That it affects our families' health and our community's health." Mr. Roman sported a green "Clean and Safe Ports" (CSP) campaign shirt, had dark skin, and wore a mustache that matched his salt-and-pepper hair. He spent his allotted three minutes speaking to the economic injustice of working as a port driver for more than twenty years. Although he acknowledged that his family was "suffering from the contamination and the impure air" caused by diesel emissions, he also told commissioners that "it's not enough for you to give us new trucks when our wages are so low, because with the wages we earn, if you give us a new truck, in three or four years we'll have the same problem." It was a common argument advanced throughout the debate by the CSP campaign. CSP members ultimately convinced the Port of Los Angeles to implement a provision within the CTP that reclassified truckers as direct-hire employees rather than independent owner-operators. They hoped that a change in employment status would make it easier for drivers to improve wages and benefits by making it easier for them to join a union.

Both drivers I mentioned earlier and other members of the CSP were trying to rewrite the parameters of green growth by including labor into the definition of sustainability. They introduced their own ecological interpretation of green capitalism into the debate by forcing commissioners to consider how their conceptions of the environment were incomplete without understanding how "biophysical nature is produced or metabolized" through labor (Prudham 2009, 1599). This ecological counternarrative was met by vehement opposition from port business owners like the American Trucking Association (ATA), who argued that the proposed employment provisions in the CTP had nothing to do with cleaning up the environment. The ATA mounted a legal challenge to the labor provision. They were ultimately successful when the United States Ninth Circuit Court of Appeals blocked the employee mandate portion of the CTP by arguing that local port authorities did not have the legal jurisdiction to reclassify transportation workers. The decision proved to be a fatal blow for the CSP's efforts to include labor into a regional sustainability model.

The CSP strategy was flawed from the beginning. Even as drivers made eloquent statements before the port commissions, leaders of the CSP campaign adopted a more moderate high-road strategy. They never explicitly made the case for why labor is a central element of an ecologically sustainable development model. In fact, they shied away from an explicit prolabor position and focused instead on framing the CTP as a progrowth policy. This mainstreaming effort was part of a national Green New Deal campaign (Herman 2001; Green New Deal Group 2008; Mattera et al. 2009). The national Green New Deal movement embraced the idea that capitalism, if guided by a neo-

Keynesian progressive state, could save workers from the ravages of the free market.

Ceding ideological control to market-based policies eventually led to a brand of Keynesian development that did not include explicit mandates for social and economic justice. In fact, debate about the CTP was dominated by rational market-based ideologies that focused on resource management. One consultant testified before the POLA harbor commission and argued that the concession plan that mandated direct-hire employees could be used as an incentive to secure a regular and available labor force. Under this framing, the concession gave regulators a way to create an adequate workforce by ensuring that the necessary labor power could be accessed when needed. The same consultant claimed that making drivers into direct-hire employees instead of independent contractors would provide new strategies for accountability and governance. It would create "reciprocal obligations, skin in the game if you will," because according to him, "the port has provided a concession. In return under this model the LMC [Licensed Motor Carrier] has committed capital—trucks—has hired employees and has them ready in service of the port." Put this way, efforts to include an employee provision in the CTP were an attempt to regularize the logistics economy by mitigating existing market externalities that subject truck drivers to precarious conditions. Commissioner Joseph Radisich argued that a more regularized market, one that reduced the number of trucks operating at the ports, would minimize congestion, increase port efficiencies, and improve economic outcomes for workers.

This notion that the CTP represented a market-correction strategy that helped to regularize the port economy is a prime example of how green capitalist proponents attempt to resolve labor and ecological problems through technical solutions. Such reasoning has ideological roots in Saint-Simonianism and Taylorism; both schools of thought argued that scientific rationalism could resolve class conflicts by increasing production and eliminating scarcity (Maier 1970). Market-based arguments stood in direct opposition to the affective moral economies expressed by truck drivers, who placed their bodies and the bodies of their loved ones at the center of logistics-based sustainability. In contrast, the green capitalism approach produced the ports as a rational system that needed effective management. Freeman illustrated this entrepreneurial management approach when he said, "We have described our objectives from the beginning to be air quality, security, and the third just as important is a stable and growing workforce." What Freeman failed to recognize was that his "stable and growing workforce" included racialized low-wage labor, without which logistics and the port economy could not expand their territorial reach. Immigrant truck drivers and their biophysical resources were thus central to the production of an urban logistics ecology. The contradictions between these two approaches could not be fixed by a policy solution that refused to reconcile this deeper connection among race, class, nature, and space.

Conclusion

The CTP was a political gamble for progressives. It marked the embrace of a green capitalism framework and an attempt to move beyond the traditional distributive models of environmental justice (Pulido 1996; Agyeman, Bullard, and Evans 2002; G. Walker 2009). Yet they were outmaneuvered by private and public port boosters who successfully deployed a green growth strategy that excluded labor and economic justice from discussions of urban ecology. What transpired with the CTP is an example of how policy approaches that fail to disrupt the engrained market and state logics of environmental racism and racial capitalism risk falling prey to the misguided notion that science and engineering can resolve capitalism's differentiated relationships of power, which systematically predispose racialized bodies to premature death (Robinson 1983; Wilson 2000; Woods 2000; Gilmore 2002). These policy solutions are often defined by market-based environmental ideologies that erase "the voice and interests of subaltern populations" (Sze et al. 2009, 836). It is therefore necessary to deploy scholarly techniques—as I have done here—that use the voices of subaltern populations (e.g., immigrant truck drivers) to wrestle with and make sense of the complex social and natural relationships that produce everyday urban life.

Cleaner trucks mitigated but did not eliminate the diesel death machine that was responsible for killing poor black and brown people by exposing them to cancer-causing toxins. Green technologies also failed to remove immigrant truck drivers from the economic and social precarity that is endemic to workers who earn their living by powering the machines of just-in-time production and distribution. Any proposed fix to the economic and environmental crisis of deadly development must account for how specific industries consume and deplete the biophysical resources of devalued communities, including racialized labor. It is not enough to mitigate or redistribute risk. Otherwise, as truckers told port commissioners, technological fixes like cleaner trucks will only provide a nicer ride into the precarious world of social and economic insecurity.

Notes

1. The statement was made at a harbor commission meeting early in his tenure as president (Schoch 2006).
2. Los Angeles, Orange, Riverside, and San Bernardino counties make up the South Coast Air Basin. The California Environmental Protection Agency has divided the state into fifteen different air basins.
3. Environmental justice advocates commonly referred to the ports as a diesel death zone. Examples include Cone (2011) and Gonzalez and Miller (n.d.).
4. The material and virtual objects that make up logistics (computer networks, ships, distribution centers, etc.) are absolutely important to this story, and I have written about them elsewhere (De Lara 2012, 2018).
5. References about the contents of the CAAP are taken directly from the policy and supporting documents (Port of Los Angeles and Port of Long Beach 2006).
6. All references to public statements are taken from the author's field notes and from transcriptions made from official meetings that took place between 2006 and 2008.
7. TEUs are a standard unit of measurement. It means twenty-foot equivalent unit. Author calculations are based on data taken from port documents.
8. Freeman was appointed by Los Angeles Mayor Antonio Villaraigosa, whose election in 2005 signaled a liberal and progressive political swing in local politics after decades of conservative and centrist administrations.
9. These comments were taken from a special joint meeting between POLA and POLB on 14 January 2008, transcribed by the author.
10. These data are taken from the 2010 U.S. Census (see www.census.gov).
11. Simon Goodal, Boston Consulting Group, presentation before the POLA Commission, 6 March 2008, author's notes.
12. Simon Goodal, Boston Consulting Group, presentation before the POLA Commission, 6 March 2008, author's notes.

References

Agyeman, J., R. D. Bullard, and B. Evans. 2002. Exploring the nexus: Bringing together sustainability, environmental justice and equity. *Space and Polity* 6 (1):77–90.

The Alameda Corridor Project: Its successes and challenges. Hearing before the Subcommittee on Government Efficiency, Financial Management and Intergovernmental Relations, Pub. L. No. 107–50, § Committee on Government Reform. 2001. Washington, DC: U.S. Government Printing Office. Accessed December 2, 2015. http://www.gpo.gov. Washington, DC: U.S. Government Printing Office.

Allen, J. P., and E. Turner. 1997. *The ethnic quilt: Population diversity in Southern California.* Northridge: The Center for Geographical Studies, California State University, Northridge.

Amin, A., and N. Thrift. 2007. Cultural-economy and cities. *Progress in Human Geography* 31 (2):143–61.

Bailey, I., and F. Caprotti. 2014. The green economy: Functional domains and theoretical directions of enquiry. *Environment and Planning A* 46 (8):1797–1813.

Blackwater, B. 2012. Two cheers for environmental Keynesianism. *Capitalism Nature Socialism* 23 (2):51–74.

Brand, U. 2016. Green economy, green capitalism and the imperial mode of living: Limits to a prominent strategy, contours of a possible new capitalist formation. *Fudan Journal of the Humanities and Social Sciences* 9 (1): 107–21.

Braz, R. 2006. Joining forces: Prisons and environmental justice in recent California organizing. *Radical History Review* 2006 (96):95–111.

Brenner, N. 2000. The urban question as a scale question: Reflections on Henri Lefebvre, urban theory and the politics of scale. *International Journal of Urban and Regional Research* 24 (2):361–78.

Conca, K., T. Princen, and M. F. Maniates. 2001. Confronting consumption. *Global Environmental Politics* 1 (3): 1–10.

Cone, M. 2011. A toxic tour: Neighborhoods struggle with health threats from traffic pollution. *Environmental Health News* October 7. Accessed January 12, 2017. http://www.environmentalhealthnews.org.

Cowen, D. 2010. A geography of logistics: Market authority and the security of supply chains. *Annals of the Association of American Geographers* 100 (3):600–620.

———. 2014. *The deadly life of logistics: Mapping violence in global trade.* Minneapolis: University of Minnesota Press.

Davis, M. 1998. *Ecology of fear: Los Angeles and the imagination of disaster.* New York: Metropolitan Books.

De Lara, J. D. 2012. Goods movement and metropolitan inequality. In *Cities, regions and flows,* ed. P. V. Hall and M. Hesse, 75–91. London and New York: Routledge.

———. 2018. *Inland shift: Race, space, and capital in inland Southern California.* Berkeley and Los Angeles: University of California Press.

Erie, S. P. 2004. *Globalizing L.A.: Trade, infrastructure, and regional development.* Stanford, CA: Stanford University Press.

Ferguson, P. 2015. The green economy agenda: Business as usual or transformational discourse? *Environmental Politics* 24 (1):17–37.

Foster, J. B. 2000. *Marx's ecology: Materialism and nature.* New York: Monthly Review Press.

Gandy, M. 2002. *Concrete and clay: Reworking nature in New York City.* Cambridge, MA: MIT Press.

———. 2004. Rethinking urban metabolism: Water, space and the modern city. *City* 8 (3):363–79.

Gilmore, R. W. 2002. Fatal couplings of power and difference: Notes on racism and geography. *The Professional Geographer* 54 (1):15–24.

Gonzalez, C., and J. Miller. n.d. Fighting for environmental justice in the diesel death zone. Accessed

January 17, 2017. http://climatehealthconnect.org/stories/fighting-for-environmental-justice-in-the-diesel-death-zone/.

Green New Deal Group. 2008. *A green new deal: Joined-up policies to solve the triple crunch of the credit crisis, climate change and high oil prices.* London: New Economics Foundation. Accessed December 4, 2016. http://www.greennewdealgroup.org.

Harvey, D. 1973. *Social justice and the city.* Baltimore: The Johns Hopkins University Press.

———. 1987. Flexible accumulation through urbanization: Reflections on "post-modernism" in the American city. *Antipode* 19 (3):260–86.

———. 1990. *The condition of postmodernity.* Cambridge, MA: Blackwell.

Hastings, J. 2002. *Natural Resources Defense Council Inc., et al. Plaintiffs and Appellants, v. City of Los Angeles, et al., Defendants.* California: Court of Appeal, Second District, Division 4.

Herman, B. 2001. How high-road partnerships work. *Social Policy* 31 (3):11–19.

Hesse, M., and J.-P. Rodrigue. 2006. Global production networks and the role of logistics and transportation. *Growth and Change* 37 (4):499–509.

Heynen, N., M. Kaika, and E. Swyngedouw, eds. 2006. *In the nature of cities: Urban political ecology and the politics of urban metabolism.* London and New York: Routledge.

Husing, J. 2004. *Logistics & distribution: An answer to regional upward social mobility.* Los Angeles: Southern California Association of Governments.

Keil, R. 2003. Urban political ecology: Progress report. *Urban Geography* 24 (8):723–38.

King, A. D. 1997. *Culture, globalization and the world-system: Contemporary conditions for the representation of identity.* Minneapolis: University of Minnesota Press.

Kyser, J. n.d. Goods movement in Southern California: How can we solve problems and generate new state sales and income tax revenues? Los Angeles: Los Angeles Economic Development Corporation.

Leachman, R. 2007. *Progress report: Port and modal elasticity study: Phase II.* Los Angeles, CA: Southern California Association of Governments.

Lefebvre, H. 1991. *The production of space.* Oxford, UK: Blackwell.

Maier, C. S. 1970. Between Taylorism and technocracy: European ideologies and the vision of industrial productivity in the 1920s. *Journal of Contemporary History* 5 (2):27–61.

Mattera, P., A. Dubro, K. Gordon, and S. Club. 2009. *High road or low road? Job quality in the new green economy.* Washington, DC: Good Jobs First.

Milkman, R. 2006. *L.A. story: Immigrant workers and the future of the U.S. labor movement.* New York: Russell Sage Foundation.

Mirpuri, A., K. P. Feldman, and G. M. Roberts. 2009. Antiracism and environmental justice in an age of neoliberalism: An interview with Van Jones. *Antipode* 41 (3):401–15.

Mongelluzzo, B. 2007. Growth seen as key to "green" U.S. ports. *Journal of Commerce Online.* Accessed December 8, 2016. http://www.joc.com.

Moore, J. W. 2000. Environmental crises and the metabolic rift in world-historical perspective. *Organization & Environment* 13 (2):123–57.

Muñoz, J. E. 1999. *Disidentifications: Queers of color and the performance of politics.* Minneapolis: University of Minnesota Press.

Pastor, M., C. Benner, and M. Matsuoka. 2009. *This could be the start of something big: How social movements for regional equity are reshaping metropolitan America.* Ithaca, NY: Cornell University Press.

Perez, C. 2015. Capitalism, technology and a green global golden age: The role of history in helping to shape the future. *The Political Quarterly* 86:191–217.

Port of Los Angeles. n.d. The Port of Los Angeles Clean Truck Program: Program overview & benefits. Accessed January 25 2017. https://www.portoflosangeles.org/.

Port of Los Angeles and Port of Long Beach. 2006. *San Pedro Bay Ports Clean Air Action Plan.* Los Angeles and Long Beach, CA: The Port of Los Angeles and the Port of Long Beach. Accessed October 1, 2010. http://www.cleanairactionplan.org.

Prudham, S. 2005. *Knock on wood: Nature as commodity in Douglas fir country.* London and New York: Routledge.

Prudham, S. 2009. Pimping climate change: Richard Branson, global warming, and the performance of green capitalism. *Environment and Planning A* 41 (7):1594–1613.

Pulido, L. 1996. A critical review of the methodology of environmental racism research. *Antipode* 28 (2):142–59.

Pulido, L. 2015. Geographies of race and ethnicity 1: White supremacy vs white privilege in environmental racism research. *Progress in Human Geography* 39 (6):809–17.

Robinson, C. J. 1983. *Black Marxism: The making of the black radical tradition.* London: Zed.

Schlosberg, D. 2004. Reconceiving environmental justice: Global movements and political theories. *Environmental Politics* 13 (3):517–40.

Schoch, D. 2006. Port panel chief has plenty to unload. *The Los Angeles Times* January 15. Accessed December 5, 2017. http://articles.latimes.com/2006/jan/15/local/me-freeman15.

Smith, E. B., M. Mendoza, and D. H. Ciscel. 2005. The world on time: Flexible labor, new immigrants and global logistics. In *The American South in a global world,* ed. J. L. Peacock, H. L. Watson, and C. R. Mathews, 23–38. Chapel Hill: University of North Carolina Press.

Smith, N. 1984. *Uneven development: Nature, capital, and the production of space.* New York: Blackwell.

Soja, E., R. Morales, and G. Wolff. 2009. Urban restructuring: An analysis of social and spatial change in Los Angeles. *Economic Geography* 59 (2):195–230.

South Coast Air Quality Management District. 2008. *Multiple air toxics exposure study in the South Coast Air Basin: MATES III.* Diamond Bar, CA: South Coast Air Quality Management District.

Southern California Association of Governments. 2005. *Southern California regional strategy for goods movement: A plan for action.* Los Angeles: Southern California Association of Governments. Accessed September 13,

2013. http://www.scag.ca.gov/goodsmove/reportsmove. htm.

Swyngedouw, E. 1996. The city as a hybrid: On nature, society and cyborg urbanization. *Capitalism Nature Socialism* 7 (2):65–80.

Swyngedouw, E., and N. C. Heynen. 2003. Urban political ecology, justice and the politics of scale. *Antipode* 35 (5):898–918.

Sze, J., J. London, F. Shilling, G. Gambirazzio, T. Filan, and M. Cadenasso. 2009. Defining and contesting environmental justice: Socio-natures and the politics of scale in the delta. *Antipode* 41 (4):807–43.

Tilly, C. 2012. Social movements as historically specific clusters of political performances. *Berkeley Journal of Sociology* 38:1–30.

The Tioga Group. 2003. *Goods movement truck and rail study executive summary.* Los Angeles: Southern California Association of Governments.

U.S. Census Bureau. 2010. American FactFinder. DP1: Profile of general population and housing characteristics. Accessed August 5, 2016. http://factfinder2. census.gov.

Walker, G. 2009. Beyond distribution and proximity: Exploring the multiple spatialities of environmental justice. *Antipode* 41 (4):614–36.

Walker, R. A. 2004. The spectre of Marxism: The return of the limits to capital. *Antipode* 36 (3):434–43.

Watts, M. J. 1992. Space for everything (a commentary). *Cultural Anthropology* 7 (1):115–29.

Wilson, B. M. 2000. *America's Johannesburg: Industrialization and racial transformation in Birmingham.* Lanham, MD: Rowman & Littlefield.

Woods, C. 2000. *Development arrested: The blues and plantation power in the Mississippi Delta.* London: Verso.

JUAN DE LARA is an Assistant Professor in the Department of American Studies & Ethnicity at the University of Southern California, Los Angeles, CA 90089–2534. E-mail: jdelara@usc.edu. His research interests include social justice and social movements, urbanization, race and ethnicity, labor, the U.S.–Mexico border, and the intersections between big data technologies and race.

22 "Wagering Life" in the Petro-City

Embodied Ecologies of Oil Flow, Capitalism, and Justice in Esmeraldas, Ecuador

Gabriela Valdivia

This article uses a political ecology approach to examine how urban residents of the refinery city of Esmeraldas "wager life" under conditions of social and chemical toxicity associated with oil capitalism. The article draws on the scholarship on affective economies and critical oil geographies to trace the knotting of social reproduction and oil capital in Esmeraldas and to illustrate how "cruel optimisms" (Berlant 2011) allow city-dwellers to make sense of everyday life amidst frontier-style petro-capitalism. Focusing on personal narratives of social reproduction, affect, and hope in the city, the article first argues that "justice" can be contradictory and politically ambivalent and, second, challenges fixed readings of resistance, refusal, or submission in resource extraction–dominated sites. Rather than presupposing resistance to petro-capitalism or submission to its workings, the article illustrates the liveliness of urban justice struggles and how attention to embodied ecologies and affective oil economies deepens scholarship on social justice. *Key Words: cruel optimisms, environmental justice, petro-capital, social reproduction, urban political ecology.*

本文运用政治生态学取径, 检视埃斯梅拉达的炼油城市中的城市居民, 如何在石油资本主义的社会与化学毒性境况下 "拿生命作赌注。" 本文运用情绪经济和批判石油地理的学术研究, 追溯埃斯梅拉达的社会再生产与石化资本间的密切结合, 并描绘 "冷酷的乐观主义" (Berlant 2011) 如何让城市居民得以在边境型的石化资本主义中应对每日生活。聚焦城市中的社会再生产、情绪和希望的个人叙事, 本文首先主张, "正义" 可以是矛盾且在政治上模棱两可的。再者, 本文挑战对资源搾取支配场域中的抵抗、拒绝或顺从的固定解读。与其预设对石化资本主义的抵抗或屈从, 本文描绘出生动的城市正义斗争, 以及对于体现的生态和情绪性的石油经济之关注, 如何能够深化有关社会正义的学术研究。 关键词: 冷酷的乐观主义, 环境正义, 石化资本, 社会再生产, 城市政治生态。

Este artículo usa un enfoque de ecología política para examinar el modo como los residentes urbanos del centro de refinería de Esmeraldas se "juegan la vida" bajo las condiciones de toxicidad social y química asociadas con el capitalismo petrolero. El artículo se apoya en el cuerpo de erudición relacionado con las economías afectivas y las geografías críticas del petróleo para seguirle el rastro al nudo de la reproducción social y el capital del petróleo en Esmeraldas, y para ilustrar cómo "los crueles optimismos" (Berlant 2011) permiten a los citadinos dar sentido a la vida cotidiana en medio de petro-capitalismo de estilo fronterizo. Enfocándonos en narrativas personales de reproducción social, afecto y de esperanza en la ciudad, el artículo arguye primero que "la justicia" puede ser contradictoria y políticamente ambivalente y, segundo, reta las lecturas fijas de resistencia, rechazo o sumisión en los sitios dominados por la extracción de recursos. Más que suponer una resistencia al petro-capitalismo o una sumisión a sus designios, el artículo ilustra la vivacidad de las contiendas por la justicia urbana y el modo como la atención a las ecologías implicadas y las economías petroleras afectivas profundiza la erudición sobre la justicia social. *Palabras clave: optimismos crueles, justicia ambiental, petro-capital, reproducción social, ecología política urbana.*

The coastal city of Esmeraldas is a strategic logistical node of oil flow in Ecuador, South America's fifth largest oil producer. Esmeraldas houses the country's most important oil processing and exporting hub, including the endpoint of the state-owned pipeline *Sistema de Oleoducto de Transporte Ecuatoriano* (SOTE); the maritime oil terminal of Balao; and the Esmeraldas refinery, which processes 110,000 barrels of oil per day. Of the US$6 billion in oil products refined in 2013 in Esmeraldas, US$1.15 billion was exported via its marine terminal (EP-Petroecuador 2013). To increase efficiency in production and energy security, between 2014 and 2017, the Ecuadorian government spent US$2.2 billion on refinery updates and, in 2016, relocated its hydrocarbon transportation firm, FLOPEC, to Esmeraldas to increase oversight over oil affairs.

The bodies of Esmeraldeños are "attuned" (Anderson and Wylie 2009) to the oil-suffused atmosphere of the city. At night, fumes from the refinery chimney assault the eyes and capacity to breathe. When the winds die down, the air is filled with the smell of sulfur and burnt plastic, which a taxi driver described as "slow death." Gastritis, diabetes, cancer, arthritis, and depression are common, particularly in areas surrounding the refinery (Valdivia and Lu 2016). Children are warned not to swim in rivers close to the city because they can develop skin rashes, which could be associated with human waste leaking from inadequate sewage systems, pesticides from agro-industrial fields upstream, or refinery by-products. The refinery is associated with emotional distress, often described as a "time bomb" that will obliterate the city.

Yet, stories about everyday life in this petro-city often do not feature oil. Instead, they refer to what Berlant (2011) called "cruel optimisms": desires for mobility, improvement, and a dignified life attached to oil's circulation that residents want to become true while being attuned to its toxicity. Optimism is a structural feeling, such that an "optimistic attachment is invested in one's own or the world's continuity" (Berlant 2011, 13). This optimism has temporal dimensions: It is like a stretch of time or holding pattern from which a person or situation moves forward slowly and that demands awareness and hypervigilance across registers of difference (e.g., class, gender, race, sex; Berlant 2011). Berlant (2011) qualified this optimism as "cruel" because it is attachment to a desired good (e.g., "the good life") that "wears out the subjects who nonetheless find their conditions of possibility within it" (27). For example, a young, politically active Afro-descendant man referred to the refinery as "wealth in a world of racism." As a state-owned entity, the refinery finances hospitals, health centers, and schools to compensate people in its zones of influence; that is, it offers goods that improve quality of life. It also regularly employs locals for the social reproduction of refinery personnel (e.g., food preparation, cleaning, housekeeping, and maintenance) via short-term subcontracts. Some Esmeraldeños even desire breakdown in operations because it means access to temporary wages. These desires, despite their associated harms, are not unfounded: 70 percent of the urban population relies on self-employment (Instituto Nacional de Estadística y Censo [INEC] 2010) and "distributive labor" (Ferguson 2015)—livelihoods that depend on distribution relations (e.g., cash payments), not "production" in Marx's sense, bound up in unequal, hierarchical, and exploitative relations of dependence—to make ends meet.

These are stories of uncertainty and urban social justice "knotted" (Haraway 2008, 4) to petro-capitalism, a form of wealth accumulation that hinges on the production, distribution, and consumption of petroleum. Petro-capitalism enables modes of knowing life associated with oil, from "democracy" and "freedom" (Watts 2004; Mitchell 2011) to cultures and aesthetics of modernity (Huber 2013; LeMenager 2014; Rogers 2015). Although oil studies focus on population scales to understand how "oil and gas move through our lives" (Appel, Mason, and Watts 2015, 27), how this flow is entangled with urban justice and with affect is less explored, however. This article begins such an examination, focusing on the cruel optimisms of life with oil in Esmeraldas to ask how social reproduction in the petro-city informs our understanding of justice.

Social reproduction broadly includes practices, relations, and conditions (e.g., political, regulatory, biological, physical, affective) through which people maintain and reproduce economically (Katz 2001; Murphy 2017). I use social reproduction as a figure of analysis in two interrelated ways. First, drawing on Marxist conceptions of living and being, "social reproduction" stands for actions and processes that enable laborers to reproduce themselves from day to day in capitalist milieus (Katz 2001). I emphasize how living and being unfolds in Esmeraldas for those whose livelihood security is tied to the logics of oil capital (but not formalized by it) to examine how oil becomes part of everyday urban endurance and resilience, beyond contamination events and toxic exposures. Esmeraldeños call this "jugarse la vida" (wagering life): the range of creative activities that people engage with to economically maintain and take care of their families, on an everyday basis.[1] Second, I use social reproduction as a genre, or tangible site of social contracts and tacit agreements, to provoke the recognition of plural conceptions of social justice. As a genre, social reproduction relates promise and ordinary life under (oil) capitalism. My aim is to show that its waning, flexibility, and endurance frame "potential openings within and beyond the impasse of adjustment that constant crisis creates" under capitalist regimes (Berlant 2011, 6–7).

Next, I situate the analysis of oil-infused urban ecologies. Then, I trace how forms of capital accumulation become the substrate of oil capitalism in Esmeraldas and draw on seven months of ethnographic observations (2012–2016) to examine the embodied political ecology of oil flow. I conclude with a discussion on how attention to the wager of life with oil brings needed attention to everyday urban justice associated with oil hegemony.

Urban Life with Oil

Critical oil scholarship has examined the ordinary geographies of oil extraction in rural areas and the inequalities produced therein (Bridge 2015). How oil flows through urban nodes remains underdeveloped, however. Notable exceptions are the works of Huber and Auyero and colleagues, which showcase two distinct analytics of how urban ecologies animate moral projects of "ordinary" life with oil. Examining oil's circulation in the United States, Huber (2013) showed how oil derivatives provide the supplementary materiality to the real subsumption of life under capital. Oil flows through civic freedoms and consumption via desired material goods (private lawns, cars, and homes) and the "good life" is valued according to the generalized mass consumption of oil products. In quite a different take on ordinary urban life with oil, Auyero and Swistun (2009) demonstrated how oil's flow reproduces not the dream of the "good life" but the disavowal of life itself. In their ethnography of a marginalized neighborhood in close proximity to YPF (now Repsol) and other petrochemical companies in Buenos Aires, they suggested that oil-enabled life is about fears regarding the origins and prognosis of infirmities and uncertainties.

In both accounts, oil capitalism becomes legible through the production of urban bodies. Huber (2013) emphasized the disciplining of abled bodies—suburban, affluent, white, and privileged—as "normal" bodies that appear emotionally and physically disconnected from the energy logistics that enable their actualization. Auyero and Swistun (2009), on the other hand, point to the production of suffering bodies ("the poor") who struggle to endure rather than thrive under oil capitalism. Their proximity to the logistics of oil, itself a product of dispossession, determines their ethical encounters in the city: "the poor do not breathe the same air, drink the same water, or play on the same playgrounds as others" (Auyero and Burbano de Lara 2012). These devalued bodies challenge oil capitalism through refusal of and confusion about toxic urban environments.

In these readings, subjects either consent or refuse to the hegemony of oil capital. The liveliness of urban living through which oil flows remains bracketed, however. How do embodied ecologies, the "fleshy, messy and indeterminate stuff of everyday life" (Katz 2001, 711), clue us into the workings of *jugarse la vida* in the petro-city? Doshi (2017) pointed out that the body is often mobilized in conceptualizations of urban infrastructure and politics, but material embodiment remains understudied.

The next two sections begin accounting for these embodied ecologies of the petro-city. First, I trace the "deep historical spatial logics" of capitalism (Heynen 2016) that became the substrate on which urban life with oil unfolds in Esmeraldas. If "capital can only be capital when it is accumulating" Melamed (2015, 77), and in urban spaces it accumulates by producing and moving through relations of inequality (Pulido 2000, 2016; Rangathan 2016), then we must recognize how historically sedimented differential access to capital has deepened the wager for life in the petro-city. Second, I trace stories of *jugarse la vida*: how people maintain attachments that sustain personal ideals of the good life, even as the day-to-day becomes riskier precisely because of the logistics of petro-capital.

From Black Freedom to Black Gold

The city of Esmeraldas (population 161,000) is flanked by the Pacific Ocean on the north, the Esmeraldas River on the east, and the low mountain ranges of the Tumbes-Chocó Magdalena bioregion on the west. These limits shaped the city's north–south growth. Newcomers with limited economic resources tend to settle the unstable edges of the city: the foothills that suffer from landslides during the rainy season; the flood-prone mangrove banks of the Esmeraldas River; or the southern margins of the city, where the urban services grid does not fully reach yet.

Fifty-five percent of the people in the Canton of Esmeraldas (where the city is located) self-define as Afro-Ecuadorian or Afro-descendent (INEC 2010), a product of colonial capitalism. Slaves escaping the Spanish Empire trade in the sixteenth century formed maroon communities in Esmeraldas, earning it the reputation of "unmanageable spaces" of black freedom (Rueda Novoa 2001). After independence from Spain in 1822, the Ecuadorian state concessioned Esmeraldas to European entrepreneurs, disavowing its sovereign black history and, as Esmeraldeño writer Juan Montaño described, "shackling our minds to racial capitalism" (personal interview,

August 2012). Private capital financed railroads that connected Esmeraldas to Andean commercial centers (Parsons 1957) but disconnected local blacks from land. Between 1948 and 1951, the Ecuadorian government invested heavily in agricultural credits, ports, and highways to expand the agrarian frontier, issuing loans to nearly 1,000 settlers to finance 25,000 acres of new bananas in Esmeraldas and surrounding areas (Parsons 1957). By 1953, more than one in every four bananas shipped from Ecuador moved through Esmeraldas (Southgate 2016). "Over 100,000 bunches of bananas ... moved through the port ... there was money everywhere" (Lopez Estupiñán 2013).

During this period, rural Afro-descendent farmers, pushed off the land by continued extractive capitalism, became workers on plantations owned by Europeans. The precarious urban margins—foothills, river banks, and the south end of the city—swelled as surplus workers moved to the city. Some found employment in the port, loading and sorting bananas and joining vessel crews. Afro-Ecuadorian poet Antonio Preciado (1992) fondly recalled his first job as a child: offering water to the workers who carried the banana bunches on their shoulders. Others recall childhood memories of black workers in saloons, unwinding from a hard day of work (La Jaula de Pájaro 2002). The banana bonanza did not last long, however. After Sigatoka leaf blight forced the abandonment of plantations in the mid-1950s, a second banana boom began in the late-1960s, but banana elites elsewhere lobbied to prohibit Esmeraldeño production. Their efforts indirectly reduced the economic activities (e.g., food preparation and entertainment, maintenance, and cleaning) that urban blacks creatively used to redistribute banana income during the boom years.

The final blow to the Esmeraldeño banana industry was dealt in 1967, when commercially viable oil reserves were identified 500 km away, in the northern Ecuadorian Amazon. Focusing on transport logistics and costs, oil firms identified Esmeraldas as the best geographically suitable and economically viable site to build an oil export terminal. Esmeraldeños waited optimistically for the promised economic boost associated with the new oil industry to be housed in Esmeraldas (Galarza 1972). The Esmeraldas refinery was built south of the city, beyond its administrative boundaries, and began operating in 1978. The film *The First Barrel of Oil* (Cuesta 1972) documents the arrival of the very first barrel of Amazonian oil to the terminal station of Balao, Esmeraldas, on 26 June 1972. Featuring footage from the events, the documentary draws attention to the ceremonial turning of the oil spigot in Esmeraldas, an act that made evident how individual bodies, nation, and oil are bound. "[U]nable to hold back their emotions," declares the narrator, nearby witnesses sought to touch the first barrel, "their calloused hands, hardened by work, stained with the black gold that symbolizes their hope." The documentary ends with the barrel's arrival in Quito, where along with other barrels it forms a squad of militant natural bodies saluting government representatives and foreign oil firm businessmen.

This affective narrative of oil-enabled life continues today—most readily in President Rafael Correa's (2007–2017) speeches on how his administration would end poverty through oil investment. The Correa administration introduced the campaign *Ecuador, ama la vida* (Ecuador, love life), its symbols emblazoned on the façade of state buildings, including the Esmeraldas refinery, to build a national imaginary of a love of life possible through oil's transformation into social programs and megainfrastructure. The state-sponsored advertisement "Your Community Is Priority" exemplifies this point. The one-minute ad shows happy children running to meet an oil refinery complex that melts away into homes, schools, and hospitals. This campaign echoes the "cruel optimisms" of Esmeraldeños: To attain a dignified life, the people must desire goods (access to urban services, homeownership, etc.) even as their materialization, conditioned by oil's circulation, introduces harms and risks to life in general.

Wagering Life in the Petro-City

To wager is to intentionally place something or oneself at risk to bear on the unfolding of an uncertain event. Achieving a dignified life under precarious economic conditions is one such anticipated event in Esmeraldas. The stories of social reproduction profiled next illustrate these wagers, where bodies and futures are put on the line to weigh in on the promise of a dignified life tied to oil capital's circulation.

Leonor

Leonor (pseudonym) is an Afro-Ecuadorian woman in her fifties, born in the city. At eight years old, her mother sent her to work as a maid in a rural parish 90 km south of the city, as she could not afford to care for her. At eighteen years old, Leonor returned to her mother's house, but city life was unbearable: The roads turned to rivers of mud that brought down hillsides when it rained, and buzzards circled the city due to exposed waste from homes and markets. Leonor moved to the capital city of Quito but soon returned to Esmeraldas. She heard that land abutting the refinery was being invaded and saw this as an opportunity to gain independence. The land Leonor and a handful of people invaded became 15 de Marzo, the neighborhood where she has resided for more than twenty-five years. Like others who could not afford to live in the city proper in the late 1980s, Leonor settled this peri-urban space, feeling that it was safe for raising her children. Despite abutting the refinery compound, it was peaceful and a contrast to the chaos of the city.

Such struggles over land at the urban edge marked shifts in the political regime of the city. The MPD (*Movimiento Popular Democrático* or Popular Democratic Movement), a self-identified revolutionary, leftist party, took root in the marginal neighborhoods of Esmeraldas in the 1980s, including 15 de Marzo. MPD militant Ernesto Estupiñan ran for the mayor's office and won in a landslide, becoming the first black mayor of Esmeraldas in 1999. Born and raised in Esmeraldas, Estupiñán was intimately aware of the land needs at the city's periphery. His administration stabilized hillsides, filled wetlands for migrants to settle, paved roads, organized commercial sectors, established a limited garbage collection system, and built limited water and sewage systems. These "underground" public works were oil-enabled: During his first term, Estupiñán took advantage of neoliberal decentralization policies and pushed for agreements with then-president Gustavo Noboa to channel oil rents directly to the city as compensation for the refinery's atmospheric externalities.

Estupiñán also understood racialized abandonment in the petro-city. 15 de Marzo residents remember Estupiñan's efforts as supporting a more "dignified way of life," and according to Leonor, "Ernesto knew how to take care of his people" (interview, January 2014). As the incumbent mayoral candidate, during reelection campaigns in 2003 and 2008, he extended urban services to areas surrounding the refinery and lobbied the refinery to financially contribute to the effort. Today, the formal industrial buffer has residential blocks with cement homes, paved roads, schools, and health centers; active neighbor associations; and diverse "economic activities" (nonwage labor). Estupiñan was reelected in 2004 and again in 2009, although by then accusations of corruption demoralized his base support.

In the late 2000s, the management of oil logistics and population in Esmeraldas changed again. President Correa, as part of his nation improvement campaign, turned his attention to securing state presence in strategic resource areas like Esmeraldas. This securitization bound oil and electoral politics, turning Esmeraldas into a political battlefield in the years preceding the 2014 municipal elections. The Esmeraldas municipality was a harsh critic of Correa's government and accused the central government of racialized meddling in local affairs. Understanding how the municipality depended on the redistribution of oil rents to manage urban living and being, Correa decreed that no oil funds would be allocated to the municipal government, even if it was entitled to these by law, to purge the opposition out of Esmeraldas.

By August 2012, the oil chokehold on the MPD-dominated municipality became evident. Neighborhoods like 15 de Marzo were in utter disrepair: Potholes, crime levels, and urban insecurity increased. For Leonor, the potholes and crime added risks to her economic activities, particularly her food vending activities. She called Estupiñán directly to fix these but received little response. In the midst of these frictions, the Ecuadorian state stepped in to facilitate urban renewal: It used oil funds to hire private firms to pave roads, install sidewalks and public lighting, and paint schools in colors associated with the state-owned oil company. By January 2014, billboards throughout the city displayed the amounts invested in urban works and Correa's support for the state-affiliated candidate, Lenin Lara. Leonor approved of the positive change she experienced: She felt safe coming back home late at night and was not worried about her food cart (her main vending means) breaking because of the potholes (Figure 1). Esmeraldeños understood and condoned the oil-enabled strategy to disarticulate Estupiñán from his constituency. In the 2014 municipal election, Lara received 46 percent of the votes in the parish where 15 de Marzo is located and the MPD received 11 percent.

Leonor lives next to the refinery but she does not relate to the refinery as a neighbor. Rather, her living and being is marked by the rhythms of oil money and oil politics, through urban services, the land to which she was eventually granted de facto possession, and the patron–client relations that secured her electoral vote. Leonor's life in 15 de Marzo is certainly tied to oil capital flow, though. Suffering from diabetes, headaches, and chronic knee pain, likely associated with long-term exposure to refinery emissions, Leonor

Figure 1. Food vending is a ubiquitous economic activity in the busy streets of Esmeraldas. Photo by author. (Color figure available online.)

moves slowly. She is the head of her household. Not formally employed, her main income comes from the *maduros* (sweet plantains) she roasts in the morning, which she takes to the nearby market or bus terminal using her "vending cart," a bicycle adapted into a three-wheel vehicle that carries a wooden platform. She used to own a larger fleet, but with sickness, rising insecurity, and poor road infrastructure, she scaled down. She brings in about $20 to $30 a week to the household. Her seventeen-year-old house, donated by a Catholic group, fell down in 2013, and she invests the lion's share of her income in rebuilding it. Leonor's teenage son is a brick layer at a school built with oil rents by the Correa administration and her husband is occasionally hired for the construction of state-sponsored, oil-financed infrastructure. Leonor's most stable income comes from a state-funded cash payment program, a form of distributive labor that she admits is the only way to keep things afloat. Although she is professionally trained as a teacher, she prefers economic activities and redistributive labor because they give her the freedom to stay close to her family and to escape the routine of wage labor.

Jaime

Jaime (pseudonym) hails from Tumaco, Colombia. He has worked as an electrician, fisher, bricklayer, and security guard. In 2009, he and his wife and three adult

daughters left Colombia and settled in Esmeraldas, convinced that life in the city "is easier." Jaime describes Esmeraldas as a place where work can be had—if you look for it. Soon after settling in Esmeraldas, Jaime met Estupiñán and asked him for work: "We became friends and he gave me a job in the municipality, as a refrigeration technician, and then he sent me to work at the slaughterhouse" (interview, July 2016).

Jaime believes that anyone in Esmeraldas who shares his last name could be kin. Part of his reasoning stems from the flow of Colombian migrants who turned to Esmeraldas to escape the economic attrition, violence, and dispossession of civil war. Some of this desire for kin also reflects how difficult it is for refugees to wager a life in Esmeraldas. Being in his sixties, Jaime did not have easy access to employment. To make ends meet, Jaime's family set up a beer business 11 km north of the refinery, on the Las Palmas beach, the city's premier tourist spot. A family member loaned him US$15 and with that money he bought and sold three crates of beer in half an hour. Everybody in the family worked: Jaime's daughters grilled brochettes while he and his wife sold the beer, every day, from 6 p.m. until late. On the best days, Jaime sold about sixty crates of beer (twenty-four bottles, at $2 each). Pooling together their joint earnings, the family purchased a home and in 2013 invested in a formal restaurant on the beach. Soon after, Jaime's wife fell ill with cancer

Figure 2. Food and beer vending on Las Palmas beach with FLOPEC building in the background. Photo by author. (Color figure available online.)

and died the following year. Jaime was forced to close the restaurant.

That same year, Lenin Lara, the state-backed mayor, was elected. Jaime experienced a noticeable shift in his wager of life in Esmeraldas. Under Estupiñán, Las Palmas was a chaotic gathering spot for night life, and small restaurants (kiosks) peppered the edge of the beach, offering food and beer with little regulation. Blending excitement and excess, gatherings at Las Palmas could escalate into violence: Drunk customers often broke into fights and empty bottles of beer flying across streets was a common sight. According to Jaime, Estupiñán did not charge taxes on economic activities like his and sponsored public events that generated more redistribution possibilities. Estupiñán likely relied more on oil rents than city taxes to administrate life in Esmeraldas and thus applied little control over the latter.

Lara, however, drew on oil-funded urban renewal and moral sanitization to change living and being in Las Palmas. As part of the oil-funded state campaign to "love life," he turned Las Palmas into an orderly public space. The sale of alcohol in public spaces was prohibited and "informal sales" (e.g., food, drinks, alcohol, etc.), such as those that supported Jaime's family, were criminalized. In addition, in 2013, Correa decreed that FLOPEC, the institution that oversees oil

derivatives marine transportation, would relocate to Esmeraldas, to be closer to its cabotage infrastructure. FLOPEC collaborated with Lara's sanitized vision of the spatial logics of the beachfront. In May 2016, it inaugurated a massive brand new office building resembling a giant white oil vessel in the place where food kiosks used to operate (Figure 2). Surrounding FLOPEC's building, a squad of vendor huts, like the barrels of oil that arrived in Quito back in 1972, sit in orderly fashion, available for lease. The idea is that food sellers like Jaime could be integrated into (and taxed as part of) the "formal" economy by renting these spaces for their business. Access to these formalized spaces is difficult, however. Rent runs from $300 per month to more than $800 per month, depending on location and size, which are prohibitive prices in a city where rent for an apartment in less desirable areas averages $100 per month.

Jaime does not typically talk about oil capital directly but emphasizes how the logistics of oil securitization become the logistics of his livelihood wager: when and where he can derive an income, how much, and with what risks. He refers to the FLOPEC building as a "white elephant" of gentrification, and he might be right. Surrounding the FLOPEC compound are announcements for the forthcoming *balnearios* (beachfront communities) that will occupy the former spaces of economic activities, which have now been pushed out of Las Palmas. This formalization of public space has deepened Jaime's wager for life. Like other sellers, Jaime has not given up on his business; it gives him a sense of freedom in the city. Jaime often comes on "off days," typically on Sundays, and only buys a limited amount of beer crates for resale—a maximum of seven or eight. His elder daughter and grandson, who live next to the refinery, walk the beach with a box full of water bottles and nonalcoholic beverages to eke out a living. "We do this hard work because we need to," Jaime's daughter says. "Even if the municipality wants to remove us, here we are, selling below the radar, that's where we are." Becoming more mobile to wager a living means becoming more flexible with less protection; to carry less and sell less, to hope for less.

Leonor and Jaime expand our understanding of social justice in the petro-city. Their actions, moral economies, and political agency complicate production-centric analyses and dualistic framings of docility and resistance vis-à-vis oil harms. The substrate of their urban vagabondage is a petro-economy that is simultaneously toxic and enabling and that knots their urban bodies with oil flow and the management of populations via patron–client relations, investments, and urban works. A particular form of urban justice unfolds in this substrate, filled with "cruel optimisms" (Berlant 2011) of petro-capital: a double bind of hope and loss that is experienced through everyday social reproduction and a desire for a "good life" (e.g., through employment, upward mobility, and social equality), under conditions of toxicity, contingency, and crisis.

Conclusion

Petro-cities like Esmeraldas showcase how the "vagabondage" of capitalism (Katz 2001) manifests in bodies, desires, and urban risks knotted with oil's circulation. Paying attention to the affective attachments of social reproduction, this article outlined cruel optimisms that both help endure the *longue durée* of capitalist devaluation and normalize the injustices and risks associated with a dignified life—a life worth living—in the petro-city. Leonor and Jaime's wagers of life provide rich insights into dimensions of social justice in petro-cities. First, they offer an expanded view of urban justice that is not limited to contamination events and toxic exposures—social or chemical. Leonor and Jaime might be targeted as objects of improvement, collateral damage, or unavoidable sacrifices of capital accumulation by those who do not know their

realities, but their living and being is not determined by such victimhood slots. Their stories exceed these toxic slots, revealing how their affective attachments with the petro-city—where they live, love, and reproduce—go beyond a politics of confusion, refusal, and recognition of harm. Theirs is a politics of dignity in the present, an optimism bound up in the lived geographies of capitalism, under conditions of "slow death." Second, Leonor and Jaime do not necessarily see themselves as agents responsible for changing the city; they see themselves as doing the best they can to achieve a life worth living. This ultimately is their challenge to normative and emancipatory conceptions of justice in the city, often organized around securing property, rights, equality before the law, and collective action. Justice, as a form of freedom, manifests in various registers and genres. People are not only divided into resisting or being subsumed by capitalism; they find room to maneuver within the toxicity of its hegemony. As both Leonor and Jaime would have it, dignity is their genre of social justice in the petro-city. Daily, affective quests for dignity—their wagers for a "dignified life"—hold the possibility for locating and learning from different forms of knowing justice within the ordinary spaces of life with oil.

Acknowledgments

My special thanks to Marcela Benavides, Kati Alvarez, Janeth Cando, Andrea Castillo, Yarita Giler, Hector Lañón, Paola Lastre, and Carlos Erazo for their friendship, insight, and collective generosity. This research would not be possible without those who opened their doors and shared their life stories in Esmeraldas. All errors of interpretation remain mine.

Funding

This research was supported by the National Science Foundation (NSF) Award 1259049 and by The Institute for the Arts and Humanities (IAH) at the University of North Carolina.

Note

1. I thank Héctor Lañón for introducing me to the phrase *jugarse la vida* in reference to the everyday risks of making a life worth living in Esmeraldas.

References

Anderson, B., and J. Wylie. 2009. On geography and materiality. *Environment and Planning A* 41 (2):318–35.

Appel, H., A. Mason, and M. Watts. 2015. *Subterranean estates: Life worlds of oil and gas.* Ithaca, NY: Cornell University Press.

Auyero, J., and A. Burbano de Lara. 2012. In harm's way at the urban margins. *Ethnography* 13 (4):531–57.

Auyero, J., and D. Swistun. 2009. *Flammable: Environmental suffering in an Argentine shantytown.* Oxford, UK: Oxford University Press.

Berlant, L. G. 2011. *Cruel optimism.* Durham, NC: Duke University Press.

Bridge, G. 2015. The hole world: Scales and spaces of extraction. *Scenario Journal.* Accessed December 1, 2016. http://scenariojournal.com/article/the-hole-world/.

Cuesta, A. 1972. *El Primer Barril* [The first barrel]. Sendip, Ecuador: Noticiero Nacional Cuestordóñez, Compañia Productora.

Doshi, S. 2017. Embodied urban political ecology: Five propositions. *Area* 49 (1):125–28.

EP-Petroecuador. 2013. *Memoria de sostenibilidad refinería esmeraldas* [Sustainability memory of the Esmeraldas refinery]. Quito, Ecuador: EP-Petroecuador.

Ferguson, J. 2015. *Give a man a fish: Reflections on the new politics of distribution.* Durham, NC: Duke University Press.

Galarza, J. 1972. *El Festín del Petróleo* [The feast of petroleum]. Quito, Ecuador: Ediciones Solitierra.

Haraway, D. J. 2008. *When species meet.* Minneapolis: University of Minnesota Press.

Heynen, N. 2016. Urban political ecology II: The abolitionist century. *Progress in Human Geography* 40 (6):839–45.

Huber, M. 2013. *Lifeblood.* Minneapolis: University of Minnesota Press.

Instituto Nacional de Estadística y Censo (INEC). 2010. Fascículo provincial de esmeraldas [Esmeraldas province datasheet]. Accessed December 1, 2016. http://www.ecuadorencifras.gob.ec/wp-content/descargas/Manulateral/Resultados-provinciales/esmeraldas.pdf.

Katz, C. 2001. Vagabond capitalism and the necessity of social reproduction. *Antipode* 33 (4):709–28.

La Jaula, de Pájaro. 2002. Ernesto Estupiñán un hombre del barrio caliente [Ernesto Estupiñán, a man from barrio caliente]. *El Universo* 31 March 2002. Accessed December 1, 2016. http://www.eluniverso.com/2002/03/31/0001/69/42C22E657B9D46829CBA5FD1D70F8756.html.

LeMenager, S. 2014. *Living oil: Petroleum culture in the American century.* New York: Oxford University Press.

Lopez Estupiñán, L. 2013. El oro verde ... [Green gold ...]. *La Hora* September 6. Accessed December 1, 2016. http://lahora.com.ec/index.php/noticias/show/1101559072/-1/El_oro_verde%E2%80%A6.html#.V7dfg5grL1s.

Melamed, J. 2015. Racial capitalism. *Critical Ethnic Studies* 1 (1):76–85.

Mitchell, T. 2011. *Carbon democracy.* New York: Verso.

Murphy, M. 2017. *The economization of life.* Durham, NC: Duke University Press.

Parsons, J. 1957. Bananas in Ecuador: A new chapter in the history of tropical agriculture. *Economic Geography* 33 (3):201–16.

Preciado A. 1992. *De Sol a Sol* [From sun to sun]. Quito, Ecuador: Libresa.

Pulido, L. 2000. Rethinking environmental racism: White privilege and urban development in Southern California. *Annals of the Association of American Geographers* 90 (1):12–40.

———. 2016. Geographies of race and ethnicity II: Environmental racism, racial capitalism and state-sanctioned violence. *Progress in Human Geography* 41 (4):524–33.

Ranganathan, M. 2016. Thinking with Flint: Racial liberalism and the roots of an American water tragedy. *Capitalism Nature Socialism* 27 (3):17–33.

Rogers, D. 2015. *The depths of Russia.* Ithaca, NY: Cornell University Press.

Rueda Novoa, R. 2001. *Zambaje y autonomía: Historia de la gente negra de la provincia de esmeraldas: Siglos XVI-XVIII* [Zambaje and autonomy: The history of black people in the province of Esmeraldas: The XVI–XVII centuries]. Esmeraldas, Ecuador: Municipalidad de Esmeraldas.

Southgate, D. 2016. *Globalized fruit, local entrepreneurs: How one banana-exporting country achieved worldwide reach.* Philadelphia: University of Pennsylvania Press.

Valdivia, G., and F. Lu. 2016. *Los impactos de la extracción petrolera, políticas regulatorias y prácticas ambientales en esmeraldas, Ecuador* [The impacts of oil extraction, regulation, and environmental practices in Esmeraldas, Ecuador]. Esmeraldas, Ecuador: Report to EP-Petroecuador.

Watts, M. 2004. Resource curse? Governmentality, oil and power in the Niger Delta, Nigeria. *Geopolitics* 9 (1):50–80.

GABRIELA VALDIVIA is an Associate Professor in the Department of Geography at The University of North Carolina, Chapel Hill, NC 27599. E-mail: valdivia@email.unc.edu. Her research interests include political ecologies of extractive industries, resource governance, and environmental justice in Latin America.

23 Decolonizing Urban Political Ecologies

The Production of Nature in Settler Colonial Cities

Michael Simpson and Jen Bagelman

This article contributes to the decolonization of urban political ecology (UPE) by centering the ongoing processes of colonization and its resistances that produce urban natures in settler colonial cities. Placing the UPE literature in conversation with scholarship on settler colonialism and Indigenous resurgence, we demonstrate how the ecology of the settler colonial city is marked by the imposition of a colonial socionatural order on existing Indigenous socionatural systems. Examining the case of Lekwungen territory, commonly known as Victoria, British Columbia, we consider how parks, property lines, and settler agriculture are inscribed on a dynamic food system maintained by the Lekwungen over millennia. The erasure of the Lekwungen socioecological system, however, has never been complete. Efforts of the Lekwungen and their allies to continue managing these lands as part of an Indigenous food system have resulted in conflict with volunteer conservationists and parks officials who assert their own jurisdictional authority over the space. Drawing on interviews and participant observation research, we argue that the seemingly quotidian and everyday acts of tending to urban greenspace by these groups are actually of central importance to struggles over the reproduction of UPEs in the settler colonial city. *Key Words: decolonization, Indigenous resurgence, settler colonialism, urban political ecology.*

本文透过聚焦生产迁佔殖民城市本质的持续的殖民及反抗过程, 对城市政治生态学 (UPE) 的去殖民作出贡献。我们将 UPE 文献置放于迁佔殖民主义与原住民復甦的文献对话之中, 展现迁佔殖民城市生态, 如何以对既存的原住民社会自然系统加诸殖民的社会自然秩序为印记。我们检视肋筐恩 (Lekwungen) 领地的案例——此地一般以英属哥伦比亚的维多利亚所为人熟知, 考量公园、产权界限与迁佔者农业, 如何拓印于肋筐恩维系上千年的动态粮食系统之上。但肋筐恩的社会生态系统却从未完全遭到抹除。肋筐恩及其同盟者持续管理这些土地作为原住民粮食系统的努力, 导致与志愿保存者和将自身管辖权投射于该空间的公园职员之间的冲突。我们运用访谈和参与式观察, 主张这些群体照料城市绿空间此般看似平凡且日常的行为, 实际上对于在迁佔殖民城市中再生产 UPE 的斗争而言至关重要。 关键词: 去殖民, 原住民復甦, 迁佔殖民主义, 城市政治生态学。

Este artículo contribuye a la descolonización de la ecología política urbana (EPU) centrando los procesos actuales de colonización y sus resistencias generadores de naturalezas urbanas en ciudades de orígenes coloniales. Colocando la literatura de la EPU en conversación con la erudición sobre colonialismo con colonos y la resurgencia indígena, demostramos el modo como la ecología de la ciudad colonial de pobladores colonos está marcada por la imposición de un orden socionatural colonial sobre los sistemas socionaturales indígenas existentes. Examinando el caso del territorio lekwungen, conocido comúnmente como Victoria, Columbia Británica, consideramos cómo los parques, las líneas de propiedad y la agricultura de los pobladores están inscritos en un sistema alimentario dinámico mantenido a través de los milenios por los lekwungen. Sin embargo, la eliminación del sistema socioecológico lekwungen nunca se ha completado. Los esfuerzos de los lekwungen y de sus aliados para continuar manejando estas tierras como parte de un sistema alimentario indígena han terminado en conflicto con conservacionistas voluntarios y oficiales de los parques, quienes reivindican su propia autoridad jurisdiccional sobre ese espacio. Con base en entrevistas y observaciones de investigación participativa, argumentamos que los actos diarios aparentemente cotidianos de ocuparse del espacio verde urbano por estos grupos son realmente de importancia central en las luchas para la reproducción de la EPU en la ciudad colonial de pobladores colonos. *Palabras clave: descolonización, resurgencia indígena, colonialismo con colonos, ecología política urbana.*

The forceful erasure of Indigenous socioecological systems is a condition of possibility for the creation of the settler colonial city. Yet, scholars of urban political ecology (UPE) have had relatively little to say about this profound social injustice that constitutes the city's foundational moment.

Centering the historical and ongoing processes of dispossession that are necessary for the making and remaking of settler colonial space in the city, this article contributes to a recent wave of scholarship that extends UPE beyond its earlier Marxian capital-centric focus by interrogating intersections with other forms of exclusions and injustices that contribute to the production of uneven urban environments (see, e.g., Heynen, Perkins, and Roy 2006; Truelove 2011; Lawhon, Ernstson, and Silver 2014; Zimmer 2015; Doshi 2016; Heynen 2016, 2017). Here, we foreground processes of settler colonialism, where the creation and continued existence of one social, political, and ecological formation is premised on the dispossession and elimination of Indigenous societies and ecologies (Wolfe 2006; Tuck and Yang 2012; Hixon 2013). Whereas scholars of UPE have tended to overlook how the spatial strategies of settler colonialism fashion the political ecologies of the city, geographers and scholars in related disciplines who do place settler colonialism at the forefront of their analyses have tended to overlook how this is an ecological project as much as it is a social or political project (Harris 1997, 2008, 2011; Clayton 1999, 2000; Blomley 2003, 2004; Edmonds 2010; Hugill 2017). By bringing UPE into conversation with scholarship on the geography of settler colonialism and Indigenous resurgence, we hope to contribute to Heynen's (2016) recent call for a more "heterodox" UPE scholarship that is "attentive to racial and colonial capitalism" (842) and to build on this call by suggesting new research directions that might reveal further pathways toward decolonizing and intersectional scholarship and practice.

Drawing on the case study of an urban greenspace on the homelands of the Lekwungen people, now known as Victoria, British Columbia, we argue that the production of the settler colonial city entails the violent imposition of a spatial order on existing Indigenous socionatural systems, both of which become inscribed in the city's political ecologies. This process remains incomplete and ongoing, however, as the spatial formation of settler power must be forcibly maintained and reproduced daily. We argue that acts of Indigenous peoples to remake the socioecologies of the city by reclaiming ancestral food systems serve to contest and disrupt the creation of smooth urban geographies of settler power and are important expressions of Indigenous resurgence and decolonization.

In the first section of this article, we begin by describing the Lekwungen socioecological system that was encountered by early European explorers and settlers.

Although remarking on the beauty and bounty of these lands, these colonizers failed to recognize that they were gazing on a complex food system that the Lekwungen produced and modified over millennia. Seeing like settlers, they were unable to recognize any forms of cultural intervention that did not resemble the particular linear order with which they were familiar in Europe (Cronon 1996; Scott 1998). Their Eurocentric frames informed them that Indigenous peoples of the Americas were incapable of producing such landscapes.

In the next section, we move on to demonstrate that this *terra nullius* view of the world was not simply a benign misunderstanding. The presumed absence of any Indigenous cultural interventions in the landscape led colonizers to believe that they could inhabit, sculpt, and build on Lekwungen lands freely and without consequence. Moreover, these Eurocentric misperceptions became the legal and discursive rationale for the removal of the Lekwungen peoples from these lands and the subsequent production of the settler colonial city.

In the third section, we move to contemporary geopolitical landscapes to demonstrate that the project of dispossession and the creation of settler colonial space is not confined to a previous historical moment but remains ongoing and incomplete. To quote Wolfe (2006), settler colonialism is "a structure, not an event" (338). Despite the violence entailed in processes of settler colonization, the creative destruction of Indigenous socioecological formations has not been totalizing; consequently, the ecological composition and dynamics of the city continue to reflect both Indigenous and settler colonial socionatural interventions. Processes of city making and the everyday reproduction of urban space are dependent on sustained practices that reproduce, maintain, and enforce settler colonial space. The spatioecological configurations of the settler colonial city therefore remain unstable and vulnerable to contestation. Our case study provides one example of how Indigenous peoples actively contest the UPEs of settler colonialism by tending to the socioecological systems of their ancestors or, perhaps more accurate, by working to create new emergent socioecological environments that disrupt the smooth imposition of a purely capitalist or settler colonial space. Far from an even topography, what results from these contestations are patchy, inconsistent socioecological landscapes. We argue that these everyday acts of tending to urban greenspace are of great importance to the struggles between competing efforts to reproduce or decolonize the UPEs of the settler colonial city.

Our call to "decolonize UPE" thereby takes on a dual meaning. First, as scholars, our call is for greater attentiveness not only to the colonial structures that shape urban ecologies but also to the ways in which colonial discourses risk being uncritically reproduced by the UPE subdiscipline itself. As Collard, Dempsey, and Sundberg (2015) argued, it is not sufficient for us to deconstruct dominant discourses of nature without also reckoning with the systemic violences of colonial capitalism that produced contemporary conditions of social and ecological impoverishment. This reckoning with both past and present structural conditions can help ensure that our radical imaginaries of a socioecological future do not impose new settler colonial conditions on Indigenous lands and waters (Tuck and Yang 2012). Second, as activists and inheritors of settler privilege, our call is to decolonize the actual material political ecologies of the city by supporting and "walking with" (Zapatistas 2005, cited in Collard, Dempsey, and Sundberg 2015) Indigenous peoples in these ongoing efforts to disrupt the settler colonial project and bring about new urban socioecological worlds rooted in social, racial, and ecological justice in the city. In conclusion, we explore several implications and tensions that arise from bringing the politics of decolonization into conversation with UPE. We hope that these implications and tensions generate new questions and research directions and contribute to ongoing efforts to build a more heterodox and intersectional UPE scholarship.

Seeing Like a Settler: Foundational Discourses of Nature

The first time that the Lekwungen people discovered Europeans in their territory was likely in 1790 when Manuel Quimper and his crew rowed to their shores and, after planting a cross on their lands, proclaimed the Lekwungen territories a Spanish possession "in the name of His Catholic Majesty Carlos IV" (Harris 1997, 31; see also Lutz 2008, 51–52). On returning two weeks later, Quimper and his crew discovered that their cross had been removed (Harris 1997).

George Vancouver and his crew on the HMS *Discovery* arrived two years later on a mission to map what is now known as Vancouver Island on behalf of the British Crown. Encountering the Lekwungen site of Camosack, Vancouver described the landscape as "almost as enchantingly beautiful as the most elegantly finished pleasure grounds in Europe ... an extensive lawn covered with luxuriant grass, and diversified with

an abundance of flowers." Vancouver noted that this landscape "seemed as if it had been planted" in a way that would have "puzzled the most ingenious designer of pleasure grounds" (Vancouver 1801, 226).

The landscape that Vancouver gazed at was clearly challenging to his frame of understanding. On the one hand, Vancouver compared the Lekwungen lands to the beauty of the most skillfully and intentionally manicured landscapes that he was familiar with in Europe, noting, "A picture so pleasing could not fail to call to our remembrance certain delightful and beloved situations in old England" (228). Vancouver wrote that it was hard to believe that "any uncultivated country had ever been discovered exhibiting so rich a picture" (227). Despite these appearances, Vancouver clearly did believe that he had encountered an uncultivated country, stating, "We had no reason to imagine that this country had ever been indebted for any of its decorations to the hand of man" (227). Even as Vancouver paused on his voyage to gather gooseberries, roses, raspberries, and currants, which "seemed to thrive excessively" and "grow very luxuriantly" (229), he refused to believe that he was standing among an expansive and skillfully crafted garden. Instead, he insisted that he was merely reaping the abundance offered by a "bounteous nature."

Yet the prairie grasslands that Vancouver admired were indeed cultured and intentional landscapes. They comprised a dynamic and highly productive food system, crafted and managed by the Lekwungen over millennia. This anthropogenic landscape produced a multitude of fruits, roots, nuts, and game that helped to sustain the Lekwungen people and culture. Perhaps none of these foods was more important to the Lekwungen than the *Kwetlal*, more commonly known today as camas (*Camassia* spp.), a lily with beautiful purple flowers and a starchy bulb that provided the Lekwungen with an important food staple, as well an important item of trade with other nearby nations (Turner and Kuhnlein 1983; Penn 2006; Corntassel and Bryce 2012). It was spring when Vancouver arrived, and the Lekwungen camas fields would have looked spectacular in full bloom.

As ethnobotanist Beckwith (2004) documented, the Lekwungen's careful planning and active maintenance of this landscape required "multifaceted cultivation activities" (210), such as weeding out the poisonous "Death Camas" (*Toxicoscordion venenosum*), thinning out and selecting camas bulbs, rotating fields, and periodic burning to prevent forests from encroaching on these grassland meadows. Controlled burns of camas fields also boosted the productivity of soils, encouraged the growth of berries on the field's edge (Turner, Davidson-Hunt, and O'Flaherty 2003), and created an inviting habitat for deer, which provided the Lekwungen with a source of meat without needing to domesticate animals (Turner 1999). Each of these activities was governed by what Beckwith described as a "camas culture"—a complex series of social, political, and economic relations that changed and developed over time, which included patterns of resource tenure, decision-making processes, and gendered divisions of labor (Beckwith 2004, 210; see also Lutz 2008; Corntassel and Bryce 2012). The landscape that Vancouver encountered in 1792 can thus rightly be described as a socio-political–economic–ecological system.

Although Vancouver's mandate on this visit was to map the island's coastline on behalf of Britain, the detailed maps of the territory that he created were nevertheless pivotal to the settler colonial project. By rendering Native lands and waters objects of colonial knowledge, and thus colonizable territory, these cartographic representations create an "imaginative geography" with "practical performative force" (Gregory 2001). These depictions anticipated colonization, serving as the basis of claims to sovereignty and setting the stage for a future British settlement. Eventually, Vancouver's simplified, abstract, and selective knowledge of Lekwungen lands became codified as scientific knowledge, drawn on to rationalize the renaming and refashioning of this geography as both "Vancouver Island" and "British Columbia" (Clayton 2000).

It was not until fifty years later in 1842, at the height of the Oregon boundary dispute, that the Hudson's Bay Company (HBC) dispatched James Douglas to build Fort Victoria, the first permanent British settlement on the Lekwungen territory (Clayton 1999). On his arrival, Douglas made assumptions similar to those of Vancouver—that the natural beauty of this land was unrelated to the people he encountered living there. Douglas described this land as a "perfect Eden" that appeared to have "dropped from the clouds into its present position," dismissing the Lekwungen as "desperate savages" demonstrating "natural barbarity" (Hargrave 1938, 420–21). This presumption shared by both Vancouver and Douglas that the Lekwungen lands were wild and natural follows in lockstep what is by now a well-known Eurocentric discourse, informed by deep-seated racist beliefs that non-Western peoples were barbaric, savage, uncivilized, and therefore incapable of creating landscapes on par with the most beautiful or productive designs of

Europe (Cronon 1996; Gregory 2001; Edmonds 2010). These explorers and settlers simply did not have the ecological literacy or the cultural understanding to recognize that the idyllic landscapes they encountered were profoundly modified by human societies. Seeing like settlers, the socioecological landscape that Vancouver and Douglas encountered in the Lekwungen territory was simply illegible.

The Production of the Settler Colonial City

This construction of the Western nature–culture dualism is not merely a theoretical conundrum. Germinating within it is the logic of dispossession and genocide. The discursive and representational erasure of people from nature justifies their political and territorial erasure in the material world (Braun 2002). This direct relationship between discourses of nature and colonial acts of dispossession is made clear in the case of the Lekwungen homelands, where Douglas was specifically instructed that native inhabitants had legal title only to those lands on which built structures already existed, or that had already been placed under cultivation, at the time of colonization (Arnett 1999). Douglas's inability to recognize the Lekwungen camas fields as a cultivated food system thus became more than an innocent ethnographic misinterpretation.

When Vancouver Island was officially made a British colony in 1849, the British government effectively outsourced the administration and governance of the colony to the HBC, a privately owned corporation with a proven track record in the colonization business. As explained by Earl Gray in the British House of Lords, the British government felt an urgency to establish a permanent colony on the island before the United States did, but the British government was also hesitant to take on the "considerable expense" that colonies required (cited in Mackie 1992, 6). Instead, the HBC was granted quasi-governmental authority over the island and was endowed with responsibility for the "advancement of colonization and the encouragement of trade and commerce" (cited in Mackie 1992, 6). The HBC would profit from parceling up the land and selling it as private lots to British settlers. In exchange, the company would be required to build colonial infrastructure, such as schools, roads, and parks.

As numerous historians have noted (Fisher 1992; Harris 1997; Clayton 1999; Edmonds 2010), this marked a significant turning point in the colonial project. For the first fifty years of the HBC's presence on

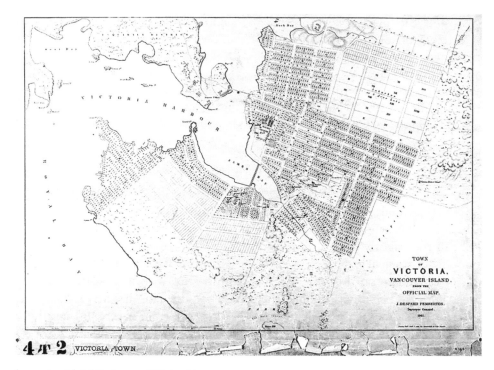

Figure 1. Joseph Pemberton's grid, 1861. *Source:* BC Archives.

Vancouver Island, the company's profits were dependent on mercantile trade with the native peoples, which required that they retain access to their lands. The shift to an economy based on the sale of land required the removal of they from these lands and the extinction of their title. As the newly appointed governor of the colony, Douglas was granted powers to facilitate this removal. He moved swiftly to sign treaties that stated that the Lekwungen lands had become "the entire property of the white people forever" (Arnett 1999). Although the legitimacy of the Douglas Treaties has been widely disputed (Duff 1969; Fisher 1992; Arnett 1999; Lutz 2008), these documents remain the legal basis for settler occupation of Lekwungen lands to this day and thus one of numerous technologies of colonial dispossession.

Shortly after signing these treaties, Joseph Pemberton was appointed the colony's Survey General and was tasked to the complete Victoria's first land survey. Imposing a cadastral grid on the Lekwungen territory, Pemberton's survey sliced the land into thousands of individual parcels, thereby introducing the notion of land as a commodity that can be bought or sold for the first time in this region. This corroborates geographers who have emphasized the role of the map, the survey, and the grid as technologies of colonial dispossession (Scott 1998; Clayton 2000; Mitchell 2002; Blomley

2004; Harris 2008). What is notable about Pemberton's survey, however, is that his own maps clearly depict the abstract lines of the grid being laid out on an existing socioecology (Figure 1).

The imposition of a private property regime not only furthered the dispossession of the Lekwungen people from their lands, separating them from the camas fields that were a major source of their sustenance, but also permitted individual private owners to run roughshod over this food system. Pemberton encouraged this behavior, advertising it as a selling point to potential buyers, suggesting:

> Open grass lands can of course be ploughed up at once, and a crop obtained. Fern lands require to be ploughed in the heat of summer, in order, by fermentation, to kill the fern, and to destroy by exposure bulbous roots, such as crocuses, kamass, etc., for which purpose pigs make admirable pioneers. (Pemberton 1860, 23)

By 1865, more than 100,000 acres had been placed under settler agricultural production in the Victoria region (Beckwith 2004), and many more acres were destroyed by overgrazing and trampling of livestock (Turner 1999; MacDougal, Beckwith, and Maslovat 2004). One settler remarked in 1868, "One of the bitterest regrets of the natives is that the encroachment of the whites is rapidly depriving them of their crops

Figure 2. Camas in *Meegan*/Beacon Hill Park, 2015. *Source:* Michael Simpson. (Color figure available online.)

of this useful and almost necessary plant [camas]" (cited in MacDougal, Beckwith, and Maslovat 2004).

Decolonizing Urban Political Ecologies in the Settler Colonial City

Processes of dispossession, displacement, and erasure of Indigenous space continued in Victoria throughout the twentieth century (Barman 2010). Whereas the Lekwungen managed thousands of acres of land as part of an intricate food system prior to colonization, today, urban development has paved over, built on, or ripped up 95 percent of that agro-ecological system. The relatively few spaces throughout the city where remnants of the camas food system can be found are mostly in parks. One such site is *Meegan*, a place of great cultural importance to the Lekwungen, which is now also Victoria's iconic greenspace known by settlers as Beacon Hill Park (Figure 2).

The management of this site as a recreational area rather than a productive food system has resulted in substantial alterations in both ecological composition and function over the past 150 years. At various times in this history, the city's parks department applied chemical pesticides to the fields, drove over them with maintenance vehicles, and chopped back native flora with lawnmowers. Parks officials also altered the hydrology of the landscape by draining water from the camas fields, making them more amenable to settler recreation activities. Settlers who grazed animals on these fields introduced pasture grasses, such as *Elymus repens* and *Poa pratensis*, which they believed to be superior to the native grasses as a feed for livestock, but which smother out native flora and food crops like camas. Other nonnative species, such as Scotch broom (*Cytisus scoparius*) were introduced into these areas for the purpose of imposing British settler aesthetic preferences. Hundreds of thousands of daffodils were planted in the camas fields to promote Victoria as a

Figure 3. Daffodils and European pasture grasses grow alongside camas in *Meegan*/Beacon Hill Park. *Source:* Michael Simpson. (Color figure available online.)

home to an international flower industry (Ringuette 2004; see Figure 3).

In sum, the present-day socioecological configuration of *Meegan*/Beacon Hill Park reflects features of both the Lekwungen socioecological system that was built and managed over millennia and the settler socioecological formations that have been forcefully imposed on these lands over the past 175 years through processes of dispossession and colonization. The ecological structure, composition, and distribution of this ecology have been directly affected by political and economic processes including imperial competition between Britain and the United States, the dispossession of the Lekwungen homelands, the imposition of a cadastral grid and a private property regime, negotiations between the interests of capital (the HBC) and empire (the Britain Crown), the introduction of European agriculture production and landscape aesthetic, legal ordinances suppressing Indigenous fire management practices, and rapid urbanization. These social, political, and economic processes are not typically thought of as ecological processes and yet they inscribe themselves in the city's landscape and structure its ecological composition.

Today, tensions between Indigenous and settler socioecological logics continue to inform the everyday practices that govern, contest, and reproduce the city's ecology. We conducted interviews and participant observation to learn about the practices of three different groups that are each involved in the maintenance of the landscape of Beacon Hill Park: the City of Victoria's Parks Department, a volunteer-run conservation group called Friends of Beacon Hill Park, and the Lekwungen Food Systems (LFS) project. Although it appears that these groups perform similar activities, such as removing invasive species and encouraging the growth of native flora, in fact they approach maintenance of this space in notably different ways informed by substantially different understandings of the land, its history, and its ecology.

These differences are reflected in the very language that they use to describe the space. For instance, whereas the Lekwungen-led group refers to this space as *Meegan* and understand it to be part of the *Kwetlal* food system, the Friends of Beacon Hill Park refer to this space as a "Garry Oak meadow" located in a "park." On their Web site, the Friends refer to the park as "one of the last refuges in Greater Victoria for some of the area's most delicate and endangered native plants and ecosystems" (Friends of Beacon Hill Park 2017). The group's work is intended to protect and preserve this native ecosystem from human intervention and to return the ecology to something that resembles species composition prior to colonization. Their conception of the precolonial landscape does

Figure 4. Volunteers with the Lekwungen Food System project pile English ivy (*Hedera helix*) removed from *Meegan*, on a nearby monument to Queen Elizabeth II. *Source:* Michael Simpson. (Color figure available online.)

not seem to involve the management by Indigenous peoples, however. In an interview conducted with one member of this group, the informant expressed concern that Lekwungen practices of harvesting camas could disturb the park's native flora.

In contrast, the LFS project, led by Lekwungen woman Cheryl Bryce, actively manages *Meegan* as a part of the Lekwungen ancestral food system. This group's Community Toolshed project involves work parties during which Indigenous and non-Indigenous volunteers remove unwanted plants and tend camas fields (Penn 2006; Corntassel and Bryce 2012). The LFS project does not ask permission from any other authority to tend to these food systems as the Lekwungen have always done because they recognize this land as the rightful territory of the Lekwungen people. Managing the space in this way has created tensions and conflicts with both local conservationists, who consider their actions to be ecologically disruptive, and parks officials, who claim sole jurisdiction. On at least one occasion, parks staff ordered LFS volunteers to halt their activities and leave the park because they were deemed to be acting in violation of the department's landscape management policies.

These differing understandings of territory, jurisdiction, and nature inform the criteria that each of these three groups employ when determining which plants belong and which are removed from the landscape. The conservation-minded Friends generally aim to keep "native" species that were present in the landscape prior to the arrival of Europeans and remove "nonnative" species. In contrast, an informant from the City of Victoria's Parks Department explained that because "nonnative" plants will never, realistically, be completely eliminated, they prioritize biological diversity and ecosystem function over returning to a precolonial landscape. Parks workers therefore do not endeavor to eradicate all nonnative species but instead target specific plants that are considered particularly problematic or invasive in accordance with an official government list of prioritized plants.

Unlike either of these other groups, the LFS project determines which plants to remove based on their practical knowledge of the Indigenous food system. Rather than relying on a strict native–nonnative categorization, this group removes plants that inhibit the growth of food plants such as camas. This entails the removal of many introduced species from the meadows (Figure 4) but also the removal of some native plants such as snowberry (*Symphoricarpos albus*). This practice of removing native plants that are hindering food production remains consistent with the practices of the Lekwungen ancestors who did so by burning these meadows to help keep species at bay and keep the fields open. The Lekwungen approach can also be differentiated from that of municipal parks officials who manage the landscape with the objective of improving overall ecological function. For instance, the Parks Department allows daffodils to remain in the landscape because they do not deem them to be an ecological nuisance. Managing the landscape as a food system, however, the Lekwungen remove daffodils because their toxic bulbs could be confused with the edible bulbs of camas on harvesting. On one occasion, a parks employee reprimanded the Lekwungen-led group for removing Spanish broom (*Spartium junceum*) from the landscape because that species was not on the municipality's list of targeted species.

Conclusion

Histories of the colonial present are inscribed in urban landscapes, legible in the ecological configurations of the settler colonial city. Indeed, settler power is predicated on a spatial arrangement that determines how and by whom urban space is to be inhabited (Blomley 2004). The spatio-ecological formations of the city that are produced by, but also uphold and legitimate, settler colonial power must be routinely maintained and enforced by local authorities. Although these arrangements are often policed with overt force, they are even more commonly maintained through everyday acts such as the tending of urban greenspace and discursive acts of representation. Attention to these subtle, embodied, and affective actions is vital to understand how urban spaces are reproduced and contested on a daily basis in the settler colonial city today. Indeed, as Mackey (2016) described, longstanding discursive representations of nature, nation, and territory act as powerfully affective or emotive "structures of feeling" that reinforce settler expectations of entitlement to the land. Even well-meaning environmentalist discourses and conservation practices remain susceptible to these affective structures of colonialization and often serve to aid and abet the reproduction of settler political ecologies by evoking these representations (Cronon 1996; Braun 2002; Safransky 2014).

The ongoing need to enforce and reproduce settler colonial space also renders this socioecological formation vulnerable, though. Conflicts over everyday practices of maintaining urban greenspaces might seem trivial; however, we argue that these quotidian practices are of central importance to the competing struggles to either reproduce or decolonize the political ecologies of the settler colonial city (Loftus 2012). Decolonization efforts do not always take the spectacular form that social movements theorists might expect and can therefore be difficult to "see" (Simpson 2011). The practices of Indigenous peoples to revitalize and strengthen the socioecologies of their ancestral food systems, as exemplified by the work of Cheryl Bryce and the LFS project, undermine the imposition of a smooth spatial logic of colonial capitalism and thereby contribute to broader struggles for decolonization led by Indigenous peoples in North America and beyond. These direct interventions into the urban landscape exemplify an Indigenous resurgence approach to decolonization, which asserts self-determination through the direct assertion of land-based practices rather than seeking permission or awaiting rights or recognition to be awarded from the settler colonial state (Alfred and Corntassel 2005; Simpson 2011; Corntassel 2012; Coulthard 2014). As Corntassel and Bryce (2012) explained, "Indigenous resurgence is about reconnecting with homelands, cultural practices, and communities, and is centered on reclaiming, restoring, and regenerating homeland relationships. ... This entails moving away from the performativity of a rights discourse geared toward state affirmation and approval toward a daily existence conditioned by place-based cultural practices" (153).

As settlers and scholars, we call for a greater attentiveness to the ways in which the political ecologies of the settler colonial city are produced through the ongoing dynamics between a racialized capitalism that seeks to entrench settler colonial power and the everyday acts of Indigenous resurgence that contest the colonial past and present, upset the spatial logic of the contemporary settler order, and contribute to the emergence of UPEs rooted in social, racial, and environmental justice in the city. Of course, historical geographies of the colonial present take different characteristics in different spaces and contexts, as do the ways in which settler colonial dynamics are contested. Consequently, the implications of our call for greater attentiveness to settler colonialism in UPE scholarship will necessarily be varied and contextual, reflecting the variegated ways in which settler colonial dynamics produce urban space. We have pointed to one example of how these dynamics operate on Lekwungen territory in Victoria, British Columbia; however, further case study examples and research are necessary to tease out the unique implications and tensions that arise when considering how settler colonial dynamics contribute to the production of urban natures in other contexts. Rather than speaking of "the settler colonial city" as a singular space,

then, we want to recognize the different ways in which settler colonial power is structured and enacted in different urban spaces, as well as identify the family resemblances across these settler colonial sites.

In the case of the Victoria, British Columbia, settler colonial processes of dispossession, urbanization, and cultural genocide have inflicted devastating violence on the Lekwungen food system, yet this violence has not been totalizing, as remnants of this socioecology remain present in the urban landscape. Further research might consider what it means to decolonize UPEs in spaces where Indigenous socioecological systems have been even more thoroughly eliminated. Moreover, we might ask how settler colonial processes have interacted with other gendered, racialized, or heteronormative logics in these spaces to produce uneven urban environments (Heynen 2016, 2017). Likewise, how do projects of urban decolonization intersect (or come into tension) with other political priorities and struggles for social and environmental justice that emerge from within these urban spaces, and how are solidarities negotiated across these differences? More work is needed to explore the tensions that might arise between decolonization projects and other intersectional approaches to UPE that decenter Marxian analysis, such as "abolition ecology" (Heynen 2016), "feminist political ecology" (Rocheleau, Thomas-Slayter, and Wangari 1996; Truelove 2011), "embodied UPE" (Doshi 2016), and the "provincialization" of UPE (Lawhon, Ernstson, and Silver 2014; Zimmer 2015). As scholars, we must also continually reconsider how to carefully navigate these tensions by working with and alongside diverse struggles for social, environmental, and racial justice in the city in respectful ways, ensuring that we are not speaking over or on behalf of people involved in these struggles.

Further still, more work is required to understand the implications that the politics of decolonization has for anticapitalist movements that aspire to reclaim or "occupy" urban space and how these movements might proceed in ways that avoid the unintentional imposition of neocolonial settler futurities that further the erasure of Indigenous socioecological systems. More broadly, we might ask what it means to "decolonize" urban landscapes that have been thoroughly and forever altered by the metabolic processes of capitalism and settler colonialism, and we might consider how the answer to this question takes different forms in different socioecological and historical contexts. In sum, the project of building a more heterodox and intersectional UPE requires a greater understanding of these variegations in settler colonial processes; how they intersect with structures of capitalism, race, and patriarchy in these different spaces; and how meaningful and respectful solidarities are built across geographical difference.

Acknowledgments

Many thanks to Cheryl Bryce, Jeff Corntassel, and Joanne Cuffe for inspiration and guidance. Thanks also to Trevor Barnes, Eric Higgs, Nathan McClintock, Nik Heynen, and the two anonymous reviewers for helpful feedback on previous versions of this article.

Funding

This article was written with support from the Social Science and Humanities Research Council of Canada.

References

Alfred, T., and J. Corntassel. 2005. Being Indigenous: Resurgences against contemporary colonialism. *Government and Opposition* 40 (4):597–614.

Arnett, C. 1999. *The terror of the coast: Land alienation and colonial war on Vancouver Island and the Gulf Islands, 1849–1863*. Burnaby, BC, Canada: Talon Books.

Barman, J. 2010. Race, greed and something more: The erasure of urban Indigenous space in early twentieth-century British Columbia. In *Making settler colonial space*, ed. T. B. Mar and P. Edmonds, 155–73. London: Palgrave Macmillan.

Beckwith, B. R. 2004. The queen root of this clime. PhD diss., University of Victoria, Victoria, BC, Canada.

Blomley, N. 2003. Law, property, and the geography of violence: The frontier, the survey, and the grid. *Annals of the Association of American Geographers* 93 (1):121–41.

———. 2004. *Unsettling the city: Urban land and the politics of property*. London and New York: Routledge.

Braun, B. 2002. *The intemperate rainforest: Nature, culture, and power on Canada's west coast*. Minneapolis: University of Minnesota Press.

Clayton, D. 1999. *Islands of truth: The imperial fashioning of Vancouver Island*. Vancouver, BC, Canada: UBC Press.

———. 2000. The creation of imperial space in the Pacific Northwest. *Journal of Historical Geography* 26 (3):327–50.

Collard, R. C., J. Dempsey, and J. Sundberg. 2015. A manifesto for abundant futures. *Annals of the Association of American Geographers* 105 (2):322–30.

Corntassel, J. 2012. Re-envisioning resurgence: Indigenous pathways to decolonization and sustainable self-determination. *Decolonization: Indigeneity, Education & Society* 1 (1):86–101.

Corntassel, J., and C. Bryce. 2012. Practicing sustainable self-determination: Indigenous approaches to cultural restoration and revitalization. *The Brown Journal of International Affairs* 18 (2):151–62.

Coulthard, G. 2014. *Red skin, white masks: Rejecting the colonial politics of recognition*. Minneapolis: University of Minnesota Press.

Cronon, W. (Ed.) 1996. The trouble with wilderness. In *Uncommon ground: Rethinking the human place in nature*, 69–90. New York: Norton.

Doshi, S. 2016. Embodied urban political ecology: Five propositions. *Area* 49 (1):125–28.

Duff, W. 1969. The Fort Victoria treaties. *BC Studies* 3:3–57.

Edmonds, P. 2010. Unpacking settler colonialism's urban strategies: Indigenous peoples in Victoria, British Columbia, and the transition to a settler-colonial city. *Urban History Review/Revue d'histoire urbaine* 38 (2):4–20.

Fisher, R. 1992. *Contact and conflict: Indian–European relations in British Columbia, 1774–1890*. 2nd ed. Vancouver, BC, Canada: UBC Press.

Friends of Beacon Hill Park. 2017. Home page. Accessed August 25, 2017. http://www.friendsofbeaconhillpark.ca/.

Gregory, D. 2001. Postcolonialism & production of nature. In *Social nature: Theory, practice, and politics*, ed. N. Castree and B. Braun, 84–111. Oxford, UK: Blackwell.

Hargrave, J. 1938. *The Hargrave correspondence, 1821–1843*. Edited by G. P. de T. Glazebrook. Toronto: Champlain Society Digital Collection.

Harris, C. 1997. *The resettlement of British Columbia: Essays on colonialism and geographical change*. Vancouver, BC, Canada: UBC Press.

———. 2008. How did colonialism dispossess? Comments from an edge of empire. *Annals of the Association of American Geographers* 94 (1):165–82.

———. 2011. *Making native space: Colonialism, resistance, and reserves in British Columbia*. Vancouver, BC, Canada: UBC Press.

Heynen, N. 2016. Urban political ecology II: The abolitionist century. *Progress in Human Geography* 40 (6):839–45.

———. 2017. Urban political ecology III: The feminist and queer century. *Progress in Human Geography*. Advance online publication. https://doi.org/10.1177/0309132517693336

Heynen, N., H. A. Perkins, and P. Roy. 2006. The political ecology of uneven urban green space: The impact of political economy on race and ethnicity in producing environmental inequality in Milwaukee. *Urban Affairs Review* 42 (1):3–25.

Hixson, W. L. 2013. *American settler–colonialism: A history*. New York: Palgrave Macmillan.

Hugill, D. 2017. What is a settler–colonial city? *Geography Compass* 11 (5):1–11.

Lawhon, M., H. Ernstson, and J. Silver. 2014. Provincializing urban political ecology: Towards a situated UPE through African urbanism. *Antipode* 46 (2):497–516.

Loftus, A. 2012. *Everyday environmentalism: Creating an urban political ecology*. Minneapolis: University of Minnesota Press.

Lutz, J. S. 2008. *Makúk: A new history of Aboriginal–white relations*. Vancouver, BC, Canada: UBC Press.

MacDougall, A. S., B. R. Beckwith, and C. Y. Maslovat. 2004. Defining conservation strategies with historical perspectives: A case study from a degraded oak grassland ecosystem. *Conservation Biology* 18 (2):455–65.

Mackey, E. 2016. *Unsettled expectations: Uncertainty, land and settler decolonization*. Halifax, NS, Canada: Fernwood.

Mackie, R. S. 1992. The colonization of Vancouver Island, 1849–1858. *BC Studies* 96:3–40.

Mitchell, T. 2002. *Rule of experts: Egypt, techno-politics, modernity.* Berkeley: University of California Press.

Pemberton, J. D. 1860. *Facts and figures relating to Vancouver Island and British Columbia: Showing what to expect and how to get there; With illustrative maps.* London: Longman, Green, Longman, and Roberts.

Penn, B. 2006. Restoring camas and culture to Lekwungen and Victoria: An interview with Lekwungen Cheryl Bryce. *Focus Magazine* June.

Ringuette, J. 2004. Beacon Hill Park history (1842–2004). Accessed June 13, 2017. http://www.beaconhillparkhistory.org/.

Rocheleau, D., B. Thomas-Slayter, and E. Wangari. 1996. *Feminist political ecology: Global issues and local experiences.* London and New York: Routledge.

Safransky, S. 2014. Greening the urban frontier: Race, property, and resettlement in Detroit. *Geoforum* 56:237–48.

Scott, J. C. 1998. *Seeing like a state: How certain schemes to improve the human condition have failed.* New Haven, CT: Yale University Press.

Simpson, L. 2011. *Dancing on our turtle's back: Stories of Nishnaabeg re-creation, resurgence, and a new emergence.* Winnipeg, MB, Canada: ARP Books.

Truelove, Y. 2011. (Re-)Conceptualizing water inequality in Delhi, India through a feminist political ecology framework. *Geoforum* 42 (2):143–52.

Tuck, E., and K. W. Yang. 2012. Decolonization is not a metaphor. *Decolonization: Indigeneity, Education & Society* 1 (1):1–40.

Turner, N. J. 1999. "Time to Burn": Traditional use of fire to enhance resource in British Columbia. In *Indians, fire, and the land in the Pacific Northwest,* ed. R. Boyd, 185–218. Corvallis: Oregon State University Press.

Turner, N. J., I. J. Davidson-Hunt, and M. O'Flaherty. 2003. Living on the edge: Ecological and cultural edges as sources of diversity for social–ecological resilience. *Human Ecology* 31 (3):439–61.

Turner, N. J., and H. V. Kuhnlein. 1983. Camas (*Camassia* spp.) and riceroot (*Fritillaria* spp.): Two liliaceous "root" foods of the Northwest Coast Indians. *Ecology of Food and Nutrition* 13 (4):199–219.

Vancouver, G. 1801. *A voyage of discovery to the North Pacific Ocean: And round the world.* Vol. 1. London: J. Stockdale.

Wolfe, P. 2006. Settler colonialism and the elimination of the native. *Journal of Genocide Research* 8 (4):387–409.

Zimmer, A. 2015. Urban political ecology "beyond the West": Engaging with South Asian urban studies. In *The international handbook of political ecology,* ed. R. L. Bryant, 591–99. London: Edward Elgar.

MICHAEL SIMPSON is a PhD Candidate in the Department of Geography at the University of British Columbia, Vancouver, BC V6T 1Z2, Canada. E-mail: m.simpson@alumni.ubc.ca. His research interests include the political ecologies of settler colonialism and conflicts over pipelines in North America.

JEN BAGELMAN is a Lecturer in the Department of Geography at the University of Exeter, Exeter EX4 4QJ, UK. E-mail: j.bagelman@exeter.ac.uk. Her work critically examines how displacement is produced through exclusionary citizenship and bordering practices and explores how anti-colonial movements enact more loving geopolitics.

24 Datafying Disaster

Institutional Framings of Data Production Following Superstorm Sandy

Ryan Burns

In the wake of disasters, communities organize to produce spatial data capturing knowledge about the disaster and to fill gaps left by formal emergency responders. The ways in which communities affect overall response efforts can produce inequalities, disempowerment, or further marginalization. Increasingly, this organizing and knowledge production occurs through digital technologies and, recently, *digital humanitarianism* has become an important suite of such technologies. Digital humanitarianism includes technologies like the crowd-sourced crisis mapping platform Ushahidi and the community of volunteers Humanitarian OpenStreetMap Team, which focuses on the amateur-generated global base map OpenStreetMap. Digital humanitarianism is shifting how needs and knowledges are captured and represented as data following disasters. These transformations raise important questions for geographers interested in the sociopolitical and institutional processes that frame data production and representation. In this article, I contribute to geographers' efforts to understand the institutional and community-based politics that frame the types of data that are produced in disaster contexts by drawing on an ethnographic project that took place in both Washington, DC, and New York City after Superstorm Sandy in 2012. I show that digital humanitarians produced data in the Rockaway Peninsula of New York in response to perceived gaps on the part of formal emergency responders. In so doing, they represented needs, individuals, and communities in ways that local community advocacy organizations found problematic. These findings shed light on the politics and struggles around why particular data sets were produced and the motives behind capturing particular disaster-related needs and knowledge as data. *Key Words: critical data studies, digital humanitarianism, disaster response, urban politics, urban technology.*

灾害过后，社区开始组织进行生产捕捉灾害知识的空间数据，并填补官方灾害急救所遗留的空缺。社区影响总体灾害回应的方式，能够生产不均、剥夺权力，抑或进一步边缘化。此般组织与知识生产，正逐渐透过数码科技而发生，而数码人道主义，并在晚近成为此般技术的重要套件。数码人道主义，包含了诸如众包危机製图平台 Ushahidi 之技术，以及"开放街道地图人道主义团队"的志愿者社群，该社群聚焦由业餘者所生产的全球底图"开放街道地图"。数码人道主义，正在改变灾后过后如何捕捉并再现需求与知识作为数据的方式。这些变迁，对于框架数据生产与再现的社会政治及制度过程感兴趣的地理学者，提出了重要的问题。我于本文中，藉由运用 2012 年珊蒂飓风侵袭后，在华盛顿特区与纽约市进行的民族志计画，对地理学者致力于理解框架在灾害脉络中生产的数据类别之制度和社区政治之努力做出贡献。我将展现，数码人道主义者在纽约的洛克威半岛生产数据，以回应官方灾害急救所意识到的阙如。他们以此呈现地方社区倡议组织发现有问题的需求、个人和社区。这些发现，为特定数据集为何被生产、以及捕捉特定的灾害相关需求与知识作为数据的动机之政治与斗争提供了洞见。 关键词： *批判数据研究，数码人道主义，灾害回应，城市政治，城市科技。*

Siguiendo los pasos a los desastres, las comunidades se organizan para producir datos espaciales que acopien conocimiento sobre el desastre, para llenar los vacíos dejados por quienes responden formalmente a la emergencia. El modo como las comunidades afectan los esfuerzos generales de respuesta puede producir desigualdades, desempoderamiento o mayor marginalización. Cada vez más, esta forma de organizar y de producir conocimiento ocurre por medio de tecnologías digitales y, recientemente, el *humanitarianismo digital* se ha convertido en una suite importante de tales tecnologías. El humanitarianismo digital incluye tecnologías como la plataforma Ushahidi para el mapeo de crisis de origen multitudinario y la comunidad de voluntariado Humanitarian OpenStreetMap Team, centrada en el mapa base global OpenStreetMap generado por aficionados. El humanitarianismo digital está derivando a cómo después de los desastres las necesidades y conocimientos son captados y representados como datos. Tales transformaciones generan preguntas importantes para los geógrafos interesados en los procesos sociopolíticos e institucionales que enmarcan la producción y representación de datos. Con este artículo, contribuyo a los esfuerzos de los geógrafos para entender las políticas institucionales y las de base comunitaria que enmarcan los tipos de datos producidos dentro de los contextos de los desastres, apoyándome en un proyecto etnográfico que se desarrolló en Washington, DC y en la Ciudad de Nueva York después de la Supertormenta Sandy en 2012. Muestro que los humanitarios digitales produjeron datos en la Península de

Rockaway en Nueva York en respuesta a vacíos percibidos en la parte de quienes respondieron formalmente a la emergencia. Haciendo esto, ellos representaron necesidades, individuos y comunidades de ciertas maneras que fueron consideradas problemáticas por las organizaciones de apoyo comunitario local. Estos descubrimientos arrojan luz sobre aspectos políticos y luchas generados alrededor del porqué se produjeron conjuntos de datos particulares, y sobre las motivaciones que indujeron la captación como datos de necesidades y conocimientos particulares relacionados con el desastre. *Palabras clave: estudios sobre datos críticos, humanitarianismo digital, respuesta al desastre, política urbana, tecnología urbana.*

Following large, disruptive events such as disasters, communities often self-organize to fill gaps in the responses of formal actors, such as through sharing resources and providing personal assistance (Stallings and Quarantelli 1985; Birch and Wachter 2006). This form of community empowerment, which has long been a topic of research (Wachtendorf and Kendra 2006), can result in the strengthening of interpersonal ties and the establishment of sociospatial networks that can be politically powerful. Such organizing efforts increasingly occur through the use of digital technologies such as social media and big data (Mayer-Schönberger and Cukier 2013), text messages on mobile phones, crisis mapping, and crowd-sourcing platforms. Despite their increasing prevalence, these technologies, collectively referred to as *digital humanitarianism* (Burns 2014, 2015; Meier 2015), have mixed and uneven impacts on community organizing in disaster contexts (Brandusescu, Sieber, and Jochems 2016; Read, Taithe, and Mac Ginty 2016). Because disasters are important moments for social and institutional upheaval and the reestablishment of new norms and relations, it is important to consider the institutional and community-based forces that frame the ways in which people use digital technologies to enact recovery processes. The indeterminate and often contradictory impacts of digital humanitarian technologies can deepen inequalities, disempower, or further marginalize, and this complexity reflects digital technologies more broadly (Boyd and Crawford 2012; Straubhaar et al. 2013; Graham 2014). Research on critical geographic information systems (GIS) and the geoweb has elucidated the forces that set limits around what is seen as possible or "legitimate" uses of technology, data, and maps (Elwood 2006; Johnson and Sieber 2011), yet these considerations have to date had limited impact on digital humanitarian and big data research (Crampton et al. 2013).

In this article, I contribute to research underscoring the institutional and social forces and processes that frame the types of data and representations used to affect social change. In so doing, I build on geographers' efforts to understand the ways in which communities use digital technologies toward achieving sociopolitical goals, such as addressing social injustices. I report on a research project conducted in 2012 and 2013 that observed the ways in which digital and physical community organizations in New York City engaged digital humanitarian technologies to fill gaps in the formal response to Superstorm Sandy. Most important, these complex relations cultivated tensions between digital humanitarians and the two other parties of community organizations and the formal responders. I argue that the contestations around Sandy-related needs and knowledge representation led to very different data being produced by different stakeholders. In fact, one cannot understand the data that were produced—and the specific languages and descriptions that were employed—without attention to these struggles around knowledge and need representation.

I begin by situating this article within research on digital labor in disasters.[1] I pay particular attention to research showing the sociospatially variegated implications of digital technologies, especially around the ability to effectively leverage technology for social change. These inequalities foreground digital technologies as a key site and means for struggles over social change and the decisions that affect communities. I follow this by describing the methodology and empirical context of this project. With institutional and community-based framings of data in mind, I make two substantive arguments related to digital humanitarian technology use in the response to Superstorm Sandy. The first is that the institutional frameworks and established workflows of formal responders elicited new data practices by community organizations, largely conducted through digital humanitarian technologies. Second, other organizations resisted these new data practices and the ways in which digital humanitarians captured and represented their needs and knowledge. In particular, the proprietary nature of much of this work—specifically, the practice of putting data behind paywalls—disenfranchised communities on the Rockaway Peninsula. This led some individuals to contest digital humanitarian approaches to data collection and representation. Finally, I conclude by briefly enumerating two recommendations for how policymakers and digital humanitarians might rethink their engagements with such technologies.

Digital Labor in Disasters

In contrast with early boosterist claims that new digital humanitarian technologies promote democracy, liberation, and empowerment (e.g., Meier and Munro 2010; Zook et al. 2010; Crowley and Chan 2011), recent research has begun to reconceptualize digital humanitarianism as an uneven, contested, and sometimes problematic sociopolitical development. Humanitarian data production and interpretation require specialized and situated knowledge and skills that are not only inaccessible to many digital humanitarian technologists (Bhroin 2015; Finn and Oreglia 2016), but they also enroll broader data relations such as gender discrepancies (Stephens 2013; Cupples 2015) and political–economic imperatives (Thatcher, O'Sullivan, and Mahmoudi 2016; Burns forthcoming). Many have shown that digital humanitarianism affects response efforts in unintended ways (Currion 2010; Jacobsen 2015) and can potentially expose vulnerable populations to increased risk (Shanley et al. 2013; Haworth and Bruce 2015; Raymond et al. 2016). Humanitarian data are produced in institutionally specific ways that mirror political imperatives, sometimes with data being locked behind proprietary restrictions (Taylor and Schroeder 2015). These data then come to influence how we know disasters and their sociopolitical foundations, in turn influencing the types of organized responses deemed "appropriate" or "legitimate" (Crawford and Finn 2015).

The digital divide thus persists here in numerous ways that delimit the technologies' potential impact on social change (Graham et al. 2014). Burns (2014) argued that internal contestations and knowledge politics underwrite the ability to access, leverage, and influence the development of digital humanitarian technologies. For example, installing, customizing, and serving an instance of the Ushahidi[2] crisis mapping platform requires skills in server administration, web scripting, and database management (Brandusescu, Sieber, and Jochems 2016). These limitations parallel Gilbert's (2010) reconceptualization of the digital divide to account for social capital. In these ways, some are always excluded from humanitarian data (Mulder et al. 2016) but, more important, asymmetrical power relations are reproduced in the digital humanitarian context (Sandvik et al. 2014; Burns 2015; Duffield 2016). A decade ago, Elwood (2006) argued that framing community organizations' GIS-based knowledge production as either activist and resistance or as cooptation is problematic, as these roles and relationships are not necessarily singular nor mutually exclusive; these sorts of slippages within digital humanitarianism have not yet been explored.

Still, many remain cautiously optimistic regarding the potential of digital labor for disaster response. Resor (2016) insisted on a blurred boundary between "digital humanitarians" and "formal responders," noting that many digital humanitarians have high degrees of professional experience and often maintain ties to formal institutions. This upholds decades of research findings about self-organization following crises (Stallings and Quarantelli 1985). Many continue to argue that volunteered geographic information (VGI) can be useful for generating knowledge of ground conditions following disasters, although its applicability to preparation, mitigation, and recovery are less clear (Heinzelman, Waters, and United States Institute of Peace 2010; Haworth, Whittaker, and Bruce 2016; Shepard et al. 2016).

To date, much research on big data tends to understand data as a direct reflection of conditions "in real life," rather than as reflecting institutional and social contexts (see, e.g., Procter, Vis, and Voss 2013). Scholars typically use the term *big data* to conjure social media, automatic sensor data, clickstream and Web behavior data, and retail purchasing information (Kitchin 2014). Analysis of social media content in particular tends to search for meaning and patterns in data as they are presented (Boyd and Crawford 2012). This approach obfuscates the institutional and community-based processes and limitations that frame the types of data produced and the representational strategies espoused (Gray 2012). Geographers have a rich history of interrogating these intersections of data and society (Dalton and Thatcher 2014), and it is in this vein that this article seeks to address that gap.

This study builds on the existing literature by exploring the ways in which community organizations and individuals engage with digital humanitarian technologies to, on the broadest level, enact social change. By acting on their local and immediate scales, they affect larger scales of data practices and representations. In doing so, they are simultaneously—either consciously or otherwise—acting on the processes by which marginalization and inequalities are reproduced. Digital technologies here are seen as an interface

Figure 1. This photo, taken in March 2013, shows remnants of the destroyed Rockaway Peninsula boardwalk and ongoing beach erosion problems. (Color figure available online.)

among multiple actors, including formal disaster response agencies, digital humanitarians, community organizations, and the individual members within these groups. As such, the current project speaks to Barkan and Pulido's (2017) encouragement to understand the ways in which claims—for resources, recognition, or participation, for example—are made, disciplined, and addressed. Barkan and Pulido (2017) touched on the ways in which cartographic knowledge production "crystallizes recognition of injustice—even

for people not interested" explicitly in injustice (38). It then becomes important for geographers to question how these claims occur through cartographic visualizations, data sets, social media, and other digital technologies. In what follows, I explore these questions in the context of research conducted around Superstorm Sandy, in which digital humanitarians, formal responders, and community organization leaders offered multiple competing interpretations of how needs and knowledge should be captured as data.

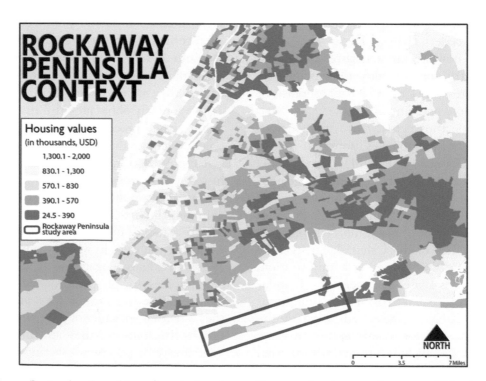

Figure 2. The Rockaway Peninsula, where this study took place, has a wide range of socioeconomic statuses. (Color figure available online.)

272

Methodology and Context

In late October 2012, Superstorm Sandy became the largest Atlantic storm on record, its hurricane-force winds stretching over 1,000 miles in diameter (National Weather Service 2013). Most of the Eastern seaboard of the United States was severely damaged, making this the second most destructive storm on record to hit the United States (Blake et al. 2013). In New York City, among the hardest hit areas were the southern Queens neighborhoods on the Rockaway Peninsula (Bloch et al. 2012), although effects were widespread due to flooding and power and public transportation outages. In the Rockaway Peninsula, however, many information and communication technology networks were disrupted, and economic infrastructures such as the peninsula's boardwalk and beachside amenities were destroyed for several years (see Figure 1).

I visited New York City in March 2013 to conduct in-depth, semistructured interviews with local community organizations, digital humanitarians, and senior administrators and policymakers. I focused this research on the Rockaway Peninsula, which was still in the initial recovery stages despite the four-month gap and despite containing a wide range of socioeconomic statuses (see Figure 2). My visit overlapped with an event hosted by the Museum of Modern Art (MoMA) PS1 to fundraise for local Sandy recovery, and this event was attended by city council members, key response organizers, and the musician Patti Smith.[3] This research was a component of a larger project primarily located at a public policy research institute in Washington, DC, where for a year I used the *extended case method* (Burawoy 1998) to understand the societal, policy, and political–economic impacts of digital humanitarianism. In addition to thirty-seven total interviews—of which seven were in New York—I worked as a participant-observer within several digital humanitarian organizations and performed archived data retrieval. Interviewees were chosen through a combination of snowball sampling and after identifying key actors as a participant-observer. These semistructured interviews ranging from half an hour to two hours sought to understand the ways in which digital humanitarian technologies have shifted data production and representation practices; in New York, the interviews focused in particular on Superstorm Sandy. All of these data were collated, transcribed, coded, and analyzed using a discourse analysis framework. In the rest of the article, I am seeking to understand the social processes that compelled some to enact particular data practices; likewise, I explore the politics and struggles behind those processes; here, communities' and individuals' perceptions of the crisis and response efforts are more important than whether or not individual members were factually correct. The data presented here have been selected because they are particularly representative of these processes that appear throughout the corpus of evidence.

(In)formal Data Collection Practices

Several digital humanitarian groups emerged with disparate data production and representation efforts following Superstorm Sandy, at times to directly confront those of formal responders such as the Federal Emergency Management Agency (FEMA) and New York's Office of Emergency Management (OEM). Among the various groups at work, multiple competing interests and goals fostered a dynamic characterized by complex actions of resistance, needs provision, and intra-institutional conflicts. At the same time, individuals and groups adopted data production and representation strategies in direct response to the established workflows, relationships between emergency management agencies, and underrepresented needs of stakeholders. That is, despite the messy relations between data producers and organizations, the data that emerged to represent disaster knowledge were framed by broad institutional processes.

FEMA, the Red Cross, and city managers in New York began intensive data collection during and immediately following Sandy, guided by institutional normative understanding that the first seventy-two hours are the most critical for effective disaster response. These efforts sought to produce "situational awareness": data intended to inform decision making and resource allocation. Simultaneously, a motley crew of entrepreneurs and technologists gathered under the ad hoc volunteer-based umbrella institution Crisis Commons[4] to similarly begin producing data and Web maps and begin acting on these knowledge outcomes. According to my interviewee Rachel, a key member of this Crisis Commons group, formal responders distributed the locations of resource centers to potentially affected households on the Rockaway Peninsula. This work marginalized families and households that were bound to their homes and unable to travel to the centers. Rachel and others in the Crisis Commons group believed that there was no formal-sector organized program to reach out to these people and that without power, these homebound families were "the most vulnerable population." She and others organized twenty volunteers who visited 500 households in a few hours, to ascertain the needs in the area, which homes had electricity available, and whether any medical emergencies were occurring. For Rachel, not only were these data not being produced by formal response agencies but this was reflective of a limitation more structurally rooted in problematic data practices. Speaking of an organization working with Crisis Commons, Rachel said:

> They needed not the same information that FEMA needs. They needed a piece of that information, and they needed it tied to a map in the same way that anyone else did. . . . [But] it evolve[s] into these fiefdoms of power. . . . FEMA I guess in some places does do door-to-door checking, but they won't share that data with anyone.

Here Rachel points to how FEMA's spatial-informational needs stem from their charter as a formal emergency management agency. Due to issues around privacy and data sharing limitations, on the one hand, and the specific responsibilities that FEMA has been delegated, on the other, the data that the organization produces can be seen to be less useful or accessible to other organizations. Invoking the notion of "fiefdoms of power," Rachel implied that data sharing can promote empowerment; this line of thinking has been central to digital humanitarianism's self-marketing and is here linked to specific data practices.

Rachel and her team took their collected data to formal response agencies to illustrate the value of their digital labor:

> They found one woman who was in need of heart and liver (or kidney) medication, and there were several other people who needed to be checked on. We reported that information. . . . I went to the police station, I talked to FEMA, I talked to Red Cross, and I went to the police station, and finally was talking with the [New York Police Department Chief Information Officer]—I showed him the walk-list, he's like, "Now I get it. I thought you were crazy at first. Now I understand what you're doing." And I was like, "Thank you, this is really important, and I have all the work here and this is the most effective way that you can protect your people. Who can I hand this over to?" And the answer was no one.

What is most important here is how Rachel and Crisis Commons–affiliated organizations collected their own data sets in direct response to perceived shortcomings of formal responders. Rachel herself did not live in or have significant ties to the Rockaway Peninsula but instead saw the communities there as in need of her group's data collection capacity. This created a tripartite relation of power between formal responders, relatively less affected individuals, and disaster-affected communities. Rachel and her group established new data collection practices without the explicit request of formal responders or communities in the Rockaway Peninsula, instead seeking to address a perceived need. As explained later, such nonresidents came to the Rockaway Peninsula en masse intending to help, yet unaware of their own incomplete understanding of the disaster's socioeconomic foundations and impacts. In this way, they were both enacting problematic representations of needs and organizing to shift resource allocation.

Rachel's anecdote speaks as well to the politics underwriting knowledge and needs inclusion and representation as data. Speaking to broader issues of legitimacy, particularly as tied to cartographic representation, in this case digital humanitarians identified gaps in the kinds of knowledges and needs considered legitimate for formal-sector intervention. Whereas past research has foregrounded the role of maps in these politics, however, Rachel pointed to the politics of data itself. Capturing needs as data enrolls a politics of exclusion, as formal actors inadvertently yet by necessity maintain gaps in their collection efforts, and here digital humanitarians contested this politics through new data practices.

Complexities in Institutionalized Spatial Data Production

Oppositional impressions of these digital humanitarian efforts emerged from formal responders and from leaders of other local organizations. These complexities pointed to the multiple competing demands, roles, and needs of organizations and the data practices they adopt or resist in relation to each of them. Those in charge of disaster-related data collection and production in some ways buttressed the digital humanitarian efforts mentioned already, but leaders of other organizations criticized the same efforts. These reactions to digital humanitarian efforts did not themselves foster new data practices, instead providing an additional framework around digital humanitarian efforts for formal responders and distancing many community organizations from digital humanitarianism writ large.

Harper, a manager for a formal GIS department in New York and advocate of digital humanitarianism, provided spatial data, analysis, and cartographic visualizations for the city during Superstorm Sandy. She expressed concern about the workflows and every-day data practices with which these efforts must align. For formal responders, this consists primarily of the Incident Command System (ICS),[5] an international structured protocol to guide and coordinate emergency response. In an interview, she estimated that her department and those with whom she maintains professional connections will draw parallels with extant spatial data platforms when evaluating the potential impact of the preceding digital humanitarian work. She spoke of the reasons why Rachel and her group likely encountered institutional challenges: "It's similar to the questions we faced with GIS: where does it fit into the ICS? Unless it gets formal adoption into this framework, it won't get 'owned.' Moving a city government takes time."

Harper asserted that attention to institutional limitations structuring disaster-related data types and data representations comprises an important part of her work as someone whose work directly contributes to the emergency management field. For her, Rachel's data collection efforts were rejected likely because of the unsolicited, amateur nature of the data sets; this contrasts with the purported authoritative, ICS-compliant data production practices of the formal emergency management sector.[6] The temporal dimension Harper underscored refers to the momentum of current established practices that digital humanitarians confront and that also frames the kinds of knowledges, needs, and places that are captured as data in disaster contexts.

This sort of reticence toward volunteer data production was echoed by managers in New York's OEM and the Department of Health and Mental Hygiene. In Superstorm Sandy, both of these agencies produced data for public consumption by publishing information to social media channels but monitored social media only to direct individuals to the proper authorities for addressing their needs. Analytics tools like GeoFeedia helped derive situational awareness from various media channels. These data production and representation techniques stemmed from formal responders' need to remain within institutional charters and workflow protocols but, in turn, generated new knowledge politics as digital humanitarians sought to address what they perceived as the ensuing gaps.

Additionally, within days of the storm, several community organizations were formed on the Rockaway Peninsula to supply labor and resources to underprivileged families in the area. Key organizers for two of these organizations—Rockaway Emergency Plan and Respond & Rebuild[7]—stated their insistence that their organizations would be directed and organized by people residing in the Rockaway Peninsula in explicit contrast with digital humanitarians such as Crisis Commons. Rowan, who worked with Respond & Rebuild, expressed to me her frustration that groups such as Crisis Commons had come to the area from distant boroughs to represent the communities' needs. She claimed that these and other digital humanitarians represented needs in ways that elevated Crisis Commons's mission and marketing over the communities' need for assistance. To illustrate the politics of needs representation, Rowan recalled a then-recent *New York Times* article that narrated an outsider's visit to the stark devastation in Rockaway Peninsula, which quoted the outsider as saying, "I'm driving my big Lexus down here. ... Thank God the car is dirty" (Nir 2012). Quinn, a leader in the Rockaway Emergency Plan, voiced similar concerns that, regardless of digital humanitarians' good intentions, they had arrived to an area under duress without a clear commission by the formal responders or by local residents. This established a politics of needs representation that made Quinn uncomfortable, given the vulnerable state of her neighborhood.

According to both Rowan and Quinn, digital humanitarians did not empower locals or accentuate residents' voices in their efforts. They reached this conclusion through two critiques of approaches toward data. First, they expressed what they claimed was widespread understanding that digital humanitarians did not solicit input from local residents on the types of information that would be collected or the ways in which information would be visualized. Indeed, Respond & Rebuild was established largely to promote local community control over needs representation and needs satisfaction; a primary goal was to gather information excluded by both formal responders and digital humanitarians—information about mold growth after the stormwater surge. Second, organizations broadly under the digital humanitarian umbrella concept arrived to the area to collect and represent data about storm damage but retained control over those data by placing it behind a paywall. Rowan was particularly critical of two organizations of volunteer first responders for following this practice. These organizations took this approach largely because they had partnered with a private software company for data collection, and that company had demanded proprietary data retention. Effectively, however, these institutional relationships excluded local communities and organizations from using the data or holding the organizations accountable for their data practices. More broadly, both interviewees expressed their concerns over implications for data ownership and usage.

Conclusion

In this article, I have argued that institutional and community-based politics frame the types of data that are produced and the ways in which those data are represented, in disaster contexts. These politics stem from the multiple competing data practices enacted by organizations digital and otherwise, formal response agencies, and individuals. Digital humanitarians play an increasingly important role in these politics, as they enroll distributed digital labor and challenge existing practices and workflows. Digital humanitarianism further elicits new data practices and contestations around how needs and knowledge will be captured as data. These findings illuminate the social and political inequalities of Big Data by foregrounding the struggles and variegations around data production practices. As the research agenda on spatial technologies continues to adapt to technological change, this research suggests that geographers should see data not as reflections of on-the-ground conditions but instead as a representational negotiation rooted in spatial inequalities.

Spatial technologies hold incredible epistemological and tactical promise, however, as demonstrated by the critical GIS literature (Kwan 2002; Sheppard 2005). To this end, digital humanitarianism is neither limited to nor necessarily characterized by exclusion and could be leveraged in work toward social justice. Some digital humanitarians use this as a guiding principle in their efforts to fill gaps left by formal response agencies. Yet, the complex politics of needs and knowledge representation I highlighted earlier insist that digital humanitarians acknowledge and account for these politics, perhaps building technology differently to promote grassroots data production and representation capacity. That is, the history of technology shows digital humanitarians multitudes of ways in which their technologies can be subverted for productive and positive social change.

To these ends, I conclude by briefly offering two recommendations to digital humanitarians and to policymakers seeking to engage these digital communities and digital technologies. First, as discussed earlier, data are neither neutral nor direct reflections of on-the-ground conditions, and one should consider not simply data presences and absences but also the contexts and forces that produce discernable spatial patterns. That is, content analysis should not take data at face value but should instead acknowledge the complex processes that lead to some data being produced and not others. Second, and most important, in this article I have identified ways in which data are always incomplete yet tell an important story. Digital humanitarian technologies hold the potential to facilitate production of data that might have marginal use to formal workflows and institutional structures, yet still convey important knowledge about a disaster. For example, digital humanitarian data might capture disrupted interpersonal networks, emotional geographies, spaces of care, new ways of thinking about and relating to urban infrastructure, or communal nonformalized knowledge. It is in these potentialities that digital humanitarianism shows the most promise for social justice work.

Acknowledgments

I thank Sarah Elwood for her long-standing and enthusiastic support of this project and for feedback on drafts of this article before it reached manuscript status. I also thank Reuben Rose-Redwood, Kristian Bredemeyer, and Ian Schulte for their feedback on an early draft. I also acknowledge and thank those who spoke with me during this research and all those working for a more socially just world through and against digital technologies.

Notes

1. I use the term *digital labor* here to invoke recent Marxist readings of crowd sourcing, attentional economies and the commodification of Web-based activities (Scholz 2012; Fuchs and Sevignani 2013; Terranova 2014), all of which underpin the processes at work in this article.
2. Ushahidi (see https://www.ushahidi.com/) is a platform that mobilizes geographically distributed labor to collect, categorize, translate, and georeference data such as social media and Short Message System (SMS) messages.
3. See http://www.momaps1.org/expo1/venue/vw-dome-2/ for more information about the event.
4. Find more information about Crisis Commons at https://crisiscommons.org/.
5. See, for more details, https://www.fema.gov/national-incident-management-system.
6. Indeed, these sorts of amateur–authoritative binaries have structured most discussions of VGI and the geoweb.
7. The official pages for these organizations are Rockaway Emergency Plan (https://www.facebook.com/rockaway help/) and Respond & Rebuild (https://www.respondan drebuild.org/).

References

Barkan, J., and L. Pulido. 2017. Justice: An epistolary essay. *Annals of the American Association of Geographers* 107 (1):33–40.

Bhroin, N. N. 2015. Social media-innovation: The case of indigenous tweets. *The Journal of Media Innovations* 2 (1):89–106.

Birch, E., and S. Wachter. 2006. Introduction: Rebuilding urban places after disaster. In *Rebuilding urban places after disaster: Lessons from Hurricane Katrina*, ed. E. Birch and S. Wachter, 1–10. Philadelphia: University of Pennsylvania Press.

Blake, E., T. Kimberlain, R. Berg, J. Cangialosi, and J. Beven, II. 2013. *Tropical cyclone report: Hurricane Sandy (AL182012)*. Miami, FL: National Hurricane Center. Accessed December 10, 2016. http://www.nhc.noaa.gov/data/tcr/AL182012_Sandy.pdf.

Bloch, M., A. McLean, A. Tse, and D. Watkins. 2012. Surveying the destruction caused by Hurricane Sandy. *New York Times* November 20. Accessed December 10, 2016. http://www.nytimes.com/newsgraphics/2012/1120-sandy/survey-of-the-flooding-in-new-york-after-the-hurricane.html.

Boyd, D., and K. Crawford. 2012. Critical questions for big data: Provocations for a cultural, technological, and scholarly phenomenon. *Information, Communication & Society* 15 (5):662–79.

Brandusescu, A., R. Sieber, and S. Jochems. 2016. Confronting the hype: The use of crisis mapping for community development. *Convergence: The International Journal of Research into New Media Technologies* 22 (6):616–32.

Burawoy, M. 1998. The extended case method. *Sociological Theory* 16 (1):4–33.

Burns, R. 2014. Moments of closure in the knowledge politics of digital humanitarianism. *Geoforum* 53:51–62.

———. 2015. Rethinking big data in digital humanitarianism: Practices, epistemologies, and social relations. *GeoJournal* 80 (4):477–90.

———. Forthcoming. "Let the private sector take care of this": The philanthro-capitalism of digital humanitarianism. In *Digital economies at the global margins*, ed. M. Graham. Cambridge, MA: MIT Press.

Crampton, J., M. Graham, A. Poorthuis, T. Shelton, M. Stephens, M. Wilson, and M. Zook. 2013. Beyond the geotag: Situating "big data" and leveraging the potential of the Geoweb. *Cartography and Geographic Information Science* 40 (2):130–39.

Crawford, K., and M. Finn. 2015. The limits of crisis data: Analytical and ethical challenges of using social and mobile data to understand disasters. *GeoJournal* 80 (4):491–502.

Crowley, J., and J. Chan. 2011. Disaster relief 2.0: The future of information sharing in humanitarian emergencies. Washington, DC: UN Foundation & Vodafone Foundation Technology Partnership. Accessed December 4, 2017. http://www.globalproblems-globalsolutions-files.org/gpgs_files/pdf/2011/DisasterResponse.pdf.

Cupples, J. 2015. Coloniality, masculinity, and big data economies. Geography/Development/Culture/Media. Accessed September 14, 2016. https://juliecupples.wordpress.com/2015/05/11/coloniality-masculinity-and-big-data-economies/.

Currion, P. 2010. "If all you have is a hammer"—How useful is humanitarian crowdsourcing? MobileActive.org. Accessed September 13, 2013. http://www.crowdsourcing.org/document/if-all-you-have-is-a-hammer—how-useful-is-humanitarian-crowdsourcing/3533.

Dalton, C., and J. Thatcher. 2014. What does a critical data studies look like, and why do we care? Seven points for a critical approach to "big data." *Society and Space Open Site*. Accessed February 9, 2017. http://societyandspace.org/2014/05/12/what-does-a-critical-data-studies-look-like-and-why-do-we-care-craig-dalton-and-jim-thatcher/.

Duffield, M. 2016. The resilience of the ruins: Towards a critique of digital humanitarianism. *Resilience* 4 (3):147–65.

Elwood, S. 2006. Beyond cooptation or resistance: Urban spatial politics, community organizations, and GIS-based spatial narratives. *Annals of the Association of American Geographers* 96 (2):323–41.

Finn, M., and E. Oreglia. 2016. A fundamentally confused document: Situation reports and the work of producing humanitarian information. In *Proceedings of the 19th ACM Conference on Computer-Supported Cooperative Work & Social Computing*, 1349–62. New York: ACM. Accessed August 29, 2016. http://www.ercolino.eu/docs/Oreglia_Pub_Fundamentally%20Confused%20Document%202016_AD.pdf.

Fuchs, C., and S. Sevignani. 2013. What is digital labour? What is digital work? What's their difference? And why do these questions matter for understanding social media?. *TripleC* 11 (2):237–93.

Gilbert, M. 2010. Theorizing digital and urban inequalities. *Information, Communication & Society* 13 (7):1000–18.

Graham, M. 2014. Internet geographies: Data shadows and digital divisions of labor. In *Society and the Internet: How networks of information and communication are changing our lives*, ed. M. Graham and M. Dutton, 99–116. Oxford, UK: Oxford University Press.

Graham, M., B. Hogan, R. Straumann, and A. Medhat. 2014. Uneven geographies of user-generated information: Patterns of increasing informational poverty. *Annals of the Association of American Geographers* 104 (4):746–64.

Gray, J. 2012. What data can and cannot do. *The Guardian* May 31. Accessed April 27, 2017. https://www.theguardian.com/news/datablog/2012/may/31/data-journalism-focused-critical.

Haworth, B., and E. Bruce. 2015. A review of volunteered geographic information for disaster management. *Geography Compass* 9 (5):237–50.

Haworth, B., J. Whittaker, and E. Bruce. 2016. Assessing the application and value of participatory mapping for community bushfire preparation. *Applied Geography* 76:115–27.

Heinzelman, J., C. Waters, and United States Institute of Peace. 2010. *Crowdsourcing crisis information in disaster-affected Haiti*. Washington, DC: U.S. Institute of Peace.

Jacobsen, K. 2015. *The politics of humanitarian technology: Good intentions, unintended consequences and insecurity.* London and New York: Routledge.

Johnson, P., and R. Sieber. 2011. Motivations driving government adoption of the Geoweb. *GeoJournal* 77 (5):667–80.

Kitchin, R. 2014. Big data, new epistemologies and paradigm shifts. *Big Data & Society* 1:1–12.

Kwan, M.-P. 2002. Feminist visualization: Re-envisioning GIS as a method in feminist geographic research. *Annals of the Association of American Geographers* 92 (4):645–61.

Mayer-Schönberger, V., and K. Cukier. 2013. *Big data: A revolution that will transform how we live, work, and think.* New York: Houghton Mifflin Harcourt.

Meier, P. 2015. *Digital humanitarians: How big data is changing the face of humanitarian response.* Boca Raton, FL: CRC.

Meier, P., and R. Munro. 2010. The unprecedented role of SMS in disaster response: Learning from Haiti. *SAIS Review* 30 (2):91–103.

Mulder, F., J. Ferguson, P. Groenewegen, K. Boersma, and J. Wolbers. 2016. Questioning big data: Crowdsourcing crisis data towards an inclusive humanitarian response. *Big Data & Society* 3 (2):1–13.

National Weather Service. 2013. *Hurricane/Post-tropical Cyclone Sandy.* Silver Spring, MD: National Oceanic and Atmospheric Administration. Accessed December 4, 2017. http://www.weather.gov/media/publications/assessments/Sandy13.pdf.

Nir, S. M. 2012. After Hurricane Sandy, Helping Hands also expose a New York divide. *The New York Times* November 16. Accessed December 14, 2017. http://www.nytimes.com/2012/11/17/nyregion/after-hurricane-sandy-helping-hands-also-expose-a-new-york-divide.html.

Procter, R., F. Vis, and A. Voss. 2013. Reading the riots on Twitter: Methodological innovation for the analysis of big data. *International Journal of Social Research Methodology* 16 (3):197–214.

Raymond, N., Z. Al Achkar, S. Verhulst, and J. Berens. 2016. *Building data responsibility into humanitarian action.* New York and Geneva: UN Office for the Coordination of Humanitarian Affairs.

Read, R., B. Taithe, and R. Mac Ginty. 2016. Data hubris? Humanitarian information systems and the mirage of technology. *Third World Quarterly* 37 (8):1314–31.

Resor, E. 2016. The neo-humanitarians: Assessing the credibility of organized volunteer crisis mappers. *Policy & Internet* 8 (1):34–54.

Sandvik, K., M. Jumbert, J. Karlsrud, and M. Kaufmann. 2014. Humanitarian technology: A critical research agenda. *International Review of the Red Cross* 86 (893):219–42.

Scholz, T., ed. 2012. *Digital labor: The Internet as playground and factory.* London and New York: Routledge.

Shanley, L., R. Burns, Z. Bastian, and E. Robson. 2013. Tweeting up a storm: The promise and perils of crisis mapping. *Photogrammetric Engineering & Remove Sensing* 79 (10):865–79.

Shepard, D., T. Hashimoto, T. Kuboyama, and K. Shin. 2016. What do boy bands tell us about disasters? The social media response to the Nepal earthquake. In *Digital humanities 2016: Conference abstracts*, ed. M. Eder and J. Rybicki, 361–64. Krakow, Poland: Jagiellonian University & Pedagogical University.

Sheppard, E. 2005. Knowledge production through critical GIS: Genealogy and prospects. *Cartographica: The International Journal for Geographic Information and Geovisualization* 40 (4):5–21.

Stallings, R., and E. Quarantelli. 1985. Emergent citizen groups and emergency management. *Public Administration Review* 45:93–100.

Stephens, M. 2013. Gender and the GeoWeb: Divisions in the production of user-generated cartographic information. *GeoJournal* 78 (6):981–96.

Straubhaar, J., J. Spence, Z. Tufekci, and R. G. Lentz, eds. 2013. *Inequity in the technopolis: Race, class, gender, and the digital divide in Austin.* Austin: University of Texas Press.

Taylor, L., and R. Schroeder. 2015. Is bigger better? The emergence of big data as a tool for international development policy. *GeoJournal* 80 (4):503–18.

Terranova, T. 2014. Free labor. In *Digital labor: The Internet as playground and factory*, ed. T. Scholz, 33–57. London and New York: Routledge.

Thatcher, J., D. O'Sullivan, and D. Mahmoudi. 2016. Data colonialism through accumulation by dispossession: New metaphors for daily data. *Environment and Planning D: Society and Space* 34 (6):990–1006.

Wachtendorf, T., and J. Kendra. 2006. Improvising disaster in the City of Jazz: Organizational response to Hurricane Katrina. *Understanding Katrina: Perspectives from the Social Sciences.* Accessed April 25, 2017. http://understandingkatrina.ssrc.org/Wachtendorf_Kendra/.

Zook, M., M. Graham, T. Shelton, and S. Gorman. 2010. Volunteered geographic information and crowdsourcing disaster relief: A case study of the Haitian earthquake. *World Medical & Health Policy* 2 (2):7–33.

RYAN BURNS is Assistant Professor of Geography at the University of Calgary, Calgary, AB T2N 1Z3, Canada. E-mail: ryan.burns1@ucalgary.ca. His work intersects GIScience and human geography. His work focuses on the social, political, and urban transformations of GIS, big data, web mapping, software, and related digital spatial phenomena. He contributes to geographers' efforts to understand the ways in which spatial technologies represent people and their knowledge, the technologies' impact on political economies, and the social inequalities sustained by new technologies.

25 Cultivating (a) Sustainability Capital

Urban Agriculture, Ecogentrification, and the Uneven Valorization of Social Reproduction

Nathan McClintock

Urban agriculture (UA), for many activists and scholars, plays a prominent role in food justice struggles in cities throughout the Global North, a site of conflict between use and exchange values and rallying point for progressive claims to the right to the city. Recent critiques, however, warn of its contribution to gentrification and displacement. With the use–exchange value binary no longer as useful an analytic as it once was, geographers need to better understand UA's contradictory relations to capital, particularly in the neoliberal sustainable city. To this end, I bring together feminist theorizations of social reproduction, Bourdieu's "species of capital" and critical geographies of race to help demystify UA's entanglement in processes of ecogentrification. In this primarily theoretical contribution, I argue that concrete labor embedded in household-scale UA—a socially reproductive practice—becomes cultural capital that a sustainable city's growth coalition in turn valorizes as symbolic sustainability capital used to extract rent and burnish the city's brand at larger scales. The valorization of UA occurs, by necessity, in a variegated manner; spatial agglomerations of UA and the ecohabitus required for its misrecognition as sustainability capital arise as a function of the interplay between rent gaps and racialized othering. I assert that ecogentrification is not only a contradiction emerging from an urban sustainability fix but is central to how racial capitalism functions through green urbanization. Like its contribution to ecogentrification, I conclude, UA's emancipatory potential is also spatially variegated. *Key Words: capital, gentrification, social reproduction, sustainability fix, urban agriculture.*

对诸多社会运动者和学者而言, 城市农业 (UA) 在整个全球北方城市的粮食正义斗争中扮演要角, 并且作为使用和交换价值之间的斗争场域, 以及有关城市权的激进宣称的号召力。但晚近的批判, 却警告城市农业将可能导致贵族化与迫迁。由于使用—交换价值的二分, 不再如同过往般提供有用的分析, 地理学者必须更佳地理解城市农业与资本之间的矛盾关系, 特别是在新自由主义的可持续城市之中。为此, 我将结合女权主义对社会再生产的理论化、布迪厄的 "资本形式", 以及种族的批判地理, 协助揭开城市农业与生态贵族化过程相互交缠的神秘面纱。在此般主要的理论贡献中, 我主张, 镶嵌于家户尺度的城市农业中的实质劳动——作为社会再生产的实践——成为了文化资本, 该资本是可持续发展的城市成长联盟回头定价作为用来在更大的尺度上榨取地租并擦亮城市品牌的象徵性可持续性资本。城市农业的定价必然以多样的方式发生; 将之误认为可持续发展的资本所需的城市农业与生态栖地的空间聚集, 浮现作为地租差异和种族化的他者化之间的互动功能。我主张, 生态贵族化并不仅只是从城市可持续性修补中浮现的矛盾, 而是种族资本主义如何透过绿色城市化运作的核心。我于结论中主张, 城市农业的解放潜能, 亦在在空间上呈现多样性。 *关键词:资本, 贵族化, 社会再生产, 可持续性修补, 城市农业。*

Para muchos activistas y eruditos, la agricultura urbana (AU) juega un prominente papel en las luchas por la justicia alimentaria de las ciudades a través de todo el Norte Global, sitio de conflicto entre el uso e intercambio de valores y punto de agitación de las reclamaciones progresistas por el derecho a la ciudad. No obstante, críticas recientes previenen sobre su contribución al aburguesamiento y el desarraigo. Con la falta de utilidad analítica que antes tenía el binario de uso-intercambio de valor, los geógrafos necesitan entender mejor las relaciones contradictorias de la AU con el capital, en particular en la ciudad neoliberal sustentable. Para este fin, traigo juntas las teorizaciones feministas de la reproducción social, las "especies de capital" de Bourdieu y las geografías críticas de raza, para ayudar a desmitificar el enredo de la AU en procesos de ecoaburguesamiento. En esta contribución primariamente teórica, sostengo que el trabajo concreto incrustado en la AU de escala hogareña—una práctica socialmente reproductiva—se convierte en capital cultural, que una coalición de crecimiento en una ciudad sustentable a su turno valora como capital de sustentabilidad simbólica usado para extraer renta y para pulir el estilo de la ciudad a escalas más grandes. La valoración de la AU, por necesidad, ocurre de una manera variada; las aglomeraciones espaciales de la AU y el ecohábito requerido para su equivocado reconocimiento como capital de sustentabilidad surgen como una función de la interacción entre las grietas de la renta y la otredad racializada. Afirmo que el ecoaburguesamiento no solo es una contradicción que surge de

Annals of the American Association of Geographers, 108(2) 2018, pp. 579–590 © 2018 by American Association of Geographers
Initial submission, November 2016; revised submission, January 2017; final acceptance, February 2017
Published by Taylor & Francis, LLC.

receta de sustentabilidad urbana, sino que es central al modo como funciona el capitalismo racial por medio de la urbanización verde. Del mismo modo que su contribución al eco-aburguesamiento, concluyo, el potencial emancipatorio de la AU también es espacialmente abigarrado. *Palabras clave: capital, aburguesamiento, reproducción social, receta de sustentabilidad, agricultura urbana.*

For many activists and scholars, urban agriculture (UA) serves as a rallying point for food justice, food sovereignty, and progressive claims to the right to the city across the Global North (Bradley and Galt 2014; Purcell and Tyman 2014). Whereas David versus Goliath standoffs between bulldozers and gardeners, use and exchange value, might have once defined UA in the public imagination, rooftop gardens growing salad greens for farm-to-table restaurants are now perhaps more representative. UA, which has come to symbolize both the environmental values undergirding urban sustainability efforts and the "local" and "artisanal" so cherished by foodies (Johnston and Baumann 2014), is also often the fruit of an "unexpected romance" (Holt 2015) between urban agriculturists and real estate developers. Critics both within and outside academia (Quastel 2009; Crouch 2012) have thus begun to scrutinize its role as "an attractive place holder on the road to gentrification" (DeLind 2015, 3).

Indeed, UA is often a temporary land use on vacant or devalued sites awaiting the next wave of investment (McClintock 2014), spaces Walker dubbed the "lumpengeography" of capital (R. Walker 1978, 32). Whereas gardens throughout history cropped up opportunistically as a coping strategy in the face of food and wage insecurity (Lawson 2005; McClintock 2010), they now more often signal the transformation of these same devalued neighborhoods (see Figure 1A). A clear connection exists between the gentrification of these lumpengeographies and the proliferation of restaurants and grocery stores that capitalize on the mainstreaming of "foodie" culture and the value it places on local, organic consumption (Burnett 2014; Anguelovski 2015). Scholars are only beginning to examine how urban food production is also entangled in processes of gentrification and capital accumulation more broadly.

Whereas earlier literature described UA in terms of gardens versus development and the incommensurability of these two use values (Schmelzkopf 2002), newer work reveals that gardening is actually quite commensurable with the market logics of development (Quastel 2009; S. Walker 2016) and, particularly in progressive urban centers, contributes to a "sustainability fix"—the "selective incorporation of environmental goals" (While, Jonas, and Gibbs 2004, 552) by growth coalitions of developers, consulting firms, nonprofits, planners, and policymakers attempting to balance the entrepreneurial imperative of economic growth and public demands for ecological regulation. Although a sustainability fix might blur the lines between environmental stewardship and economic growth, it is nevertheless a fix, temporary and prone to fracture. Gentrification—"the production of space for progressively more affluent users" (Hackworth 2002, 815)—or, more specifically, ecogentrification resulting from "the implementation of an environmental agenda driven by an environmental ethic" (Dooling 2009, 41)—is perhaps the

Figure 1. (A) A market garden surrounded by new-build condos in a rapidly gentrifying area of Portland, Oregon. (B) A front yard garden in a gentrifying neighborhood of inner Portland. Photos by the author. (Color figure available online.)

central contradiction arising from an urban sustainability fix. A growing body of scholarship reveals how green infrastructure fuels rising property values and rents and how green space becomes commodified for consumption (Pearsall 2010; Checker 2011; Rosol 2013; Gould and Lewis 2016), generating surplus value not only for developers, rentiers, and the growth coalitions to which they belong but also for global finance capital (Knuth 2016). As these scholars have demonstrated, sites of ecogentrification are also becoming central sites of contestation in the sustainable city (Lubitow and Miller 2013; Pearsall and Anguelovski 2016).

The use–exchange value binary is thus no longer as useful an analytic as perhaps it once was, and there is significant work to be done to clarify the dynamics by which UA contributes to capitalist accumulation in the neoliberal green city. Moreover, how noncommodified forms of UA actually produce value—particularly under racial capitalism (Robinson 2000; Pulido 2017)—begs closer examination. To this end, I attempt to break some new ground with this article, not by shoring up the jejune claim that UA is a bellwether of gentrification but by sketching out a theory of UA's uneven valorization within racialized processes of capitalist urbanization. I work through UA's valorization process in two steps. First, I situate UA within the realm of social reproduction, drawing on Bourdieu to explain how UA, as a socially reproductive practice, becomes cultural capital that the sustainable city's "green growth machine" (Gould and Lewis 2016, 35–36) mobilizes as symbolic *sustainability capital*, both to extract rent and to burnish the city's brand at larger scales. Second, I describe how such valorization occurs, by necessity, in a spatially variegated manner. Agglomerations of both UA and the ecohabitus required for its misrecognition as cultural capital, I assert, arise as a function of the interplay between ground rent and racialized othering. Ecogentrification, therefore, is not only a contradiction emerging from an urban sustainability fix but is fundamental to how racial capitalism works through green urbanization. I conclude by reflecting very briefly on UA's emancipatory potential.

Valorizing Social Reproduction through Misrecognition

Decentering orthodox readings of Marx, which define value as the "abstract human labor objectified or materialized" in a commodity as measured by the socially necessary labor time required to produce it (Marx 1976, 129), feminist political economists and geographers have recast the production of value to encompass the entirety of the production–reproduction dialectic (Dalla Costa and James 1972; McDowell 1999; Federici 2004; Mitchell, Marston, and Katz 2004; Bezanson and Luxton 2006; Meehan and Strauss 2015). The capitalist mode of production, from this perspective, does not require "commodities all the way down" (Fraser 2014), nor does it simply subsume other modes; rather, it relies on them for its own reproduction, transforming them (or not) in the process. As Henderson (1998) observed, "The social relations of production at any point in time will lie along a continuum" (78) of capitalist subsumption. Moreover, individuals are enmeshed in multiple relations at once, some of which are more capitalist than others (Gibson-Graham 2006; Meehan and Strauss 2015).

Most labor dedicated to UA in the Global North lies at the reproductive end of such a continuum. With the exception of a small cadre of market gardeners, urban agriculturalists (a majority of whom are women) for the most part only produce food for household use. Scholars of agrarian political economy have argued that the fruits of the home gardener's labor serve as a subsidy to capital, where self-provisioning lowers the cost of living (i.e., social reproduction), thus lowering the real cost of labor power and thereby justifying lower wages paid, ultimately lowering the cost of production (McClintock 2010; Minkoff-Zern 2014; Weissman 2015). A household-scale subsidy, from this perspective, would produce value only when aggregated across the workforce. Although more than a third of the U.S. population engages in food production (National Gardening Association 2014), the impact of homegrown produce on grocery costs is nevertheless minimal (CoDyre, Fraser, and Landman 2015). I argue, therefore, that another kind of valorization of social reproduction is taking place, where the practice of UA contributes to accumulation not only by subsidizing the cost of labor but also symbolically, as I now explain.

In theorizing social space, Bourdieu (1986) described how relative stocks of different "species" of capital confer social status or "distinction" within a given social field. Whereas economic capital is defined in monetary terms, cultural capital includes a combination of values and tastes (the embodied state), cultural goods (the objectified state), and qualifications or valorizations (the institutionalized state). Recognizing these various forms allows us to "reintegrate and reconceive use value within circuits of cultural and economic capital" (Beasley-Murray 2000, 107), while "enabling us to see consumption outside the workplace as likewise production and not simply as need-driven utility" (Beasley-Murray 2000, 113). When "perceived and recognized as legitimate" (Bourdieu 1989, 17), these forms of cultural capital serve as symbolic capital that can be leveraged to foster accumulation of economic capital. Bourdieu's species of capital therefore allow us to place social reproduction squarely within production, much as feminist reworkings of political economy do, rather than sidelining it as part of a consumption fund that subsidizes capital accumulation (e.g., Harvey 1989).

I maintain that although commercial UA generates substantial exchange value in the sustainable city through commodity production, the valorization of socially reproductive forms of UA generates even more surplus value—when mobilized as a particular symbolic form of cultural capital I call *sustainability capital*. In progressive, green cities, gardening carries a certain cachet at the household and neighborhood scales; as a performative act, it signals an awareness of and adherence to environmental values (Naylor 2012; Lebowitz and Trudeau 2017). UA holds no intrinsic value as cultural capital, though, so how is it valorized?

Following Bourdieu, valorization of UA as symbolic capital requires its *misrecognition* as cultural capital in both objectified and embodied forms; that is, the garden and the gardener.[1] A "symbolic logic of distinction" marks the practitioner from others living less sustainably, for "any given cultural competence (e.g., being able to read in a world of illiterates) derives a scarcity value from its position in the distribution of cultural capital and yields profits of distinction to its owner" (Bourdieu 1986, 245). Misrecognition of UA, I submit, depends on the predominance of an ecohabitus or set of practices and dispositions undergirded by green values. A "re-articulation of the field of high-class consumption, fostered by a more general social valorization of environmental consciousness" (Carfagna et al. 2014, 3), an ecohabitus also fuses concern for environmental sustainability with the "valorization of the local" and "revalorization of manual labor" (Carfagna et al. 2014, 15) that undergird foodie emphases on "quality, rarity, organic, hand-made, creativity, and simplicity" (Johnston and Baumann 2014, 3).

When and where an ecohabitus predominates, it is therefore use value itself—rather than an abstracted form of value based on socially necessary labor time—that is valorized via misrecognition. Whereas the latter

form of value "grasps only immediate time, time as present" (Postone 1978, 770), use value is a function of "the accumulation of past knowledge and labor time which ... finds no expression in the value-determined forms of appearance" (770). Rather than a measure of the socially necessary labor time required to till, sow, tend, and harvest a garden, UA's symbolic value is instead derived from the concrete time and labor spent by an urban agriculturalist both coaxing use values from the soil and acquiring the skills, practices, and dispositions—mastering the "feel for the game" (Bourdieu 1998, 98)—that allow her to do so.

With the rare exception of the gardener selling produce or teaching workshops, however, urban agriculturalists themselves do not "cash out" this cultural capital. Rather, rentiers, boosters, and financiers exchange these "free gifts of culture" (Stehlin 2016, 483) for economic capital at different scales. Urban growth coalitions mobilize this symbolic capital to competitively promote a city's reputation as a hotbed of sustainability and livability in hopes of attracting new investment, skilled labor, and green consumers, just as developers exchange it at the neighborhood scale for differential ground rent. At the global scale, these concrete transactions become "a collection of abstract, intangible, and interchangeable assets" (Knuth 2016, 628) that are themselves commodified, generating profits for financial institutions and shareholders.

For example, in Portland, Oregon—among the "top ten cities in the US for urban farming" (Renner 2016)—a growth coalition of municipal agencies, nonprofits, developers, and other businesses tout the city's green amenities (including UA) via tours, conferences, exchanges, and other forms of "policy boosterism" (McCann 2013) that burnish Portland's international renown as a paradigmatic sustainable city, attracting both investors and the highly educated consumers willing to work for lower real wages (relative to other metropolitan areas) in exchange for the green lifestyle that Portland has to offer (Jurjevich and Schrock 2012). As the director of Portland's Bureau of Planning and Sustainability admitted, sustainability is entrepreneurial as much as it is rooted in environmental values:

> We're not doing it just to be altruistic. Part of the reason we're doing a lot of this is there's money to be made, to be crass. ... And most of these things are things we want to do to create better, healthier places, anyway. But by doing that, you create a place where people want to live and have businesses. (quoted in D. M. Smith 2012)

In the case of noncommercial UA, then, surplus value accrued at these larger scales is extracted from household-scale social reproduction, and a garden's valorization as economic capital depends less on the habitus of the individual gardener and more on its wider misrecognition as cultural capital and eventual mobilization as symbolic capital at these larger scales. As I argue in the next section, the process of misrecognition—and valorization across scales—is fundamentally spatial, both arising from and contributing to racialized processes of uneven development.

Uneven Valorization, Rent, and the Racialized Recoding of Urban Space

For misrecognition to occur, two types of spatial agglomeration are necessary. First, there simply must be enough UA happening in a given place to be noticed; only when multiple households practice UA does it become visible enough to accrue any real symbolic value. Second, there also has to be a sufficient aggregation of people with similar values and tastes—with "identical categories of perception and appreciation" (Bourdieu 1998, 100)—who perceive UA as a marker of distinction, more common in cities such as Vancouver or Portland than in cities where UA might still be viewed as backward (Naylor 2012; McClintock et al. 2016). In other words, an ecohabitus must be sufficiently predominant within an urban population—whether spatially clustered or networked via affinity groups and social media (see Tarr 2015)—to affect the various individuals and institutions (e.g., government agencies, nonprofits, media, real estate markets) that mediate cultural capital's symbolic exchange value.[2]

UA's ability to function as sustainability capital is therefore spatially variegated. Where there are agglomerations of gardens and chicken coops and of ecominded residents who misrecognize these agglomerations as cultural capital, UA gains symbolic value, generating both profits of localization for those in close spatial proximity and profits of position for those whose labor is misrecognized as cultural capital (Bourdieu 2000). Conversely, where gardens or ecominded individuals are more dispersed, disconnected, or distant from the core, their relative distance limits the potential aggregative capacity to convert social reproduction into sustainability capital. Likewise, where an ecohabitus does not predominate—that is, where environmental concerns do not drive food production or, alternately, where existing gardens are perceived as backward or unsightly—UA accrues little or no

value as cultural capital. Even if practiced widely, it essentially remains invisible at the urban scale. Indeed, the UA practices of longtime residents, often cultivated for food security or cultural reasons, are often simply overlooked (Cheung 2016; Reynolds and Cohen 2016); in other cases, they are actually undermined by new restrictions on garden forms (e.g., trellises, greenhouses, coops) or increased fees for community garden plots, for example (Eizenberg 2012; Ghose and Pettygrove 2014).

I further maintain that for valorization to occur, such spatial variegation is necessary rather than contingent. As Stehlin and Tarr (2016) explained, "Creating spaces that are *unusually* livable can be completely congruent with property-based accumulation, which depends on qualitative differences between spaces that prompt flows of capital between them" (14–15). Labor-intensive practices, whether socially reproductive or artisanal, have symbolic value—and can therefore contribute to an urban sustainability fix—only when they can help distinguish spaces deemed sustainable or livable from those less so. Put simply, the symbolic value of UA depends to a certain extent on its scarcity value. In the sustainable city, a raised bed in the front yard (see Figure 1B) distinguishes the ecominded gardener from the typical North American, whose chemically fertilized lawn has come to signify unsustainable living. In aggregate, these gardens distinguish a city's hip, livable, green neighborhoods—or the promise of those to come—from the anomie and sprawl of mainstream suburbia and from the not-yet-gentrified lumpengeographies of the postindustrial inner core and devalued inner-ring suburbia. UA's symbolic value (and its misrecognition as such) is tied to these qualitative differences and therefore to the perpetual see-sawing motion of capital that creates them (N. Smith 2008).

For the better part of four decades, the rent gap (N. Smith 1979) between capitalized and potential ground rent—between actual market prices and unrealized market potential—has been the linchpin of production-side explanations of gentrification and of uneven urban development more broadly (Harvey 1989; Hackworth 2002). Moving beyond the false dichotomy of the production–consumption debate, however, requires recognizing that rent gaps open and close in dialectical tension with the movement of more affluent and educated and predominantly white populations into devalued neighborhoods (Slater 2006). With new arrivals come new forms of habitus and attendant changes in amenity and commodity consumption, which, in aggregate, raise potential ground rents, attracting investors. The budding agglomeration's gravitational pull on additional gentrifiers depends on the availability of cheap housing (or land) at first but eventually on the suite of amenities (including UA) making a name for the neighborhood (Zukin 1987; N. Smith 1996). Longtime residents might garden, but only when enough newcomers practice UA in way that is visible—in their front yards, community garden plots, or vacant lots—does UA function as an amenity that signals to future gentrifiers and investors that the neighborhood is on the road to being livable and green. Agglomerations of UA, then, emerge disproportionately in those areas where a rent gap can be closed.

Our research in Portland illustrates such a pattern (McClintock et al. 2016).[3] Front yard gardens, visible and performative, are disproportionately concentrated in inner Portland's single-family residential neighborhoods, particularly in those areas deemed most livable given their walkability and proximity to green amenities, as well as in neighborhoods that have been gentrified or where gentrification is underway (see Figure 2). A commitment to sustainability drives UA here; a whopping 91 percent of front yard gardeners—overall more affluent than the surrounding population—are motivated by a desire "to live in a more environmentally sustainable way" (McClintock et al. 2016, 10). Clusters of backyard gardens, on the other hand, exist throughout the city, including in low-income, highly diverse East Portland, where gardeners are less affluent than the surrounding population and are motivated more by concerns over food security. Misrecognized as sustainability capital by affluent, predominately white, ecominded residents, inner Portland's home gardens are valorized as economic capital by the city's green growth coalition at larger scales, contributing to higher home values at the city scale and to Portland's brand as a sustainable city nationally and globally. Conversely, gardens in the lumpengeography of East Portland—an area that ranks lowest in terms of livability by all city metrics (Goodling, Green, and McClintock 2015)—remain largely invisible, their use value unvalorized beyond the household scale. A quick scan of real estate listings containing the term *garden beds*, for example, revealed twenty-nine listings, all but four located in Portland's inner core or the newly gentrifying areas of East Portland just east of 82nd Avenue (Zillow 2016). Listings with the term *urban farm* were likewise situated in gentrification hot spots such as Cully, the most ethnically diverse neighborhood in the state. Although longtime residents have gardened

in Cully's spacious lots, its reputation as a UA hot spot rests more on the successes of younger (mostly white) newcomers, several of whom are engaged in highly visible, commercial UA. One market gardener recalls looking for land here: "So I drove down here and started cruising around. ... I'd just heard that this is the urban farming, or homesteading district of Portland" (interview with the author 1 September 2015). As the gentrification frontier pushes eastward past 82nd Avenue, a critical agglomeration of old and new UA—and the habitus necessary to misrecognize it as sustainability capital—is sure to follow.

Pulido (2017), however, observed that just as capital depends on spatial unevenness, "human difference is essential to the production of differential value" (527). The "racialization of space and spatialization of race" (Lipsitz 2007, 12) are therefore codetermined (Pulido 2000; Woods 2000; Gilmore 2002; Barraclough 2009). Recent scholarship further clarifies how a sustainability fix articulates with the larger racial project of capitalist accumulation and illustrates McKittrick's (2011) assertion that "the process of uneven development calcifies the seemingly natural links between blackness, underdevelopment, poverty, and place" (951). Situating gentrification's pioneer imaginary (see N. Smith 1996) within persistent settler colonial logics of dispossession, Safransky (2014) described how investments in green infrastructure in Detroit turn on "frontier narratives," wherein potential redevelopment sites are described as "empty and underutilized ... awaiting inhabitants and transformation, [thus] nullifying existing ways of life" (238). Similarly, Dillon (2014) described how the othering of urban spaces—as "under-utilized, economically unproductive lands" (1214)—that precedes redevelopment is tightly bound to the discursive and material ghettoization of surrounding neighborhoods of color. The characterization of one space as sustainable, green, or livable—in the media, in advertising, in government reports, and in everyday speech—thus renders another unlivable or uninhabitable (McKittrick 2006), a swath of *"urbs nullius"* (Coulthard 2014, 176) awaiting pioneer (re)settlement (Blomley 2003; Safransky 2014). It becomes clear, then, that ecogentrification is not only a contradiction emerging from an urban sustainability fix but is central to how racial capitalism works through green urbanization.

At its core, distinguishing the livable from the uninhabitable—indeed, delimiting the lumpengeography of capital and spatial contours of the rent gap—depends on "shared cultural ideals and moral

284

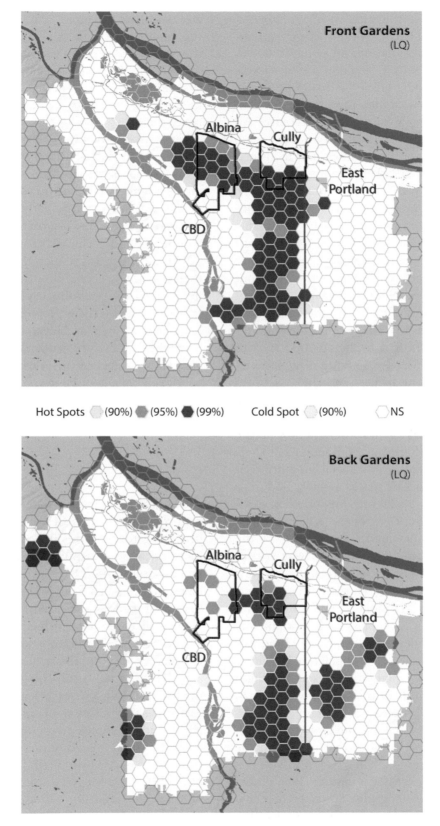

Figure 2. Hot spot analysis of front gardens (top) and back gardens (bottom) in Portland, based on a location quotient calculated for each 1.52 km² hexagon. Front yard gardens, visible and performative, are concentrated in the city's livable inner core, including Albina—ground zero of gentrification—and Cully, a neighborhood in the early stages of gentrification known for its vibrant urban agriculture scene. Clusters of less visible backyard gardens exist in these same areas but can also be found in lower income areas east of 82nd Avenue (indicated by the vertical red line). Map by Dillon Mahmoudi. Data from McClintock et al. (2016). (Color figure available online.)

geographies based on a romance with pure spaces" (Lipsitz 2007, 12). Such conceptions of purity have long underwritten ecohabitus and can be implicated in processes of racial othering (Cronon 1996; Moore, Pandian, and Kosek 2003; Finney 2014). Alternative agrifood practices, too, with their roots in the environmental movement, invoke such purity discourse. An explicit rejection of the industrial agrifood system often turns on the demonization of certain foods and body types (Guthman 2011), appealing to agrarian imaginaries that efface people of color (Carlisle 2014). Moreover, many food system activists trumpet consumption-oriented solutions that require disposable income; for example, "voting with your fork" by purchasing unprocessed, organic vegetables at a farmers market (Alkon and McCullen 2011). Other solutions require free time, such as volunteering with a nonprofit garden project (Pudup 2008). Even the most well-intentioned food justice activists are culpable of othering via their missionary zeal to "bring good food to others" (Guthman 2008). The embrace of such purity narratives, a "feel for the game," and the "viscosity" (Slocum 2007) or attraction of white bodies sharing an ecohabitus together contribute to the coding of these spaces as white and to the alienation of people of color (Henson and Munsey 2014; Ramírez 2015).

Ground rent, ecohabitus, and racial viscosity thus work together to render UA spaces as white and in so doing contribute to the racial recoding of urban space at ecogentrification's frontier, where a succession of racially coded descriptors marks the settling of the uninhabitable landscape: Sketchy gives way to up-and-coming and hip and eventually to family-friendly, livable, and green. Both an amenity that attracts foodie activists and hipster pioneers and a manifestation of their ecohabitus, UA plays an important role in distinguishing "new development, rising home values, and a whiter residential population" from a neighborhood's "racially marginalized past" (Dillon 2014, 1211). In the words of one young white urban farmer in Vancouver, "If there are community gardens or a street that looks like people are growing food on it, I think for the people I know, that plays into the 'Oh, this is up-and-coming' or 'It's okay to live here' type thing" (interview with the author 14 October 2016). Conversely, for longtime residents of these neighborhoods, lower income and often non-white, the new gardens symbolize their impending fate. An African American business leader in Portland lamented, "I knew black people were fucked as soon as I saw the bike lanes. That's when we knew we weren't welcome here anymore." His colleague added, "And the community gardens. That's another bad sign for the African American community. We always gardened. We always shared our gardens and our food. We didn't need 'community gardens.' That's a white invention" (quoted in Hern 2016, 10). At the same time, their own histories of gardening, canning, and other artisanal foodways—like the spatial clusters of UA found in East Portland and other uninhabitable areas devoid of ecohabitus—are rendered invisible by media attention lavished on young, white, and affluent foodies working in particularly photogenic, often capital-intensive gardens, on rooftops and other high-profile, performative locations (Reynolds and Cohen 2016). Given the agglomerations of white bodies and their underlying ecohabitus, these performances, while functioning as symbolic capital to increase ground rents and burnish the entrepreneurial sustainable city's brand, can alienate and ultimately exclude. Describing a nearby community garden, an African American community organizer in Portland averred, "You ain't seeing any people of your kind, and you're kinda like, 'Oh, this ain't for me'" (interview with the author 11 August 2016).

Conclusion

My goal here, rather than viewing UA within the context of pitched battles over conflicting use values between gardeners and bulldozers, has been to theorize its valorization as capital within the ecogentrification process itself, which I see as one of the primary urban manifestations of racial capitalism. I want to underscore that gardens and other forms of food production alone do not drive gentrification. Rather, as I have argued here, their uneven valorization is sociospatially dependent on both the agglomeration of ecohabitus and the racialized historical–geographical factors giving rise to rent gaps in particular neighborhoods. The growing agglomeration of white bodies in formerly uninhabitable spaces gives a spatial dimension to the ecohabitus that misrecognizes concrete time and labor invested in UA as sustainability capital. Moreover, a postpolitical, color-blind discourse of sustainability obscures the resulting whitening of urban space but nevertheless functions tacitly within the larger racial project of gentrification arising from an urban sustainability fix. Understanding UA's valorization this way reveals how practices of social reproduction can contribute to such a fix and "connects the discursive construction of race to the structural, material, and corporeal production of white racial hegemony" (Bonds and Inwood 2016, 720), especially in green cities where white people pride themselves on their liberal or progressive values—and their color-blindness.

Notwithstanding its subsumption by green entrepreneurial logics, however, social reproduction is also a sphere in which political organizing can occur, as feminist geographers have long asserted (Mitchell, Marston, and Katz 2004; Gibson-Graham 2006). The everyday space of noncapitalist UA can thus serve as a potential site of change, a rallying point for social justice and self-determination (White 2011; Reynolds and Cohen 2016). Indeed, the threat to community gardens has mobilized diverse, cross-class coalitions in the past (Staeheli, Mitchell, and Gibson 2002; Irazábal and Punja 2009). Ecogentrification, too, in dialectical fashion, sows the seeds of resistance (Pearsall and Anguelovski 2016; Safransky 2017; McClintock, Miewald, and McCann 2017), and the presence of community, collective, and commercial gardens, old and new, led by people of color in gentrifying communities offers a hopeful example of UA serving as a tool for racial justice, empowerment, and economic development (White 2011; Ramírez 2015; Pearsall and Anguelovski 2016; Reynolds and Cohen 2016; Sbicca 2016) and helps to resist the processes I have described in this article.

Like its contribution to ecogentrification, though, UA's emancipatory potential arguably also differs across time and space, for it is in the lumpengeography of the sustainable city—where an ecohabitus has not yet taken hold—that UA resists subsumption by green entrepreneurial logics. Foregrounding already-existing gardens in such spaces (e.g., those clusters in East Portland) is one simple but crucial way for scholars, journalists, and activists alike to unsettle the disproportionate misrecognition of eco-oriented, hipster-led UA as sustainability capital. Reynolds and Cohen (2016), for example, took stock of dozens of UA initiatives led by people of color in New York City and beyond, emphasizing the radical nature of their work, as well as the real need for material support to keep it going. Highlighting and honoring these invisible gardens—and, when possible, providing material support—also serves as an entry point for understanding ecogentrification within the broader context of white supremacy and the settler colonial present, a necessary prerequisite, I would argue, for fostering and strengthening alliances with activists engaged in a wider range of social and racial justice movements.

Acknowledgments

I am grateful to the many graduate students and research assistants whose labor helped bring this research to fruition: Dillon Mahmoudi, Mike Simpson, Erin Goodling, Jamaal Green, Amy Coplen, Anthony Levenda, and Jacinto Santos. Warm thanks also to Eugene McCann and Christiana Miewald for helping me take it in new directions and to Geoff Mann, Harold Perkins, Nathan Sayre, Evan Weissman, and audience members at the Tampa, Chicago, and Vancouver meetings of the American Association of Geographers and Canadian Association of Geographers for their thoughtful feedback on earlier versions. Finally, I am grateful to John Stehlin, Alex Tarr, and two anonymous reviewers for their close readings and incisive comments, all of which strengthened the final version of the article. All errors are entirely my own.

Funding

This research was supported by grants from the National Science Foundation (#1539750: "Urban agriculture, policymaking, and sustainability") and Portland State University's Institute for Sustainable Solutions.

ORCID

Nathan McClintock ⓘ http://orcid.org/0000-0002-3634-3799

Notes

1. *Misrecognition* for Bourdieu speaks to the arbitrary nature of assigning value to something, rather than implying flawed or mistaken recognition.
2. Whereas Bourdieu saw the state as the "central bank of symbolic credit" (Beasley-Murray 2000, 115) mediating the valorization of the various forms of capital, a range of institutions are today involved in the valorization of the concrete time invested in cultivating cultural capital. Given the broader neoliberal context of state retrenchment and attendant entrenchment of market logics through privatization and public–private partnerships, the private sector plays a growing role in the misrecognition of institutionalized cultural capital (see Lave 2012). Institutional recognition for UA (e.g., master gardener certification) continues to play an important role in misrecognizing UA as sustainability capital, but the state is hardly the sole arbiter.
3. My research assistants and I have been conducting mixed-methods research (including mapping, spatial analysis, mail surveys, interviews, media analysis, and

archival work) in Portland for the past five years (Good-ling, Green, and McClintock 2015; McClintock et al. 2016; McClintock and Simpson 2016). An ongoing relational comparative study of UA policy and practice in Portland and Vancouver builds on this work (McClintock, Miewald, and McCann 2017).

References

Alkon, A. H., and C. G. McCullen. 2011. Whiteness and farmers markets: Performances, perpetuations . . . con-testations? *Antipode* 43 (4):937–59.

Anguelovski, I. 2015. Healthy food stores, greenlining and food gentrification: Contesting new forms of privilege, displacement and locally unwanted land uses in racially mixed neighborhoods. *International Journal of Urban and Regional Research* 39 (6):1209–30.

Barraclough, L. R. 2009. South Central farmers and Shadow Hills homeowners: Land use policy and relational raci-alization in Los Angeles. *The Professional Geographer* 61 (2):164–86.

Beasley-Murray, J. 2000. Value and capital in Bourdieu and Marx. In *Pierre Bourdieu: Fieldwork in culture*, ed. N. Brown and I. Szeman, 100–119. Lanham, MD: Row-man & Littlefield.

Bezanson, K., and M. Luxton. 2006. *Social reproduction: Feminist political economy challenges neo-liberalism*. Mon-treal, Canada: McGill-Queen's University Press.

Blomley, N. 2003. *Unsettling the city: Urban land and the poli-tics of property*. London and New York: Routledge.

Bonds, A., and J. Inwood. 2016. Beyond white privilege: Geographies of white supremacy and settler colonial-ism. *Progress in Human Geography* 40 (6):715–33.

Bourdieu, P. 1986. The forms of capital. In *Handbook of the-ory and research for the sociology of education*, ed. J. Richardson, 241–58. Westport, CT: Greenwood.

———. 1989. Social space and symbolic power. *Sociological Theory* 7 (1):14–25.

———. 1998. *Practical reason: On the theory of action*. Palo Alto, CA: Stanford University Press.

———, ed. 2000. Site effects. In *The weight of the world: Social suffering in contemporary society*, 123–29. Stanford, CA: Stanford University Press.

Bradley, K., and R. E. Galt. 2014. Practicing food justice at Dig Deep Farms & Produce, East Bay Area, Cali-fornia: Self-determination as a guiding value and intersections with foodie logics. *Local Environment* 19 (2):172–86.

Burnett, K. 2014. Commodifying poverty: Gentrification and consumption in Vancouver's Downtown Eastside. *Urban Geography* 35 (2):157–76.

Carfagna, L. B., E. A. Dubois, C. Fitzmaurice, M. Y. Ouim-ette, J. B. Schor, M. Willis, and T. Laidley. 2014. An emerging eco-habitus: The reconfiguration of high cul-tural capital practices among ethical consumers. *Journal of Consumer Culture* 14 (2):158–78.

Carlisle, L. 2014. Critical agrarianism. *Renewable Agriculture and Food Systems* 29 (2):135–45.

Checker, M. 2011. Wiped out by the "greenwave": Environ-mental gentrification and the paradoxical politics of urban sustainability. *City & Society* 23 (2):210–29.

Cheung, C. 2016. Meet East Vancouver's original urban farmers. *Vancouver Courier*. Accessed January 17, 2017. http://www.vancourier.com/news/meet-east-vancouver-s-original-urban-farmers-1.2320486.

CoDyre, M., E. D. G. Fraser, and K. Landman. 2015. How does your garden grow? An empirical evaluation of the costs and potential of urban gardening. *Urban Forestry & Urban Greening* 14 (1):72–79.

Coulthard, G. S. 2014. *Red skin, white masks: Rejecting the colonial politics of recognition*. Minneapolis: University of Minnesota Press.

Cronon, W. 1996. The trouble with wilderness: Or, getting back to the wrong nature. *Environmental History* 1 (1):7–28.

Crouch, P. 2012. Evolution or gentrification: Do urban farms lead to higher rents? *Grist*. November 1. Accessed September 8, 2017. http://grist.org/food/evolution-or-gentrification-do-urban-farms-lead-to-higher-rents/.

Dalla Costa, M., and S. James. 1972. *The power of women and the subversion of community*. Bristol, UK: Falling Wall Press.

DeLind, L. B. 2015. Where have all the houses (among other things) gone? Some critical reflections on urban agricul-ture. *Renewable Agriculture and Food Systems* 30:3–7.

Dillon, L. 2014. Race, waste, and space: Brownfield redevel-opment and environmental justice at the Hunters Point shipyard. *Antipode* 46 (5):1205–21.

Dooling, S. 2009. Ecological gentrification: A research agenda exploring justice in the city. *International Jour-nal of Urban and Regional Research* 33 (3):621–39.

Eizenberg, E. 2012. The changing meaning of community space: Two models of NGO management of community gardens in New York City. *International Journal of Urban and Regional Research* 36 (1):106–20.

Federici, S. 2004. *Caliban and the witch: Women, the body and primitive accumulation*. New York: Autonomedia.

Finney, C. 2014. *Black faces, White spaces: Reimagining the relationship of African Americans to the great out-doors*. Chapel Hill: The University of North Caro-lina Press.

Fraser, N. 2014. Can society be commodities all the way down? Post-Polanyian reflections on capitalist crisis. *Economy and Society* 43 (4):541–58.

Ghose, R., and M. Pettygrove. 2014. Actors and networks in urban community garden development. *Geoforum* 53:93–103.

Gibson-Graham, J. K. 2006. *The end of capitalism (as we knew it): A feminist critique of political economy*. Minne-apolis: University of Minnesota Press.

Gilmore, R. W. 2002. Fatal couplings of power and differ-ence: Notes on racism and geography. *The Professional Geographer* 54 (1):15–24.

Goodling, E. K., J. Green, and N. McClintock. 2015. Uneven development of the sustainable city: Shifting capital in Portland, Oregon. *Urban Geography* 36 (4):504–27.

Gould, K. A., and T. L. Lewis. 2016. *Green gentrification: Urban sustainability and the struggle for environmental jus-tice*. London and New York: Routledge.

Guthman, J. 2008. Bringing good food to others: Investigat-ing the subjects of alternative food practices. *Cultural Geographies* 15 (4):431–47.

Guthman, J. 2011. *Weighing in: Obesity, food justice, and the limits of capitalism.* Berkeley: University of California Press.

Hackworth, J. 2002. Postrecession gentrification in New York City. *Urban Affairs Review* 37 (6):815–43.

Harvey, D. 1989. *The urban experience.* Baltimore, MD: The Johns Hopkins University Press.

Henderson, G. 1998. Nature and fictitious capital: The historical geography of an agrarian question. *Antipode* 30 (2):73–118.

Henson, Z., and G. Munsey. 2014. Race, culture, and practice: Segregation and local food in Birmingham, Alabama. *Urban Geography* 35 (7):998–1019.

Hern, M. 2016. *What a city is for: Remaking the politics of displacement.* Cambridge, MA: MIT Press.

Holt, S. 2015. An unexpected romance: Urban farmers and real estate developers. Accessed October 4, 2016. http://www.citylab.com/cityfixer/2015/10/the-newest-odd-couple-real-estate-developers-and-urban-farmers/409060/.

Irazábal, C., and A. Punja. 2009. Cultivating just planning and legal institutions: A critical assessment of the South Central farm struggle in Los Angeles. *Journal of Urban Affairs* 31 (1):1–23.

Johnston, J., and S. Baumann. 2014. *Foodies: Democracy and distinction in the gourmet foodscape.* London and New York: Routledge.

Jurjevich, J. R., and G. Schrock. 2012. *Is Portland really the place where young people go to retire? Migration patterns of Portland's young and college-educated, 1980–2010.* Portland, OR: Population Research Center. Accessed February 11, 2015. http://pdxscholar.library.pdx.edu/prc_pub/5/.

Knuth, S. 2016. Seeing green in San Francisco: City as resource frontier. *Antipode* 48 (3):626–44.

Lave, R. 2012. Bridging political ecology and STS: A field analysis of the Rosgen Wars. *Annals of the Association of American Geographers* 102 (2):366–82.

Lawson, L. J. 2005. *City bountiful: A century of community gardening.* Berkeley: University of California Press.

Lebowitz, A., and D. Trudeau. 2017. Digging in: Lawn dissidents, performing sustainability, and landscapes of privilege. *Social & Cultural Geography* 18 (5):706–31.

Lipsitz, G. 2007. The racialization of space and the spatialization of race: Theorizing the hidden architecture of landscape. *Landscape Journal* 26 (1):10–23.

Lubitow, A., and T. R. Miller. 2013. Contesting sustainability: Bikes, race, and politics in Portlandia. *Environmental Justice* 6 (4):121–26.

Marx, K. 1976. *Capital: A critique of political economy.* Vol. 1. London: Penguin.

McCann, E. 2013. Policy boosterism, policy mobilities, and the extrospective city. *Urban Geography* 34 (1):5–29.

McClintock, N. 2010. Why farm the city? Theorizing urban agriculture through a lens of metabolic rift. *Cambridge Journal of Regions, Economy and Society* 3 (2):191–207.

———. 2014. Radical, reformist, and garden-variety neoliberal: Coming to terms with urban agriculture's contradictions. *Local Environment* 19 (2):147–71.

McClintock, N., D. Mahmoudi, M. Simpson, and J. P. Santos. 2016. Socio-spatial differentiation in the sustainable city: A mixed-methods assessment of residential gardens in metropolitan Portland, Oregon, USA. *Landscape and Urban Planning* 148:1–16.

McClintock, N., C. Miewald, and E. McCann. 2017. The politics of urban agriculture: Sustainability, governance, and contestation. In *The Routledge handbook on spaces of urban politics,* ed. A. E. G. Jonas, B. Miller, K. Ward, and D. Wilson. London: Routledge.

McClintock, N., and M. Simpson. 2016. Cultivating in Cascadia: Urban agriculture policy and practice in Portland, Seattle, and Vancouver. In *Cities of farmers: Problems, possibilities and processes of producing food in cities,* ed. J. Dawson and A. Morales, 59–82. Iowa City: University of Iowa Press.

McDowell, L. 1999. *Gender, identity and place: Understanding feminist geographies.* Minneapolis: University of Minnesota Press.

McKittrick, K. 2006. *Demonic grounds: Black women and the cartographies of struggle.* Minneapolis: University of Minnesota Press.

———. 2011. On plantations, prisons, and a black sense of place. *Social & Cultural Geography* 12 (8):947–63.

Meehan, K., and K. Strauss, eds. 2015. *Precarious worlds: Contested geographies of social reproduction.* Athens: University of Georgia Press.

Minkoff-Zern, L.-A. 2014. Hunger amidst plenty: Farmworker food insecurity and coping strategies in California. *Local Environment* 19 (2):204–19.

Mitchell, K., S. A. Marston, and C. Katz, eds. 2004. *Life's work: Geographies of social reproduction.* New York: Wiley.

Moore, D. S., A. Pandian, and J. Kosek, eds. 2003. *Race, nature, and the politics of difference.* Durham, NC: Duke University Press.

National Gardening Association. 2014. *Garden to table: A 5-year look at food gardening in America.* Williston, VT: National Gardening Association. Accessed June 23, 2015. http://goo.gl/lf4xSD.

Naylor, L. 2012. Hired gardens and the question of transgression: Lawns, food gardens and the business of "alternative" food practice. *Cultural Geographies* 19 (4):483–504.

Pearsall, H. 2010. From brown to green? Assessing social vulnerability to environmental gentrification in New York City. *Environment and Planning C: Government and Policy* 28 (5):872–86.

Pearsall, H., and I. Anguelovski. 2016. Contesting and resisting environmental gentrification: Responses to new paradoxes and challenges for urban environmental justice. *Sociological Research Online* 21 (3):1–6.

Postone, M. 1978. Necessity, labor, and time: A reinterpretation of the Marxian critique of capitalism. *Social Research* 45 (4):739–88.

Pudup, M. 2008. It takes a garden: Cultivating citizen-subjects in organized garden projects. *Geoforum* 39 (3):1228–40.

Pulido, L. 2000. Rethinking environmental racism: White privilege and urban development in Southern California. *Annals of the Association of American Geographers* 90 (1):12–40.

———. 2017. Geographies of race and ethnicity II: Environmental racism, racial capitalism and state-sanctioned violence. *Progress in Human Geography* 41 (4):524–33.

Purcell, M., and S. K. Tyman. 2014. Cultivating food as a right to the city. *Local Environment* 20 (10):1132–47.

Quastel, N. 2009. Political ecologies of gentrification. *Urban Geography* 30 (7):694–725.

Ramírez, M. M. 2015. The elusive inclusive: Black food geographies and racialized food spaces. *Antipode* 47 (3):748–69.

Renner, S. 2016. Top 10 cities in the U.S. for urban farming. Accessed November 17, 2016. http://inhabi tat.com/top-10-cities-in-the-us-for-urban-farming/.

Reynolds, K., and N. Cohen. 2016. *Beyond the kale: Urban agriculture and social justice activism in New York City.* Athens: University of Georgia Press.

Robinson, C. J. 2000. *Black Marxism: The making of the black radical tradition.* 2nd ed. Chapel Hill: The University of North Carolina Press.

Rosol, M. 2013. Vancouver's "EcoDensity" planning initiative: A struggle over hegemony? *Urban Studies* 50 (11):2238–55.

Safransky, S. 2014. Greening the urban frontier: Race, property, and resettlement in Detroit. *Geoforum* 56:237–48.

———. 2017. Rethinking land struggle in the postindustrial city. *Antipode* 49 (4):1079–1100.

Sbicca, J. 2016. These bars can't hold us back: Plowing incarcerated geographies with restorative food justice. *Antipode* 48 (5):1359–79.

Schmelzkopf, K. 2002. Incommensurability, land use, and the right to space: Community gardens in New York City. *Urban Geography* 23 (4):323–43.

Slater, T. 2006. The eviction of critical perspectives from gentrification research. *International Journal of Urban and Regional Research* 30 (4):737–57.

Slocum, R. 2007. Whiteness, space and alternative food practices. *Geoforum* 38:520–33.

Smith, D. M. 2012. Breaking: Portland sustainability chief admits "Portlandia" isn't really a parody. Accessed November 17, 2016. http://grist.org/cities/breaking-port land-sustainability-chief-admits-portlandia-isnt-really-a-parody/.

Smith, N. 1979. Toward a theory of gentrification: A back to the city movement by capital, not people. *Journal of the American Planning Association* 45 (4):538–48.

———. 1996. *The new urban frontier: Gentrification and the ravanchist city.* London and New York: Routledge.

———. 2008. *Uneven development: Nature, capital, and the production of space.* Athens: University of Georgia Press.

Staeheli, L. A., D. Mitchell, and K. Gibson. 2002. Conflicting rights to the city in New York's community gardens. *GeoJournal* 58 (2–3):197–205.

Stehlin, J. G. 2016. The post-industrial "shop floor": Emerging forms of gentrification in San Francisco's innovation economy. *Antipode* 48 (2):474–93.

Stehlin, J. G., and A. R. Tarr. 2016. Think regionally, act locally?: Gardening, cycling, and the horizon of urban spatial politics. *Urban Geography.* Advance online publication. doi:10.1080/02723638.2016.1232464

Tarr, A. R. 2015. Have your city and eat it too: Los Angeles and the urban food renaissance. PhD diss., University of California, Berkeley.

Walker, R. 1978. Two sources of uneven development under advanced capitalism: Spatial differentiation and capital mobility. *Review of Radical Political Economics* 10 (3):28–37.

Walker, S. 2016. Urban agriculture and the sustainability fix in Vancouver and Detroit. *Urban Geography* 37 (2):163–82.

Weissman, E. 2015. Brooklyn's agrarian questions. *Renewable Agriculture and Food Systems* 30 (1):92–102.

While, A., A. E. G. Jonas, and D. Gibbs. 2004. The environment and the entrepreneurial city: Searching for the urban "sustainability fix" in Manchester and Leeds. *International Journal of Urban and Regional Research* 28 (3):549–69.

White, M. M. 2011. Sisters of the soil: Urban gardening as resistance in Detroit. *Race/Ethnicity: Multidisciplinary Global Contexts* 5 (1):13–28.

Woods, C. 2000. *Development arrested: The blues and plantation power in the Mississippi Delta.* London: Verso.

Zillow. 2016. Accessed November 23, 2016. http://www.zil low.com.

Zukin, S. 1987. Gentrification: Culture and capital in the urban core. *Annual Review of Sociology* 13:129–47.

NATHAN McCLINTOCK is an Associate Professor in the Toulan School of Urban Studies and Planning at Portland State University, Portland, OR 97207. E-mail: n.mcclintock@pdx.edu. His research interests include urban political ecology, critical urbanism, and food systems planning.

26 From "Rust Belt" to "Fresh Coast"

Remaking the City through Food Justice and Urban Agriculture

Margaret Pettygrove and Rina Ghose

Rising levels of urban food insecurity and diet-related disease have led to many inquiries into the urban food environment and its relation to health. Community-based food activism and urban agriculture (UA) provide alternatives to conventional food systems and promote food justice. Forms of food activism include community gardens, farmers' markets, antihunger initiatives, legislative advocacy, food literacy campaigns, and organic food consumption. Although many benefits are noted, scholars also contend that food activism often serves to bolster neoliberal structures by encouraging neoliberal citizen subjectivities or engaging in localized activities that do not directly challenge broader structural injustices. To the extent that neoliberalization is a racist (and racialized) process, the reproduction of neoliberal structures contributes to reproducing racial difference. This article examines the complexities of food activism within the context of neoliberal governance, with particular attention to the role of the local entrepreneurial state and its interactions with nonstate actors. City government and private development agencies promote UA as a means of neoliberal economic development that operates via public–private partnership to revitalize and generate value from central city neighborhoods. In so doing, these actors appropriate discourses from community-based UA organizations to legitimize their political–economic interests. Community-based organizations in turn recognize these interests and engage strategically with the city and private agencies to survive in the context of heightened resource competition and performance pressures within the nonprofit sector. Our research is based on seven years of fieldwork in Milwaukee, collecting data through intensive semistructured interviews, participant observations, and documents analysis. *Key Words: community gardens, food justice, neoliberal urbanism, urban agriculture.*

城市粮食不安全程度的加剧，以及与饮食相关的疾病，已引发诸多有关城市粮食环境及其与健康的关系之探问。以社区为基础的粮食行动主义与城市农业 (UA)，提供了传统粮食系统之外的另类选择，并提倡粮食正义。粮食行动主义的形式，包含社区花园、农夫市场、反饥饿运动、立法倡议、粮食知识运动，以及有机食品消费。尽管诸多益处已受注意，但学者仍主张粮食行动主义经常通过鼓励新自由主义的公民主体，抑或参与无法直接挑战更为广阔的结构性不正义之在地化活动，因而经常强化了新自由主义结构。如同新自由主义化作为种族歧视(和种族化)的过程，新自由主义结构的再生产导致了种族差异的再生产。本文检视新自由主义治理脉络中的粮食行动主义，并特别聚焦企业型地方政府的角色，及其与非政府行动者的互动。市政府与私人发展机构，提倡 UA 作为透过公私伙伴关系操作的新自由主义经济发展的方式，以此复兴并创造市中心邻里的价值。这些行动者藉由这麼做来挪用以社区为基础的UA组织之论述，以正当化其政治经济利益。以社区为基础的组织，从而认识到这些利益，并与市政单位和私人行动者进行策略性合作，以在非盈利部门紧缩的资源竞争与表现压力之脉络中生存。我们的研究是根据在密尔沃基为期七年的田野工作，并透过密集的半结构式访谈、参与式观察和档案分析来搜集资料。*关键词：社区花园，粮食正义，新自由主义城市主义，城市农业。*

Los niveles crecientes de inseguridad alimentaria urbana y de enfermedades relacionadas con la dieta han conducido a muchas indagaciones dentro del entorno alimentario urbano y su relación con la salud. El activismo alimentario de base comunitaria y la agricultura urbana (AU) proporcionan alternativas a los sistemas alimentarios convencionales y promueven la justicia alimentaria. Las formas de activismo alimentario incluyen huertas comunales, mercados de granjeros, iniciativas contra el hambre, apoyo legislativo, campañas de concientización alimentaria y consumo de productos orgánicos. Si bien al respecto se notan muchos beneficios, los eruditos también sostienen que el activismo alimentario a menudo sirve para apuntalar estructuras neoliberales estimulando las subjetividades ciudadanas neoliberales o comprometiéndose en actividades localizadas que no retan directamente las injusticias estructurales de mayor envergadura. En la medida en que la neoliberalización es un proceso racista (y racializado), la reproducción de estructuras neoliberales contribuye a reproducir la diferencia racial. Este artículo examina las complejidades del activismo alimentario dentro del contexto de la gobernanza neoliberal, con particular atención sobre el papel del estado empresarial local y sus interacciones con actores no estatales. El gobierno de la ciudad y las agencias privadas de desarrollo promueven la AU como medio de desarrollo económico neoliberal que opera a través de la asociación público–privada para revitalizar y generar valor desde los vecindarios de la ciudad central. Haciendo esto, estos actores se apropian de los discursos de las organizaciones de AU de base comunitaria para legitimar sus intereses político–económicos. Las organizaciones de base comunitaria a su vez reconocen estos intereses y se comprometen estratégicamente con la ciudad y las agencias privadas para sobrevivir dentro del contexto de competencia exacerbada por los recursos y por presiones de desempeño dentro del sector de ánimo no lucrativo. Nuestra investigación se basa en siete años de trabajo de campo en Milwaukee, durante el cual se recogieron datos por medio de entrevistas semiestructuradas intensivas, observaciones participativas y análisis de documentos. *Palabras clave: huertas comunales, justicia alimentaria, urbanismo neoliberal, agricultura urbana.*

In 2013, the City of Milwaukee launched HOME GR/OWN (HG), an initiative endeavoring to increase fresh produce consumption and reduce obesity rates citywide through the development of urban agriculture (UA) and other local food system infrastructure. In doing so, the city joined a growing number of government and community-based organizations seeking to promote food justice[1] by improving access to healthy foods (Alkon and Agyeman 2011; Guthman 2012). Simultaneously, food justice and UA are framed by Milwaukee city leaders as an economic development approach that will stimulate revitalization through sustainable green infrastructure development, productive reuse of vacant city land, and job creation. Accordingly, UA development efforts have tended to focus on the working-class, African American neighborhoods of Milwaukee's Northside, which have disproportionately borne the consequences of white flight, disinvestment, and economic recession. The case of UA development in Milwaukee thus provides an opportunity to examine the intersections of food justice, green space production, and neoliberal urban economic development.

In this article, we examine the complexities of food activism within the context of neoliberal governance, with particular attention to the role of the local entrepreneurial state and its interactions with nonstate actors. We ask how UA-based revitalization efforts take shape and what it means for UA (and related local food system development) to be used by a municipal government as a neoliberal economic development strategy. We argue that local government agencies both constrain and exploit community-based food justice organizing to advance neoliberal interests. City government and private development agencies promote UA as a means of neoliberal economic development that operates via public–private partnership to revitalize and generate value from central city neighborhoods. In so doing, these actors appropriate discourses from community-based UA organizations to legitimize their political economic interests. Community-based organizations in turn recognize these interests and engage strategically with the city and private agencies to survive in the context of heightened resource competition and performance pressures within the nonprofit sector.

Further, because the city now actively champions and collaborates with these organizations, neoliberal development activities paradoxically create openings for these organizations to advance their own interests. This has led to a flourishing of UA-centered food projects coalescing around a dominant narrative that frames food justice as economic development.

We situate this project relative to existing research on neoliberalization and economic development. To the extent that neoliberalization is a racist (and racialized) process, the reproduction of neoliberal structures contributes to reproducing racial difference. Thus, we attend to the fundamental structuring role of race and racism, noting that efforts to develop UA in Milwaukee often reproduce racializing discourses that constitute "inner-city" black communities and spaces as "unhealthy" and thus viable to be leveraged in the interest of (and even standing to benefit from) neoliberal economic development. Food and dietary health inequities that arise from systemic poverty and racism are addressed through place-based narratives of "health and wellness," prompting localized land use interventions through UA and community wellness programs. *Unhealthy* might thus stand in for more politically loaded terms (e.g., *blight*) that have traditionally been used to justify urban renewal and gentrification.

Using a mix of methodologies, we have addressed these complex questions through seven years of case study research in Milwaukee, Wisconsin. We have examined food and dietary health inequities through spatial analysis. We have also drawn on seven years of in-depth qualitative fieldwork to examine food discourses and the UA movement. This includes approximately fifty-five open-ended interviews with city officials, community organizers, and activists, along with content analysis of policy documents, promotional materials, organizational correspondence, social media text, and news articles. The following sections elaborate on our theoretical approaches and examine key research findings.

Neoliberalism, Economic Development, and Food Activism

Shaped by the ideologies of market liberalization, entrepreneurial governance, and retrenchment of state welfare, neoliberalism has been a dominant policy influence at all levels of U.S. government. Public–private partnership approaches to economic and community development are increasingly common in the context of neoliberalization, as local governments have sought to shift social service management to voluntary and private-sector actors and as community organizations have been compelled to cope with precarious funding (Newman and Lake 2006; Perkins 2009; Ghose and Pettygrove 2014b). Increases in poverty and dependence on cheap (but nutritionally poor) food are some of the effects of neoliberalization. Neoliberal welfare reforms have also contributed significantly to rising food insecurity and declining health, by reducing welfare benefits and pushing more individuals into temporary and low-wage employment (Lightman, Mitchell, and Herd 2008; Cook 2012).

Although food justice organizing and other efforts to improve food access in disinvested urban areas are often conceived of as responses to processes of neoliberalization and uneven development (Staeheli, Mitchell, and Gibson 2002; Baker 2004; Eizenberg 2012), research indicates that these efforts might serve to reproduce neoliberal governmentalities, by encouraging reforms centered on volunteerism and consumer choice (Guthman 2008; Pudup 2008; Ghose and Pettygrove 2014a). These forms of neoliberal governance could also be spatialized through urban land use processes, where environmental design is used to incentivize or compel idealized behaviors (Shannon 2014; Carter 2015). These are often centered on narratives about "place-based" health and wellness programs (Carter 2015, 375). Efforts to alleviate dietary health inequities through highly localized neighborhood-scale environmental modification (e.g., developing a grocery store in a low-income neighborhood), however, can produce a form of neoliberal paternalism that nudges individuals into healthy behaviors while disconnecting the problem from larger political economic systems (Shannon 2014).

Further, local governments could promote activities associated with improving food accessibility and UA development as a neoliberal economic development strategy or a type of sustainability fix that resolves crises of accumulation by building sustainable infrastructure (e.g., green space) and positioning the city as innovative to attract corporate investment (While, Jonas, and Gibbs 2004; Castree 2008; Quastel 2009; Draus, Roddy, and McDuffie 2014; Walker 2015). UA can thus contribute to processes of eco-gentrification, wherein green space and community gardens become amenities that elevate property values (Quastel 2009). Simultaneously, municipalities continue to engage in land use conflicts in which urban community gardens and farms are heavily regulated or evicted in favor of revenue-generating uses (Domene and Saurí 2007; Barraclough 2009; Irazabal and Punja 2009; Perkins 2009; Rosol 2012). Neoliberal urbanism has thus tended to favor forms of urban green space development and management that are economically productive.

It is also important to consider how neoliberal political economies are both racializing and fundamentally structured by race (Pulido 2000; Barraclough 2009; Wilson 2009; Lai 2012; Bonds 2013b). Contemporary efforts to revitalize or remake urban spaces are "imagined through and embedded within" histories of white supremacist land use practices, which include rules about property ownership and the right of states to dictate land use (Bonds 2013a, 1392). Neoliberal discourses about individual responsibility and the supposed color-blindness of market-based systems have served to simultaneously obscure and reproduce race and racism as organizing principles of society (Melamed 2006; Roberts and Mahtani 2010). Whiteness and white supremacy are reproduced through various practices and policies, including land use planning, economic development and revitalization, and mortgage lending practices (Pulido 2000; Delaney 2002; Lawson, Jarosz, and Bonds 2010; Bonds 2013a; Feagin 2013).

The racialization of space that occurs with the production of the inner city or areas targeted for redevelopment functions in different ways to reproduce racial hierarchies, in part by providing boundaries that "demarcate devaluation" (McClintock 2011, 95) or, conversely, enabling urban renewal and gentrification (Lai 2012). Racialization of space can, in a sense, be considered to reflect the differential valuation of space and its inhabitants (Pulido 2000). What is identified as valuable or productive land use is inextricably tied to racial constructions. In the case of development initiatives tied discursively to public health, it might also be useful to consider how racialization has occurred often through discourses that delineate more or less "healthy" populations and spaces on the basis of bodily or environmental norms (Guthman and DuPuis 2006; Marvin and Medd 2006; Brown 2009; Keil 2009; Shannon 2014).

Indeed, in the case of food justice and UA organizing, many scholars have drawn attention to the role of race. Some have noted that the pervasive whiteness of many organizations contributes to reinforcing racialized exclusion through the creation and defense of white spaces (Slocum 2007; Alkon and McCullen 2009; Ramirez 2015). Particularly in the context of UA development, the presence of white organizers might racialize spaces in particular ways, reinforcing notions of struggling black neighborhoods needing

developers and the state to revitalize them and thus "normalize processes of black dispossession" (Ramirez 2015, 762; see also McKittrick 2011). Further, by using development of marginalized neighborhoods to generate value for municipal entities (e.g., by stimulating gentrification or attracting commercial activity and tourism), such activities exploit marginalized populations (Lai 2012; Bonds 2013a).

In the case of HG, however, the ways in which the City of Milwaukee is "intervening" simultaneously provide openings for black organizations to reshape space according to their own imaginaries and reproduce neoliberal racialized discourses. As Ramirez (2015) noted, "Race, power and privilege emerge through community food spaces; they either reify existing inequalities or challenge them, depending on how the food space is being produced" (752). We emphasize that both can occur within the same space,

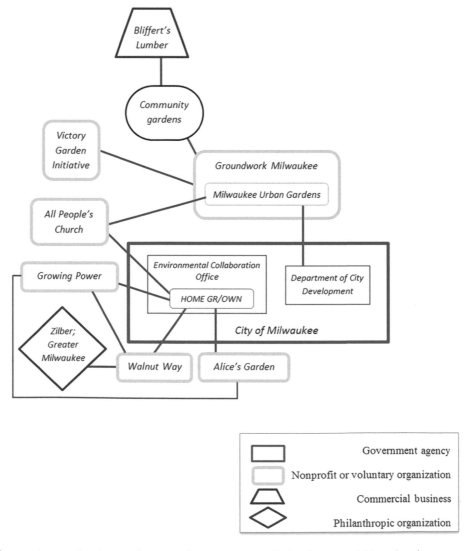

Figure 1. Actors and networks in Milwaukee's urban agriculture movement. (Color figure available online.)

as with neoliberalization more broadly (Roy 2011). Further, as Ramirez (2015) showed, black UA and food justice projects can "use land as a tool of liberation, drawing from practices of resistance that stem from plantation survival strategies" (751; referencing McKittrick 2013).

These theorizations of how racialization structures political–economic processes add an important dimension to ongoing discussions about neoliberal economic development and its relationship to sociospatial inequities. If racialization of space entails valuing spaces differently, then we should attend to the racial narratives entwined with (and potentially propelling) revitalization and gentrification projects, especially where they involve decision making about best or appropriate land uses. For example, the decision to allow or disallow UA in a neighborhood cannot be fully understood without reference to the historical

and contemporary division of space in a particular locale (Barraclough 2009). Accordingly, we consider how HG's UA initiative racializes the spaces of its interventions and contributes to potential dispossession and displacement through environmental production.

Food Organizing and Urban Agriculture in Milwaukee, Wisconsin

The sociospatial landscape of Milwaukee and its surrounding metropolitan area reflects many characteristic urbanization processes that have occurred since the mid-1960s in cities across the United States. As a city historically fueled by a robust industrial manufacturing economy, the period of deindustrialization and post-Fordist political–economic restructuring

Filling in Milwaukee's Food Deserts:
Turning the Vacant Lots in Food Deserts into Community Gardens

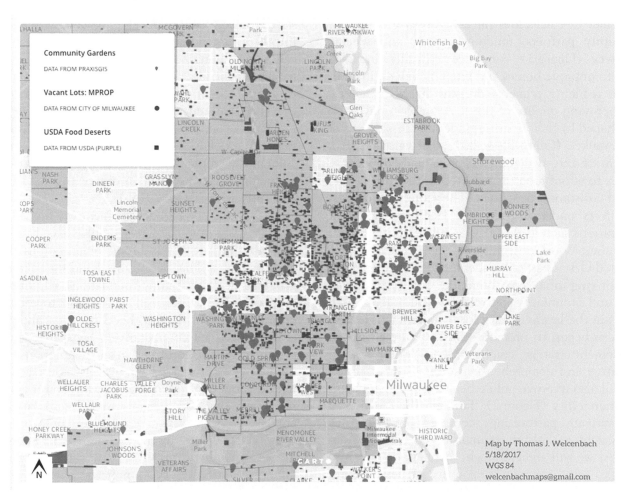

Figure 2. USDA-defined food deserts and urban agriculture in Milwaukee. (Color figure available online.)

led to particularly significant impacts via employment loss, urban disinvestment, suburbanization, and white flight. Milwaukee is also significantly racially unequal, in terms of housing patterns, employment, transportation access, and incarceration rates (Quinn and Pawasarat 2013; Rast 2015). In the wake of economic recession and a housing and mortgage crisis that began in 2006, home foreclosures and land vacancy have increased significantly (Derus 2007; Pawasarat and Quinn 2007).

Since the early 1980s, the City of Milwaukee has pursued various forms of neoliberal economic development, including downtown redevelopment initiatives and defunding municipal public services such as parks management (Ghose 2007; Zimmerman 2008; Perkins 2009). In response, voluntary organizations have proliferated to fill social service needs (Roy 2011).

UA development efforts in Milwaukee have emerged as largely community-led projects against a backdrop of historical and ongoing revitalization initiatives, many of which have aimed for neighborhood-level economic and community development on the Northside. UA activities have involved a variety of community partners, funding sources, and specific objectives (Figure 1) but have typically included efforts to improve homes and property values, in addition to reducing crime, addressing health concerns, and improving quality of life. UA projects have also been shaped by the disparities in food access that exist in Milwaukee (Gibbs-Plessl 2012; Pettygrove 2016), in terms of the availability of food and individuals' abilities to afford food (as reflected in the distribution of food assistance benefits).[2] Therefore, the earliest UA efforts undertaken in the 1990s by nonprofit organizations such as Alice's Garden, Growing Power, and Walnut Way Conservation Corps (WWCC) have promoted UA as part of broader food justice organizing goals, linking urban environmental quality to dietary health and community control over food systems.

Within the last five years, local government leaders have also begun to promote UA as a key economic development strategy, emphasizing its potential as an innovative and comprehensive form of revitalization that will also improve public health and environmental sustainability (Figure 2). Through the creation of public–private partnerships centered on UA, they have sought to bolster community-led UA development. As we will argue, though, these efforts have also served to reinforce neoliberal forms of development that operate through the racialization of urban space. As the city promotes the redevelopment of vacant city

lots into productive uses that generate value for the city, it does so in part by framing predominantly black neighborhoods as "unhealthy" and in need of environmental improvements (in the form of UA) to expose residents to healthier foods.

Economic Development and Urban Agriculture

In 2013, the City of Milwaukee launched HG as part of its comprehensive sustainability plan.[3] This coincided with the launch of the Mayor's Strong Neighborhoods Plan, and both were heralded as efforts to revitalize inner-city neighborhoods through the redevelopment of vacant, "blighted" spaces (Anonymous, personal communication, 2014; City of Milwaukee 2013, 2015). Whereas the Strong Neighborhoods Plan has focused on preventing and mitigating impacts of home foreclosure, HG has targeted vacant lot improvement through reuse. HG, housed in the Environmental Collaboration Office (ECO), operates through a public–private partnership model, with the stated objective of redeveloping vacant public lots into UA spaces that will come under the long-term care of community organizations. HG has also worked to implement land use and building code changes to enable a broader range of UA activities, in the hope that this will encourage more community-led UA projects beyond the scope of HG. This represents a substantial policy shift from the previous tolerance of community gardens to the active promotion and financing of UA projects on city-owned land (Ghose and Pettygrove 2014a, 2014b).

HG's initial phase of development was funded with a $75,000 grant from the Greater Milwaukee Foundation and other private philanthropic organizations (Northwestern Mutual Foundation, the Fund for Lake Michigan, and the Zilber Family Foundation). HG distributes part of these funds through small grants to community organizations and neighborhood groups. Funded projects are designed collaboratively between community groups and ECO. Construction and maintenance during the first year is provided through city-contracted labor from organizations including Growing Power and Walnut Way's Blue Sky Landscaping. Community groups are then encouraged to raise money or apply for funding (e.g., via the Neighborhood Improvement Development Corporation [NIDC]) to cover costs in the years following. To date, HG's projects primarily include parks and fruit orchards, along with some production farms and community gardens.

As a public–private partnership, HG works with a variety of local nonprofit and private organizations. HG's first UA development project, Ezekiel Gillespie Park (opened in 2014), received financial and in-kind support from a variety of city agencies and private entities, including University of Wisconsin–Milwaukee's Community Design Solutions (CDS), which provided design labor. The construction labor for the park was contracted to the lead community group Walnut Way's Blue Sky Landscaping, an employment training program run by the organization. City of Milwaukee administrators have been particularly explicit about the strategic advantage to the city of collaborating with nongovernmental actors in their efforts to develop UA and improve healthy food access. The participation of community organizations in HG has been facilitated by a well-established and highly interconnected network of local UA and food justice organizations, in which the city has positioned itself as simply a participant (Figure 1).

Because they are central to the flow and exchange of organizational resources and a source of political legitimacy, partnerships are a normal part of how both community-based and government food projects in Milwaukee function. HG has positioned itself as a continuation of Milwaukee's long history of community-based food organizing, drawing on popular perceptions of Milwaukee as a UA center. HG accordingly highlights how the local government simultaneously coopts community narratives to further its own interests and supports these narratives to the benefit of community activists.

In 2016, despite initial enthusiasm, HG began to shift its focus away from UA development toward more conventional forms of neighborhood revitalization. HG now focuses principally on redeveloping vacant lots (of which there are more than 5,000 in Milwaukee) into parks and greenscape projects. As HG came to understand the challenges and financial costs associated with urban food production—particularly the need for greenhouses to allow growing through Milwaukee's long, cold winters—they became reluctant to invest. Community groups, however, continue to build and maintain community gardens on the Northside, their efforts made easier by HG's land use policy changes.

Thus, although HG has lowered barriers to community gardening in Milwaukee, the work of developing community gardens remains with residents. Although framed as a panacea for marginalized neighborhoods, HG has prioritized the interests of the City of Milwaukee, treating UA as a sustainability fix intended to generate increased property values in inner-city neighborhoods and increase the city's attractiveness to investors. This is reflected in their open reliance on Growing Power[4] (an organization that is high-profile nationally and focused on the commercial production of organic produce) as the model for UA, without considering local factors critical to sustainable economic development through UA (e.g., the need for agricultural training and the relatively small acreage offered by vacant lots). With HG's movement away from UA, Employ Milwaukee (through its EARN and LEARN program), along with the City of Milwaukee Common Council and several state legislators, has taken interest in continuing to promote UA as economic development, with a dedicated focus on creating employment.

UA development in Milwaukee is thus tied to broader discourses surrounding neoliberal economic development and place making. The City and other organizations actively position UA as a means of economic revitalization—particularly in the inner city, where land vacancy, in the wake of the home foreclosure crisis, has increased pressure to find productive reuses of vacant space. The City of Milwaukee has expressed increasing concern about an abundance of vacant land concentrated on the Northside of the city since 2012, when the foreclosure crisis first became apparent (although many date the origins of the crisis to 2006 and the subprime mortgage crisis; Derus 2007; Quinn and Pawarasat 2007; Wisla n.d). Although the city has permitted community gardens on vacant public residential lots since 2000, it was only in 2012, with the initial conception of HG, that the city began actively pursuing development of UA and community gardens.

Urban Agriculture and Dietary Health

HG frames local food systems development, centered on UA, not only as a form of economic development but as a solution to inequities in healthy food access.[5] UA development is typically described as most needed in neighborhoods where large chain grocery stores and farmers' markets are scarce and convenience stores abound and where residents are thus generally characterized as being in poor dietary health.[6] Such conditions are nearly always ascribed to low-income, predominantly black, inner-city neighborhoods (synonymous colloquially with the Northside). Dietary health issues are commonly attributed to the abundance of convenience stores and fast food restaurants on the Northside, which are typically characterized as unhealthy, threatening, or aesthetically unpleasant; one community organizer describes corner stores as "creepy, dark, nasty" (Anonymous, personal communication, 2014). These narratives of unhealthy places then become discursively linked with long-standing tropes about blight and disinvested neighborhoods.

Despite emphasizing that dietary inequities are structural (a function of the environment), food organizers and city representatives suggest that individuals need to be persuaded to engage in healthier behaviors, in many cases by being acculturated into healthy foodways. It casts people of color as relatively unaware and seems to pathologize the Northside, implicitly constituting residents of these neighborhoods as unhealthy and framing dietary health as a matter of personal choice. According to some organizers, successfully changing residents' dietary health will also require a shift in cultural values surrounding food. As a member of the Milwaukee Food Council (MFC) explained, in discussing strategies for improving healthy food access in low-income neighborhoods, "We can't just put veggies there [in the store] and expect people to eat them" (Anonymous, personal communication, 2014). Broadly, organizers and city leaders tend to construct the built environment and cultural systems as interlinked contexts that shape individual behavior. As one organizer explained, individuals make "bad decisions" not out of apathy but because "they are in an environment—in a culture, in a system—that pushes them to make the wrong decisions" (Anonymous, personal communication, 2015). A Milwaukee Health Department staff member attributed unhealthy eating habits on the Northside, where he is involved in developing healthy corner stores, as a function of

> lack of knowledge or just not being exposed to it . . . kids are not used to getting an apple handed to them to eat . . . people don't know how to cook, so we tried to have some cooking demonstrations . . . it was marginally successful . . . it's a systematic thing. (Anonymous, personal communication, 2015)

Thus, describing the problem as systemic in this case appears to refer primarily to cultural systems, rather than political–economic systems that structure resource access. Constructing unhealthy dietary behaviors in this manner contributes to racializing the subjects of this intervention according to long-standing tropes that pathologize black cultural systems, often without explicitly mentioning race (Guthman 2008; Slocum 2011).[7] It also simultaneously provides justification for interventions that center on physically redeveloping urban space.

In this context, UA is positioned by many groups (including the city) as an essential strategy to address dietary inequities, in part because UA is understood as an effective environmental "nudge" that promotes consumption of foods (fresh produce, in particular) valued as healthy, in contrast to more "conventional," clinical forms of nutrition education. UA encourages cultural shifts by inscribing particular food values into the urban landscape.

This linkage between public health and environmental conditions highlights city leaders' fundamental interest in UA as a revitalization strategy that works through remaking urban space. In describing HG, staff indicate that creating healthy food access is not the initiative's primary or most important goal. Although official descriptions of HG in the city's sustainability plan and on its Web site present reducing dietary health inequities as a principal objective, discussions about it elsewhere indicate that public health and food access are tangential to the basic motivation for development of HG. In public presentations, local media reports, and personal interviews, HG is framed as being first and foremost about rejuvenating vacant lots into green spaces (whether that includes food or not). HG's first pilot project, a "pocket park" in the Lindsay Heights neighborhood, completed in 2015, contains fruit trees and perennial fruits but also contains areas of lawn and other features intended to enable its use as a park. Other HG developments have included fruit orchards and community garden spaces, as well as more traditional parks. Thus, although framed in significant part through the discourse of UA as a means to improve healthy food access, HG appears to be driven by broader economic development interests. As the director of HG explained, the initiative is

> really just trying to solve the multitude of problems that we have, in the central city, of urban blight, neighborhood destabilization, poor access to healthy food, and joblessness. . . . You know, if you impact one vacant lot . . . you may create income for someone working on the site . . . you help solve urban blight . . . lower city operating costs . . . crime tends to drop in greener neighborhoods. . . . We touch this piece of land, we're going to get

7 to 9 benefits out of it ... that's a pretty darn good return on investment. (Anonymous, personal communication, 2014)

The city thus contends that it is not only reducing city operating costs (by shifting costs of lot maintenance to community or private groups) but also stabilizing property values and creating employment (through construction and maintenance). This discourse—that UA promotes economic development—is echoed by nongovernmental economic and community development organizations, including philanthropic foundations.

UA development is also part of the city's effort to stimulate economic development by remaking the inner city (and Milwaukee overall); that is, by reconstructing vacant and blighted places as sites of greening, the city positions itself as innovative, sustainable, and thus economically competitive. Mayor Barrett has described this as part of his campaign to shift the image of Milwaukee from "rust belt" to "fresh coast."[8] This emphasis on UA as a means to improve the city's economic appeal and put vacant lots back into economically productive uses underscores the city's support for UA as it fits into neoliberal economic development agendas.

As we note in subsequent sections, the city's framing of UA as economic development relies in part on the resonance of this narrative with community groups, many of whom view UA as a strategy to generate resources and economic self-sufficiency on a local scale. In seeking to remake Milwaukee, city leaders emphasize the existing discourse of Milwaukee as a UA "hub" and pioneer, built on a history of local community organizing around this activity (iconized by Will Allen, who established Growing Power in 1993). A city staff member noted that the idea for HG began to circulate in 2012 when "Will Allen had just been named one of the 100 most influential people. Food was starting to get hot, and we decided ... let's ride this wave" (Anonymous, personal communication, 2012). Thus, the city appropriates what has been a largely community-driven UA movement (led by many black organizations) to advance its neoliberal economic development interests. The Bloomberg competition, where HG first emerged as an idea, is important because it drew attention to Milwaukee and served to demonstrate the appeal of the HG idea to funders (even though HG was ultimately not selected to receive an award) and to the general public.

That HG is understood as being able to address multiple issues simultaneously also appears to help explain its strategic appeal to the city, in that this enables HG to be adaptable to constraints. As HG's director stated,

> We're really working on multiple issues simultaneously, and that helps, because ... sometimes things are tough to do in city government and you hit a roadblock, and we've got the ability to veer off in a different direction. (Anonymous, personal communication, 2014)

The extent of popular support (whether perceived or real) for UA has been central to the city's view of UA as a viable economic development tool, as it has enabled the city to see UA as a low-cost effort, with the bulk of the work carried out by voluntary or private organizations. HG is consciously discussed as an initiative that is merely supporting existing community efforts toward UA, which already has legitimacy and is best left to the responsibility of civil society. So, for example, the director of HG consistently frames the initiative as responding to and emerging from the community, and thus as inherently collaborative.

Although the city has provided financial support for UA—in the form of a funded staff position responsible for coordinating HG projects, labor donated from agencies like the Department of Public Works, and financing construction costs through a Greater Milwaukee Fund grant—the focus of HG work since its inception has been coordinating projects carried out by nonprofit organizations or private entities (even while retaining ownership of the land). As the director of HG explained, in the case of a pocket park developed as part of the program, the space will continue to be city property maintained by the partnering community organization in the immediate future, but the city is "hoping that engaged residents will take over the management of the property, which actually helps us lower our city operating costs" (Anonymous, personal communication, 2014). The cost of UA implementation continues to reside within the civil society.

In these ways, UA development generates value for the city, putting places and people "to work." By drawing on and reinforcing social constructions of Milwaukee's black inner-city neighborhoods as blighted, unproductive, and unhealthy (environmentally and culturally), the city positions UA development in these neighborhoods as a necessary and beneficial intervention. City-sponsored UA development is then simultaneously reproducing neoliberal structures, which are themselves fundamentally racialized and racist (Roberts and Mahtani 2010).

This discourse—that UA promotes economic development—is echoed by nongovernmental economic and community development organizations, including philanthropic foundations. The Zilber Family Foundation, for example, frames its decision to give $500,000 for Walnut Way's Innovation and Wellness Commons as an economic investment that will draw more investment. The prominence of the narrative equating locally grown food with health in Milwaukee might be at least partially due to the popularity of this narrative with funders who have a history of funding community development activities locally. Medical College of Wisconsin's Healthier Wisconsin Partnership Program (HWPP), for example, has funded collaborative UA projects among seven of the organizations considered here.

Many community organizations also support the discourse of UA as economic development, albeit with an emphasis on the economic opportunities that local food systems provide in addition to its potential to improve land value. In part, this appears to reflect the idea that linking UA and local food to economic processes will enable its long-term sustainability and viability. Beyond this, however, there are groups that construct UA as a means to economic survival and autonomy for groups that tend to be excluded from or marginalized within broader economic processes on the basis of race and class. The production of green space for UA, or other local food spaces (e.g., farmers' markets), often dovetails with practices centered on food and farming business and labor development, particularly efforts to train and support individuals (often youth) in farming or other food-related entrepreneurial skills, provide gainful employment, and develop markets for those entrepreneurs. For many organizations, these practices use farming or local food business management as a means to employment and skills development for youth or adults who face structural barriers to formal, living-wage employment.

Although the development of UA here certainly aligns with and possibly reinforces elements of the city's economic development agenda (and many of these projects have directly benefited from partnering with HG to obtain land or other resources), these projects emphasize economic development via UA as a direct benefit to marginalized groups. To the extent that the dearth of adequate economic opportunities in African American neighborhoods is a function of structural racism expressed through historical land use processes (including redlining, home mortgage lending policies, urban renewal projects, white flight, deindustrialization, and disinvestment from central cities), efforts to develop community-based economies via

food systems are antiracist because they enable communities on the receiving end of racial discrimination to promote their own sustenance and survival (see Ramirez 2015).

UA and local food system development are thus also positioned as part of efforts by marginalized groups to assert political control by making food systems directly meet local needs and controlling the production of space. In this context, community ownership of space is central to food system localization and UA development. A Northside farmers' market director noted that the market is important in part because it "fills niches ... that big agriculture can't, like, sweet potato greens—those are popular especially among African immigrants" (Anonymous, personal communication, 2015). Revitalization, as a central narrative of the city's economic agenda, has been embraced by Northside UA groups, to the extent that they are able to leverage this narrative in support of projects to reclaim and transform vacant lots for a variety of purposes. For example, the director of a prominent black-led UA organization describes particular gardens in Northside neighborhoods as sites of healing in the wake of violent deaths of residents.

This challenges straightforward distinctions between "activists" and opponents, as negotiations over ownership and development of city space for UA are not a simple conflict between the city (and commercial interests) and community groups, but between different groups with distinct interests. The city, in other words, exists simultaneously as an ally and an antagonist in its governance of space. Although urban economic development is often synonymous with neoliberalization, there might in practice exist multiple economic development agendas. In the context of Milwaukee UA, where city agencies promote UA as part of a neoliberal economic agenda centered on reducing municipal operating costs and stimulating investment (directly and through boosting the city's image), multiple Northside community organizations frame UA as a means of economic development that will build wealth directly for local communities. Wealth for these groups includes financial resources, land ownership, and capacity to self-sustain outside of formal economic systems.

Conclusion

In Milwaukee, different conceptions of food and health (with distinct rationales and goals) have coalesced to support UA development and related practices. The tendency of groups to link UA and dietary health and the prominent role of UA across Milwaukee food projects appear to a significant degree to reflect the resonance of the discourse of UA as a comprehensive form of economic or community development that can address numerous concerns. UA is also a low-cost, politically feasible form of organizing and economic development. It is thus a form of neighborhood revitalization (framed as sustainable development and public health promotion) that appeals to the local government leaders and community groups. Perhaps most significant, UA has been framed and leveraged by the city as a means of creating value from vacant lots and thus remaking the inner city. The foreclosure and land vacancy crisis has created an economic incentive for the city to promote UA development on vacant lots, and the existence of strong, thriving UA and food movements within Milwaukee has facilitated this strategy of economic development. In this way, environmental sustainability, revitalization, and dietary health seem to function as mutually reinforcing discourses.

The City of Milwaukee's promotion of dietary health in this manner aligns with its broader neoliberal interest in ensuring productive uses of land and promoting public–private partnerships. The City uses UA to leverage value from disinvested neighborhoods and remake the image of Milwaukee according to sustainability narratives. Yet municipal policy changes to promote healthy food access have also facilitated the work of community organizations engaged in UA development. There is thus tension surrounding these activities, as they simultaneously contest and reinforce white supremacist capitalist economic development agendas.

Although the development of UA sites in this context does appear to facilitate black-led community organizations' access to resources (especially land) and to provide tangible benefits for residents of these neighborhoods (e.g., green space, fruit trees), the city's focus on this particular strategy of investment does not address the various structural processes that produce racial inequities in Milwaukee. Further, by reinforcing the narrative of the inner city as a space to be (re)developed and made productive by (and for) the state—supported by the narrative that black residents need to be exposed to and educated about healthy foodways—this initiative reproduces white supremacist (and colonial)

imaginaries. Again, although race is often unspoken (at least in text), the framing of UA as a means of encouraging cultural change in the communities where it is developed seems to racialize revitalization as whiteness, in that it is a process meant to improve neighborhoods understood to be black.

Acknowledgments

We thank the anonymous reviewers for their insightful comments and suggestions. We also thank Dr. Nik Heynen and Jennifer Cassidento at the *Annals* for their invaluable assistance throughout the review and publication process.

Notes

1. We use *food justice* in this article to refer to activities associated with pursuit of (more) equitable food systems, including, primarily, equitable access to nutritious foods and community control over food production.
2. In 2013, North-Central Milwaukee contained the highest density of Supplemental Nutrition Assistance Program (administered in Wisconsin via the FoodShare program) benefit dollars distributed per person.
3. HG is the first municipal program devoted solely to food and UA in Milwaukee.
4. In 1993, Will Allen established Growing Power on a 2-acre farmland in Milwaukee, to practice urban agriculture. Today, Growing Power has multiple farm sites in Wisconsin, including a 40-acre rural farm, and sites in Chicago. Although it supports community gardening endeavors, Growing Power functions as a commercial agricultural enterprise where products are sold for a profit.
5. They also emphasize creating an explicit, unified set of priorities, goals, and values to guide food activism. This discursively configures the networked space of food activism around the ideal of consensus and cooperation. Although there has been apparent alignment around particular narratives and practices among Milwaukee food organizations involved in shared networks, this might in part reflect the practical necessity of participating in these networks, as a result of which actors might opt to alter their priorities to maintain support. In this way, the alignment of narratives could serve to construct a political space that delineates the appropriate scope of action, the actors that are included, and the appropriate arrangement of actor relationships.
6. Milwaukee actors often use chain grocery stores as a referent for ideal (quality) retail food sources, although development of such stores is rarely pursued as a solution to food access concerns. When one organization assessed local healthy food accessibility, they counted stores as healthy food sources, "if they had a good, decent selection, or a really nice, like, Pick 'n' Save kind of thing" (Anonymous, personal communication, 2014).

7. This is not unlike "culture of poverty" discourses that seem to shift blame from individuals but still amount to constructing particular racialized groups as abnormal or inherently destructive (Goode and Maskovsky 2001).

8. On 8 December, Mayor Tom Barrett was in Washington, DC, where Milwaukee

> was one of 16 cities featured in case studies as part of a recent report by the Federal Reserve titled 'The Enduring Challenge of Concentrated Poverty in America.' Barrett cited initiatives such as the new Urban Entrepreneur Partnership of Milwaukee ... set up to encourage economic development. Taking issue with the report's use of the term 'rust belt' to describe Milwaukee, Barrett (said) that he prefers to think of the city being on the 'fresh coast,' a reference to Lake Michigan's fresh water. (Marrerro 2008)

HG is part of such initiatives, and the city has emphasized innovations associated with UA, such as storm water management, cisterns, and porous paving stones.

References

Alkon, A. H., and J. Agyeman, eds. 2011. *Cultivating food justice: Race, class, and sustainability.* Cambridge, MA: MIT Press.

Alkon, A. H., and C. G. McCullen. 2009. Whiteness and farmers markets: Performances, perpetuations ... contestations? *Antipode* 43 (4):937–59.

Baker, L. E. 2004. Tending cultural landscapes and food citizenship in Toronto's community gardens. *Geographical Review* 94 (3):305–25.

Barraclough, L. R. 2009. South central farmers and shadow hills homeowners: Land use policy and relational racialization in Los Angeles. *The Professional Geographer* 61 (2):164–86.

Bonds, A. 2013a. Economic development, racialization, and privilege: "Yes in my backyard." Prison politics and the reinvention of Madras, Oregon. *Annals of the American Association of Geographers* 103 (6):1389–1405.

———. 2013b. Racing economic geography: The place of race in economic geography. *Geography Compass* 7 (6):398–411.

Brown, M. 2009. 2008 urban geography plenary lecture—Public health as urban politics, urban geography: Venereal biopower in Seattle, 1943–1983. *Urban Geography* 30 (1):1–29.

Carter, E. D. 2015. Making the blue zones: Neoliberalism and nudges in public health promotion. *Social Science and Medicine* 133:374–88.

Castree, N. 2008. Neoliberalising nature: Processes, effects, and evaluations. *Environment and Planning A* 40 (1):153–73.

City of Milwaukee. 2013. ReFresh Milwaukee: City of Milwaukee sustainability plan. Accessed October 9, 2017. http://city.milwaukee.gov/ImageLibrary/Groups/cityGreenTeam/documents/2013/ReFreshMKE_PlanFinal_Web.pdf.

———. 2015. City of Milwaukee STRONG Neighborhoods Plan: Impact report—Q4 2015. Accessed October 9, 2017. http://city.milwaukee.gov/ImageLibrary/Groups/citySNP/ImpactReportQ4-Final.pdf.

Cook, K. 2012. Neoliberalism, welfare policy and health: A qualitative meta-synthesis of single parents' experience of the transition from welfare to work. *Health: An Interdisciplinary Journal for the Social Study of Health, Illness & Medicine* 16 (5):507–30.

Delaney, D. 2002. The space that race makes. *The Professional Geographer* 54 (1):6–14.

Derus, M. 2007. Faces of foreclosure 53206. *Milwaukee Journal Sentinel* May 6.

Domene, E., and D. Saurí. 2007. Urbanization and class-produced natures: Vegetable gardens in the Barcelona metropolitan region. *Geoforum* 38:287–98.

Draus, P. J., J. Roddy, and A. McDuffie, 2014. "We don't have no neighbourhood": Advanced marginality and urban agriculture in Detroit. *Urban Studies* 51 (12):2523–38.

Eizenberg, E. 2012. Actually existing commons: Three moments of space of community gardens in New York City. *Antipode* 44:764–82.

Feagin, J. 2013. *The white racial frame.* 2nd ed. London and New York: Routledge.

Ghose, R. 2007. Politics of scale and networks of association in public participation GIS. *Environment and Planning A* 39 (8):1961–80.

Ghose, R., and M. Pettygrove. 2014a. Actors and networks in urban community gardens. *Geoforum* 53:93–103.

———. 2014b. Urban community gardens as spaces of citizenship. *Antipode* 46:1092–1112.

Gibbs-Plessl, T. 2012. *Twinkies, tomatoes, and tomatillos: A quantitative assessment of healthy food accessibility in Milwaukee County.* Milwaukee, WI: Hunger Task Force. Accessed October 10, 2017. https://www.hungertaskforce.org/wp-content/uploads/2015/05/Final_HTF_Healthy_Food_Access_Report.pdf.

Goode, J., and J. Maskovsky, eds. 2001. *New poverty studies: The ethnography of power, politics, and impoverished people in the United States.* New York: New York University Press.

Guthman, J. 2008. "If they only knew": Color blindness and universalism in California alternative food institutions. *The Professional Geographer* 60 (3):387–97.

Guthman, J. 2012. Opening up the black box of the body in geographical obesity research: Toward a critical political ecology of fat. *Annals of the American Association of Geographers* 102 (5):951–57.

Guthman, J., and M. DuPuis. 2006. Embodying neoliberalism: Economy, culture, and the politics of fat. *Environment and Planning D: Society and Space* 24 (3):427–48.

Irazábal, C., and A. Punja. 2009. Cultivating just planning and legal institutions: A critical assessment of the South Central farm struggle in Los Angeles. *Journal of Urban Affairs* 31 (1):1–23.

Keil, R. 2009. Urban politics and public health: What's urban, what's politics? *Urban Geography* 30 (1):36–39.

Lai, C. 2012. The racial triangulation of space: The case of urban renewal in San Francisco's Fillmore District. *Annals of the American Association of Geographers* 102 (1):151–70.

Lawson, V., L. Jarosz, and A. Bonds. 2010. Articulations of place, poverty, and race: Dumping grounds and unseen grounds in the rural American Northwest. *Annals of the American Association of Geographers* 100 (3):655–77.

Lightman, E. S., A. Mitchell, and D. Herd. 2008. Globalization, precarious work, and the food bank. *Journal of Sociology & Social Welfare* 35 (2):9–28.

Marrerro, D. 2008. Milwaukee on the "fresh coast," not "rust belt," says Milwaukee Mayor Tom Barrett in D.C. speech. *Journal Sentinel* December 3. Accessed October 9, 2017. http://archive.jsonline.com/blogs/news/35494089.html.

Marvin, S., and W. Medd. 2006. Metabolisms of obe-city: Flows of fat through bodies, cities and sewers. In *In the nature of cities: Urban political ecology and the politics of urban metabolism*, ed. N. Heynen, M. Kaika, and E. Swyngedouw, 137–49. London and New York: Routledge.

McClintock, N. 2011. From industrial garden to food desert: Demarcated devaluation in the flatlands of Oakland, California. In *Cultivating food justice: Race, class, and sustainability*, ed. A. H. Alkon and J. Agyeman, 89–120. Cambridge, MA: MIT Press.

McKittrick, K. 2011. On plantations, prisons, and a black sense of place. *Journal of Social and Cultural Geography* 12 (8):947–63.

———. 2013. Plantation futures. *Small Axe* 17:1–15.

Melamed, J. 2006. The spirit of neoliberalism from racial liberalism to neoliberal multiculturalism. *Social Text* 89 (4):1–24.

Newman, K., and R. W. Lake. 2006. Democracy, bureaucracy and difference in US community development politics since 1968. *Progress in Human Geography* 30 (1):44–61.

Perkins, H. A. 2009. Turning feral spaces into trendy places: A coffee house in every park? *Environment and Planning A* 41 (11):2615–32.

Pettygrove, M. 2016. *Food inequities, urban agriculture and the remaking of Milwaukee, Wisconsin*. Doctoral dissertation, University of Wisconsin–Milwaukee, Milwaukee, WI.

Pudup, M. B. 2008. It takes a garden: Cultivating citizen-subjects in organized garden projects. *Geoforum* 39 (3):1228–40.

Pulido, L. 2000. Rethinking environmental racism: White privilege and urban development in Southern California. *Annals of the American Association of Geographers* 90 (1):12–40.

Quastel, N. 2009. Political ecologies of gentrification. *Urban Geography* 30 (7):694–725.

Quinn, L. M., and J. Pawasarat. 2007. *Milwaukee's housing crisis: Subprime mortgages, foreclosures, evictions and affordability*. Milwaukee: Employment and Training Institute, University of Wisconsin–Milwaukee. Accessed October 9, 2017. http://www4.uwm.edu/eti/2007/HMDACensusUpdate.pdf.

———. 2013. *Wisconsin's mass incarceration of African American males: Workforce challenges for 2013*. Milwaukee: Employment and Training Institute, University of Wisconsin–Milwaukee. Accessed October 9, 2017. http://www4.uwm.edu/eti/2013/BlackImprisonment.pdf.

Ramírez, M. M. 2015. The elusive inclusive: Black food geographies and racialized food spaces. *Antipode* 47:748–69.

Rast, J. 2015. *Public transit and access to jobs in the Milwaukee metropolitan area, 2001–2014*. Milwaukee: University of Wisconsin–Milwaukee, Center for Economic Development. Accessed October 9, 2017. https://www4.uwm.edu/ced/publications/Transit2015_FINAL-1.pdf.

Roberts, D. J., and M. Mahtani. 2010. Neoliberalizing race, racing neoliberalism: Placing "race" in neoliberal discourses. *Antipode* 42 (2):248–57.

Rosol, M. 2012. Community volunteering as neoliberal strategy? Green space production in Berlin. *Antipode* 44:239–57.

Roy, P. 2011. Analyzing empowerment: An ongoing process of building state–civil society relations—The case of Walnut Way in Milwaukee. *Geoforum* 41 (2):337–48.

Shannon, J. 2014. Food deserts: Governing obesity in the neoliberal city. *Progress in Human Geography* 38 (2):248–66.

Slocum, R. 2007. Whiteness, space, and alternative food practice. *Geoforum* 38 (3):520–33.

———. 2011. Race in the study of food. *Progress in Human Geography* 35 (3):303–27.

Staeheli, L. A., D. Mitchell, and K. Gibson. 2002. Conflicting rights to the city in New York's community gardens. *GeoJournal* 58:197–205.

Walker, S. 2015. Urban agriculture and the sustainability fix in Vancouver and Detroit. *Urban Geography* 39 (2):163–82.

While, A., A. E. G. Jonas, and D. Gibbs. 2004. The environment and the entrepreneurial city: Searching for the urban "sustainability fix" in Manchester and Leeds. *International Journal of Urban and Regional Research* 28 (3):549–69.

Wilson, D. 2009. Introduction: Toward a refined racial economy perspective. *The Professional Geographer* 61 (2):139–49.

Wisla, M. n.d. Special report: Foreclosures, block by block. *Milwaukee Neighborhood News Service*. Accessed October 9, 2017. http://milwaukeenns.org/milwaukee-foreclosures/.

Zimmerman, J. 2008. From brew town to cool town: Neoliberalism and the creative city development strategy in Milwaukee. *Cities* 25:230–42. doi:10.1016/j.cities.2008.04.006.

MARGARET PETTYGROVE has a doctorate in geography from the University of Wisconsin–Milwaukee, Milwaukee, WI 53211. E-mail: petygr2@uwm.edu. She is currently a geographic information systems (GIS) analyst with the U.S. Forest Service. Her research centers on urban food systems, political ecology, and health geographies. She has published articles on these topics in a variety of research journals.

RINA GHOSE is Professor in the Department of Geography, University of Wisconsin–Milwaukee, Milwaukee, WI 53211. E-mail: rghose@uwm.edu. Her research interests intersect critical GIS, urban governance, political ecology, and health geographies. She has published many research articles in acclaimed journals on these topics.

Index

T - #0083 - 311024 - C324 - 276/219/15 - PB - 9780367663551 - Gloss Lamination